STATISTICS AND RELATED TOPICS

STATISTICS AND RELATED TOPICS

Proceedings of the International Symposium on
Statistics and Related Topics
held in Ottawa, Canada, May 5-7, 1980

Edited by

M. CSÖRGÖ
D. A. DAWSON
J. N. K. RAO
A. K. Md. E. SALEH

Department of Mathematics and Statistics
Carleton University
Ottawa, Canada

NORTH-HOLLAND PUBLISHING COMPANY – AMSTERDAM • NEW YORK • OXFORD

ISBN: 0 444 86293 5

Publishers:

NORTH-HOLLAND PUBLISHING COMPANY
AMSTERDAM • NEW YORK • OXFORD

Sole distributors for the U.S.A. and Canada:

ELSEVIER NORTH-HOLLAND INC.
52 VANDERBILT AVENUE, NEW YORK, N.Y. 10017

Library of Congress Cataloging in Publication Data

International Symposium on Statistics and Related
Topics (1980 : Ottawa, Ont.)
Statistics and related topics.

1. Mathematical statistics--Congresses.
I. Csörgő, M. II. Title.
QA276.A1I57 1980 519.5 81-14023
ISBN 0-444-86293-5 AACR2

PRINTED IN THE NETHERLANDS

DEDICATED TO THE MEMORY OF

HERMAN OTTO HARTLEY

(1912 - 1980)

PREFACE

An International Symposium on Statistics and Related Topics was held at Carleton University, Ottawa, Canada, May 5-8, 1980 in conjunction with an International Symposium on Current Topics in Survey Sampling. This volume contains invited papers presented at the Symposium on Statistics and Related Topics; a separate volume (to be published by Academic Press) will consist of the invited papers given at the latter symposium. Several well-known statisticians from Canada, U.S.A. and other countries presented invited papers and more than three hundred statisticians and scientists in other disciplines interested in applications of statistics participated in the Symposia.

The selection of topics for the Symposium on Statistics and Related Topics was somewhat biased by our own personal research interests, but the papers here do cover a wide range of areas of current interest including linear models and optimal designs, parametric inference and goodness-of-fit, biostatistics, nonparametric methods and robust inference, stochastic differential equations and Markov processes. The emphasis at the symposium was on both theory and applications. The papers cover a variety of applications including the analysis of evoked response experiments, hazard rate models in carcinogenic testing, lotteries and the analysis of hail data. An important feature of the book is that it contains, in addition to the many research papers which will not be published elsewhere, authoritative review papers of recent developments on variance components, optimal designs, quantile processes and goodness-of-fit, dose response models for quantal response data, adaptive robust inference, robust tests and Erdös-Rényi laws of large numbers. This volume should be of interest to both mathematical and applied statisticians as well as scientists in other disciplines interested in applications of statistics.

Two of us (M. Csörgő and D.A. Dawson) received Killam Foundation Senior Research Fellowships which provided the seed money for the symposium. Our thanks are also due to Natural Sciences and Engineering Research Council Canada (NSERC) for additional financial support. One of us (A.K.Md.E. Saleh) acted as the Chairman of the Organizing Committee and administered the Symposium. We also wish to thank Dan Krewski and Mike Hidiroglou for their help in coordinating the two symposia and the local arrangements committee: Karol Krotki (Chairman), Jack Graham, Mike Bankier and Andy McCleod for their excellent job. Our appreciation is also due to our colleagues Amit Bose, Jean-Paul Dionne, Roger Fischler, Jack Graham, Ed Hughes, Gale Ivanoff, Lutful Kabir, and S. Raman for presiding over the sessions.

We wish to thank Dr. K.S. Williams, Chairman, Department of Mathematics and Statistics and Dr. George Skippen, Dean of Science at Carleton University, for their opening remarks and Dr. William G. Madow for the special lecture on "Statistical Models and Statistics" before the wine and cheese party. Our special thanks are, of course, due to the invited speakers for giving excellent talks and for their excellent cooperation in preparing the papers for publication. We also thank the speakers in the contributed papers sessions and the following invited speakers who could not prepare their papers for inclusion in the volume (for unavoidable reasons): R.R. Bahadur: "Combining Independent Tests of Significance"; C.R. Blyth: "Cramer-Rao Inequality without Regularity Conditions"; C.M. Deo: "Probabilities of Moderate Deviations in Banach Space"; J.P. Dionne: "Generating Function in Branching Processes with Application to Maximum Likelihood Estimation"; G.B. Kallianpur: "Some Remarks on the Feynman Integrals"; R. Marcus: "Stochastic Evolution Equations"; W. Philip: "Approximation Theorems for Sums of C(S) Valued Random Variables with Applications to Random Fourier Series and Empirical Characteristic Processes"; R. Rishel: "Control of Completely Observed and Partially Observed Jump Markov Processes"; R.J. Tomkins: "Limiting Behaviour of Normed Sums of Independent Random Variables".

While preparing this volume for publication, we were deeply saddened to hear that Professor H.O. Hartley passed away on December 30, 1980, in Durham, North Carolina. He underwent open heart surgery in September, 1980 and in his characteristic style for tackling problems tried valiantly for three months to overcome post-operative complications but succumbed in the end. We are most grateful to him for having presented two important papers, one dealing with hazard rate models for carcinogenic testing (jointly with H.D. Tolley and R.L. Sielken, Jr) appearing in this volume and the other on response errors in sample surveys which will appear in the Sampling Volume. Anyone who attended the symposia will agree that his very active participation was an important contributing factor in the success of the symposia. Professor Hartley's contributions to statistics and related areas are monumental. His enthusiasm for solving problems in diverse areas of statistical application is astounding. He was a great teacher and a mentor to many graduate students who were fortunate enough to have studied under him (including one of us, J.N.K. Rao). Above all, he was a kind and most considerate human being, always ready to help anyone. As a token of our deep respect and appreciation, we are dedicating this volume to the memory of Professor Hartley. We also wish to convey our sincere condolences to his wife Grace, son Michael and daughter Jennifer

Finally we wish to express our appreciation to the North-Holland Publishing Company for their cooperation and Gill Murray for excellent typing help.

Miklós Csörgő
D.A. Dawson
J.N.K. Rao
A.K.Md.E. Saleh

TABLE OF CONTENTS

DEDICATION v

PREFACE vii

PART I LINEAR MODELS

RECENT DEVELOPMENTS IN DESIGNS AND ESTIMATORS FOR VARIANCE COMPONENTS 3

R.L. Anderson

1. Introduction 3
2. Use of Iterated Least Squares (ITLS) to Estimate Variance Components 3
3. Comparison of Designs and Estimators for Variance Components: Two-stage Nested Designs 6
4. Extensions to Multistage Nested Designs 7
5. Non-balanced Designs for Two-way Classification Models 11
6. Compositing 18
7. References 21

AN ANALYSIS OF THE MASSACHUSETTS NUMBERS GAME 23

Herman Chernoff

1. Introduction 23
2. The Payoff System 25
3. The Data 25
4. The Systems 27
5. A More Elaborate Model 33
6. Summary 36
7. Bibliography 37

STUDY OF OPTIMALITY CRITERIA IN DESIGN OF EXPERIMENTS 39

A. Hedayat

1. Preliminary 39
2. Some Well-known Optimality Criteria 42
3. S-optimality and (M,S)-optimality 44
4. Φ_p-criteria 45
5. Universal Optimality 47
6. Type 1 and Type 2 Criteria 51
7. Schur Optimality 52
8. References 54

EQUALITIES AND INEQUALITIES FOR CONDITIONAL AND PARTIAL CORRELATION COEFFICIENTS 57

Michael C. Lewis and George P.H. Styan

1. Introduction and Summary 57
2. Inequalities for $\rho(UV|W)$ when $\rho(UV:W) = \rho(UV\cdot W)$ and both Regressions are Linear 58
3. Inequalities for $\rho(UV\cdot Z)$ when $\rho(UV|Z) = \rho(UV:Z)$ and both Regressions are Linear in $1/Z$ 62
4. References 65

PART II PARAMETRIC INFERENCE AND GOODNESS-OF-FIT

QUANTILE PROCESSES AND SUMS OF WEIGHTED SPACINGS FOR COMPOSITE GOODNESS-OF-FIT 69

Miklós Csörgő and Pál Révész

1. Introduction and Legend 69
2. Discussion of a New Result 79
3. Nuisance Parameter Free Goodness-of-Fit Statistics for the Shift and Scale Family 82
4. On Weiss' Estimate of the Scale Parameter of the Shift and Scale Family 84
5. References 85

ASYMPTOTIC DISTRIBUTIONS OF FUNCTIONS OF THE EIGENVALUES OF THE REAL AND COMPLEX NONCENTRAL WISHART MATRICES 89

C. Fang and P.R. Krishnaiah

1. Introduction 89
2. Perturbation Technique 89
3. Asymptotic Joint Distribution of Functions of the Roots of Noncentral Wishart Matrix 91
4. Applications in Investigation of the Structures of Interactions 96
5. Applications in Cluster Analysis 99
6. Asymptotic Distributions of Functions of the Roots of the Complex Wishart Matrix 104
7. References 107

ON EFFICIENT INFERENCE IN SYMMETRIC STABLE LAWS AND PROCESSES 109

Andrey Feuerverger and Philip McDunnough

1. Introduction 109
2. Some Properties of Stable Distributions 110
3. Maximum Likelihood By Inversion 111
4. Maximum Likelihood Simulation Study 112
5. Fourier Methods for Inference 113
6. Fourier Gridpoints for the Symmetric Stable Laws 117
7. A Simulation Study for AR(1) Stable Processes 117
8. References 121

SOME COMMENTS ON THE MINIMUM MEAN SQUARE ERROR AS A CRITERION OF ESTIMATION 123

C. Radhakrishna Rao

1. Introduction 123
2. Estimation of a Single Parameter 124
3. Estimation of Variance 126
4. Direct and Inverse Regression 130
5. Simultaneous Estimation of Two Parameters 132
6. Estimation of Several Parameters 134
7. References 141
8. Appendix 143

ESTIMATING QUANTILES OF EXPONENTIAL DISTRIBUTION 145

A.K.Md. Ehsanes Saleh

1. Introduction 145
2. Estimation of Quantiles of Exponential Distribution Based on k Arbitrary Order Statistics 145
3. Estimation of Quantiles Based on k Selected Order Statistics 147
4. ARE of $\hat{\hat{x}}_\xi$ Relative to $\bar{x}_\xi$ and $\dot{x}_\xi$ 149
5. References 150

PART III BIOSTATISTICS

SOME ASPECTS OF THE ANALYSIS OF EVOKED RESPONSE EXPERIMENTS 155

David R. Brillinger

1. Introduction 155
2. Some Formalization 156
3. Investigation of the AER for Model 1 158
4. Uses of Model 1 159
5. A Formal (Linear) Approach 160
6. An Alternative Viewpoint 161
7. Superposability 161
8. Robust/Resistant Estimates 162
9. A Recursive Procedure 163
10. Resistance (Frequency Domain) 164
11. A Partially Parametric Model 165
12. A Fully Parametric Model 165
13. References 166

MEAN RESIDUAL LIFE 169

W.J. Hall and Jon A. Wellner

1. Introduction 169
2. Bounds for MRL 170
3. Characterizations of MRL; the Inversion Formula 170
4. Residual Moment Formulas and Some Characterizations 173
5. Applications of the Inversion Formula 177
6. MRL 'at Great Age' 179
7. Use of MRL in Modelling 181
8. References 182

THE PRODUCT FORM OF THE HAZARD RATE MODEL IN CARCINOGENIC TESTING 185

H.O. Hartley, H.D. Tolley and R.L. Sielken, Jr.

1. Introduction 185
2. The Dynamic, Compartment-Analytic Model 186
3. The "Multiple Attack" Model 192
4. References 194
5. Appendix 1 196
6. Appendix 2 197
7. Appendix 3 197

DOSE RESPONSE MODELS FOR QUANTAL RESPONSE TOXICITY DATA 201

Daniel Krewski and John Van Ryzin

1. Introduction 201
2. Dose-Response Models 202
3. Estimation Procedures 207
4. Empirical Results 211
5. Discussion 223
6. Appendix: Maximum Likelihood Estimation 224
7. References 229

PART IV NONPARAMETRIC METHODS AND ROBUST INFERENCE

THE CONDITIONAL APPROACH TO ROBUSTNESS 235

George A. Barnard

1. Conditional Approach 235
2. Appendix 240
3. References 241

THE STRONG CONVERGENCE OF EMPIRICAL NEAREST NEIGHBOR ESTIMATES OF INTEGRALS 243

Luc Devroye

1. Introduction 243
2. Proofs 244
3. The Nearest Neighbor Rule 248
4. Refinements 250
5. References 252

ON ADAPTIVE ROBUST INFERENCES 253

Robert V. Hogg

1. Introduction 253
2. Estimation of the Center 254
3. Distribution-free Methods 257
4. The Regression Situation 259
5. Conclusions 262
6. References 262

AN APPLICATION OF RANK INVARIANT MULTIPLE REGRESSION AND VARIABLE SELECTION 265

D. L. McLeish

1. Examples 267
2. Numerical Procedures 269
3. Tests for Partial Correlation 270
4. Application to the Hail Data 271
5. Appendix 273
6. References 276

A NONPARAMETRIC TEST FOR EQUALITY AGAINST ORDERED ALTERNATIVES IN THE CASE OF SKEWED DATA, WITH A BIOMEDICAL APPLICATION 279

A.R. Padmanabhan, M.L. Puri and A.K.Md. Ehsanes Saleh

1. Introduction 279
2. The Proposed Test 279
3. Asymptotic Relative Efficiency (ARE) of V-test 281
4. Application to Lung Cancer Data 282
5. References 283

RANK ANALYSIS OF COVARIANCE UNDER PROGRESSIVE CENSORING, II 285

Pranab Kumar Sen

1. Introduction 285
2. Preliminary Notions 286
3. The Proposed PCS Tests 287
4. Asymptotic Properties of the Proposed Tests 289
5. Some General Remarks 294
6. References 295

ROBUST TWO-SAMPLE TEST AND ROBUST REGRESSION AND ANALYSIS-OF-VARIANCE VIA MML ESTIMATORS 297

M.L. Tiku

1. Introduction 297
2. Joint Efficiency of μ_c and σ_c 298
3. Robust Two-Sample Test 301
4. Robust Regression Based on Grouped Data 304
5. Efficiency of θ^* 305
6. Robust Regression 307
7. Robust Analysis-of-Variance 308
8. Bibliography 312

PART V STOCHASTIC PROCESSES

GALERKIN APPROXIMATION OF NONLINEAR MARKOV PROCESSES 317

D.A. Dawson

1. Introduction 317
2. Background Information on Nonlinear Markov Systems 317
3. The Gauss-Galerkin Numerical Approximation 321
4. Gauss-Galerkin Approximation of Function Space Integrals 331
5. The Question of Convergence 334
6. References 338

DONSKER CLASSES OF FUNCTIONS 341

R.M. Dudley

1. Introduction and Preliminaries 341
2. Sequences of Functions 346
3. Metric Entropy with Bracketing 346
4. References 351

CAUSAL CALCULUS OF BROWNIAN FUNCTIONALS, AND ITS APPLICATIONS 353

Takeyuki Hida

1. Background 353
2. Generalized Brownian Functionals 354
3. Differential Calculus 356
4. Stochastic Partial Differential Equations 358
5. References 360

WHITE NOISE ANALYSIS AND AN APPLICATION TO STOCHASTIC DIFFERENTIAL EQUATIONS IN HILBERT SPACE 361

Yoshio Miyahara

0. Introduction 361
1. Stochastic Integrals and Multiple Wiener Integrals 362
2. Application to Stochastic Differential Equations 364
3. Concluding Remarks 373
4. References 373

ON THE INCREMENTS OF STOCHASTIC PROCESSES AND THE RECONSTRUCTION OF THEIR DISTRIBUTIONS 375

Josef Steinebach

1. The Stochastic Geyser Problem 375
2. The Erdős-Rényi Law of Large Numbers 376
3. Some other Erdős-Rényi Type Laws 378
4. Convergence Rates in Erdős-Rényi Laws 382
5. References 384

TITLES OF CONTRIBUTED PAPERS 387

PART I
LINEAR MODELS

STATISTICS AND RELATED TOPICS
M. Csörgö, D.A. Dawson, J.N.K. Rao, A.K.Md.E. Saleh (eds.)

RECENT DEVELOPMENTS IN DESIGNS AND ESTIMATORS FOR VARIANCE COMPONENTS

R. L. Anderson

University of Kentucky

1. INTRODUCTION

At a conference on Statistical Design and Linear Models at Colorado State University in 1973, I prepared a survey paper on Designs and Estimators for Variance Components; this was published in the proceedings of the Conference (Anderson, 1975). Since that time, I have participated in a number of research projects in this area. Papers with W. O. Thompson (1975), H. D. Muse (1978) and W. W. Stroup and J. W. Evans (1980) have been published. Excerpts from a dissertation by B. Thitakomal (1977) have been submitted for publication. A dissertation by J. H. Schwartz (1980) has contributed additional information.

2. USE OF ITERATED LEAST SQUARES (ITLS) TO ESTIMATE VARIANCE COMPONENTS

In 1961 and again in 1967, I proposed an iterated weighted least squares procedure for estimating variance components, using the single degree-of-freedom contributions to the sums of squares in the analysis of variance as the dependent variables and the variance components as the regression coefficients. In general, one would combine all such sums of squares with the same expected values in the computations. Let us assume that we have t sums of squares, SS_i, each with f_i degrees of freedom (d.f.) and expectation E_i, where E_i is a linear function of the variance components, σ_j^2:

$$E_i = \sum_{j=1}^{p} X_{ij}\,\sigma_j^2.$$

Since the SS_i are multiples of χ^2-variates, they have unequal variances:

$$V(SS_i) = 2E_i^2/f_i.$$

The estimating equations are

$$\underline{\hat{\sigma}}^2 = (X'\Sigma^{-1}X)^{-1}X'\Sigma^{-1}\underline{SS},$$

where $\underline{\sigma}^2$ is the p-vector of variance components, X is the t×p matrix of coefficients in the E_i and Σ is the t×t variance-covariance matrix of the SS_i. If the SS_i are independent, Σ is a diagonal matrix; if the SS_i are not independent, the set of sufficient statistics to estimate $\underline{\sigma}^2$ must also include some cross-product terms. Since Σ is a function of the unknown variance components, an iterated computing procedure is required.

Originally ITLS utilized only sums of squares adjusted for the mean; hence, the estimates corresponded to those obtained by Restricted Maximum Likelihood (REML). The Thompson-Anderson (1975) article considered two-stage nested designs with k_1 classes having one sample per class and k_2 classes having two samples per class;

the model is

$$y_{ij} = \mu + a_i + b_{j(i)},$$

with variance components σ_a^2 and σ_b^2. The sums of squares are presented in Table 1 (assuming $k_1 \geq 1$). In Table 1, $k = k_1 + k_2$ and $N = k_1 + 2k_2$, SSA_1 and SSA_2 are the sums of squares between classes within the two groups, SSA_3 the sum of squares between the two groups and SSB the within-classes sum of squares.

Table 1: Usual Sums of Squares for (k_1,k_2) Design

Sum of squares (SS_i)	D.F. (f_i)	Expectation (E_i)	Variance ($2E_i^2/f_i$)
SSA_1	k_1-1	$(k_1-1)(\sigma_a^2+\sigma_b^2)$	$2(k_1-1)(\sigma_a^2+\sigma_b^2)^2$
SSA_2	k_2-1	$(k_2-1)(2\sigma_a^2+\sigma_b^2)$	$2(k_2-1)(2\sigma_a^2+\sigma_b^2)^2$
SSB	k_2	$k_2\ \sigma_b^2$	$2k_2\ \sigma_b^4$
SSA_3	1	$(2k\sigma_a^2/N+\sigma_b^2)$	$2(2k\sigma_a^2/N+\sigma_b^2)^2$

As with maximum likelihood (ML), the estimates from the ITLS procedure cannot be negative; hence, a truncation rule was adopted which was consistent with the likelihood principle in a restricted parameter space: if at any stage of the iteration, $\hat{\sigma}_a^2 < 0$, one might set $\hat{\sigma}_a^2 = 0$ and use $\sigma_b^2 = SS_t/(N-1)$, where $SS_t=SSA_1+SSA_2+SSA_3+SSB$. A more efficient computing procedure (when $\hat{\sigma}_a^2 < 0$) would be to set $\hat{\sigma}_a^2 = \varepsilon$, a small positive number, and continue the iteration until $\hat{\sigma}_b^2$ converges; if $\hat{\sigma}_a^2$ is still negative, then set it equal to zero. Since ML estimators have lower mean squared errors than does REML for this situation[1/], ITLS based on the four SS_i in Table 1 is inferior to ML.

Recently, I realized that one could obtain complete ML estimates by including μ, the population mean, in the iteration process. The complete ITLS procedure for the (k_1,k_2) design is presented in Table 2. An example of this procedure is included in a technical report by Jennrich and Moore (1975, p. 24) in which they consider g classes, each with n_i samples. The statistics used are presented in Table 3. It is interesting to observe that the procedure in Table 3 can be modified to estimate μ, σ_a^2 and $\sigma_{b_i}^2$, where the sampling variance within classes is not assumed to be constant from class to class. For this case in Table 3, $n_i\sigma_a^2+\sigma_b^2$ is replaced by $n_i\sigma_a^2+\sigma_{b_i}^2$ and $(n_i-1)\sigma_b^2$ by $(n_i-1)\sigma_{b_i}^2$. For all of these examples, the Σ-matrix is diagonal.

[1/] See Harville (1975).

Table 2: ITLS Procedure for (k_1,k_2)-Design

(i) Let y_{1i} represent the observations for the classes with one sample per class:

$$\sum_{i=1}^{k_1} y_{1i} = G;\ SSA_1^* = \sum_i (y_{1i}-\mu)^2 = SSA_1 + (G_1-k_1\mu)^2/k_1 ;$$

(ii) Let y_{2ij} represent the observations for the classes with two samples per class $(i=1,2,\ldots,k_2;\ j=1,2)$:

$$y_{2i1} + y_{2i2} = G_{2i};\ \sum_i G_{2i} = G_2;$$

(iii) The statistics used are

Statistic	Expectation (E_i)	Variance (V_i)
$\bar{y}_1 = G_1/k_1$	μ	$(\sigma_a^2+\sigma_b^2)/k_1$
$\bar{y}_2 = G_2/2k_2$	μ	$(2\sigma_a^2+\sigma_b^2)/2k_2$
SSA_1^*	$k_1(\sigma_a^2+\sigma_b^2)$	$2k_1(\sigma_a^2+\sigma_b^2)^2$
SSA_2^*	$k_2(2\sigma_a^2+\sigma_b^2)$	$2k_2(2\sigma_a^2+\sigma_b^2)^2$
SSB	$k_2\ \sigma_b^2$	$2k_2\ \sigma_b^4$

Table 3: Statistics Used in ITLS Estimation: General Two-stage Nested Design

Statistic	Expectation	Variance
$\bar{y}_1$	μ	$(n_1\sigma_a^2+\sigma_b^2)/n_1$
$n_1(\bar{y}_1-\mu)^2$	$n_1\sigma_a^2+\sigma_b^2$	$2(n_1\sigma_a^2+\sigma_b^2)^2$
SSB_1	$(n_1-1)\sigma_b^2$	$2(n_1-1)\sigma_b^4$
.	.	.
.	.	.
.	.	.
$\bar{y}_g$	μ	$(n_g\sigma_a^2+\sigma_b^2)/n_g$
$n_g(\bar{y}_g-\mu)^2$	$n_g\sigma_a^2+\sigma_b^2$	$2(n_g\sigma_a^2+\sigma_b^2)^2$
SSB_g	$(n_g-1)\sigma_b^2$	$2(n_g-1)\sigma_b^4$

3. COMPARISON OF DESIGNS AND ESTIMATORS FOR VARIANCE COMPONENTS: TWO-STAGE NESTED DESIGNS

In Thompson-Anderson (1975), it was shown that for balanced two-stage nested models, a modified maximum likelihood (MML) estimator[2/] has an appreciably smaller mean squared error (MSE) of $\hat{\sigma}_a^2$ when N is small and $\rho = \sigma_a^2/\sigma_b^2$ is less than 1. Some pertinent comparisons of MSE are presented in Table 4 for N = 30, $\sigma_b^2 = 1$ and selected ρ and k.

Table 4: Mean Squared Errors of $\hat{\sigma}_a^2$ for Balanced Designs for N = 30

ρ	k	NTAOV[a/]	TAOV[a/]	ML	MML
0.1	6	.0393	.0294	.0176	.0125
	5	.0378	.0293	.0163	.0115
	3	.0407	.0345	.0143	.0100
	2	.0559	.0501	.0134	.0096
0.5	15	.176	.155	.136	.126
	10	.165	.155	.129	.119
	6	.199	.192	.143	.129
1.0	15	.355	.347	.313	.301
	10	.406	.402	.343	.326
2.0	15	.926	.925	.937	.812
	10	1.221	1.220	1.044	.999
4.0	15	2.926	2.926	2.643	2.564
	10	4.184	4.184	3.579	3.425

[a/] AOV = Analysis of Variance; NT = non-truncated; T = truncated.

Table 4 also includes the results for the usual non-truncated Analysis-of-Variance (AOV) and the truncated AOV estimators. Note that for $\rho \geq 0.5$, the optimal balanced design is the same for all estimators. In all cases, the MSE for ML-estimators is much less than for even the truncated AOV estimators. MML offers only a slight improvement when $\rho \geq 1$ if the optimal design is used; however, its MSE is much smaller for small k.

The substantial decrease in MSE for large ρ when k increases from 10 to 15 (sample size per class decreases from 3 to 2) indicates a need for an unbalanced design. Table 5 presents some estimated MSE based on simulations and asymptotic variances for selected (k_1,k_2) designs [see Tables 3 and 5 in Thompson-Anderson (1975)].

Table 5: Comparisons of Estimators of σ_a^2 for Selected (k_1,k_2)-Designs[b/]

	ρ				
	0.5	1.0	2.0	4.0	8.0
			$(k_1,k_2)=(10,10)$		
AV (ML)	.218	.377	.835	2.34	7.75
MSE (ML)	.166	.354	.842	2.33	7.72
MSE (UAOV)	.207	.406	.907	2.52	8.31
MSE (WAOV)	.180	.370	.882	2.59	8.89

[2/] Introduced by Klotz, et. al.(1969).

Table 5: (Continued)

	ρ				
	0.5	1.0	2.0	4.0	8.0
			$(k_1,k_2)=(50,25)$		
AV (ML)	.0810	.129	.258	.670	2.13
MSE (ML)	.0785	.126	.254	.631	1.93
MSE (UAOV)	.0928	.141	.271	.646	1.95
MSE (WAOV)	.0809	.123	.259	.684	2.12

b/ AV = Asymptotic Variance; MSE = Mean Squared Error; AOV = Analysis of Variance (U = Unweighted; W = Weighted). MSE based on 12,000 simulated experiments for (10,10) and 3,000 for (50,25).

The tendency of the simulated MSE for N = 100 to be smaller than the asymptotic variances for large ρ (when there would be few truncations) is somewhat disconcerting. It may be that there was a slight non-normality in the simulated observations. It is obvious that ML is a consistently better estimator than AOV; however, the unweighted AOV is quite good for large ρ. There is a slight reduction in MSE by use of a nonbalanced design for ρ large, but the gain is not large.

Thitakomal (1977) has investigated the use of a modified ML estimator for the (k_1,k_2) designs, in which $k_1+\lambda_1$ degrees of freedom were used for SSA_1^* and $k_2+\lambda_2$ degrees of freedom for SSA_2^* (Table 2). Using the restriction $\lambda_1/k_1 = \lambda_2/k_2$, he found the optimal value of $\lambda_1 + \lambda_2 \doteq 2$ for μ known; and $\lambda_1 + \lambda_2 \doteq 1$ when μ had to be estimated. For N = 30, the mean squared error was reduced by about 10% for large ρ and 20% for ρ = 0.5 when μ was known; the reductions were only 3% and 10% when μ had to be estimated. There was a trivial reduction, as expected, when N = 100.

4. EXTENSIONS TO MULTISTAGE NESTED DESIGNS

In the Colorado Conference paper, I discussed a number of non-balanced multistage designs. The Bainbridge (1963) staggered design seemed to be the best, when AOV estimators were used. The model for three stages is

$$y_{ijk} = \mu + a_i + b_{j(i)} + c_{k(ij)},$$

with variance components σ_a^2, σ_b^2 and σ_c^2. The Bainbridge design consists of a repetitions of this design:

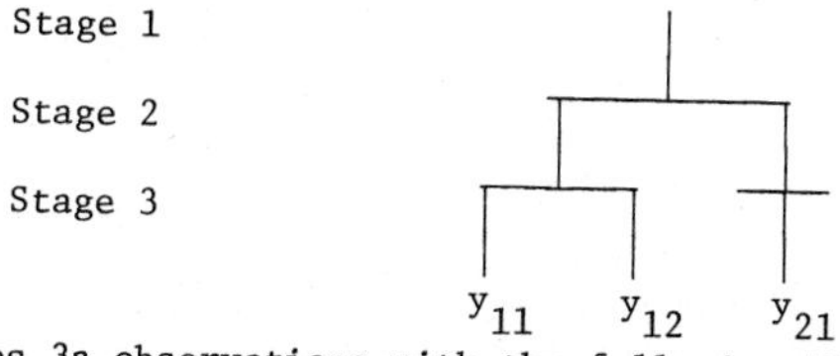

Stage 1

Stage 2

Stage 3

y_{11} y_{12} y_{21}

This produces 3a observations with the following AOV:

	D.F.	SS	MS	E(MS)
Stage 1	a-1	SSA	MSA	$\sigma_c^2 + 5/3\sigma_b^2 + 3\sigma_a^2$
Stage 2	a	SSB	MSB	$\sigma_c^2 + 4/3\sigma_b^2$

	D.F.	SS	MS	E(MS)
Stage 3	a	SSC	MSC	σ_c^2

The estimators are $\hat{\sigma}_c^2$ = MSC, $\hat{\sigma}_b^2$ = 3 (MSB−MSC)/4 and $\hat{\sigma}_a^2$ = (4MSA−5MSB+MSC)/12. The usual truncation procedures are followed to make all $\hat{\sigma}_i^2 \geq 0$.

The terms needed for ITLS are presented in Table 6 and the variance-covariance matrix is given in Table 7. The results in Table 7 are based on the formula:

$$\text{Cov}(Z_iZ_j,\ Z_{i'}\ Z_{j'}) = \sigma_{ii'}\ \sigma_{jj'} + \sigma_{ij'}\ \sigma_{i'j}.$$

Table 6: Statistics Used in ITLS Estimation: Bainbridge Three-Stage Design

Statistic	Expectation
$\bar{y}_1 = \sum_i (y_{i11}+y_{i12})/2a$	μ
$\bar{y}_2 = \sum_i (y_{i21})/a$	μ
$SSA^* = \sum_i (Z_{i1}-3\mu)^2$	$a(9\sigma_a^2 + 5\sigma_b^2 + 3\sigma_c^2)$
$SCP = \sum_i Z_{i1}Z_{i2}$	$2a\sigma_b^2$
$SSB = \sum_i Z_{i2}^2$	$a(8\sigma_b^2 + 6\sigma_c^2)$
$SSC = \sum_i Z_{i3}^2$	$2a\sigma_c^2$

where $Z_{i1} = y_{i11}+y_{i12}+y_{i21}$; $Z_{i2} = y_{i11}+y_{i12}-2y_{i21}$;

$Z_{i3} = y_{i11}-y_{i12}$.

Table 7: Variance-Covariance Matrix for Statistics in Table 6

$$\Sigma = \begin{bmatrix} V_1(2\times2) & \\ & V_2(4\times4) \end{bmatrix}$$

where

$$V_1 = \begin{bmatrix} (2\sigma_a^2+2\sigma_b^2+\sigma_c^2)/2a & \sigma_a^2/a \\ \sigma_a^2/a & (\sigma_a^2+\sigma_b^2+\sigma_c^2)/a \end{bmatrix}$$

$$V_2 = \begin{bmatrix} a_{11} & a_{12} & a_{13} & 0 \\ a_{12} & a_{22} & a_{23} & 0 \\ a_{13} & a_{23} & a_{33} & 0 \\ 0 & 0 & 0 & 8a\sigma_c^4 \end{bmatrix}$$

Table 7: (Continued)

$$a_{11} = 2a(9\sigma_a^2+5\sigma_b^2+3\sigma_c^2)^2; \quad a_{12} = 4a\sigma_b^2(9\sigma_a^2+5\sigma_b^2+3\sigma_c^2);$$
$$a_{13} = 8a\sigma_b^4; \quad a_{22} = a[(9\sigma_a^2+5\sigma_b^2+3\sigma_c^2)(8\sigma_b^2+6\sigma_c^2)+4\sigma_b^4];$$
$$a_{23} = 4a\sigma_b^2(8\sigma_b^2+6\sigma_c^2); \quad a_{33} = 2a(8\sigma_b^2+6\sigma_c^2)^2.$$

The ITL equations are

$$\underline{\theta} = (X'\Sigma^{-1}X)^{-1} X'\Sigma^{-1}\underline{S},$$

where $\underline{\theta}' = (\mu, \sigma_a^2, \sigma_b^2, \sigma_c^2)$, X is the 6×4 coefficient matrix in the expectations and $\underline{S}$ is the vector of statistics.

It is easy to show that this is an ML-procedure. The log likelihood is proportional to

$$L = -[a\ell n\ |C| + Q]/2,$$

where

$$C = \begin{bmatrix} C_1 & 2\sigma_b^2 & 0 \\ 2\ {}_b^2 & C_2 & 0 \\ 0 & 0 & 2\sigma_c^2 \end{bmatrix}; \quad C_1 = 9\sigma_a^2+5\sigma_b^2+3\sigma_c^2 \quad C_2 = 8\sigma_b^2+6\sigma_c^2 .$$

$$Q = \frac{C_2\ SSA^* + C_1\ SSB - 4\sigma_b^2\ SCP}{C_1C_2 - 4\sigma_b^4} + \frac{SSC}{2\sigma_c^2};$$

$$\hat{\mu} = [\hat{C}_2\Sigma Z_{i1} - 2\hat{\sigma}_b^2\ \Sigma Z_{i2}]/3a\hat{C}_2$$
$$= \bar{y} - 2\hat{\sigma}_b^2\ \bar{z}_2/3\hat{C}_2,$$

where $\bar{z}_2$ is an ancillary statistic. The variance components would be estimated by

$$\text{ITLS:}\quad \underline{\hat{\sigma}}^2 = (X^{*\prime}V_2^{-1}X^*)^{-1}X^{*\prime}V_2^{-1}\underline{S}^*$$

where

$$X^* = \begin{bmatrix} 9a & 5a & 3a \\ 0 & 2a & 0 \\ 0 & 8a & 6a \\ 0 & 0 & 2a \end{bmatrix}; \quad \underline{S}^* = \begin{bmatrix} SSA^* \\ SCP \\ SSB \\ SSC \end{bmatrix}.$$

In the Anderson-Bancroft text (1952, page 334), I proposed an alternative type of non-balanced five-stage nested design. The Anderson three-stage design is

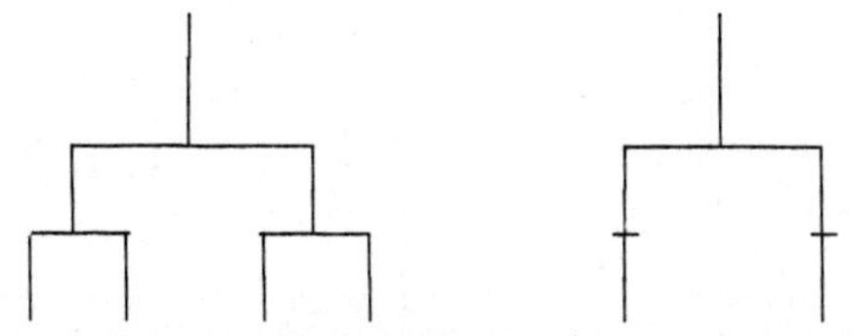

To make the size of the design comparable to that of Bainbridge's, one would have a/2 repetitions of the basic design. The statistics needed for ITLS are presented in Table 8.

Table 8: Statistics Used in ITLS for Anderson Three-stage Nested Design

Statistic[a]	Expectation	Variance
$\bar{y}_1$	μ	$(\sigma_c^2+2\sigma_b^2+4\sigma_a^2)/2a$
$\bar{y}_2$	μ	$(\sigma_c^2+\sigma_b^2+2\sigma_a^2)/a$
$SSA_1^* = \sum_{i=1}^{a/2} (z_{1i}-4\mu)^2/4$	$\frac{a}{2}(\sigma_c^2+2\sigma_b^2+4\sigma_a^2)$	$a(\sigma_c^2+2\sigma_b^2+4\sigma_a^2)^2$
$SSA_2^* = \sum_{i=1}^{a/2} (z_{2i}-2\mu)^2/2$	$\frac{a}{2}(\sigma_c^2+\sigma_b^2+2\sigma_a^2)$	$a(\sigma_c^2+\sigma_b^2+2\sigma_a^2)^2$
SSB_1	$\frac{a}{2}(\sigma_c^2+2\sigma_b^2)$	$a(\sigma_c^2+2\sigma_b^2)^2$
SSB_2	$\frac{a}{2}(\sigma_c^2+\sigma_b^2)$	$a(\sigma_c^2+\sigma_b^2)^2$
SSC_1	$a\sigma_c^2$	$2a\sigma_c^4$

[a] The subscript 1 refers to the design with 4 observations per A-class; 2 refers to the other design.

The Bainbridge design has been recommended, mainly because it is amenable to a simple AOV estimation procedure. The Anderson design requires an unrealistic pooling of sums of squares for AOV. If one uses ITLS, the Anderson design has one distince advantage: the Σ-matrix is diagonal, since $\bar{y}_1$ and $\bar{y}_2$ and the various SS are independent of one another. Therefore, one does not need to determine any cross-product statistics and $X'\Sigma^{-1}X$ is a simple matrix.

Schwartz (1980) conducted 5000 simulated experiments for each parameter set to compare the performance of the Bainbridge, Anderson and Balanced designs in estimating the variance components in a three-stage nested process with a=20 (60 observations). The traces of the mean square errors of these estimates are presented below ($\sigma_c^2 = 1$):

Variance Components σ_a^2	σ_b^2	Anderson	Bainbridge	Balanced
0.1	0.1	0.1632	0.1765	0.0885
0.1	1.0	0.5037	0.5034	0.3505
0.1	10.0	14.22	13.42	12.73
1.0	0.1	0.3554	0.3706	0.3268
1.0	1.0	0.8230	0.8295	0.7896
1.0	10.0	14.99	15.13	14.08
10.0	0.1	10.85	10.89	13.69
10.0	1.0	12.37	11.74	15.63
10.0	10.0	37.64	36.56	46.38

There is very little to choose between the Bainbridge and Anderson designs; the balanced design is best when σ_a^2 is small and least efficient when σ_a^2 is large.

5. NON-BALANCED DESIGNS FOR TWO-WAY CLASSIFICATION MODELS

Muse and Anderson (1978) considered two-way classification models with no interaction:

$$y_{ij} = n_{ij}(\mu + r_i + c_j + e_{ij}),$$

where $i=1,2,\ldots,r$; $j=1,2,\ldots,c$; $n_{ij}=0$ or 1 and $\sum_{ij} n_{ij}=n$. The respective variance components are σ_r^2, σ_c^2 and σ_e^2. Gaylor (1960) proposed an L-design and a disconnected Balanced Disjoint Rectangle (BD)-design; Bush and Anderson (1963) proposed S-and C-designs. Muse compared Bush's S-design, a modified MS-design, Gaylor's L-design, a balanced B-design, an off-diagonal OD-design and four variations of Gaylor's BD-design. The incidence X-matrices for these designs with 36 or 37 observations are presented in Table 9.

Table 9: Incidence Matrices for the Muse Designs [a]

S-37

$$\begin{bmatrix} \text{Let } i,j = 1,2,\ldots,13 \text{ where} \\ n_{11}=n_{12}=n_{13,12}=n_{13,13}=1; \\ n_{ij}=1, \text{ if } i-1 \le j \le i+1, \text{ for} \\ i=2,\ldots,12;\ n_{ij}=0, \text{ otherwise.} \end{bmatrix}$$

MS-36

$$\begin{bmatrix} \text{Let } i,j = 1,2,\ldots,12 \text{ where} \\ n_{11}=n_{12}=n_{1,12}=1; \\ n_{12,1}=n_{12,11}=n_{12,12}=1, \\ n_{ij}=1, \text{ if } i-1 \le j \le i+1, \\ \text{for } i=2,\ldots,11;\ n_{ij}=0, \text{ otherwise.} \end{bmatrix}$$

OD3-36

$$\begin{bmatrix} \underset{6\times 6}{I} \otimes \begin{bmatrix} 0 & 1 & 1 \\ 1 & 0 & 1 \\ 1 & 1 & 0 \end{bmatrix} \end{bmatrix}$$

L-36

$$\begin{bmatrix} \text{Let } i,j=1,2,\ldots,10 \\ n_{ij}=0 \text{ if } i \ge 3 \text{ and } j \ge 3; \\ n_{ij}=1, \text{ otherwise} \end{bmatrix}$$

BD3-36

$$\begin{bmatrix} \underset{4\times 4}{I} \otimes \begin{bmatrix} \underline{J}_3\ \underline{J}_3' \end{bmatrix} \end{bmatrix}$$

BD2-36

$$\begin{bmatrix} \underset{9\times 9}{I} \otimes (\underline{J}_2 \underline{J}_2') \end{bmatrix}$$

BD2×3-36

$$\begin{bmatrix} \underset{6\times 6}{I} \otimes (\underline{J}_2 \underline{J}_3') \end{bmatrix}$$

BD3×2-36

$$\begin{bmatrix} \underset{6\times 6}{I} \otimes (J_3 J_2') \end{bmatrix}$$

B-36

$$\begin{bmatrix} \underline{J}_6\ \underline{J}_6' \end{bmatrix}$$

[a] $\underline{J}_n$ denotes an n-vector of ones.

The asymptotic variances of $\hat{\mu}$ and $\hat{\underline{\sigma}}^2$ are presented by Searle (1970):

$$AV(\hat{\mu}) = [\underline{J}_n' \, V^{-1} \, \underline{J}_n]^{-1},$$

$$AV(\hat{\underline{\sigma}}^2) = 2\{tr(V^{-1} \frac{\partial V}{\partial \sigma_i^2} V^{-1} \frac{V}{\partial \sigma_j^2}) \text{ for } i,j=e,r,c,\}^{-1},$$

where $V = \sigma_e^2 I_n + \sigma_r^2 V_1 + \sigma_c^2 V_2 = Var(\underline{y})$ and $\underline{J}_n$ is an n-vector of ones. Determination of $AV(\hat{\underline{\sigma}}^2)$ was simplified by use of a two-step orthogonal transformation method, whereby $\underline{y}$ is transformed to a vector $\underline{y}^* = (X^*)' \, \underline{y}$, where X^* consists of orthogonal vectors. The elements of X^* are not functions of $\underline{\sigma}^2$. Step one yields $c_1 (0 \le c_1 \le n)$ specific orthogonal vectors denoted by

$$X_1^* = [\underline{X}_1^*, \underline{X}_2^*, \ldots, \underline{X}_{c_1}^*],$$

such that $V_1\underline{X}_i^* = j_i\underline{X}_i^*$ and $V_2\underline{X}_i^* = k_i\underline{X}_i^*$, where j_i and k_i are arbitrary constants, ≥ 0. The variance-covariance matrix (V_i^*) for X_1^* is diagonal.

If $c_1 < n$, step two is employed to obtain a second orthogonal matrix

$$X_2^* = [\underline{X}_{c_1+1}^*, \ldots, \underline{X}_n^*]$$

such that its variance $V_2^* = X_2^{*\prime}\, V\, X_2^*$ has a block diagonal structure.

The application of steps one and two yields an orthogonal matrix $X^* = [X^*_1 | X_2^*]$. If the design is disconnected, the procedure is applied to each block of the design (OD and BD-designs).

The method is illustrated for one block of the OD3-36 design. In this case

$$V = \sigma_e^2 I_6 + \sigma_r^2 R_1 R_1' + \sigma_c^2 C_1 C_1',$$

where $R_1 = I_3 \otimes \underline{J}_2$ and $C_1 = \begin{bmatrix} 0 & 0 & 1 & 0 & 1 & 0 \\ 1 & 0 & 0 & 0 & 0 & 1 \\ 0 & 1 & 0 & 1 & 0 & 0 \end{bmatrix}$.

Step one produced two vectors:

$$X_1^{*\prime} = \begin{bmatrix} 1 & -1 & -1 & 1 & 1 & -1 \\ 1 & 1 & 1 & 1 & 1 & 1 \end{bmatrix}.$$

Note that $V_1\underline{X}_1^* = V_2\underline{X}_1^* = 0$; $V_1\underline{X}_2^* = V_2\underline{X}_2^* = 2\underline{X}_2^*$. No other orthogonal vector meets the requirements of step one.

$$V_1^* = \text{Diag}[6\sigma_e^2;\ 6(\sigma_e^2 + 2\sigma_r^2 + 2\sigma_c^2)].$$

The second step provides

$$X_2^{*\prime} = \begin{bmatrix} 1 & 0 & -1 & -1 & 0 & 1 \\ 1 & -2 & 1 & 1 & -2 & 1 \\ 1 & 2 & -1 & 1 & -2 & -1 \\ 1 & 0 & 1 & -1 & 0 & -1 \end{bmatrix}; \quad V_2^* = \begin{bmatrix} A_2 & 0 \\ 0 & A_3 \end{bmatrix}$$

where

$$A_2 = \begin{bmatrix} 2(2\sigma_e^2 + 3\sigma_1) & -6\sigma_2 \\ -6\sigma_2 & 6(2\sigma_e^2 + \sigma_1) \end{bmatrix};$$

$$A_3 = \begin{bmatrix} 6(2\sigma_e^2 + 3\sigma_1) & 6\sigma_2 \\ 6\sigma_2 & 2(2\sigma_e^2 + \sigma_1) \end{bmatrix}$$

with $\sigma_1 = \sigma_r^2 + \sigma_c^2$ and $\sigma_2 = \sigma_r^2 - \sigma_c^2$. Since V_2^* has two sets of off-diagonal terms, the set of sufficient statistics will be

$$\Sigma Z_{1i}^2,\ \Sigma(Z_{2i} - 6\hat{\mu})^2,\ \Sigma Z_{3i}^2,\ \Sigma Z_{3i}Z_{4i},\ \Sigma Z_{4i}^2,\ \Sigma Z_{5i}^2,\ \Sigma Z_{5i}Z_{6i},\ \Sigma Z_{6i}^2,$$

where $\hat{\mu} = \bar{Y}$ for this design, the summations are over the six blocks and $Z_{hi} = X_{hi}^{*\prime}\underline{y}_i$.

Muse and Anderson (1978) compare the asymptotic variances for the designs of Table 9 for σ_r^2 and σ_c^2 ranging from 0 to 8. The criteria used in the comparisons were trace (AV), Var($\hat{\sigma}^2$), Var($\hat{\sigma}_r^2$) and Var($\hat{\sigma}_c^2$), where $\sigma^2 = \sigma_r^2 + \sigma_c^2 + \sigma_e^2$. For Var($\hat{\sigma}^2$), the OD3-design is generally superior. For trace (AV), the results were as follows:

Condition	Preferred Design
σ_e^2 is dominant	B-36
$\max(\sigma_r^2/\sigma_e^2, \sigma_c^2/\sigma_e^2) \doteq 1$	BD3
$1 \leq \sigma_r^2/\sigma_e^2 \doteq \sigma_c^2/\sigma_e^2 < 2$	MS
$2 \leq \sigma_r^2/\sigma_e^2 \doteq \sigma_c^2/\sigma_e^2$	S or OD3
$\sigma_r^2 \neq \sigma_c^2$ and one larger than σ_e^2	BD2
$\sigma_r^2 \neq \sigma_c^2$ and both larger than σ_e^2	OD3

The L-design should not be used. When σ_e^2 is not the dominant variance component, there is little to choose among the other five non-balanced design, except that OD3 is superior when σ_r^2 and σ_c^2 are quite large.

Muse and Anderson (1978) also compared the small sample (SS) properties of BD2-36 and B-36 and Thitakomal (1977) computed results for OD3-36, using $\sigma_r^2 = \sigma_c^2 = 0.5$ and 8.0 and $\sigma_r^2 = 0.5$, $\sigma_c^2 = 8.0$. These comparisons were based on 5,000 simulated experiments for each parameter set. A comparison of large and small sample results is presented in Table 10.

Table 10: Ratios of Small Sample MSE Estimates (SS) and Ratios of Asymptotic Variances (LS) for the BD2- and OD3-Designs Relative to the B-Design ($\sigma_e^2 = 1$)

σ_r^2	σ_c^2	Ratio	Type	$R(\hat{\sigma}_r^2)$	$R(\hat{\sigma}_c^2)$	$R(\hat{\sigma}_e^2)$	R(trace)	$R(\hat{\sigma}^2)$
0.5	0.5	BD2:B	SS	1.21	1.21	2.17	1.42	0.706
			LS	1.18	1.18	2.62	1.46	0.701
		OD3:B	SS	1.25	1.25	2.58	1.55	1.13
			LS	1.21	1.21	3.04	1.56	0.703
8.0	8.0	BD2:B	SS	0.653	0.653	2.81	0.655	0.436
			LS	0.523	0.523	2.78	0.527	0.433
		OD3:B	SS	0.510	0.510	5.56	0.520	0.446
			LS	0.442	0.442	4.14	0.448	0.404
0.5	8.0	BD2:B	SS	1.68	0.385	2.41	0.403	0.378
			LS	1.53	0.377	2.77	0.395	0.375
		OD3:B	SS	1.62	0.395	3.02	0.414	0.402
			LS	1.95	0.374	3.98	0.399	0.373

There is not as close agreement between LS and SS results as one would like; however, these results agree insofar as the selection between designs is concerned, when the trace criterion is used. Furthermore, these results support the viewpoint that LS results provide a reasonable indication of design performance provided the asymptotic ratio of interest for the two designs is not too close to one. I should mention that Muse obtained closed form solutions for the ML-equations for the B-36 and BD2-36 designs.

Thitakomal (1977) has also studied the more general two-way random model:

$$y_{ijk} = \mu + r_i + c_j + (rc)_{ij} + e_{k(ij)},$$

with variance components σ_r^2, σ_c^2, σ_{rc}^2 and σ_e^2. He compared the designs described in Table 11.

Table 11: Description of Thitakomal Designs

Design	N*	Descriptions
BI	50	B-25 with 2 observations per cell
BII	72	B-36 with 2 observations per cell
BDI	60	BD2-12 with 2 observations per cell BD2-36 with 1 observation per cell
BDII	60	BD2-24 with 2 observations in the upper-left and lower-right cells and 1 observation in each of the remaining cells; BD2-24 with 1 observation per cell
BDIII	60	BD2-48 with 2 observations in the upper-left cell and 1 observation in each of the remaining cells
BDIV	64	BD4-32 with 2 observations per cell
OD	60	OD3-12 with 2 observations per off-diagonal cell; OD3-36 with 1 observation per off-diagonal cell

*N = Total number of observations.

Based on asymptotic variances, the trace criterion produced these results:

Condition		Preferred Design[a/]		
		σ_{rc}^2		
σ_r^2	σ_c^2	.5	1	2
.5	.5	BI	BII	BII
≤ 1	1	IV	IV	BI or BII
≤ 1	2	I	I	IV
≤ 1	8	I	I	I
2	2	I	I	I
2	8	OD	OD	I
8	8	OD	OD	OD

a/ BDI = I, BDII = II, etc.; $\sigma_e^2 = 1$.

As before, OD is desirable when both σ_r^2 and σ_c^2 are large; BDI is the most useful when either σ_r^2 or σ_c^2 is large; a balanced design is best when σ_r^2 and σ_c^2 are small.

The statistics needed for ITLS estimation for those non-balanced designs in Table 11 with diagonal Σ are presented in Table 12.

Table 12: Statistics for BDI and BDIV Designs

Statistic	Expectation	Variance
	BDI-Design	
$\bar{y}_2$	μ	$(4\sigma_r^2+4\sigma_c^2+2\sigma_{rc}^2+\sigma_e^2)/24$
$\bar{y}_1$	μ	$(2\sigma_r^2+2\sigma_c^2+\sigma_{rc}^2+\sigma_e^2)/36$
SSB_2^*	$3(4\sigma_r^2+4\sigma_c^2+2\sigma_{rc}^2+\sigma_e^2)$	$6(4\sigma_r^2+4\sigma_c^2+2\sigma_{rc}^2+\sigma_e^2)^2$
SSR_2	$3(4\sigma_r^2+2\sigma_{rc}^2+\sigma_e^2)$	$6(4\sigma_r^2+2\sigma_{rc}^2+\sigma_e^2)^2$
SSC_2	$3(4\sigma_c^2+2\sigma_{rc}^2+\sigma_e^2)$	$6(4\sigma_c^2+2\sigma_{rc}^2+\sigma_e^2)^2$
$SS(RC)_2$	$3(2\sigma_{rc}^2+\sigma_e^2)$	$6(2\sigma_{rc}^2+\sigma_e^2)^2$
SSE_2	$12\sigma_e^2$	$24\sigma_e^2$
SSB_1^*	$9(2\sigma_r^2+2\sigma_c^2+\sigma_{rc}^2+\sigma_e^2)$	$18(2\sigma_r^2+2\sigma_c^2+\sigma_{rc}^2+\sigma_e^2)^2$
SSR_1	$9(2\sigma_r^2+\sigma_{rc}^2+\sigma_e^2)$	$18(2\sigma_r^2+\sigma_{rc}^2+\sigma_e^2)^2$
SSC_1	$9(2\sigma_c^2+\sigma_{rc}^2+\sigma_e^2)$	$18(2\sigma_c^2+\sigma_{rc}^2+\sigma_e^2)^2$
$SS(RC)_1$	$9(\sigma_{rc}^2+\sigma_e^2)$	$18(\sigma_{rc}^2+\sigma_e^2)^2$
	BD-IV Design	
$\bar{y}$	μ	$(8\sigma_r^2+8\sigma_c^2+2\sigma_{rc}^2+\sigma_e^2)/64$
SSB*	$2(8\sigma_r^2+8\sigma_c^2+2\sigma_{rc}^2+\sigma_e^2)$	$4(8\sigma_r^2+8\sigma_c^2+2\sigma_{rc}^2+\sigma_e^2)^2$
SSR	$6(8\sigma_r^2+2\sigma_{rc}^2+\sigma_e^2)$	$12(8\sigma_r^2+2\sigma_{rc}^2+\sigma_e^2)^2$
SSC	$6(8\sigma_c^2+2\sigma_{rc}^2+\sigma_e^2)$	$12(8\sigma_c^2+2\sigma_{rc}^2+\sigma_e^2)^2$
SSRC	$18(2\sigma_{rc}^2+\sigma_e^2)$	$36(2\sigma_{rc}^2+\sigma_e^2)^2$
SSE	$32\sigma_e^2$	$64\sigma_e^4$

For BDII, BDIII and OD, there are off-diagonal terms in Σ; hence, some cross-product terms are needed in the set of sufficient statistics. For BDII the part with one observation per cell has the same estimators as that part of BDI, with the coefficients of expectations and variances multiplied by 2/3. The part with two observations in the diagonal cells involves two correlated estimators of μ: $\bar{y}_1$ with a variance of $(2\sigma_r^2+2\sigma_c^2+2\sigma_{rc}^2+\sigma_e^2)/24$ and $\bar{y}_2$ with a variance of $(\sigma_r^2+\sigma_c^2+\sigma_{rc}^2+\sigma_e^2)/12$; the covariance of $\bar{y}_1$ and $\bar{y}_2$ is $(\sigma_r^2+\sigma_c^2)/12$. The quadratic forms are based on the following transformation matrix:

$$X^{*\prime} = \begin{bmatrix} 1 & -1 & 0 & 0 & 0 & 0 \\ 0 & 0 & 0 & 0 & 1 & -1 \\ 1 & 1 & 1 & 1 & 1 & 1 \\ 1 & 1 & -2 & -2 & 1 & 1 \\ 1 & 1 & 0 & 0 & -1 & -1 \\ 0 & 0 & 1 & -1 & 0 & 0 \end{bmatrix} ; \quad V^* = \begin{bmatrix} V_1^* & & \\ & V_2^* & \\ & & V_3^* \end{bmatrix}$$

where $V_1^* = 2\sigma_e^2 \; I(2\times 2)$;

$$V_2^* = \begin{bmatrix} 18(\sigma_r^2+\sigma_c^2) + 2(5\sigma_{rc}^2+3\sigma_e^2) & 4\sigma_{rc}^2 \\ 4\sigma_{rc}^2 & 2(8\sigma_{rc}^2+6\sigma_e^2) \end{bmatrix}$$

$$V_3^* = \begin{bmatrix} 4(2\sigma_r^2+2\sigma_c^2+2\sigma_{rc}^2+\sigma_e^2) & 4(\sigma_r^2-\sigma_c^2) \\ 4(\sigma_r^2-\sigma_c^2) & 2(\sigma_r^2+\sigma_c^2+\sigma_{rc}^2+\sigma_e^2) \end{bmatrix} .$$

Hence the additional statistics from this part of BDII are ΣZ_{1i}^2, ΣZ_{2i}^2, ΣZ_{3i}^2, $\Sigma Z_{3i}Z_{4i}$, ΣZ_{4i}^2, ΣZ_{5i}^2, $\Sigma Z_{5i}Z_{6i}$ and ΣZ_{6i}^2, where the summation is $i=1,\ldots,6$.

For BDIII, there will be 12 blocks of 5 observations each. The estimator will be based on $(\bar{y}_{11},\bar{y}_{12},\bar{y}_{21},\bar{y}_{22})$ with (24,12,12,12) observations and variance-covariance matrix.

$$V(\bar{y}) = \begin{bmatrix} (2\sigma_r^2+2\sigma_c^2+2\sigma_{rc}^2+\sigma_e^2)/24 & \sigma_r^2/12 & \sigma_c^2/12 & 0 \\ \sigma_r^2/12 & \sigma^2/12 & 0 & \sigma_c^2/12 \\ \sigma_c^2/12 & 0 & \sigma^2/12 & \sigma_r^2/12 \\ 0 & \sigma_c^2/12 & \sigma_r^2/12 & \sigma^2/12 \end{bmatrix}$$

where $\sigma^2 = \sigma_r^2+\sigma_c^2+\sigma_{rc}^2+\sigma_e^2$. The X^* and V^*-matrices are

$$X^{*\prime} = \begin{bmatrix} 1 & -1 & 0 & 0 & 0 \\ 1 & 1 & 1 & 1 & 1 \\ -3 & -3 & 2 & 2 & 2 \\ 0 & 0 & 1 & 1 & -2 \\ 0 & 0 & 1 & -1 & 0 \end{bmatrix} ; \quad V^* = \begin{bmatrix} 2\sigma_e^2 & 0 \\ 0 & V_2^* \end{bmatrix}$$

$$V_2^* = \begin{bmatrix} T_{11} & T_{12} & T_{13} & T_{14} \\ T_{12} & T_{22} & T_{23} & T_{24} \\ T_{13} & T_{23} & T_{33} & T_{34} \\ T_{14} & T_{24} & T_{34} & T_{44} \end{bmatrix} \quad \begin{aligned} T_{11} &= 13(\sigma_r^2+\sigma_c^2)+7\sigma_{rc}^2+5\sigma_c^2; \\ T_{22} &= 32(\sigma_r^2+\sigma_c^2)+48\sigma_{rc}^2+30\sigma_e^2; \\ T_{33} &= 2(\sigma_r^2+\sigma_c^2)+6(\sigma_{rc}^2+\sigma_e^2); \\ T_{44} &= 2(\sigma_r^2+\sigma_c^2+\sigma_{rc}^2+\sigma_e^2); \end{aligned}$$

$$T_{12} = -(4\sigma_r^2+4\sigma_c^2+6\sigma_{rc}^2);\ T_{13} = \sigma_r^2+\sigma_c^2;\ T_{14} = \sigma_r^2-\sigma_c^2;$$

$$T_{23} = -8(\sigma_r^2+\sigma_c^2);\ T_{24} = -8(\sigma_r^2-\sigma_c^2);\ T_{34} = 2(\sigma_r^2-\sigma_c^2).$$

Hence the quadratic statistics are: ΣZ_{1i}^2, ΣZ_{21i}^2, $\Sigma Z_{21i}Z_{22i}$, $\Sigma Z_{21i}Z_{23i}$, $\Sigma Z_{21i}Z_{24i}$, ΣZ_{22i}^2, $\Sigma Z_{22i}Z_{23i}$, $\Sigma Z_{22i}Z_{24i}$, ΣZ_{23i}^2, $\Sigma Z_{23i}Z_{24i}$ and ΣZ_{24i}^2, where i=1,...,12.

For the OD-design the part with one observation per occupied cell has the same statistics as Muse's OD-3 design except that σ_e^2 is replaced by $\sigma_e^2+\sigma_{rc}^2$. The part with two observations per occupied cell has A_2 and A_3 the same except they are multiplied by 4 and $2\sigma_e^2$ is replaced by $2\sigma_{rc}^2+\sigma_e^2$, $V_1^*=\text{Diag}[2\sigma_e^2 I_{6\times 6};\ 12(2\sigma_{rc}^2+\sigma_e^2);\ 12(4\sigma_r^2+4\sigma_c^2+2\sigma_{rc}^2+\sigma_e^2)]$. The summation of the squares and products is for i=1,2.

I have developed a new design, whose properties have not been investigated. The basic incidence matrix is

$$\begin{bmatrix} 2 & 1 & 0 \\ 1 & 1 & 1 \\ 0 & 1 & 2 \end{bmatrix}$$

There will be four means of interest: $(\bar{y}_{111}+\bar{y}_{112}+\bar{y}_{331}+\bar{y}_{332})/4$, $(\bar{y}_{12}+\bar{y}_{32})/2$, $(\bar{y}_{21}+\bar{y}_{23})/2$ and $\bar{y}_{22}$. The transformation matrix and variances are presented in Table 13. Hence for this design, the quadratic statistics will be $[\Sigma_i Z_{hi}^2$, h=1,2,...,9], $\Sigma Z_{3i}Z_{4i}$, $\Sigma Z_{3i}Z_{5i}$, $\Sigma Z_{3i}Z_{6i}$, $\Sigma Z_{4i}Z_{5i}$, $\Sigma Z_{4i}Z_{6i}$, $\Sigma Z_{5i}Z_{6i}$, $\Sigma Z_{7i}Z_{8i}$, $\Sigma Z_{7i}Z_{9i}$ and $\Sigma Z_{8i}Z_{9i}$.

Table 13: Transformation Matrices and Variances for New Design

$$X^{*\prime} = \left[\begin{array}{ccccccccc} + & - & 0 & 0 & 0 & 0 & 0 & 0 & 0 \\ 0 & 0 & 0 & 0 & 0 & 0 & 0 & + & - \\ \hline + & + & -2 & -2 & 4 & -2 & -2 & + & + \\ + & + & + & + & + & + & + & + & + \\ + & + & + & -2 & -2 & -2 & + & + & + \\ + & + & -2 & + & -2 & + & -2 & + & + \\ \hline 0 & 0 & + & - & 0 & + & - & 0 & 0 \\ + & + & - & - & 0 & + & + & - & - \\ + & + & + & + & 0 & - & - & - & - \end{array}\right] = \left[\begin{array}{c} X_1^{*\prime} \\ \hline X_2^{*\prime} \\ \hline X_3^{*\prime} \end{array}\right]$$

$$V^* = \begin{bmatrix} V_1^* & & \\ & V_2^* & \\ & & V_3^* \end{bmatrix};\qquad V_1^* = 2\sigma_e^2\ I(2\times 2);$$

$$V_3^* = \begin{bmatrix} 2(\sigma_r^2+\sigma_c^2)+4(\sigma_{rc}^2+\sigma_e^2) & 2(\sigma_r^2-\sigma_c^2) & 6(\sigma_r^2-\sigma_c^2) \\ 2(\sigma_r^2-\sigma_c^2) & 2(\sigma_r^2+\sigma_c^2)+12\sigma_{rc}^2+8\sigma_e^2 & 4\sigma_{rc}^2+6(\sigma_r^2+\sigma_c^2) \\ 6(\sigma_r^2-\sigma_c^2) & 4\sigma_{rc}^2+6(\sigma_r^2+\sigma_c^2) & 18(\sigma_r^2+\sigma_c^2)+12\sigma_{rc}^2+8\sigma_e^2 \end{bmatrix}$$

V_2^* has all off-diagonal terms of $4\sigma_{rc}^2$ and diagonal terms:

$$40\sigma_{rc}^2+36\sigma_e^2;\ 27(\sigma_r^2+\sigma_c^2)+13\sigma_{rc}^2+9\sigma_e^2;\ 54\sigma_r^2+22\sigma_{rc}^2+18\sigma_e^2;\ 54\sigma_c^2+22\sigma_{rc}^2+18\sigma_e^2.$$

Stroup, et al (1980) have developed an ITLS procedure to estimate the variance components for completely random Balanced Incomplete Block Designs; it is demonstrated that ITLS is superior to the usual Analysis of Variance and Symmetric Sums estimating procedures. In an article recently submitted for publication, BIB designs are compared with the Balanced and S, MS, L, BD and OD designs. The BIB design with 9 blocks, 12 treatments and 4 treatments per block compares favorably with the BD2 and OD3 designs.

6. COMPOSITING

If the sampling and measurement process is very time consuming (e.g. if it involves conducting a chemical analysis), the optimal number of measurements per sample often will be less than 1. Cameron (1951) discusses a number of compositing plans for estimating the components of variance when sampling baled wool for percent clean content. This is one example of the probelms encountered in bulk sampling; see for example Duncan (1962) and Bicking (1964). A comparison of various sampling plans is presented by Kussmaul and Anderson (1967), in which is it assumed that the measurement variance component σ_m^2 is so small that it can be neglected. As far as I know, all of these articles use ANOVA estimating procedures. Kussmaul's research should be extended to designs for which σ_m^2 can be estimated and in which other estimating procedures, such as ITLS, are considered.

Cameron presents some data for which 7 bales of wool had one core per bale and 7 bales had 4 cores per bale; hence, there were 14 measurements and 35 samples. Unfortunately these data do not provide information to separate σ_b^2 and σ_m^2. The data and ANOVA are presented in Table 14. Note the standard errors for both σ_s^2 and $\sigma_b^2 + \sigma_m^2$ exceed the estimators, indicating that many more than 14 bales were needed to estimate the components of variance. Because the mean based on composites of four samples per bale is more precise than the one based on only one sample per bale, the two means should be weighted inversely to their variances in computing $\hat{\eta}$. Unfortunately these weights are not too precise because each is based on only six degrees of freedom.

Table 14: Average Percent Clean Content of Composited Samples of Wool [Cameron (1951)].

	1 Sample/Bale	4 Samples/Bale
	58.31	55.48
	58.46	57.81
	54.71	57.23
	60.23	57.33
	61.52	58.86
	62.23	58.78
	60.88	60.78
Average	59.48	58.04

Model: $y_{1i} = \eta + b_i + s_i + m_i$; $y_{4i} = \eta + b_i + \bar{s}_i + m_i$

where $i=1,2,\ldots,7$ and $\bar{s}_i = \sum_{j=1}^{4} s_{ij}/4$.

ANOVA

Source	DF	SS	MS	E(MS)
Batches (1S/B)	6	39.409	6.577	$\sigma_b^2 + \sigma_m^2 + \sigma_s^2$
Batches (4S/B)	6	16.494	2.749	$\sigma_b^2 + \sigma_m^2 + \sigma_s^2/4$

$$\hat{\sigma}_s^2 = \frac{4}{3}(6.577-2.749) = 5.104$$

$$\hat{\sigma}_b^2+\hat{\sigma}_m^2 = \frac{1}{3}[4(2.749)-6.577] = 1.473$$

$$\hat{\eta} = \frac{\frac{59.48}{6.577} + \frac{58.04}{2.749}}{\frac{1}{6.577} + \frac{1}{2.749}} = 58.46.$$

$$\hat{V}(\hat{\sigma}_s^2) = (\frac{4}{3})^2 \cdot \frac{2}{6}[(6.577)^2 + (2.749)^2] = 30.11$$

$$\hat{V}(\hat{\sigma}_b^2+\hat{\sigma}_m^2) = \frac{1}{9} \cdot \frac{2}{6}[16(2.749)^2 + (6.577)^2] = 6.080$$

$$\hat{V}(\hat{\eta}) = [\frac{7}{6.577} + \frac{7}{2.749}]^{-1} = 0.2770$$

S.E.: 5.487 for $\hat{\sigma}_s^2$; 2.466 for $\hat{\sigma}_b^2 + \hat{\sigma}_m^2$; 0.526 for $\hat{\eta}$.

In Table 15, I present a suggested compositing plan, which would enable me to estimate all three components. It is recommended that the ITLS estimation procedure be used, because there are more sums of squares in the analysis of variance than components of variance. Also it is desirable to estimate η and the variance components using a single computing algorithm, because $\hat{\eta}$ will be a weighted linear function of the three means ($\bar{y}_1$, $\bar{y}_2$ and $\bar{y}_3$). Cameron (1951) presents other compositing procedures than the one used in Table 14; however, in none of these is an optimal estimating procedure such as ITLS proposed. Since bulk sampling is so important and measurement costs are high, it appears that some research should be devoted to both optimal sampling plans and estimating procedures for experiments in which compositing is necessary.

Table 15: A Suggested Compositing Plan for Which All Variance Components Can Be Estimated

1) k_1 batches, each with 1 measurement based on a composite of n_1 samples.

2) k_2 batches, each with 1 measurement based on a composite of $n_2(>n_1)$ samples.

3) k_3 batches, each with 2 measurements, each measurement based on a composite of n_3 samples.

Total numbers: $K = k_1 + k_2 + k_3$ batches;

$M = k_1 + k_2 + 2k_3$ measurements; $N = n_1k_1 + n_2k_2 + 2n_3k_3$ samples.

ITLS Statistic	Expectation	Variance
$\bar{y}_1$	η	$\frac{\sigma_b^2+\sigma_m^2}{k_1} + \frac{\sigma_s^2}{n_1k_1}$
$\bar{y}_2$	η	$\frac{\sigma_b^2+\sigma_m^2}{k_2} + \frac{\sigma_s^2}{n_2k_2}$
$\bar{y}_3$	η	$\frac{\sigma_b^2}{k_3} + \frac{\sigma_m^2}{2k_3} + \frac{\sigma_s^2}{2n_3k_3}$
$SSB_1 + k_1(\bar{y}_1-\eta)^2$	$k_1(\sigma_b^2+\sigma_m^2+ \frac{\sigma_s^2}{n_1})$	$2k_1(\sigma_b^2+\sigma_m^2+ \frac{\sigma_s^2}{n_1})^2$
$SSB_2 + k_2(\bar{y}_2-\eta)^2$	$k_2(\sigma_b^2+\sigma_m^2+ \frac{\sigma_s^2}{n_2})$	$2k_2(\sigma_b^2+\sigma_m^2+ \frac{\sigma_s^2}{n_2})^2$
$SSB_3 + 2k_3(\bar{y}_3-\eta)^2$	$k_3(2\sigma_b^2+\sigma_m^2+ \frac{\sigma_s^2}{n_3})$	$2k_3(2\sigma_b^2+\sigma_m^2+ \frac{\sigma_s^2}{n_3})^2$
SSM_3	$k_3(\sigma_m^2 + \frac{\sigma_s^2}{n_3})$	$2k_3(\sigma_m^2 + \frac{\sigma_s^2}{n_3})^2$

7. REFERENCES

[1] Anderson, R. L. (1961): Designs for estimating variance components. Proceedings of the Seventh Conf. on the Design of Experiments in Army Research Development and Testing. Fort Monmouth, New Jersey, October 18-20, 781-823. Inst. Stat. Mimeo Ser. 310.

[2] Anderson, R. L. (1975): Designs and estimators for variance components. Chapter 1 in Statistical Design and Linear Models. Edited by J.N. Srivastava, North Holland Publishing Co., Amsterdam.

[3] Anderson, R. L. and Bancroft, T. A. (1952): Statistical Theory in Research. McGraw-Hill, New York.

[4] Anderson, R. L. And Crump, P. P. (1967): Comparison of designs and estimation procedures for estimating parameters in a two-stage nested process. Technometrics 9, 499-516.

[5] Bainbridge, T. R. (1963): Staggered nested designs for estimating variance components. ASQC Annual Conference Transactions, 93-103.

[6] Bicking, C. A. (1964): Bibliography on Sampling of raw materials and products in bulk. Tappi 47(5), 147A-170A.

[7] Bush, N. and Anderson, R. L. (1963): A comparison of three different procedures for estimating variance components. Technometrics 5, 421-440.

[8] Cameron, J. M. (1951): The use of components of variance in preparing schedules for sampling of baled wool. Biometrics 7, 83-96.

[9] Duncan, A. J. (1962): Bulk sampling: problems and lines of attack. Technometrics 4, 319-344.

[10] Gaylor, D. W. (1960): The construction and evaluation of some designs for the estimation of parameters in random models. Unpublished Ph.D. dissertation, North Carolina State University at Raleigh. Inst. Stat. Mimeo Ser. 256.

[11] Harville, D. A. (1975): Maximum likelihood approaches to variance component estimation and to related problems. Technical Report ARL TR 75-0175, Aerospace Research Laboratory, Wright-Patterson AFB, Ohio.

[12] Jennrich, R. I. and Moore, R. H. (1975): Maximum likelihood estimation by means of non-linear least squares. Statistical Computing Section Proceedings, American Statistical Association, 57-65.

[13] Klotz, J. H., Milton, R. C. and Zacks, S. (1969): Mean square efficiency of estimators of variance components. Journal American Statistical Association, 64, 1383-1402.

[14] Kussmaul, K. and Anderson, R. L. (1967): Estimation of variance components in two-stage nested designs with composite samples. Technometrics 9, 373-389.

[15] Muse, H. D. and Anderson, R. L. (1978): Comparison of designs to estimate variance components in a two-way classification model. Technometrics 20, 159-166.

[16] Schwartz, J. H. (1980): Optimal allocation of resources to reduce product or process variability; a comparison of some designs to estimate the allocation parameters and the variance components. Unpublished Ph.D. Dissertation, University of Kentucky.

[17] Searle, S. R. (1970): Large sample variances of maximum likelihood estimators of variance components. Biometrics 26, 505-524.

[18] Stroup, W. W., Evans, J. W. and Anderson, R. L. (1980): Maximum likelihood estimation of variance components in a completely random BIB design. Commun. Statist. - Theor. Meth., A9(7), 725-756.

[19] Thitakomal, B. (1977): Extension of previous results on properties of estimators of variance components. Unpublished Ph.D. dissertation, University of Kentucky.

[20] Thompson, W. O. and Anderson, R. L. (1975): A comparison of designs and estimators for the two-staged nested design. Technometrics 17, 37-44.

STATISTICS AND RELATED TOPICS
M. Csörgö, D.A. Dawson, J.N.K. Rao, A.K.Md.E. Saleh (eds.)

AN ANALYSIS OF THE MASSACHUSETTS NUMBERS GAME

Herman Chernoff[1,2,3]

Massachusetts Institute of Technology
Department of Mathematics
Cambridge, Massachusetts 02139

The Massachusetts Numbers Game is operated on a pari-mutuel system where the winners divide the pot after the state takes 40%. Thus, it would be possible to develop a winning strategy if one could determine sufficiently unpopular numbers. A study of the winning numbers and payoffs over the four year history of the game shows that it has become more difficult to obtain winning numbers. Prospective gamblers must make allowance for Gambler's Ruin and Regression to the Mean. A moderately complex model allowing for the observed drift in behaviour of the payoffs fails to produce a reasonably useful winning strategy. An Empirical Bayes approach is used.

1. INTRODUCTION

The Massachusetts Numbers Game began on April 10, 1976. After almost two years of its operation, a newspaper article, reporting on it, commented that, as was to be expected, none of the randomly selected four digit numbers had been repeated. This article aroused some interest because the probability of no duplication of a 4-digit number in 500 trials is very small. In a letter to the Editor of the Boston Globe, the Commissioner of the State Lottery corrected the original report which had been based on a careless comment and pointed out that there had been several duplications.

In the meantime, the potential use of the data from the lottery as an example in courses in Statistics, and the possibility that it might furnish an interesting application in pattern recognition led me to request the publicly available data which consisted of the winning numbers and the winning amounts.

The Numbers Game operates on a pari-mutuel system where, each day, the state divides up the total amount bet (the *pot*) among the winners, after subtracting 40% for expenses and revenue. Thus it is conceivable that the player, who can choose his numbers, will have a positive expectation if he bets on numbers which are sufficiently unpopular. In that case, when one of these numbers is *hit*, there will be relatively few winners to divide the pot and the resulting payoff may be high enough to compensate for the amount subtracted by the state and the low probability of a hit. The role of pattern recognition might be that of determining, from a limited experience, the rules which define unpopular numbers.

Upon examining the data from the first few weeks (see Table 1) it became clear that no sophisticated pattern recognition methods would be required to see that zeroes and nines were extremely unpopular and that ones were too popular. Indeed, the simple naive strategy of betting on all numbers whose first three digits did not contain 1, 2, 3, 4 applied over a trial period of the last two months for which I then had data (days 553-600), turned out to be profitable.

Table 1. Winning Numbers and Payoffs

For First 60 plays of Massachusetts Numbers Game

Number	Payoff*	Number	Payoff	Number	Payoff	Number	Payoff
5057	$ 5,017	0681	$10,631	2180	$ 4,033	9543	$ 6,610
6712	3,982	9702	13,286	9227	10,110	9825	8,733
5253	5,804	4256	2,535	8604	9,154	7427	4,235
6549	4,867	0237	6,289	9837	8,005	7474	4,621
0420	9,222	5871	3,609	7712	5,949	4247	4,592
9943	39,170	0661	12,439	0888	7,911	7029	4,897
4716	2,465	5196	3,385	8415	3,846	8257	2,476
5421	2,015	4575	4,143	0249	5,645	4502	5,697
5930	6,766	1598	4,935	2007	7,655	2079	3,751
4962	5,364	5782	5,895	8702	7,316	8363	6,284
6784	5,804	5991	12,661	0955	8,708	0867	8,512
2147	1,586	7088	13,725	3741	3,411	1133	3,082
3509	4,456	0746	8,856	5237	2,183	5571	6,772
7234	3,029	7273	5,797	7521	2,886	4218	3,262
0488	8,278	2933	5,454	2908	6,254	8813	6,913

* Payoff is return for $1 bet on the four-digit winning number.

My interest in the Numbers Game resulted in visits to my office by several students who had tried to beat the game and had failed. Basically, there were two reasons for their failures. They may be somewhat inaccurately labeled, *Gambler's Ruin* and *Regression*.

Gambler's Ruin or Variability refers to the fact that even with a positive expectation on each bet, the gambler may lose his entire fortune before the law of large numbers, which assures him of a positive reward in the long run, takes effect. Specifically if one bets on 10% of all numbers for 300 days a year, the number of wins or hits has expectation 30 and standard deviation 5.2, which is about 17% of the expectation. Thus to be reasonably sure of coming out ahead by the end of the year, the numbers selected must compensate for an additional 30 or more percent due to the possible random fluctuation in the number of hits in addition to the *take* of the state. The relative magnitude of this random component is reduced as one bets on more numbers each day, but then one may have to include numbers which are not so favorable. Related to this variability phenomenon is the above-mentioned success of the naive strategy. Upon analysis, it turned out that the reason for the success was that these numbers appeared 15 times which was more often than the expected 10.4. However, the average winning per hit was only 92% of the amount required to feel that this strategy would succeed in the long run.

Regression refers to the fact that numbers, which were selected because they look good by virtue of their success in the past, are not as good as they look. Their past success, which led to their selection was partly due to their goodness and partly due to good luck which cannot be depended upon in the future. Thus, a correction, essentially related to the "regression to the mean" phenomenon of Galton or Empirical Bayes is required to compensate for an overly optimistic estimate of future performance based on their past performance.

Both of these phenomena, though well appreciated by statisticians, are seldom understood by the untrained, and they were responsible for the sad stories of my visitors, who seemed to be more interested in recounting those stories than in learning my recipe for a winning strategy.

The payoff system is described in more detail in Section 2 and the data consisting of 1320 results of the games are described in Section 3. In Section 4 several betting systems, using predictions based on simple adjustments for regression to the mean and the first 851 results, are proposed. These are evaluated with the data obtained subsequently and found to be over-optimistic. A more complex model is proposed and the 1320 observations are used to derive a new system for which the predicted success is not overwhelmingly optimistic.

2. THE PAYOFF SYSTEM

The Numbers Game operates on a pari-mutuel system which is complicated by the fact that bettors are permitted to select various types of bets. A four digit number is to be selected by a random mechanism each evening. Earlier that day or on previous days, the bettors are permitted to choose a four digit number or to bet on four not necessarily distinct digits in any order. They may also bet on the first three digits in exact or in any order. They may also bet on the last three digits, the first two digits, etc. or even, on any one of the digits.

These separate types of bets are not treated as separate games. All of the money is put into one pot from which the state subtracts 40%. Then the money is disbursed according to a system which can be described as follows. Assign 5000 points for each dollar bet on a 4-digit number in exact order, 700 for each dollar bet on a 3-digit number in exact order, 60 and 6 for 2 and 1-digit numbers, respectively. A four-digit number in any order is treated as 24 separate smaller bets (in exact order) corresponding to each permutation, if the four digits are distinct. If they are not distinct, the appropriate accommodation is made; e.g., 1233 is treated as 12 bets. Then the money left in the pot is assigned to the winning bettors in proportion to their winning points.

If the bettor's choices were made in a random fashion, this payoff system would yield a negative expected gain for each bet made. However, the disadvantage would depend on the type of bets with the three-digit numbers being the least unfavorable. I do not know the reason for this favoritism. Its effect is to give the 3-digit number bettor an advantage over the others; so much so that if almost everyone else bet on 4-digit numbers, he would expect a return of \$0.84 instead of \$0.60 on each dollar bet (in a random betting environment).

A simplified analysis assuming that N_4 dollars are uniformly distributed on 4-digit bets and N_3 dollars on 3-digit bets and no other bets are made gives an expected payoff per dollar bet on a 3-digit number, of

$$E_3 = \frac{(0.6)(N_4 + N_3)\cdot\frac{700}{1000}}{500\,\frac{N_4}{10{,}000} + \frac{700\,N_3}{1000}} = \frac{0.6[1+\lambda]}{1+\frac{5}{7}\lambda} \tag{2.1}$$

where $\lambda = N_4/N_3$ is the ratio of the amounts invested in these bets. Then E_3 ranges from 0.60 to 0.84 as λ varies from 0 to ∞.

Because of the payoff system an anticipated payoff of \$10,000 or more for a \$1 bet on a four-digit number in exact order is required for that number to be worthwhile investing in. However, a payoff of (5/7) of \$10,000 = \$7,143 (in return for \$1 bet on a four-digit number) would be adequate if the bettor chose a three-digit number.

3. THE DATA

The data consists of 1,320 winning numbers and associated payoffs dating from April 10, 1976 thru July 7, 1980. These correspond, with a few minor exceptions, to the results for six days per week, excluding Sundays. The payoffs for the winning numbers in the first few weeks varied considerably from very large numbers to some which were quite small. Afterwards the larger numbers occurred much less often and the fluctuations seemed much reduced. To study the trend, the 1,320 payoffs were arranged in groups of 60 in chronological order, and the means and standard deviations of these 22 groups were computed. First, however, the payoffs were normalized by dividing the payoff per dollar bet on a four digit number by $7,143. This *normalized payoff* is, except for a round-off error, one-thousandth of the payoff per dollar bet on a three-digit number. It represents an estimate of the average return on a dollar bet on one of the corresponding three-digit numbers. Thus a normalized payoff of over $1.00 corresponding to a four-digit number *seems* to indicate that the first three digits and the last three digits represent favorable numbers on which to bet.

On the sixth day of the game, the winning amount for $1.00 bet on the four-digit number 9943, was $39,170 which was an extraordinarily large value, the next highest was $13,725. To reduce the impact of that outlier, it was trimmed down to $13,950 in all calculations.

The means and standard deviations of the groups of 60 normalized payoffs are presented in chronological order in Table 2. Both the means and standard deviations have declined considerably over the four-year period recorded. This decline seems to have been great in the first couple of years and to have stopped more or less in the last two years. These declines are a matter of serious concern to the prospective bettor. The decline in means implies that the general level of payoff is now lower. The decline in standard deviation, even without the decline in mean, would imply a lower frequency of high profitable payoffs. Both of these make it more difficult to come up with a winning system.

Table 2. Means and Standard Deviations of Normalized Payoffs of 22 Groups of 60 in Chronological Order

Group	Mean	Standard Deviation
1	0.863	0.426
2	0.820	0.279
3	0.801	0.293
4	0.788	0.270
5	0.731	0.248
6	0.722	0.223
7	0.766	0.230
8	0.757	0.240
9	0.769	0.287
10	0.756	0.215
11	0.755	0.193
12	0.751	0.234
13	0.727	0.169
14	0.659	0.148
15	0.735	0.209
16	0.700	0.162
17	0.713	0.183
18	0.703	0.175
19	0.729	0.157
20	0.719	0.170
21	0.690	0.212
22	0.716	0.173
Mean	0.744	0.223
Pooled Standard Deviation		0.231

The reduction in the mean suggests that more people are betting on three-digit numbers thereby reducing the gambler's opportunity to take advantage of them. The reduction in the standard deviation suggests that more people are betting on the numbers which were formerly neglected. In short it seems that there are more skillful bettors playing. Since a successful system must attempt to take advantage of the lack of skill of the other bettors, there seems to be a considerably reduced opportunity to profit.

4. THE SYSTEMS

In principle, one way to find favorable numbers is to select those four-digit numbers for which the average payoff in the past has been very large. Unfortunately there are 10,000 four-digit numbers and most of them have not appeared and cannot be evaluated. An alternative is to group these into a smaller number of non-overlapping "homogeneous" sets and to evaluate these sets. For example, we may confine attention to groups for which the first three digits are the same. By homogeneous, we mean that the members of the group would be expected to be more equally advantageous or disadvantageous than randomly selected non-members. The advantage of representing the group by the average performance of those members who have been hit depends on how homogeneous the group is, and is measured by the within group variability.

Using the second three digits for grouping is also a possibility. But a little analysis reveals that the within group variability is higher for these groups than for those formed from the first three digits.

Thus we have two reasons to select and compare numbers by their first three digits. Bets on three-digit numbers are advantageous and the grouping by the first three digits is comparatively homogeneous.

It is possible to group or aggregate still further. For example, there would be greater reliability in using larger groups which would permit us to average over more hits in assessing groups. This advantage is offset by the loss of homogeneity that results from the use of larger groups. One natural way to aggregate further the groups using the first three digits is, to put together in one group, all numbers with the same first three digits in arbitrary order. Such a group can be represented by the element where the three digits are in increasing order. Thus, 122, 212, 221 are all represented by 122. We shall let I_0 represent the four digits, I_1, the first three digits and I_2, the first three digits in arbitrary order. Thus, for $I_0 = 2{,}124$, $I_1 = 212$ and $I_2 = 122$.

In general let $I = I(I_0)$ be an index which identifies the group to which I_0 belongs. For each possible value i of I there is a possibly null set of $N(i) \geq 0$ values of X_{it}, which are the observed normalized payoffs X corresponding to the $I = i$ at the t-th play of the game. Let

$$X = E(X|I) + \varepsilon = Y + \varepsilon \tag{4.1}$$

where

$$Y = \tilde{Y}_I = E(X|I) \tag{4.2}$$

is uncorrelated with ε which has mean 0 and variance σ_ε^2. Then

$$\sigma_X^2 = \sigma_Y^2 + \sigma_\varepsilon^2 \tag{4.3}$$

where $\mu = E(X)$, σ_X^2 and σ_ε^2 are estimated by

$$\bar{X} = \sum_t X_{it}/n\,, \qquad n = \sum_i N(i)$$

$$s_X^2 = \sum_t (X_{it} - \bar{X})^2/(n-1)$$

and

$$s_\varepsilon^2 = \sum_t (X_{it} - \bar{X}(i))^2/\Sigma^*[N(i)-1]$$

with Σ^* the sum over $\{i : N(i) > 1\}$, $\bar{X}(i) = \Sigma_i X_{it}/N(i)$, and Σ_i the sum over $\{t : I=i\}$.

If we assume that $Y = \tilde{Y}_I$ is normally distributed, i.e., $L(Y) = N(\mu, \sigma_Y^2)$ and that ε is normally distributed and independent of I, then

$$L(\bar{X}(I) \mid I = i) = N(\tilde{Y}_i, \frac{\sigma_\varepsilon^2}{N(i)})$$

and for $N(i) > 0$, the Empirical Bayes [2] approach estimates $\tilde{Y}_i$ by

$$\hat{Y}(i) = \frac{\hat{\mu}\hat{\sigma}_Y^{-2} + \bar{X}(i)N(i)\hat{\sigma}_\varepsilon^{-2}}{\hat{\sigma}_Y^{-2} + N(i)\hat{\sigma}_\varepsilon^{-2}} = w\hat{\mu} + (1-w)\,\bar{X}(i) \quad (4.4)$$

where

$$w = \left[1 + \frac{N(i)\,\hat{\sigma}_Y^2}{\hat{\sigma}_\varepsilon^2}\right]^{-1}, \quad (4.5)$$

$\hat{\sigma}_\varepsilon^2 = s_\varepsilon^2$, $\hat{\mu} = \bar{X}$, and $\hat{\sigma}_Y^2 = s_X^2 - s_\varepsilon^2$.

Thus $\hat{Y}(i)$ is a weighted average of the overall mean $\hat{\mu}$ and $\bar{X}(i)$ where the weight bringing $\hat{Y}(i)$ toward $\hat{\mu}$ depends on $N(i)$ as well as on $\hat{\sigma}_Y^2$ and $\hat{\sigma}_\varepsilon^2$. Note that a simple minded estimate of $\tilde{Y}_i$ would be $\bar{X}(i)$ but that for large $\bar{X}(i)$, $\hat{Y}(i)$ comes closer to the overall mean and plays the role of the regression to the mean.

Several betting strategies were devised on the basis of the first 851 observations (which was the data set available to one when I first carried out some non-trivial calculations). We list and discuss several of these.

S_1: Bet on all first three-digit numbers in exact order, i_1, for which $\hat{Y}_1(i_1) > 1.00$ where $\hat{Y}_1(i_1)$ is the estimate of $\tilde{Y}_{it}$ derived by the method described in the last section for $I = I_1$.

S_2: Bet on all first three-digit numbers in arbitrary order, i_2, for which $\hat{Y}_2(i_2)$ is estimated with $I = I_2$, and $\hat{Y}_2(i_2) > 1.00$.

These two systems ignore the declining behavior with time of the means and standard deviations of X and s_X. At the time of the first 851 observations, the estimate of the rate of decline of $E(X)$ as a function of time t, was $-\hat{\beta} = 0.0001441$. Then X was replaced by $X^* = X - \hat{\beta}(t-851)$. As a result S_1 and S_2 were modified to S_1^* and S_2^* using estimates $\hat{Y}_1^*$ and $\hat{Y}_2^*$.

Simple computations yield $\bar{X} = 0.7613$, $s_X = 0.257$, $\hat{\beta} = -0.0001441$, and $s_{X*} = 0.254$. For $I = I_1$, $s_\varepsilon = 0.152$ and $s_{\varepsilon*} = 0.152$ yielding $\hat{\sigma}_Y = 0.207$ and $\hat{\sigma}_{Y*} = 0.203$. For $I = I_2$, $s_\varepsilon = 0.181$, $s_{\varepsilon*} = 0.181$, $\hat{\sigma}_Y = 0.182$ and

$\hat{\sigma}_{Y*} = 0.178$. These calculations permit us to estimate $\tilde{Y}_{I1}$ for $I_1 = i_1$ and $\tilde{Y}_{I_2}$ for $I_2 = i_2$ by $\hat{Y}_1(i_1)$, $\hat{Y}_1^*(i_1)$, $\hat{Y}_2(i_2)$, and $\hat{Y}_2^*(i_2)$.

In Table 3, we list in order of increasing i_1, four groups of i_1 values with the corresponding predicted value of $\tilde{Y}_{I_1}$ and some other information. In detail we list for each of these values of i_1, $\bar{X}_1$ the average of the normalized payoffs for $1 \leq t \leq 851$, $N_1(i_1)$, the number of hits for $1 \leq t \leq 851$, $\hat{Y}_1$ and $\hat{Y}_1^*$. These are followed by the group number G_1, to be described, and N_1' and N_1'' the number of hits for the period $851 < t \leq 1061$ and $1061 < t \leq 1320$ respectively, and the corresponding average payoffs $\bar{X}_1'$ and $\bar{X}_1''$. The four groups are characterized as follows:

Group 1: $\hat{Y}_1^* > 1.00$

Group 2: $\hat{Y}_1^* \leq 1.00 \leq \hat{Y}_1$

Group 3: $0.90 \leq \hat{Y}_1^* \leq 1.00$,

Group 4: Relatives: Values of i_1 for which there were no hits for $t \leq 851$ but which resemble values with large $\hat{Y}_1^*$.

In Table 4, the same presentation is made for I_2, i.e., for three digit numbers in any order.

The betting System S_1^* consists of betting \$1 a day on each of the three digit numbers in group 1, because these are the numbers for which the estimates of the predicted values exceed 1.00. A related system is S_1^{**} where group 1 is supplemented by the members of group 4 which have been selected in a rather subjective fashion as described above.

System S_1^* based on the data in the first 851 results, was tested against the data accumulated in the next 210 games and did very well, returning \$44.19 per day for \$33 bet. This amount was a profit of 33.9% and even exceeded the estimate of the predicted return \$37.70 by 17%. The System S_2^* did not do so well during that same period. S_2^*, which consisted of betting 50 cents or one dollar a day on each of the three-digit numbers in group 1 (depending on whether or not the three digits had a repetition), cost 9(1/2) + 2 or \$6.50 per day and returned only \$5.58 per day which was a loss of 15% and was 20% less than \$6.96, the estimate of the predicted daily return (which should have been a profit of 7.1%).

An analysis of the data reveals that the main source of variation was the variability in the number of hits. For S_1^* there were 10 hits compared to 6.43 expected and for S_2^* there were only 8 when 8.19 were expected. Moreover, when the returns are compared to predicted amounts for those numbers that were hit, there is a deficit of 14.5% for S_1^* and 15.3% for S_2^*. These observed deficits exceed the estimated profits for these systems and suggest that the calculations leading to these systems were over-optimistic for some reason and that the systems will not be profitable. Table 5 presents more detail on the outcomes for the systems S_1^*, S_2^{**}, S_2^*, S_2^{**} for the two successive time periods $851 < t \leq 1061$ and $1062 \leq t \leq 1320$.

These deficits indicate two major tendencies. The deficit tends to be a little greater for the second period than for the first. The deficit is much larger for group 1 than for the other two groups. In fact the deficit is quite small for groups two and three. It seems that the regression to the mean correction was inadequate for the I_1 and I_2 values with the largest Y^*. Incidentally, the deficits for I_1 and I_2 behave in very similar fashions.

Table 3

Predictions Based on $T \leq 851$ and Subsequent Results for Four Groups of I_1 Values

i_1	$\overline{X}_1$	N_1	$\hat{Y}_1$	$\hat{Y}^*_1$	G_1	N'_1	$\overline{X}'_1$	N''_1	$\overline{X}''_1$
001	1.11	2	1.03	0.99	2			1	1.04
002					4				
003					4				
004					4				
005					4			1	1.13
006					4			1	.92
007	1.22	2	1.12	1.07	1			1	.83
008	1.67	1	1.35	1.29	1				
009	1.18	2	1.09	1.02	1				
020	1.43	1	1.20	1.11	1				
030					4				
040	1.34	1	1.14	1.10	1				
042	1.30	1	1.11	1.00	2				
049	1.11	1	.99	.93	3				
050					4			1	1.07
054	1.20	2	1.10	1.04	1				
060					4				
064	1.04	2	.98	.94	3			2	.69
066	1.74	1	1.40	1.29	1				
068	1.49	1	1.24	1.13	1				
070	1.60	2	1.42	1.35	1				
072	1.07	1	.96	.91	3				
074	1.24	1	1.07	.97	2			1	.79
079	1.20	2	1.10	1.06	1	2	.99		
080	1.48	1	1.23	1.13	1				
081	1.27	1	1.09	.99	2	1	1.07	1	.92
086	1.15	3	1.09	1.02	1				
089	1.21	1	1.09	.97	2			1	.88
090	1.15	3	1.09	1.05	1				
095	1.07	4	1.03	.97	2				
299	1.07	1	.96	.90	3				
300	1.58	1	1.29	1.23	1				
330	1.03	1	.94	.90	3				
466	1.45	1	1.21	1.17	1				
494	1.09	1	.97	.90	3			3	.92
498	1.21	1	1.05	.96	2				
533	1.38	1	1.16	1.07	1				
566	1.07	1	.96	.93	3			1	.57
588	1.12	2	1.04	.98	2				
598	1.11	1	.99	.92	3			1	.76
599	1.77	1	1.42	1.31	1	1	.93		
655	1.34	1	1.14	1.07	1	2	1.06		
661	1.04	1	.98	.96	3	1	.76		
679	1.20	1	1.04	.95	2			1	.75
688	1.07	1	.96	.92	3			1	.97
698	1.31	1	1.12	1.03	1	1	1.07		
699	1.34	1	1.14	1.05	1				
708	1.92	1	1.52	1.41	1				
760	1.14	1	1.01	.93	2				
779	1.27	1	1.09	1.01	1				

Table 3 continued...

i_1	$\overline{X}_1$	N_1	$\hat{Y}_1$	$\hat{Y}_1^*$	G_1	N_1'	$\overline{X}_1'$	N_1''	$\overline{X}_1''$
848	1.36	2	1.23	1.15	1	1	.75		
859	1.69	1	1.36	1.29	1				
860	1.28	1	1.10	1.00	2	1	.82	1	1.00
866	1.04	2	.98	.91	3	1	.81		
877	1.26	2	1.15	1.06	1				
882	.99	2	.94	.90	3				
884	1.18	2	1.09	1.01	1	1	.76		
886	1.56	1	1.28	1.19	1				
922	1.17	2	1.08	1.01	1				
940	1.32	1	1.13	1.03	2	1	.85		
944	1.11	2	1.04	.97	2	2	1.05		
950	1.11	2	1.03	.95	2				
956	1.09	3	1.04	.98	2	1	.61		
959	1.06	2	1.00	.95	2				
966	1.24	2	1.14	1.07	1				
970	1.86	1	1.47	1.36	1				
979	1.27	1	1.09	.99	2				
980					4				
982	1.09	2	1.02	.93	2				
983	1.13	1	1.00	.90	2				
989	1.39	1	1.17	1.09	1			1	.96
990	1.11	2	1.04	.98	2			1	1.00
994	1.95	1	1.54	1.43	1				
996	1.04	1	.94	.90	3	1	1.13		

$\hat{Y}_1$ = Prediction assuming no trend

$\hat{Y}_1^*$ = Prediction assuming trend

G_1 = Group no.

N_1 = No. of hits for $1 < t \leq 851$

N_1' = No. of hits for $851 < t \leq 1061$

N_1'' = No. of hits for $1061 < t \leq 1320$

$\overline{X}_1$ = Average for hits in $1 \leq t \leq 851$

$\overline{X}_1'$ = Average for hits in $851 < t \leq 1061$

$\overline{X}_1''$ = Average for hits in $1061 < t \leq 1320$

Table 4. Predictions Based on $T \leq 851$ and Subsequent Results for Four Groups of I_2 Values

i_2	$\bar{X}_2$	N_2	$\hat{Y}_2$	$\hat{Y}_2^*$	G_2	N_2'	$\bar{X}_2'$	N_2''	$\bar{X}_2''$
002	1.25	2	1.09	1.00	2				
003	1.58	1	1.17	1.10	1				
004	1.34	1	1.05	1.00	2				
005					4			2	1.10
007	1.41	4	1.28	1.22	1			1	.83
008	1.58	2	1.31	1.23	1				
009	1.16	5	1.09	1.04	1			1	.69
049	1.06	4	1.00	.93	2	2	.98	1	.69
059	1.06	8	1.02	.94	2	2	.90		
066	1.18	3	1.08	1.00	2			1	.54
067	1.00	4	.96	.91	3			3	.84
068	1.16	7	1.11	1.02	1	1	.90	2	.90
078	1.14	5	1.08	1.00	2	1	1.21	2	1.02
079	1.14	6	1.09	1.02	1	2	.99	1	.74
229	1.17	2	1.03	.96	2				
289	1.00	6	.97	.90	3	2	.81	2	.91
335	1.06	3	.99	.91	3			1	.74
449	1.10	3	1.02	.95	2	3	1.05	3	.92
488	1.20	5	1.13	1.05	1	2	.76		
499	1.21	3	1.10	1.02	1	2	.86	1	.82
556	1.34	1	1.05	.99	2	2	1.07		
559					4				
588	1.12	2	1.00	.93	2	1	.88		
589	.99	6	.96	.90	3			3	.73
599	1.30	3	1.17	1.10	1	1	.93	1	.85
669	1.24	2	1.08	1.01	1			1	.68
688	1.32	2	1.13	1.07	1			2	.97
699	1.19	3	1.05	.98	2	1	1.13		
778	1.05	4	.99	.91	3	1	.81	1	1.34
779	1.00	3	.94	.90	3				
799	1.27	1	1.02	.92	2				
899	1.39	1	1.08	1.00	2	1	.69	3	.84

$\hat{Y}_2$ = Prediction assuming no trend

$\hat{Y}_2^*$ = Prediction assuming trend

G_2 = Group no.

N_2 = No. of hits for $1 < t \leq 851$

N_2' = No. of hits for $851 < t < 1061$

N_2'' = No. of hits for $1061 < t \leq 1320$

$\bar{X}_2$ = Average for hits in $1 < t \leq 851$

$\bar{X}_2'$ = Average for hits in $851 < t \leq 1061$

$\bar{X}_2''$ = Average for hits in $1061 < t \leq 1320$

Table 5. Comparison Between Predictions and Results for the Betting Systems For the Two Time Periods

$P_1 : 851 < t \leq 1061$ and $P_2 : 1061 < t \leq 1320$

System	S_1^*		S_1^{**}		S_2^*		S_2^{**}	
Period	P_1	P_2	P_1	P_2	P_1	P_2	P_1	P_2
Amt. bet per day	33.00	33.00	42.00	42.00	6.50	6.50	7.50	7.50
Pred. return	37.70	37.70	--	--	6.96	6.96	--	--
Observed return	44.19	6.91	44.19	18.96	5.58	5.37	5.58	6.78
Pred. no. of hits	6.93	8.55	8.82	10.88	8.19	10.10	9.45	11.66
Observed no. of hits	10.00	2.00	10.00	5.00	8.00	10.00	8.00	12.00
Observed/pred.	1.44	0.23	1.13	0.46	0.98	0.99	0.85	1.03
Pred. return/amt. bet	1.14	1.14	--	--	1.07	1.07	--	--
Observed return/amt.bet	1.34	0.21	1.05	0.45	0.85	0.83	0.74	0.90
Pred.return per hit	1.14	1.14	--	--	1.07	1.07	--	--
Observed/pred.	0.81	0.79	--	--	0.82	0.78	--	--
Pred. return per hit given the numbers hit	1.09	1.08	--	--	1.04	1.06	--	--
Observed/pred.	0.86	0.83	--	--	0.85	0.79	--	--

5. A MORE ELABORATE MODEL

Part of the difficulty giving rise to the "deficits" of the previous section can be attributed to the inadequacy of the model implicit in the regression corrections. This had two major shortcomings. First it was assumed that the variance of ε was independent of I. An investigation of the sample standard deviations for fixed I reveals that the variance of ε given I seems to be roughly proportional to $\tilde{Y}$. Thus the values of I with higher $\bar{X}$ should have their predictions "regressed" more heavily toward the mean.

The second shortcoming was the use of the assumptions of no trend in the X values or of a linear trend. A view of a graph of the means and standard deviations of the 22 groups of 60 observations (see Table 2) indicates that means and standard deviations seemed to decline over the first 720 games and then to stabilize.

Assuming stability for $720 \leq t \leq 1320$ one could base predictions on this period alone and ignore the previous data. If, as seems likely, the prospects for finding many good numbers on which to bet is slight, it seems preferable to use all of the data with a more complicated model.

Indeed the use of the last period alone, assuming no trend and making no allowance for the dependence of ε on I, yields no values of I_1 and I_2 for which the prediction $\hat{Y}$ exceeds one. The most optimistic predictions are $\hat{Y}_1 = 0.95$ for $I_1 = 787$ and $\hat{Y}_2 = 0.97$ for $I_2 = 078$. Thus if any good betting numbers are to be discovered, the more elaborate model and the entire data set will be required.

The following model allows for a trend over the period $1 \leq t \leq 720$ in the conditional mean and standard deviation of X given I and assumes that the standard deviation is roughly proportional to the mean. Let the normalized payoff corresponding to $I = i$ at time t be

$$X_{it} = \tilde{Y}_{it} + \varepsilon_{it} \tag{5.1}$$

where the ε_{it} are independent $N(0, \tilde{\sigma}^2_{it})$ random variables,

$$\tilde{Y}_{it} = \mu + \alpha_{1i}\alpha_{3it}, \quad \tilde{Y}_i = \mu + \alpha_{1i} \tag{5.2}$$

$$\tilde{\sigma}_{it} = (\mu + \alpha_{1i})\beta_{3it}^{-\frac{1}{2}} \tag{5.3}$$

$$\alpha_{3it} = 1 - \alpha_2 h(t) \tag{5.4}$$

$$\beta_{3it} = \beta_1 - \beta_2 h(t) \tag{5.5}$$

and

$$\begin{aligned} h(t) &= 721 - t \quad &\text{for} \quad t \leq 720 \\ &= 0 \quad &\text{for} \quad t > 721. \end{aligned} \tag{5.6}$$

The maximum likelihood estimates of the parameters μ, α_2, β_1 and β_2 and α_{1i}, for each i represented in the 1320 observations, may be obtained by minimizing

$$S = \sum_t [\tilde{\sigma}^{-2}_{it}(X_{it} - \tilde{Y}_{it})^2 + \log \tilde{\sigma}^2_{it}]. \tag{5.7}$$

For $I = I_1$ the estimate of β_2 was negative which suggests that $\tilde{\sigma}$ increases with t which seems counter-intuitive. Hence, in the case of $I = I_1$, estimates of the other parameters were also derived under the restriction $\beta_2 = 0$. The estimates of μ, α_2, β_1 and β_2 are listed in Table 6.

Table 6. Estimates of the Parameters of the Complex Model

	$I = I_1$		$I = I_2$
	Restricted*	Unrestricted	Unrestricted
μ	0.660	0.686	0.625
α_2	-0.00151	-0.00174	-0.00135
β_1	68.9	57.4	32.9
β_2	0.0	-0.0776	0.0248
$\bar{X}^{(0)} = 0.7710$			
$s_X^{(0)} = 0.1773$			
$\hat{\sigma}_2^2$	.0237	.0222	.0155
$\hat{\sigma}_2$	.1541	.1491	.1245

* Restricted refers to the restriction of $\beta_2 = 0$
$\bar{X}^{(0)}$, $s_X^{(0)}$ refer to the period $720 < t \leq 1320$
$\hat{\sigma}_2^2$ is an estimate of the variance of $\tilde{Y}_I$

Having estimated the parameters by $\hat{\mu}$, $\hat{\alpha}_{1i}$, $\hat{\alpha}_2$, $\hat{\beta}_1$ and $\hat{\beta}_2$ we obtain preliminary predictors $\hat{Y}^0(i) = \hat{\mu} + \hat{\alpha}_{1i}$ of $\tilde{Y}_i = \tilde{Y}_{it}$ for $t > 720$. But these predictions are subject to the regression to the mean criticism and we refine these estimates to

$$\hat{Y}^{(1)}(i) = w_i \bar{X}^{(0)} + (1-w_i)\hat{Y}^{(0)}(i) \tag{5.8}$$

where $\bar{X}^{(0)} = 0.711$ is the average of the X_t for $721 \leq t \leq 1320$ and

$$w_i = (1 + \sigma_2^2/\sigma_{1i}^2)^{-1} \tag{5.9}$$

is derived in the following argument.

The mean $\bar{X}^{(0)}$ based on 600 observations is a relatively reliable estimate of the mean of the $\tilde{Y}_I$. Then σ_2^2/σ_{1i}^2 should represent the ratio of the variance of $\tilde{Y}_I$ to the variance of $\hat{Y}^{(0)}(i)$, to fit in with the standard adjustment for regression to the mean.

To estimate σ_2^2 using the data for $720 < t \leq 1320$, we note that

$$X_t = \tilde{Y}_I + \varepsilon_{It} = \tilde{Y}_I[1 + \eta_{It}]$$

where the variance of η_{It}, given $I = i$, is β_1^{-1}. Hence

$$EX_t = E\tilde{Y}_I$$

and

$$EX_t^2 = [E\tilde{Y}_I^2][1 + \beta_1^{-1}].$$

Thus we have

$$E\tilde{Y}_I^2 = \sigma_2^2 + [E\tilde{Y}_I]^2 = (EX_t^2)(1 + \beta_1^{-1})^{-1}$$

$$\sigma_2^2 = [\sigma_{X_t}^2 + \{E(X_t)\}^2][1 + \beta_1^{-1}]^{-1} - \{E[X_t]\}^2 \tag{5.10}$$

which may be estimated by using the sample mean and standard deviation of the X_t over $720 < t \leq 1320$ to estimate $E(X_t)$ and $\sigma_{X_t}^2$ and $\hat{\beta}_1$ to estimate β_1. These estimates appear in Table 6.

The estimates of μ, α_2, β_1 and β_2 are based on 1320 observations while those of α_{1i} are based on very few. Thus we may regard the estimates of μ, α_2, β_1 and β_2 as relatively exact. In that case $\hat{\alpha}_{1i}$ is obtained from a few observations where $I = i$, each of which yields a Fisher Information (according to our model) of

$$(2 + \alpha_{3it}^2\beta_{3it})/\tilde{Y}_i^2 .$$

Thus we estimate

$$\sigma_{1i}^{-2} = \frac{2N(i) + \sum_{I=i} \alpha_{3it}^2\beta_{3it}}{\tilde{Y}_i^2} \tag{5.11}$$

by substituting our estimates of $\hat{\alpha}_{1i}$, $\hat{\alpha}_2$, $\hat{\beta}_1$, $\hat{\beta}_2$. A derivation appears in [1]. Equations (5.8) and (5.9) yield the refined estimate $\hat{Y}^{(1)}(i)$. However, the weight

w_i used the unrefined estimate $\hat{Y}^{(0)}(i)$ in (5.11). By replacing the $\tilde{Y}_i$ in (5.11) by the refined estimate $\hat{Y}^{(1)}$, we start an iterative process which quickly converges to a final estimate $\hat{Y}(i)$.

In Table 7, we list those values of I_1 and I_2 for which $\hat{Y}(i) > 1.0$. Only two values of I_2 appear. These are 007 and 008 and the anticipated profit is only 4%. On the other hand, the predicted payoffs for I_1 look more promising. Here there are 23 values for which the "restricted" prediction $\hat{Y}_1^r$, assuming $\beta_2 = 0$, exceed 1.0, forecasting an average profit of 8.6%.

Table 7. Large Predicted Payoffs for I_1 and I_2

I_1	N_1	$\hat{Y}_1$	$\hat{Y}_1^r$	I_1	N_1	$\hat{Y}_1$	$\hat{Y}_1^r$	I_2	N_2	$\hat{Y}_2$
007	3	1.01	1.00	466	1	1.09	1.12	007	5	1.04
008	1	1.18	1.18	599	2	1.12	1.10	008	2	1.04
020	1	1.04	1.03	655	3	1.06	1.06			
040	1	1.04	1.07	709	1	1.19	1.18			
066	1	1.12	1.11	848	3	1.03	1.02			
068	1	1.02	1.01	859	1	1.17	1.16			
070	2	1.20	1.20	886	1	1.09	1.08			
079	4	1.08	1.07	966	2	1.05	1.03			
080	1	1.04	1.04	970	1	1.16	1.15			
087	1	0.99	1.01	989	2	1.02	1.01			
090	3	1.03	1.03	994	1	1.19	1.18			
300	1	1.14	1.14							

$\hat{Y}_1^r$ is prediction based on estimates with β_2 restricted to 0.

6. SUMMARY

Preliminary estimates based on the first 851 plays of the lottery and assumptions of stability or linear trend and homoscedasticity predict that certain numbers will be profitable for betting. Subsequent experience proved disappointing. The predictions were over-optimistic for these "good" numbers by about 18% while for somewhat less desirable numbers these predictions seemed to be much closer (over-optimistic by about 5%).

A more elaborate model which assumes (1) that there is no basic shift in bettor preferences but that there is a gradual trend in payoffs and variability of payoffs for the first 720 games and stability thereafter, and (2) that the standard deviation of the payoff is proportional to the mean payoff for various numbers, leads to a new choice of desirable numbers.

Using these we could play 23 numbers with a predicted profit margin of 8.6%. However, 23 numbers are few and over a period of 300 games, the expected number of hits is 6.9 but the coefficient of variation is .38. This implies that even if the forecasts are sound and stable, one would have to play for many years before being reasonably certain of making a profit.

More detailed tables are presented in [1].

7. BIBLIOGRAPHY

[1] Chernoff, Herman. An Analysis of the Massachusetts Numbers Game: MIT Tech. Report No. 23.

[2] Efron, B. and Morris, C. (1975). Data Analysis Using Stein's Estimator and its Generalizations, Journal of the American Statistical Association, Vol. 70, No. 350, 311-319.

FOOTNOTES

[1] Work done with the partial support of NSF Grant MCS80-058483.

[2] Key Words. Lottery, Numbers Game, Regression to the Mean, Gambler's Ruin, Empirical Bayes

[3] AMS 1980 Subject Classification: Primary 62-70, Secondary 62C12

STATISTICS AND RELATED TOPICS
M. Csörgö, D.A. Dawson, J.N.K. Rao, A.K.Md.E. Saleh (eds.)

STUDY OF OPTIMALITY CRITERIA IN DESIGN OF EXPERIMENTS

A. Hedayat

Department of Mathematics
University of Illinois, Chicago
Chicago, Illinois
U.S.A.

In this paper we have rigorously studied various optimality criteria currently adopted by design specialists in choosing a best design for performing an experiment. These optimality criteria include: G-optimality, D-optimality, L-optimality, E-optimality, S-optimality, (M,S)-optimality, Φ_p-criteria, Universal optimality, type 1 and 2 criteria and Schur optimality.

1. PRELIMINARY

We perform experiments mainly to estimate or test hypotheses about some specified unknown parameters of a given model efficiently. Different considerations lead us to different criteria for the choice of the "best" design. Although Definition 2.1 is a response function criterion, most criteria in design theory are directly related to parameter estimation. Hence the information matrices play an important role and thus by Caratheodory theorem we can limit our search to discrete designs which are supported on sets consisting of finite number of points.

To see how the optimality criteria in design theory arose, we first give an example of the very basic motivation: Let d be a design and let Y be the vector of observations obtained under d. Assume

$$E(Y) = X\underline{\theta}\ ,\quad Cov(Y) = \sigma^2 I\ , \tag{1.1}$$

where Y is an $n \times 1$ vector of observations, X is an $n \times k$ matrix with known entries specified by d, $\underline{\theta}$ is a $k \times 1$ vector of unknown constants, and I denotes the identity matrix of order n. In many cases we are only interested in the subvector $\underline{\theta}_1$ of $\underline{\theta}$. With no loss of generality we can write $\underline{\theta}' = (\underline{\theta}_1' \,\vdots\, \underline{\theta}_2')$, where $\underline{\theta}_1$ is a $v \times 1$ vector, $1 \leq v \leq k$. According to the partition $\underline{\theta}' = (\theta_1' \,\vdots\, \underline{\theta}_2')$ the Model (1.1) can be written as

$$E(Y) = (X_1 \,\vdots\, X_2)\begin{pmatrix}\underline{\theta}_1\\ \cdots\\ \underline{\theta}_2\end{pmatrix},\quad Cov(Y) = \delta^2 I\ . \tag{1.1$'$}$$

The information matrix of $\underline{\theta}_1$ under d and the Model (1.1)' is $X_1'X_1 - X_1'X_2(X_2'X_2)^{-}X_2'X_1$. We shall denote this by M_d. Note that $M_d = X'X$ when $v = k$, i.e., $\underline{\theta}_1 = \underline{\theta}$. Now we consider four cases:

(i) <u>To estimate each component of</u> $\underline{\theta}$:

Assume $X'X$ is nonsingular, and suppose we want to estimate each of the individual parameters. By Gauss-Markov Theorem, the best linear unbiased estimator (b.l.u.e.) $\hat{\underline{\theta}}$ of θ is given by

$$\hat{\underline{\theta}} = (X'X)^{-1}X'Y \tag{1.2}$$

with

$$\text{Cov}(\hat{\underline{\theta}}) = \sigma^2(X'X)^{-1}. \tag{1.3}$$

Let x_i be the ith column of X and c_j be the jth column of $X(X'X)^{-1}$, then from (1.2) and (1.3) it follows that

$$\hat{\theta}_i = c_i'Y \tag{1.4}$$

with

$$\text{Var}(\hat{\theta}_i) = \sigma^2(c_i'c_i) . \tag{1.5}$$

Since $(X'X)^{-1}(X'X) = I_k$, we have $c_i'x_j = \delta_{ij}$, where δ_{ij} is the Kronecker delta. Applying the Schwarz inequality, we obtain

$$(x_i'x_i)(c_i'c_i) \geq (x_i'c_i)^2 = 1 \tag{1.6}$$

hence

$$\text{Var}(\hat{\theta}_i) \geq \sigma^2/x_i'x_i . \tag{1.7}$$

Usually, the experimenter has some amount of freedom in the choice of the k vectors x_i. If possible, we would like to select a design which estimates each of the parameters with minimum variance. Observe that the equality in (1.6) holds if and only if $c_i = cx_i$ for a constant c , which implies that $X'X$ is a diagonal matrix. Hence, theoretically speaking, the "best design" is a design in which $X'X$ is a diagonal matrix with diagonal entries as large as possible. For example, if $x_{ij} \in \{0,1,-1\}$ then $x_i'x_i \leq n$. Thus the best design is the one for which $X'X = nI_k$. But such a design does not always exist, see Hedayat and Wallis (1979). When such designs do not exist, the question arises to how a best design should be defined. A reasonable approach is to minimize the average variance of each of the estimated parameters or to minimize the generalized variance, etc.

(ii) <u>To estimate linear functions of a subvector $\underline{\theta}_1$ of $\underline{\theta}$:</u>

Suppose we want to estimate linear functions of $\underline{\theta}_1$ in the form $q_1'\underline{\theta}_1$. The b.l.u.e. of $q_1'\underline{\theta}_1$ is $q_1'\hat{\theta}_1$ with

$$\text{Var}(q_1'\hat{\theta}_1) = \sigma^2 q_1'M_d^- q_1 , \tag{1.8}$$

where

$$\hat{\theta}_1 = M_d^- Q_d , \tag{1.9}$$

and

$$Q_d = [X_1' - X_1'X_2(X_2'X_2)^- X_2']Y , \tag{1.10}$$

while M_d^- is any generalized inverse of M_d .

In choosing a design for estimating $q_1'\underline{\theta}_1$ there are many criteria. One of them is based on the following inequality

$$\mu_{min} \leq \frac{q_1'M_d^- q_1}{q_1'q_1} \leq \mu_{max} , \tag{1.11}$$

where μ_{max} and μ_{min} are the maximum and minimum (non-zero) eigenvalues of M_d^-

respectively. This inequality gives a bound for the variance of $\underline{q}_1'\hat{\underline{\theta}}_1$:

$$\mu_{min}\, q_1'q_1\sigma^2 \le Var(\underline{q}_1'\hat{\underline{\theta}}_1) \le \mu_{max}\, q_1'q_1\sigma^2 . \tag{1.12}$$

(iii) To test hypotheses:

Suppose in addition Y is multivariate normal and we want to test $\theta_1 = \theta_2 = \ldots = \theta_v = 0$ $(v \le k)$. (Assume M_d is nonsingular). Then the usual F test has a power function depending monotonically (increasing) on a parameter λ where

$$\lambda = \sigma^{-2}\underline{\theta}_1' M_d \underline{\theta}_1 \tag{1.13}$$

and thus by (1.11) and (1.13)

$$\sigma^{-2}\bar{\mu}_{min}\underline{\theta}_1'\underline{\theta}_1 \le \lambda \le \sigma^{-2}\bar{\mu}_{max}\underline{\theta}_1'\underline{\theta}_1 \tag{1.14}$$

where $\bar{\mu}_{max}$ and $\bar{\mu}_{min}$ are the maximum and the minimum eigenvalues of M_d.

(iv) To construct confidence region:

Again assume Y is multivariate normal and M_d is nonsingular. A $1-\alpha$ joint confidence region for $\underline{\theta}_1$ is a solid ellipsoid:

$$(\underline{\theta}_1-\hat{\underline{\theta}}_1)' M_d\, (\underline{\theta}_1-\hat{\underline{\theta}}_1) \le \sigma^2\chi_\alpha^2(v)\ , \quad \text{if } \sigma^2 \text{ is known,} \tag{1.15}$$

where $\chi_\alpha^2(v)$ is the $1-\alpha$ percentile of the χ^2 distribution with v degrees of freedom. Or

$$(\underline{\theta}_1-\hat{\underline{\theta}}_1)' M_d\, (\underline{\theta}_1-\hat{\underline{\theta}}_1) \le vS^2F_\alpha(v,n-r)\ , \quad \text{if } \sigma^2 \text{ is unknown,} \tag{1.16}$$

where $F_\alpha(v,n-r)$ is the $1-\alpha$ percentile of the F distribution with v and n-r degrees of freedom, and $S^2 = Y'[I - X(X'X)^- X']Y/(n-r)$ is an unbiased estimator of σ^2 (assume rank $(X'X) = r$).

We observe that:

(a) The volume (expected volume, if σ^2 is unknown) of the above ellipsoid is proportional to the square root of $\det M_d^{-1}$.

(b) The semi-axes (expected semi-axes, if σ^2 is unknown) of the above ellipsoid is proportional to the square roots of the eigenvalues of M_d^{-1}.

In Section 2 we shall study some well-known optimality criteria. Sections 3-7 will be some generalization of those in Section 2, or some recent developments in the determination of optimal designs. Throughout this paper we write the optimality criteria as a class of convex nonincreasing functionals Φ on the set of information matrices rather than a class of convex nondecreasing functionals ψ on the set of covariance matrices, since the former is more general than the latter. For instance, when the covariance matrix of interest is equal to M_d^- (as in (iii)), we have $\Phi(M_d) = \psi(M_d^-)$ which is convex in M_d if ψ is convex in M_d^- but not on the other hand. The strict inclusion of one class in the other is illustrated by the fact that, if $\lambda_1(M_d^-) \ge \ldots \ge \lambda_v(M_d^-)$ are the eigenvalues of M_d^-, then $\Sigma\lambda_i^{\frac{1}{2}}(M_d^-) = \Sigma\lambda_i^{-\frac{1}{2}}(M_d)$ is convex in M_d but $\Sigma\lambda_i^{\frac{1}{2}}(M_d^-)$ is not convex in M_d^-.

Notations used in the rest of this paper are listed below:

$\mathcal{B}_v$ = the class of all $v \times v$ nonnegative definite matrices.

$\mathcal{B}_{v,o}$ = the class of all $v \times v$ nonnegative definite matrices with zero row and column sums.

$\mathcal{D}$ = the class of designs under consideration.

$\mathcal{C}$ = $\{M_d,\ d \in \mathcal{D}\}$.

Also, let $\mu_{d1} \geq \mu_{d2} \geq \cdots \geq \mu_{dv}$ be the eigenvalues of M_d. Note that if $\mathcal{C} \subseteq \mathcal{B}_{v,o}$, $\mu_{dv} = 0$, for all $d \in \mathcal{D}$. If necessary, we let ξ denote an approximate design (a probability measure on the experimental space) and $M\xi$ be the associated information matrix.

To avoid messy expressions, the dimensions of matrices should be deduced from the context if they are not explicitly specified.

2. SOME WELL-KNOWN OPTIMALITY CRITERIA

Assume $\mathcal{C} \subseteq \mathcal{B}_v$.

I. G-optimality

Smith (1918) introduced a response function criterion which can be stated as follows:

Definition 2.1. A design $\xi^* \in \mathcal{D}$ is G-optimal if and only if

$$\min_{\xi \in \mathcal{D}} \max_{x \in \chi} \operatorname{var}_{\xi} \widehat{EY}_x = \max_{x \in \chi} \operatorname{var}_{\xi^*} \widehat{EY}_x ,$$

where $\widehat{EY}_x$ is the b.l.u.e. of EY_x and χ is the experimental space. Kiefer called it G-optimal (for global or minimix), since we are minimizing the maximum variance of any predicted value over the experimental space.

II. D-optimality

Definition 2.2. A design $d^* \in \mathcal{D}$ is D-optimal if and only if M_{d^*} is non-singular and $\min_{d \in \mathcal{D}} \det(M_d^{-1}) = \det(M_{d^*}^{-1})$. Here, "D-" stands for determinant. The concept introduced and studied by Wald (1943) and applied by Mood (1946). This criterion has many appealing properties;

(1) under normality, if d^* is D-optimal, d^* minimizes:

(a) The volume (or expected volume, if σ^2 is unknown, and rank of (M_d) is invariant under d) of the smallest invariant confidence region on $\theta_1, \theta_2, \ldots, \theta_v$ for any given confidence coefficient.

(b) The generalized variance of the estimators of parameters. (See remark below).

(2) In the class of approximate designs, D-optimality <=> G-optimality whenever $v = k$, i.e., $\underline{\theta}_1 = \underline{0}$.

(3) The design remains D-optimal if one changes the scale of the parameters: Let $\theta_1', \theta_2', \ldots, \theta_v'$ be related to $\theta_1, \theta_2, \ldots, \theta_v$ by a non-singular linear transformation. If d^* is D-optimal for $\theta_1, \ldots, \theta_v$, then d^* is also D-optimal for $\theta_1', \ldots, \theta_v'$. The analogue for other criteria is false in even the simplest settings.

Remark: Suppose $X = (X_1, X_2, \ldots, X_n)'$ is distributed as multivariate $N(\underline{\mu}, V)$. The determinant of V is called the generalized variance of X as defined by Wilks (1932).

In the theory of linear regression, under normal assumption, $\hat{\underline{\theta}}_1 = (\hat{\theta}_1, \hat{\theta}_2, \ldots, \hat{\theta}_v)'$ is distributed as $N(\underline{\theta}_1, M_d^{-1}\sigma^2)$, so the generalized variance of $(\hat{\theta}_1, \hat{\theta}_2, \ldots, \hat{\theta}_v)$ is equal to the determinant of $M_d^{-1}\sigma^2$ which is the product of σ^{2v} and $\det M_d^{-1}$. (Assume M_d is non-singular.)

III. L-optimality

Definition 2.3. A design $d^* \in \mathcal{D}$ is linear optimal (L-optimal) if and only if $\min_{d \in \mathcal{D}} L(M_d^{-1}) = L(M_{d^*}^{-1})$ where L is a nonnegative linear functional on $\mathcal{C}$.

One of the most useful linear criteria of optimality if A-optimality defined when

$$L(M_d^{-1}) = \mathrm{Tr}(M_d^{-1}) .$$

Definition 2.4. A design $d^* \in \mathcal{D}$ is A-optimal if and only if M_{d^*} is non-singular and $\min_{d \in \mathcal{D}} \mathrm{Tr}(M_d^{-1}) = \mathrm{Tr}(M_{d^*}^{-1})$. "A-" stands for average. In a statistical sense, if d^* is A-optimal, it minimizes the average variance of $\hat{\theta}_1, \hat{\theta}_2, \ldots, \hat{\theta}_v$. This criterion was introduced and studied by Elfving (1952) and Chernoff (1953).

IV. E-optimality

Definition 2.5. A design $d^* \in \mathcal{D}$ is E-optimal if and only if $\min_{d \in \mathcal{D}} \mu_{dv}^{-1} = \mu_{d^*v}^{-1}$. E-optimality was first considered in hypothesis testing (Wald (1943), Ehrenfield (1955)). "E-" stands for eigenvalue. It has the following properties:

(1) In hypothesis testing. Under the normality assumption, an E-optimal design maximizes the minimum power of the associated F-test of size α on the contour $\underline{\theta}_1'\underline{\theta}_1 = c$ for every α and c. (See (1.14)).

(2) In point estimation. An E-optimal design minimizes the maximum variance of the b.l.u.e. of the $q_1'\underline{\theta}_1$ over all $v \times 1$ vectors q_1 with $q_1'q_1 = 1$. (See (1.12)).

(3) In interval estimation. An E-optimal design minimizes the largest semi-axis of the (hyper) ellipsoid when normality assumptions are made on the observations.

Now it seems natural to specify some optimality functional Φ on $\mathcal{C}$ and to pose the problem: Find d to minimize $\Phi(M_d)$. We call Φ an optimality criterion. The above well-known criteria are then:

$$\text{D-optimality:} \quad \Phi_D(M_d) = \det(M_d^{-1}) = \prod_{i=1}^{v} \mu_{di}^{-1} \tag{2.1}$$

$$\text{L-optimality:} \quad \Phi_L(M_d) = L(M_d^{-1}) \tag{2.2}$$

$$\text{A-optimality:} \quad \Phi_A(M_d) = \mathrm{Tr}(M_d^{-1}) = \sum_{i=1}^{v} \mu_{di}^{-1} \tag{2.3}$$

$$\text{E-optimality:} \quad \Phi_E(M_d) = \mu_{dv}^{-1} . \tag{2.4}$$

(2.1), (2.3) and (2.4) are regarded as infinite if M_d is singular.

Note, in case $\mathcal{C} \subseteq \mathcal{B}_{v,o}$, the definitions of D-, A-, E-optimality are similar, one can simply replace the index v in (2.1),(2.2) and (2.4) by v-1.

3. S-OPTIMALITY AND (M,S)-OPTIMALITY

Assume $\mathcal{C} \subseteq \mathcal{B}_v$. When $Tr(M_d) = \sum_i \mu_{di} = A$ is a constant, for all $d \in \mathcal{D}$, the D-, A-, E-optimalities are attained when all the μ_{di}'s are equal (we call such a design a symmetric design). Unfortunately, symmetric designs do not always exist. Intuitively, in the absence of a symmetric design, we may want to believe that the "closest" design to the hypothetical symmetric design is a reasonable design to use. Shah (1960) proposed the Euclidean distance between the vector of eigenvalues of the designs as the measure of distance between the corresponding designs. Thus, according to Shah (1960) if there is no symmetric design in D, we should use the design d which minimizes the Euclidean distance between $(\mu_{d1},\ldots,\mu_{dv})$ and the vector of eigenvalues of the hypothetical symmetric design $(A/v,\ldots,A/v)$, i.e.

$$[\Sigma\, \mu_{di}^2 - (\Sigma\, \mu_{di})^2/v]^{1/2} . \tag{3.1}$$

Clearly, this is only a heuristic approach with no statistical justification. However, it has the merit that when $Tr(M_d)$ is a constant, the minimization of (3.1) is equivalent to that of $TrM_d^2 = \sum_{i,j} m_{dij}^2$ which is easier to handle.

Define $\Phi : \mathcal{B}_v \Rightarrow [0, +\infty]$ such that

$$\Phi(M_d) = TrM_d^2 = \sum_i \mu_{di}^2 = \sum_{i,j} m_{dij}^2 . \tag{3.2}$$

Formally, we have:

Definition 3.1. Suppose $Tr(M_d) = A$ is a constant for all $d \in \mathcal{D}$. A design $d^* \in \mathcal{D}$ is called S-optimal if and only if d^* minimizes $\Phi(M_d)$ (as in 3.2) for all $d \in \mathcal{D}$.

Motivated by Shah's criterion, Eccleston and Hedayat (1974) proposed a similar procedure in the case when TrM_d is not a constant.

Let $\mathcal{C}' \subseteq \mathcal{C}$ be such that the matrices in $\mathcal{C}'$ have maximum trace.

Definition 3.2. A design $d^* \in \mathcal{D}$ is (M,S)-optimal if and only if $M_{d^*} \in \mathcal{C}'$ and d^* minimizes $\Phi(M_d)$ (as in (3.2)), for all $d \in \mathcal{D}'$, where $\mathcal{D}' = \{d \in \mathcal{D};\ M_d \in \mathcal{C}'\}$.

A geometric interpretation of (M,S)-optimality can be given as follows. Set

$$S_A = \{(\mu_{d1},\ldots,\mu_{dv});\ \mu_{di} > 0,\ \sum_i \mu_{di} = A\} ,$$

and

$$S_{AB} = \{(\mu_{d1},\ldots,\mu_{dv});\ \mu_{di} > 0,\ \sum_i \mu_{di} = A;\ \sum_i \mu_{di}^2 = B\} .$$

Then S_A is an open simplex and S_{AB} is part of a (v-2)-dimensional sphere with $(A/v,\ldots,A/v)$ as the center and the quantity

$$P = [\sum_i \mu_{di}^2 - (\sum_i \mu_{di})^2/v]^{1/2} \quad \text{as the radius, when}$$

$B \geq A^2/v$. The procedure of finding an (M,S)-optimal design is the same as to choose a simplex S_A as far away from the origin as possible, and then find a design with the vector of eigenvalues on S_A which is closest to the center of the simplex in the Euclidean sense.

In the $\mathcal{B}_{v,o}$ context, same arguments hold except replacing v by $v-1$.

4. Φ_p-CRITERIA

In Kiefer (1974), the following family of criteria was introduced. We shall describe it in the $\mathcal{B}_v$ context.

Let

$$\Phi_p(M_d) = [\frac{1}{v} \operatorname{Tr}(M_d^{-p})]^{1/p}$$

$$= [\frac{1}{v} \sum_{i=1}^{v} \mu_{di}^{-p}]^{1/p}, \quad 0 < p < \infty \quad . \tag{4.1}$$

Definition 4.1. A design $d^* \in \mathcal{D}$ is Φ_p-optimal if and only if d^* minimizes $\Phi_p(M_d)$, $d \in \mathcal{D}$.

When $C \subseteq \mathcal{B}_v$, we may restrict ourself to d with M_d nonsingular. The following theorem will give a connection between D-, A-, E-criterion and the Φ_p-criterion.

Theorem 4.1. (i) $\Phi_1(M_d) = \frac{1}{v} \operatorname{Tr}(M_d^{-1}) = \frac{1}{v}(\sum_{i=1}^{v} \mu_{di}^{-1})$

$$\text{(ii)} \quad \Phi_0(M_d) = \lim_{p \to 0} \Phi_p(M_d) = (\prod_{i=1}^{v} \mu_{di}^{-1})^{\frac{1}{v}} \tag{4.2}$$

$$\text{(iii)} \quad \Phi_\infty(M_d) = \lim_{p \to \infty} \Phi_p(M_d) = \mu_{dv}^{-1} \; .$$

Proof: (i) is clear

$$\text{(ii)} \quad \Phi_p(M_d) = [\frac{1}{v} \sum_{i=1}^{v} \mu_{di}^{-p}]^{\frac{1}{p}}$$

$$\log \Phi_p(M_d) = \frac{1}{p} \log [\frac{1}{v} \sum_{i=1}^{v} \mu_{di}^{-p}] \; .$$

As p tends to zero, the right hand side goes to $\frac{0}{0}$, so by applying L'Hospital's rule, we obtain

$$\lim_{p \to 0} \log \Phi_p(M_d) = \lim_{p \to 0} \frac{\frac{1}{v}[\sum_{i=1}^{v} \mu_{di}^{-p} \log \mu_{di}^{-1}]}{\frac{1}{v}[\sum_{i=1}^{v} \mu_{di}^{-p}]}$$

$$= \frac{1}{v} \sum_{i=1}^{v} \log \mu_{di}^{-1}$$

$$= \frac{1}{v} \log \prod_{i=1}^{v} \mu_{di}^{-1} .$$

Hence
$$\lim_{p \to 0} \Phi_p(M_d) = (\prod_{i=1}^{v} \mu_{di}^{-1})^{\frac{1}{v}} .$$

(iii) Let $\mu'_{di} = \mu_{dv}\mu_{di}^{-1}$.

Then

$$\log \Phi_p(M_d) = \frac{1}{p} \log [\frac{1}{v} \sum_{i=1}^{v} (\mu'_{di}\mu_{dv}^{-1})^p]$$

$$= \frac{1}{p} \log [\frac{1}{v} \mu_{dv}^{-p} \sum_{i=1}^{v} \mu_i'^p]$$

$$= \frac{1}{p} \log \frac{1}{v} + \log \mu_{dv}^{-1} + \frac{1}{p} \log (\sum_{i=1}^{v} \mu_{di}'^p) .$$

Since $\mu'_{di} \leq 1$, for all i , we conclude

$$0 \leq \log (\sum_{i=1}^{v} \mu_{di}'^p) \leq \log v .$$

Hence
$$\lim_{p \to \infty} \frac{1}{p} \log (\sum_{i=1}^{v} \mu_{di}'^p) = 0 .$$

Therefore
$$\lim_{p \to \infty} \log \Phi_p(M_d) = \log \mu_{dv}^{-1} ,$$

and consequently

$$\lim_{p \to \infty} \Phi_p(M_d) = \mu_{dv}^{-1} .$$

Corollary 4.1.

(i) When $p = 1$, Φ_p-criterion is equivalent to A-optimality.

(ii) When p approaches to 0, the limiting case of Φ_p-criterion is equivalent to D-optimality.

(iii) When p approaches to ∞, the limiting case of Φ_p-criterion is equivalent to E-optimality.

Remark: The Φ_p-criterion in the $\mathcal{B}_{v,o}$ context is

$$\Phi_p(M_d) = [\frac{1}{v-1} \sum_{i=1}^{v-1} \mu_{di}^{-p}]^{\frac{1}{p}} .$$

5. UNIVERSAL OPTIMALITY

In Kiefer (1975), a strong optimality criterion was considered. Here, we restrict ourself in $\mathcal{B}_{v,o}$. (Since in $\mathcal{B}_v$ context, it is easier.)

Definition 5.1. We say $d^* \in \mathcal{D}$ is a universally optimal design, if d^* minimizes $\Phi(M_d)$, $d \in \mathcal{D}$ for any $\phi : \mathcal{B}_{v,o} \Rightarrow (-\infty, +\infty]$ satisfying:

(i) Φ is convex,

(ii) $\Phi(bM)$ is nonincreasing in the scalar $b \geq 0$ for each $M \in \mathcal{B}_{v,o}$. (5.1)

(iii) Φ is invariant under each permutation of rows and (the same on) columns.

Remark: Unlike the preceeding optimality criteria, not every design setting contains a universally optimal design.

Since $-\mathrm{Tr}(M)$ satisfies (5.1), immediately we have the following theorem:

Theorem 5.1. If $d^* \in \mathcal{D}$ is universally optimal, then $\mathrm{Tr}M_{d^*}$ is maximum.

Definition 5.2. A matrix M is called a completely symmetric (c.s.) matrix if $M = \alpha I_v + \beta J_v$ where α, β are scalars, I_v is the identity matrix and J_v is the $v \times v$ matrix consists of all 1's.

Lemma 5.1. If M_1 and M_2 are two completely symmetric matrices in $\mathcal{B}_{v,o}$, then there exists an h such that $M_2 = hM_1$.

Proof: Suppose

$$M_1 = \alpha_1 I_v + \beta_1 J_v$$

$$M_2 = \alpha_2 I_v + \beta_2 J_v$$

$$M_i \cdot \underline{1} = 0 \Rightarrow \alpha_i + v\beta_i = 0 \quad \text{for } i = 1,2.$$

So

$$M_i = -v\beta_i I_v + \beta_i J_v, \quad i = 1,2.$$

Let $h = \beta_2/\beta_1$.

Then

$$M_2 = -v\beta_2 I_v + \beta_2 J_v = h(-v\beta_1 I_v + \beta_1 J_v) = hM_1 .$$

The following theorems are simple tools in determining such an optimal design.

Theorem 5.2. Suppose $\mathcal{C} \subseteq \mathcal{B}_{v,o}$ contains a M_{d^*} for which

(a) M_{d^*} is c.s. (5.2)

(b) $\mathrm{Tr}M_{d^*} = \max_{d \in \mathcal{D}} \mathrm{Tr}M_d$.

Then d^* is universally optimal in $\mathcal{D}$.

Proof: From Theorem 5.1 it suffices to show that $\Phi(M_{d^*})$ minimizes $\Phi(M_d)$ for all Φ satisfies (5.1), $M_d \in \mathcal{C}'$ where $\mathcal{C}' \subseteq \mathcal{C}$ consists of the matrices which have maximum trace.

For any $M_d \in \mathcal{C}'$, let τM_d be obtained from M_d by permuting rows and columns according to τ , and let $\bar{M}_d = \sum_\tau \tau M_d/v!$, the symmetrized version of M_d. By (5.1) (a) and (c) we have

$$\Phi(\bar{M}_d) \leq \sum_\tau \frac{1}{v!} \Phi(\tau M_d) = \Phi(M_d) , \tag{5.3}$$

for any Φ satisfying (5.1). Of course $\bar{M}_d$ need not be in $\mathcal{C}$, but $\bar{M}_d$ is c.s. and in $\mathcal{B}_{v,o}$. By Lemma 5.1, $\bar{M}_d$ is of the form bM_{d*} for some $b \geq 0$. Now $Tr(\bar{M}_d) = Tr(M_d)$. But $Tr(M_d) = Tr(M_{d*})$ by assumption. This implies $b = 1$ and hence $\bar{M}_d = M_{d*}$. By (5.3), $\Phi(M_{d*}) = \Phi(\bar{M}_d) \leq \Phi(M_d)$ for all Φ satisfying (5.1) and $M_d \in \mathcal{C}'$. Therefore M_{d*} is universally optimal.

Remark: It would be very interesting if another sufficient condition could be established for universal optimality.

Theorem 5.3. Suppose an M_{d*} satisfying (5.2) exists. Let $\Phi : \mathcal{B}_{v,o} \Rightarrow (-\infty, +\infty]$ be a function satisfying (5.1). If, in addition, Φ is strictly convex (and hence also "nonincreasing" in property (ii) is replaced by "decreasing"), then every Φ-optimal d' has $M_{d'} = M_{d*}$. (i.e., d' is also universally optimal).

Proof: Let $\bar{M}_{d'} = \sum_\tau M_{d'}/v!$. Since Φ is strictly convex, we have

$$\Phi(\bar{M}_{d'}) < \sum_\tau \frac{1}{v!} \Phi(\tau M_{d!}) = \Phi(M_{d'}) . \tag{5.4}$$

Again $\bar{M}_{d'}$ is c.s. and in $\mathcal{B}_{v,o}$, this implies that $\bar{M}_{d'} = bM_{d*}$ for some $b \geq 0$. Since M_{d*} satisfies (5.2), $Tr(M_{d*}) \geq Tr(M_{d'})$ which implies $b \leq 1$. But if $b < 1$

$$\Phi(M_{d'}) > \Phi(\bar{M}_{d'}) = \Phi(bM_{d*}) > \Phi(M_{d*}) . \tag{5.5}$$

This contradicts the assumption that d' is Φ-optimal. From (5.4) and (5.5) we can conclude that

$$b = 1 \quad \text{and} \quad M_{d'} = \bar{M}_{d'}$$

i.e.,

$$M_{d'} = M_{d*} .$$

And d' is indeed universally optimal. □

Let Φ_1 and Φ_2 be two convex functions satisfying (5.1). Suppose $d*$ is Φ_1-optimal, the following theorem gives a sufficient condition for $d*$ to be Φ_2-optimal.

Theorem 5.4. If $\Phi_1 \leq \Phi_2$ on $\mathcal{C}$ and if $\Phi_1(M_{d*}) = \Phi_2(M_{d*})$, then $d*$ is Φ_2-optimal if $d*$ is Φ_1-optimal.

Proof: Assume $d*$ is Φ_1-optimal, then $\Phi_1(M_{d*}) \leq \Phi_1(M_d)$ for all $d \in \mathcal{D}$.

By assumption

$$\Phi_2(M_{d*}) = \Phi_1(M_{d*}) \leq \Phi_1(M_d) \leq \Phi_2(M_d) .$$

Hence the result.

Example 5.1. A useful family of criteria in the $\mathcal{B}_{v,o}$ context is the Φ_p-criteria, for $0 < p < \infty$, with the limiting values

$$\Phi_0(M_d) = \prod_i \mu_{di}^{-\frac{1}{v-1}} \quad \text{and} \quad \Phi_\infty(M_d) = \mu_{d(v-1)}^{-1} .$$

Here $p < q \Rightarrow \Phi_p(M_d) \leq \Phi_q(M_d)$ with equality if and only if all μ_{di} are equal. Hence from Theorem 5.4 if M_{d*} is c.s. and $d*$ is Φ_p-optimal $\Rightarrow$ $d*$ is Φ_q-optimal for all $q > p$.

In the absence of universal optimality, some weaker optimality results which have some useful statistical implications (for instance, include A-, E-, D-criteria and all ϕ_p-criteria, $0 < p < \infty$) has been discussed by Kiefer (1974).

Observe that (4.1) and (4.2) are equivalent to the following:

(a) $\Phi_p^*(M_d) = \sum_i \mu_{di}^{-p}$, $0 < p < \infty$;

(b) $\Phi_0^*(M_d) = - \sum_i \log \mu_{di}$ (5.6)

(c) $\Phi_\infty(M_d) = \mu_{dv}^{-1}$.

Let

$$\Phi^*(M_d) = \sum_i f(\mu_{di}) , \tag{5.7}$$

where f is convex and $[0, +\infty)$. We want to find conditions under which a design d is Φ^*-optimal.

Lemma 5.2. If f is a convex function on $[0, +\infty)$, then

$$\sum_{i=1}^{v-1} f(\mu_{di}) \geq \frac{v-1}{v} \sum_{j=1}^{v-1} f\left(\frac{v}{v-1} m_{djj}\right) \tag{5.8}$$

for any M_d in $\mathcal{B}_{v,o}$, with equality if all the μ_{di}'s are equal or M_d is c.s.

Proof: Let P be the $(v-1) \times v$ orthonormal matrix such that

$$PM_dP' = \Lambda_d = \begin{bmatrix} \mu_{d1} & & & 0 \\ & \cdot & & \\ & & \cdot & \\ & & & \cdot \\ 0 & & & \mu_{d(v-1)} \end{bmatrix} .$$

Argument P with $(\frac{1}{\sqrt{v}}, \frac{1}{\sqrt{v}}, \ldots, \frac{1}{\sqrt{v}})$ and call the resulting matrix P^* .

$$P^*M_dP^{*\prime} = \begin{bmatrix} PM_dP' & 0 \\ 0 & 0 \end{bmatrix} = \begin{bmatrix} \Lambda_d & 0 \\ 0 & 0 \end{bmatrix} .$$

Assume $P^* = (p_{ij}^*)$, and let $e_{ij} = p_{ij}^{*2}$.

Then $\sum_{j=1}^{v} e_{ij} = 1$ and $\sum_{i=1}^{v-1} = 1 - \frac{1}{v} = \frac{v-1}{v}$.

Also $M_d = P'\Lambda_d P \Rightarrow m_{djj} = \sum_{i=1}^{v-1} e_{ij}\mu_{di}$.

Thus
$$\frac{v-1}{v} f(\frac{v}{v-1} m_{djj}) = \frac{v-1}{v} f(\frac{v}{v-1} \sum_{i=1}^{v-1} e_{ij}\mu_{di})$$

$$= \frac{v-1}{v} f(\sum_{i=1}^{v-1} \frac{v}{v-1} e_{ij}\mu_{di}) .$$

Since $\sum_{i=1}^{v-1} e_{ij} = \frac{v-1}{v} \Rightarrow \sum_{i=1}^{v-1} \frac{v}{v-1} e_{ij} = 1$ and $e_{ij} \geq 0$.

The convexity of f

$$\Rightarrow \frac{v-1}{v} f(\sum_{i=1}^{v-1} \frac{v}{v-1} e_{ij}\mu_{di}) \leq \frac{v-1}{v} \sum_{i=1}^{v-1} \frac{v}{v-1} e_{ij} f(\mu_{di})$$

$$= \sum_{i=1}^{v-1} e_{ij} f(\mu_{di}) .$$

Hence we have

$$\frac{v-1}{v} f(\frac{v}{v-1} m_{djj}) \leq \sum_{i=1}^{v-1} e_{ij} f(\mu_{di}) . \tag{5.9}$$

Summing on j , we obtain

$$\frac{v-1}{v} \sum_{j=1}^{v} f(\frac{v}{v-1} m_{djj}) \leq \sum^{v-1}_{i=1} f(\mu_{di}) .$$

If $\mu_{d1} = \mu_{d2} = \ldots = \mu_{d(v-1)} = \mu_d$ (i.e., M_d is c.s.)

$$m_{djj} = \sum_{i=1}^{v-1} e_{ij}\mu_{di} = \mu_d(\frac{v-1}{v}) .$$

Then

$$\frac{v-1}{v} \sum_{j=1}^{v} f(\frac{v}{v-1} m_{djj}) = \frac{v-1}{v} \sum_{j=1}^{v} f(\mu_d) = (v-1) f(\mu_d)$$

$$= \sum_{i=1}^{v-1} f(\mu_d) = \sum_{i=1}^{v-1} f(\mu_{di}) . \tag{5.10}$$

Theorem 5.5. If Φ^* is given by (5.7) with f convex, and if $d^* \in \mathcal{D}$ satisfies:

(i) M_d^* is c.s.

(ii) d^* minimizes $\sum_{j=1}^{v-1} f(\frac{v}{v-1} m_{djj})$, (5.11) then d^* is Φ^*-optimal.

Proof: Follows directly from (5.10).

Example: In the case of (5.6), we obtain,

(a) If M_d^* is c.s. and minimizes $\sum_j m_{djj}^{-p} \Rightarrow d^*$ is Φ_p^*-optimal.

(b) If M_d^* is c.s. and maximizes $\sum_j \log m_{djj} \Rightarrow d^*$ is Φ_p^*-optimal (i.e., it is D-optimal).

(c) If M_d^* is c.s. and maximizes $\min_j m_{djj} \Rightarrow d^*$ is Φ_∞-optimal (i.e., it is E-optimal).

Also, from Theorem 5.2,

(d) If M_d^* is c.s. and maximizes $\sum_j m_{djj} \Rightarrow d^*$ is Φ_p-optimal, $0 \le p \le \infty$ and more.

6. TYPE 1 AND TYPE 2 CRITERIA

Cheng (1978) refined Kiefer's criteria and defined a larger class of optimality criteria that includes A-, E-, D-, all Φ_p-criteria, $0 < p < \infty$, and more.

Again, let $\mathcal{C} \subseteq \mathcal{B}_{v,o}$. (In the $\mathcal{B}_v$ context, similar arguments hold.) Let $t_{\mathcal{D}} = \max_{d \in \mathcal{D}} \text{Tr} M_d$.

Definition 6.1. A design $d^* \in \mathcal{D}$ satisfies optimality criteria of type 1 if d^* minimizes $\Phi_f(M_d) = \sum_{i=1}^{v-1} f(\mu_{di})$ where f is a real-valued function defined on $[0, t_{\mathcal{D}})$ such that

(a) f is continuous, strictly convex, and strictly decreasing on $[0, t_{\mathcal{D}}]$. We include here the possibility that

$$f(0) = \lim_{x \to 0^+} f(x) = +\infty . \tag{6.1}$$

(b) f is continuously differentiable on $(0, t_{\mathcal{D}})$, and f' is strictly concave on $(0, t_{\mathcal{D}})$, i.e., $f' < 0$, $f'' > 0$, and $f''' < 0$ on $(0, t_{\mathcal{D}})$.

Definition 6.2. A design $d^* \in \mathcal{D}$ satisfies optimality criteria of type 2, if d^* minimizes $\Phi_f(M_d) = \sum_{i=1}^{v-1} f(\mu_{di})$ where f has the same property as in Definition 6.1. Except that the strict concavity of f' is replaced by strict convexity, i.e., $f''' > 0$ on $(0, t_{\mathcal{D}})$.

Also, a generalized optimality criterion of type i $(i = 1,2)$ is defined to be the pointwise limit of a sequence of type i criteria.

From (4.2) and (5.6), the A-, D-, and Φ_p-criterion are of type 1 and the E-criterion is a generalized criterion of type 1 (being the limit of Φ_p-criteria, as $p \to \infty$). Note that the A- and D-criteria correspond to the choices of $f(x) = x^{-1}$ and $-\log x$ respectively.

Remarks: (i) There do exist functions satisfying the requirements for a type 2 criterion. For example, let $f(x) = \varepsilon x^3 - ax$ over the interval $[0,t_{\mathcal{D}}]$ of interest, when $\varepsilon > 0$, $a > 0$ and ε compared with a, is small.

(ii) From Section 4 if there is a symmetric design which maximizes $\mathrm{Tr}M_d$ over $\mathcal{D}$, then it is optimal with respect to a very general class of criteria including both generalized type 1 and type 2 criteria. □

It appears that most optimality criteria (universal optimality is an exception) which place equal emphasis on all the parameters can be formulated in terms of the eigenvalues of the information matrix. In Section 7 we shall introduce another optimality criterion of the form $\Phi(\mu_{d1},\ldots,\mu_{d(v-1)})$ with Φ Schur convex symmetric.

7. SCHUR OPTIMALITY

The concept of Schur optimality was introduced by Magda (1979). To see how it was defined, let us recall the following:

Definition 7.1. A matrix with nonnegative entries is called doubly stochastic if the sum of the entries is 1 in every row and every column.

Definition 7.2. Let I be an interval on the real line. A function $\Phi : I^n \to R$ is called Schur convex (after Schur (1923)) if

$$\Phi(Sx) \leq \Phi(x)$$

for all $x \in I^n$ and every doubly stochastic matrix S. A Schur convex function is not necessarily convex, e.g., $\phi(x_1,x_2) = |x_1 - x_2|^{\frac{1}{2}}$. Any Schur convex function is symmetric, because for any permutation matrix P we have

$$\Phi(Px) \leq \Phi(x) = \Phi(P^{-1}Px) \leq \phi(Px) .$$

Hence $\Phi(Px) = \Phi(x)$ as desired. We have used the fact that a permutation matrix and its inverse are examples of doubly stochastic matrices.

While symmetry is a necessary condition to have Schur convexity it is by no means sufficient. When convexity is added to symmetry we can insure Schur convexity. This is seen as follows: By Birkhoff (1946) every doubly stochastic matrix S can be written as a convex sum of permutation matrices. Let $S = \Sigma\lambda_i P_i$, $(\Sigma\lambda_i = 1)$. Then

$$\Phi(Sx) = \underbrace{\Phi(\Sigma\lambda_i P_i x) \leq \Sigma\lambda_i\Phi(P_i x)}_{\text{convexity of } \phi(\cdot)} \qquad \underbrace{= \Sigma\lambda_i\phi(x)}_{\text{symmetry of } \phi(x)}$$

$= \Phi(x)$ and this proves Schur convexity.

Assume $\mathcal{C} \subseteq \mathcal{B}_{v,o}$, let $I = [0,t_{\mathcal{D}}]$ and n be the smallest integer for which $\mu_{d(n+1)} = \mu_{d(n+2)} = \ldots = \mu_{dv} = 0$ for all $d \in \mathcal{D}$.

Define $\sigma(M_d)$ to be the following vector in I^n:

$$\sigma(M_d) = \begin{pmatrix} \mu_{d1} \\ \vdots \\ \mu_{dn} \end{pmatrix} . \tag{7.1}$$

For $d \in \mathcal{D}$ and any Schur convex function Φ defined on I^n and nonincreasing in its arguments, set

$$\Phi(M_d) = \Phi(\sigma(M_d)) . \qquad (7.2)$$

Schur optimality is now defined as follows:

Definition 7.3. A design $d^* \in \mathcal{D}$ is called Schur-optimal if d^* minimizes $\Phi(M_d)$, for all $d \in \mathcal{D}$, and all Schur convex functions Φ nonincreasing in their arguments.

Remark: Not every design setting contains a Schur optimal design.

Note that, if $\phi : I \to R$ is convex, then

$$\Phi(x) = \sum_{i=1}^{n} \phi(x_i), \; x' = (x_1, x_2, \ldots, x_n) \qquad (7.3)$$

is Schur convex on I^n because $\Phi(\cdot)$ is symmetric and convex. From (5.6), D-, A-, and all Φ_p-criteria defined so far on the eigenvalues of the information matrices are instances of Schur functions. As a symmetric and convex function on I_n

$$E(x_1, \ldots, x_n) = - \min_{1 \le i \le n} \{x_1, x_2, \ldots, x_n\}$$

is also Schur convex. This function is associated with E-optimality. Note that E-optimality is no longer a limiting case when dealt with as a Schur convex function. To prove Schur optimality, we state the following very useful tool.

Theorem 7.1. (Derived from Ostrowski (1952)).

Let $F(x_1, \ldots, x_n)$ be a Schur convex and nonincreasing function in its arguments on I^n. Let

$$y_1 \ge y_2 \ge \cdots \ge y_n; \quad x_1 \ge x_2 \ge \cdots \ge x_n \qquad (7.4)$$

satisfy the following

$$y_1 + \ldots + y_\ell \le x_1 + \ldots + x_\ell \quad \text{for all} \quad 1 \le \ell \le n . \qquad (7.5)$$

Then

$$F(y_1, \ldots, y_n) \le F(x_1, \ldots, x_n) .$$

For convenience, when two vectors x and $y \in I^n$, satisfy (7.4) and (7.5) we write $y \le x$.

Applying the above result we can immediately conclude:

Theorem 7.2. d^* is Schur optimal if $\sigma(M_d^*) \le \sigma(M_d)$ for all $d \in \mathcal{D}$.

It should be pointed out that the ordered partial sums in (7.4) are examples of Schur convex functions. Further useful results can be obtained in Hardy and Littlewood (1967).

Lemma 7.1. Let $M_d \in \mathcal{C} \subseteq \mathcal{B}_{v,o}$ and $P_i (1 \le i \le n)$ be n orthogonal matrices such that $M_d^{(i)} = P_i^{-1} M_d P_i$ also satisfies $M_d^{(i)} \underline{1} = 0$ for all $1 \le i \le n$. Set

$\bar{M}_d = \frac{1}{n} \sum_{i=1}^{n} M_d^{(i)}$. Then for any Schur convex function Φ nonincreasing in its arguments we have $\Phi(\bar{M}_d) \leq \Phi(M_d)$.

Proof: Since the P_i's are orthogonal, we have $\sigma(M_d^{(i)}) = \sigma(M_d)$ and hence $\Phi(M_d^{(i)}) = \Phi(M_d)$ for all $1 \leq i \leq n$. Moreover, let $\{\mu_{di}\}$ and $\{\bar{\mu}_{di}\}$ denote the eigenvalues of M_d and $\bar{M}_d$ respectively (and let them be ordered non-increasingly.) Then it is known (see Bellman (1970)) that

$$\sum_{i=1}^{\ell} \bar{\mu}_{di} \leq \sum_{i=1}^{\ell} \mu_{di} \quad \text{for } \ell = 1,2,\ldots,v-1.$$

By Theorem 7.1 we obtain $\Phi(\bar{M}_d) \leq \Phi(M_d)$.

Remark: We call $\bar{M}_d$ (defined in Lemma 7.1), an averaged version of M_d.

Verifying the requirements of Theorem 7.2 is difficult because of the large variety of information matrices M_d. It is practically impossible to find $\sigma(M_d)$. When averaging M_d properly, however, it is easily seen that finding $\sigma(\bar{M}_d)$ is a tractable task. Hence comparing $\sigma(M_{d*})$ and $\sigma(\bar{M}_d)$ (in view of Theorem 7.3) is often time possible.

Theorem 7.3. d^* is Schur optimal if $\sigma(M_{d*}) \leq \sigma(\bar{M}_d)$ for all $d \in \mathcal{D}$, where $\bar{M}_d$ is some average version of M_d.

Proof: $\phi(M_{d*}) \leq \phi(\bar{M}_d) \leq \phi(M_d)$ where the first inequality holds from the assumption $\sigma(\bar{M}_{d*}) \leq \sigma(\bar{M}_d)$ and the latter from Lemma 7.1.

Closing Remarks: We refer the reader to "Special Issue on Optimal Design Theory", No. 14, Vol. A7 (1978) of Communications in Statistics (edited by this author) for further ideas, results and references. Currently we are preparing a book on the subject of optimal design of experiments. The book should be available for distribution within a year or so. Meanwhile, the interested reader can obtain preliminary versions of some chapters of the book.

Acknowledgement: This work was supported by Grant AFOSR 76-3050C. The help of H. Hwang in gathering relevant material was essential in the preparation of the paper.

8. REFERENCES

[1] Bellman, R., Introduction to Matrix Analysis. 2nd edition, McGraw-Hill, (1970).

[2] Birkhoff, G., Tres observations sobre el algebra linear, Univ. nac. Tucuman, Revista Ser. A.5, (1946) 147-150.

[3] Box, M.J. and Draper, N.R., Factorial designs, the |X'X| criterion, and some related matters. Technometrics 13, (1971) 731-742.

[4] Cheng, C.-S., A note on (M,S)-optimality. Commun. Statist. A7 (14), (1978) 1327-1338.

[5] Cheng, C.-S., Optimality of certain asymmetrical experimental designs. Ann. Math. Statist., 6, (1978) 1239-1361.

[6] Chernoff, H., Locally optimal designs for estimating parameters. Ann. Math. Statist. 24, (1953) 586-602.

[7] Cramer, H., Mathematical Methods of Statistics. 13th printing, Princeton University Press, (1974).

[8] Eccleston, J.A. and Hedayat, A., On the theory of connected designs: characterization and optimality. Ann. Statist. 2, (1974) 1238-1255.

[9] Ehrenfeld, S., On the efficiency of experimental designs. Ann. Math. Statist. 26, (1955) 247-255.

[10] Elfving, G., Optimal allocation in linear regression theory. Ann. Math. Statist. 23, (1952) 255-262.

[11] Fedorov, V.V., Theory of Optimal Experiments, New York: Academic Press,(1972).

[12] Hardy, G.H., Littlewood, J.E. and Polya, G., Inequalities, Cambridge University Press, (1967).

[13] Hedayat, A. and Wallis, W.D., Hadamard matrices and their applications. Ann. Statist. 6, (1978) 1184-1238.

[14] Hedayat, A., Theory of Optimal Design of Experiments. A book in preparation, (1980).

[15] Kiefer, J., On the nonrandomized optimality and randomized nonoptimality of symmetrical designs. Ann. Math. Statist. 29, (1958) 675-699.

[16] Kiefer, J., Optimum experimental designs. J. R. Statist. Soc. B21, (1959) 272-319.

[17] Kiefer, J., General equivalence theory for optimum designs (approximate theory). Ann. Statist. 2, (1974) 849-879.

[18] Kiefer, J., Construction and optimality of generalized Youden designs. In A Survey of Statistical Design and Linear Models. (J. N. Srivastava, ed.). Amsterdam: North-Holland Publishing Co., (1975) 333-353.

[19] Magda, C.G., On E-optimal block designs and Schur optimality. Ph.D. Thesis. University of Illinois at Chicago Circle. (1979).

[20] Mood, A.M., On Hoteling's weighing problem. Ann. Math. Statist. 17, (1946) 432-446.

[21] Ostrowski, A., Sur quelques applications des fonctions convexes et concaves au sens de I. Schur, Journ. de Math. Pures et Appliq., 31, (1952) 253-292.

[22] Seber, G.A.F., Linear Regression Analysis. Wiley and Sons, (1977).

[23] Shah, K.R., Optimality criteria for incomplete block designs. Ann. Math. Statist. 31, (1960) 791-794.

[24] Smith, K., On the standard deviations of adjusted and interpolated values of an observed polynomial function and its constants and the guideance they give towards a proper choice of the distribution of observations. Biometrika 12, (1918) 1-85.

[25] Wald, A., On the efficient design of statistical investigations. Ann. Math. Statist. 14, (1943) 134-140.

[26] Wilks, S.S., Certain generalizations in the analysis of variance. Biometrika 24, (1932) 471-494.

STATISTICS AND RELATED TOPICS
M. Csörgö, D.A. Dawson, J.N.K. Rao, A.K.Md.E. Saleh (eds.)

EQUALITIES AND INEQUALITIES FOR CONDITIONAL AND PARTIAL CORRELATION COEFFICIENTS

Michael C. Lewis and George P.H. Styan

Internal Audit Department
Canadian International Paper Company
Montréal, Québec
Canada

Department of Mathematics
McGill University
Montréal, Québec
Canada

The conditional correlation coefficient $\rho(UV|Z)$ is the correlation between U and V when Z is held fixed. The partial correlation coefficient may be defined as either the correlation $\rho(UV:Z)$ between the residuals $U-\mathbb{E}(U|Z)$ and $V-\mathbb{E}(V|Z)$, or as the correlation $\rho(UV\cdot Z)$ between the residuals after linear regressions of U and V on Z. Following Fleiss and Tanur (1971), Gokhale (1976), and Lawrance (1976), we obtain various inequalities and equalities concerning $\rho(UV|Z)$, $\rho(UV:Z)$, $\rho(UV\cdot Z)$, and $\mathbb{E}[\rho(UV|Z)]$.

1. INTRODUCTION AND SUMMARY

The partial correlation coefficient between two random variables U and V has been defined in a variety of ways in the literature. Let Z be a $k\times 1$ random vector of one or more random variables, and let

$$\rho(UV|Z) = \mathbf{corr}(U, V \mid Z),$$

$$\rho(UV:Z) = \mathbf{corr}[U - \mathbb{E}(U|Z), V - \mathbb{E}(V|Z)],$$

where $\mathbf{corr}(.)$ denotes the usual product-moment correlation coefficient. We define $\rho(UV|Z)$ to be the *conditional correlation coefficient between* U *and* V *given* Z, and $\rho(UV:Z)$ to be the *partial correlation coefficient between* U *and* V *adjusted for* Z. This distinction between *conditional* and *partial* correlation coefficients seems only to have been made quite recently, cf. Fleiss and Tanur (1971), Gokhale (1976), and Lawrance (1976). Nevertheless these authors all assume that the two regressions $\mathbb{E}(U|Z)$ and $\mathbb{E}(V|Z)$ are linear. Our more general definition $\rho(UV:Z)$ appears only to have been given by Sverdrup (1967, p.18).

We may denote the linear regression of U on Z as

$$\mathbb{LR}(U\cdot Z) = \mathbb{E}(U) + \mathbf{Cov}(U, Z)[\mathbf{Var}(Z)]^{-1}[Z - \mathbb{E}(Z)],$$

where $\mathbf{Cov}(U, Z)$ is the $1\times k$ row vector of covariances between U and the k components of Z, and $\mathbf{Var}(Z)$ is the $k\times k$ covariance matrix of Z. We assume that the joint covariance matrix of U, V and Z is positive definite. Then

$$\begin{aligned}\rho(UV\cdot Z) &= \mathbf{corr}[U - \mathbb{LR}(U\cdot Z), V - \mathbb{LR}(V\cdot Z)] \\ &= (\rho_{uv} - \rho_{uz}'R_z^{-1}\rho_{vz})(1 - \rho_{uz}'R_z^{-1}\rho_{uz})^{-\frac{1}{2}}(1 - \rho_{vz}'R_z^{-1}\rho_{vz})^{-\frac{1}{2}} \qquad (1.1)\end{aligned}$$

is the partial correlation coefficient most often considered in the literature. Yule (1897) introduced $\rho(UV\cdot Z)$ as the *net correlation coefficient;* the term *partial correlation coefficient* was introduced by Pearson (1902). We will call $\rho(UV\cdot Z)$ the *Yule-Pearson partial correlation coefficient* and $\rho(UV:Z)$ the *Sverdrup partial correlation coefficient.* The joint correlation matrix of U, V and Z is

$$R = \begin{pmatrix} 1 & \rho_{uv} & \rho'_{uz} \\ \rho_{uv} & 1 & \rho'_{vz} \\ \rho_{uz} & \rho_{vz} & R_z \end{pmatrix}.$$

When the two regressions $\mathbb{E}(U|Z)$ and $\mathbb{E}(V|Z)$ are linear then

$$\rho(UV:Z) = \rho(UV\cdot Z).$$

When U, V and Z follow a multivariate normal distribution, then the regressions are linear, the variances and covariances are free of Z, and

$$\rho(UV:Z) = \rho(UV\cdot Z) = \rho(UV|Z), \tag{1.2}$$

cf. e.g., Anderson (1958, p.29). Fleiss and Tanur (1971) and Gokhale (1976) claimed that (1.2) holds whenever the two regressions $\mathbb{E}(U|Z)$ and $\mathbb{E}(V|Z)$ are linear. Lawrance (1976), however, showed that this is not necessarily so since $\rho(UV|Z)$ is then not necessarily free of Z, cf. (2.3) below.

Gokhale (1976) has shown that $\rho(UV\cdot Z) = 1 \Rightarrow \rho(UV|Z) = 1$ for all Z, and gives an example where $\rho(UV\cdot Z) = 0$ but $\rho(UV|Z) = \pm 1$. Furthermore he has shown that $\rho(UV|Z)$ could equal zero while $\rho(UV\cdot Z)$ may be made arbitrarily close to $+1$. He also has given examples where the univariate marginal distributions are all normal but in which $\rho(UV|Z)$ and $\rho(UV\cdot Z)$ differ.

In Section 2 of this paper we consider a situation, following Lawrance (1976), where $\rho(UV:Z) = \rho(UV\cdot Z)$ and show that this may exceed or be less than $\mathbb{E}[\rho(UV|Z)]$. In Section 3 we extend the example suggested by Fleiss and Tanur (1971), and find that there $\rho(UV:Z) = \rho(UV|Z) \le \rho(UV\cdot Z)$; we note that $\rho(UV:Z) = \rho(UV|Z)$ and $\rho(UV\cdot Z)$ may have opposite signs. In a further paper, Lewis and Styan (1981) construct a situation where $\rho(UV|Z)$, $\rho(UV:Z)$, and $\rho(UV\cdot Z)$ all differ.

2. INEQUALITIES FOR $\rho(UV|W)$ WHEN $\rho(UV:W) = \rho(UV\cdot W)$ AND BOTH REGRESSIONS ARE LINEAR

Let U, V and W be three random variables. Then, following Lawrance (1976), let us consider the construction

$$U = A + BW, \qquad V = C + DW, \tag{2.1}$$

where the random variables A, B, C, and D are distributed independently of W. Then the two regressions $\mathbb{E}(U|W)$ and $\mathbb{E}(V|W)$ are both linear, and so

$$\rho(UV|W) = \frac{\mathrm{cov}(A,C) + [\mathrm{cov}(A,D)+\mathrm{cov}(B,C)]W + \mathrm{cov}(B,D)W^2}{[(\mathrm{var}(A)+2\mathrm{cov}(A,B)W+\mathrm{var}(B)W^2)(\mathrm{var}(C)+2\mathrm{cov}(C,D)W+\mathrm{var}(D)W^2)]^{\frac{1}{2}}}. \tag{2.2}$$

When A, B, C, and D all have the same variance σ^2 and common correlation coefficient ρ then

$$\mathrm{cov}(A,C) = \mathrm{cov}(A,D) = \mathrm{cov}(B,C) = \mathrm{cov}(B,D) = \sigma^2\rho \ge -\sigma^2/3,$$

since the common correlation coefficient ρ between 4 random variables must be at least $-1/3$, cf. Anderson (1958, p.43). Hence

$$\mathrm{cov}(U,V \mid W) = \sigma^2\rho(1 + 2W + W^2),$$

while

$$\mathbf{var}(U \mid W) = \mathbf{var}(V|W) = \sigma^2(1 + 2\rho W + W^2)$$

is positive unless $\rho = +1$ and $W = -1$, or $\rho = -1$ and $W = +1$. Excluding these two cases we may then write

$$\rho(UV|W) = \frac{\sigma^2\rho(1 + 2W + W^2)}{\sigma^2(1 + 2\rho W + W^2)} = \frac{\rho(1 + W)^2}{1 + 2\rho W + W^2} \quad . \tag{2.3}$$

This is a function of W unless $\rho = +1$ and so $\rho(UV|W)$ cannot, in general, be equal to $\rho(UV{:}W)$ or to $\rho(UV{\cdot}W)$ as these partial correlation coefficients do not depend on W.

Since the two regressions $\mathbb{E}(U|W)$ and $\mathbb{E}(V|W)$ are linear, we find that

$$\rho(UV{:}W) = \rho(UV{\cdot}W) = \frac{\rho[1 + 2\mathbb{E}(W) + \mathbb{E}(W^2)]}{1 + 2\rho\mathbb{E}(W) + \mathbb{E}(W^2)} = \frac{\rho\mathbb{E}(1 + W)^2}{1 + 2\rho\mathbb{E}(W) + \mathbb{E}(W^2)} \quad , \tag{2.4}$$

using (1.1). We have, therefore, established the following:

THEOREM 2.1. *Let* $U = A+BW$ *and* $V = C+DW$, *where the random variables* $A, B, C,$ *and* D *are distributed independently of the random variable* W. *If* $A, B, C,$ *and* D *all have the same variance* σ^2 *and common correlation coefficient* ρ *then the conditional correlation coefficient* $\rho(UV|W)$ *is as given by* (2.3) *unless* $\rho = +1$ *and* $W = -1$, *or* $\rho = -1$ *and* $W = +1$, *and* $\rho(UV|W)$ *depends on* W *unless* $\rho = +1$, *while the Sverdrup and Yule-Pearson partial correlation coefficients are equal with value as given by* (2.4) *unless* $\rho = +1$ *and* $W = -1$ *with probability one, or* $\rho = -1$ *and* $W = +1$ *with probability one.*

It is of interest to compare (2.3) and (2.4). We notice first that $\rho(UV|W)$, $\rho(UV{\cdot}W)$ and ρ all have the same sign. Moreover,

$$\left.\begin{aligned} -1 \le \frac{2\rho}{1+\rho} \le \rho(UV|W) \le 0; &\qquad -\tfrac{1}{3} \le \rho \le 0, \\ 0 \le \rho(UV|W) \le \frac{2\rho}{1+\rho} < 1; &\qquad 0 \le \rho < 1, \end{aligned}\right\} \tag{2.5}$$

since

$$\frac{2\rho}{1+\rho} - \rho(UV|W) = \frac{2\rho}{1+\rho} - \frac{\rho(1+W)^2}{1 + 2\rho W + W^2} = \frac{\rho(1-\rho)(1-W)^2}{(1+\rho)(1 + 2\rho W + W^2)} \quad .$$

Equality in (2.5) may be characterized as follows:

$$\left.\begin{aligned} \rho(UV|W) = \frac{2\rho}{1+\rho} &\iff W = +1, \\ \rho(UV|W) = 0 \quad &\iff W = -1. \end{aligned}\right\} \tag{2.6}$$

Similarly

$$\left.\begin{aligned} -1 \le \frac{2\rho}{1+\rho} \le \rho(UV{\cdot}W) \le 0; &\qquad -\tfrac{1}{3} \le \rho \le 0, \\ 0 \le \rho(UV{\cdot}W) \le \frac{2\rho}{1+\rho} < 1; &\qquad 0 \le \rho < 1, \end{aligned}\right\} \tag{2.7}$$

since

$$\frac{2\rho}{1+\rho} - \rho(UV\cdot W) = \frac{\rho(1-\rho)[\mathrm{E}(1-W)^2]}{(1+\rho)[1+2\rho\mathrm{E}(W)+\mathrm{E}(W^2)]}$$

has the same sign as that of ρ. Equality in (2.7) is characterized by, cf. (2.6),

$$\rho(UV\cdot W) = \frac{2\rho}{1+\rho} \iff W = +1 \text{ with probability } 1,$$

$$\rho(UV\cdot W) = 0 \iff W = -1 \text{ with probability } 1.$$

§2.1 THE CASE OF $\mathrm{E}(W) = 0$.

If $\mathrm{E}(W) = 0$ then (2.4) reduces to $\rho(UV{:}W) = \rho(UV\cdot W) = \rho$, and if $\rho > 0$ then

$$\left.\begin{aligned} \rho(UV|W) &\geq \rho(UV{:}W) = \rho \iff W \geq 0, \\ \rho(UV|W) &\leq \rho(UV{:}W) = \rho \iff W \leq 0. \end{aligned}\right\} \quad (2.8)$$

Equality holds in (2.8) $\iff W = 0$. If $\rho < 0$ then the inequalities in (2.8) go the other way. To illustrate (2.8) we offer a plot in Figure 2.1 of

$$\rho(UV|W) = \frac{\rho(1+W)^2}{1+2\rho W+W^2} = \rho(UV|W;\ \rho)$$

say, for $\rho = +0.3$ and for $\rho = -0.3$. Also shown is $\mathrm{E}[\rho(UV|W)]$ when W is uniformly distributed on the interval $(-4, +4)$. We notice from Figure 2.1 that

$$\rho(UV\cdot W) = \rho > \mathrm{E}[\rho(UV|W)] = \mathrm{E}\left[\frac{\rho(1+W^2)}{1+2\rho W+W^2}\right]$$

for both $\rho = +0.3$ and $\rho = -0.3$. This leads to:

THEOREM 2.2. *Let the random variables* U, V *and* W *be defined as in Theorem 2.1. If* W *is distributed uniformly on* $(-k, +k)$ *with* $k>0$ *then* $\mathrm{E}(W) = 0$ *and*

$$\mathrm{E}[\rho(UV|W)] = \rho\mathrm{E}\left[\frac{(1+W)^2}{1+2\rho W+W^2}\right] < \rho = \rho(UV{:}W) = \rho(UV\cdot W)$$

for all $\rho \neq 0$ *such that* $-1/3 \leq \rho < 1$.

For a proof see Lewis (1980, Theorem 3.2).

§2.2 THE CASE OF $\mathrm{E}(W) \neq 0$.

Now let $\mathrm{E}(W) \neq 0$. If $\rho > 0$ then $\rho(UV|W) \geq \rho(UV{:}W) = \rho(UV\cdot W)$ if and only if

$$\left.\begin{aligned} \tfrac{1}{2}[\nu - (\nu^2-4)^{\frac{1}{2}}] \leq W \leq \tfrac{1}{2}[\nu + (\nu^2-4)^{\frac{1}{2}}]; &\quad \mathrm{E}(W) > 0, \\ W \leq \tfrac{1}{2}[\nu - (\nu^2-4)^{\frac{1}{2}}] \text{ or } W \geq \tfrac{1}{2}[\nu + (\nu^2-4)^{\frac{1}{2}}]; &\quad \mathrm{E}(W) < 0, \end{aligned}\right\} \quad (2.9)$$

where $\nu = [1 + \mathrm{E}(W^2)]/\mathrm{E}(W)$. Moreover $\rho(UV|W) = \rho(UV{:}W)$ if and only if $W = \frac{1}{2}[\nu \pm (\nu^2-4)^{\frac{1}{2}}]$. To prove (2.9) we write

$$\rho(UV|W) - \rho(UV{:}W) = \frac{\rho(1+W)^2}{1+2\rho W+W^2} - \frac{\rho(2+\nu)}{2\rho+\nu} = \frac{2\rho(1-\rho)(\nu W - 1 - W^2)}{(1+2\rho W+W^2)(2\rho+\nu)}$$

and so (2.9) follows at once. We note that $2\rho + \nu = \mathbf{var}[U - \mathrm{E}(U|W)]/[\sigma^2\mathrm{E}(W)]$ has the same sign as $\mathrm{E}(W)$.

Let us now compare $\rho(UV{:}W)$ with the expectation of $\rho(UV|W)$. If $\rho \neq 0$ and if $\rho < 1$ then

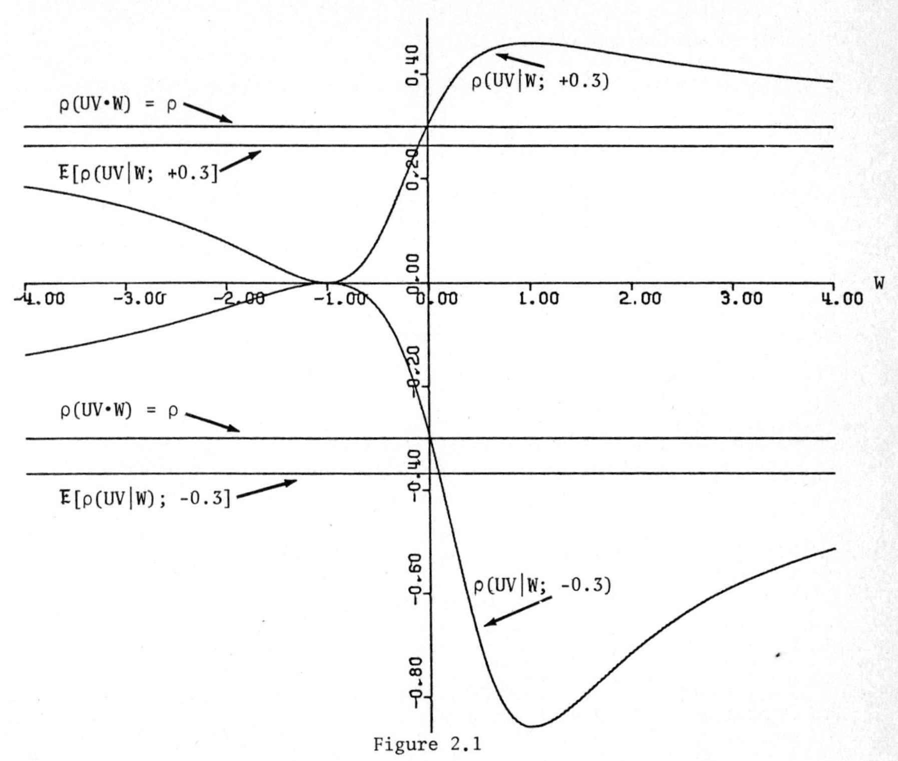

Figure 2.1

Plot of $\rho(UV{:}W) = \rho = \rho(UV{\cdot}W)$ *and of* $\rho(UV|W) = \rho(UV|W;\ \rho)$ *and its expectation when* W *is unformly distributed on the interval* $(-4, +4)$ *for* $\rho = +0.3$ *and for* $\rho = -0.3$.

$$\frac{E[\rho(UV|W)] - \rho}{2\rho(1-\rho)} = E\left[\frac{W}{1 + 2\rho W + W^2}\right],$$

while

$$\frac{\rho(UV{:}W) - \rho}{2\rho(1-\rho)} = \frac{E(W)}{1 + 2\rho E(W) + E(W^2)} \le \frac{E(W)}{1 + 2\rho E(W) + E^2(W)} \tag{2.10}$$

when $E(W) \ge 0$ because $E(W^2) \ge E^2(W)$. If $E(W) > 0$ then equality holds in (2.10) if and only if W is constant (with probability one). If $\rho > 0$ and $E(W) \ge 0$ then $E[\rho(UV|W)] \ge \rho(UV{:}W)$ if

$$E\left[\frac{W}{1 + 2\rho W + W^2}\right] \ge \frac{E(W)}{1 + 2\rho E(W) + E^2(W)} \tag{2.11}$$

and (2.11) holds provided $f(W) = W/(1 + 2\rho W + W^2)$ is convex (Jensen's inequality). The second derivative $f''(W) = 2(W^3 - 3W - 2\rho)/[(1 + 2\rho W + W^2)^3] > 0$ whenever $W \ge 2$ and $\rho < 1$. We have, therefore, proved:

THEOREM 2.3. *Let the random variables* U, V *and* W *be defined as in Theorem 2.1. If* $W \ge 2$ *with probability one and* $0 < \rho < 1$ *then* $E[\rho(UV|W)] \ge \rho(UV{:}W) = \rho(UV{\cdot}W)$, *with equality if and only if* W *is a constant with probability one.*

3. INEQUALITIES FOR $\rho(UV \cdot Z)$ WHEN $\rho(UV|Z) = \rho(UV:Z)$ AND BOTH REGRESSIONS ARE LINEAR IN $1/Z$

Fleiss and Tanur (1971) considered the following two situations which lead to linear regressions:

$$\text{(I)}\quad U = aX + bW,\quad V = cY + dW; \qquad \text{(II)}\quad U = b + aXW,\quad V = d + cYW,$$

where a, b, c, and d are constants with $ac > 0$. The random variables X and Y are distributed independently of W but are correlated between themselves with correlation ρ. Lawrance's construction (2.1) is a generalization of (I) and (II) with

(2.1)	A	B	C	D
(I)	aX	b	cY	d
(II)	b	aX	d	cY

and so we obtain $\rho(UV|W) = \rho$ for both (I) and (II) as $ac > 0$. Furthermore, $\rho(UV:W) = \rho(UV \cdot W)$ since the regressions are linear and $\rho(UV:W) = \mathbf{corr}[U - \mathrm{E}(U|W), V - \mathrm{E}(V|W)] = \rho$ so that $\rho(UV|W) = \rho(UV:W) = \rho(UV \cdot W) = \rho$.

Now let us suppose that $Z \neq 0$ with probability 1 and put $W = 1/Z$ in (II). Then $U = b + aX/Z$ and $V = d + cY/Z$, and since $ac > 0$ we have

$$\rho(UV|Z) = \rho(UV|W) = \rho,$$

$$\rho(UV:Z) = \mathbf{corr}(a\bar{X}/Z,\ c\bar{Y}/Z) = \mathbf{corr}(\bar{X}, \bar{Y}) = \rho = \rho(UV|Z),$$

where $\bar{X} = X - \mathrm{E}(X|W)$ and $\bar{Y} = Y - \mathrm{E}(Y|W)$ have zero means and are independent of $Z = 1/W$. Furthermore, using (1.1) we obtain

$$\rho(UV \cdot Z) = (\rho_{uv} - \rho_{uz}\rho_{vz})(1 - \rho_{uz}^2)^{-\frac{1}{2}}(1 - \rho_{vz}^2)^{-\frac{1}{2}} \tag{3.1}$$

with

$$\rho_{uv} = \mathbf{corr}(X/Z,\ Y/Z),\quad \rho_{uz} = \mathbf{sgn}(a)\cdot\mathbf{corr}(X/Z,\ Z),\quad \rho_{vz} = \mathbf{sgn}(c)\cdot\mathbf{corr}(Y/Z,\ Z).$$

We will use the following:

LEMMA 3.1. *Let* X, Y *and* Z *be random variables with* Z *distributed independently of* X *and* Y *and with* $Z \neq 0$ *with probability one. Then*

$$\mathbf{cov}(X/Z,\ Y/Z) = \mathbf{cov}(X,\ Y)\mathrm{E}(1/Z^2) + \mathrm{E}(X)\mathrm{E}(Y)\mathbf{var}(1/Z),$$

$$\mathbf{var}(X/Z) = \mathbf{var}(X)\mathrm{E}(1/Z^2) + \mathrm{E}^2(X)\mathbf{var}(1/Z),$$

$$\mathbf{cov}(X/Z,\ Z) = \mathrm{E}(X)\mathbf{cov}(Z,\ 1/Z).$$

The proof is straightforward (cf. Lewis, 1980, p.53) and, therefore, omitted.

Using Lemma 3.1 we may now evaluate (3.1). Since $ac > 0$ both a and c have the same sign and this sign leaves (3.1) unchanged. To make the algebra more tractable we follow Fleiss and Tanur (1971) in part and let

$$\mathrm{E}(X) = \mathrm{E}(Y) = \mu, \qquad \mathbf{var}(X) = \mathbf{var}(Y) = \sigma^2.$$

Then

$$\rho(UV \cdot Z) = (\rho_{uv} - \rho_{uz}^2)/(1 - \rho_{uz}^2)$$

with

$$\rho_{uv} = (\rho + \alpha_1^2\alpha_2^2)/(1 + \alpha_1^2\alpha_2^2), \quad \rho_{uz}^2 = \alpha_1^2\alpha_3^2/(1 + \alpha_1^2\alpha_2^2) \ ,$$

where

$$\alpha_1^2 = \frac{\mu^2}{\sigma^2}, \quad \alpha_2^2 = \frac{\mathbf{var}(1/Z)}{\mathbb{E}(1/Z^2)}, \quad \alpha_3^2 = \frac{\mathbf{cov}^2(Z,\ 1/Z)}{\mathbf{var}(Z)\mathbb{E}(1/Z^2)} = \mathbf{corr}^2(Z,\ \tfrac{1}{Z})\cdot\alpha_2^2. \tag{3.2}$$

Hence

$$\rho(UV\cdot Z) = [\rho + \alpha_1^2(\alpha_2^2-\alpha_3^2)]/[1 + \alpha_1^2(\alpha_2^2-\alpha_3^2)] \geq \rho = \rho(UV|Z) = \rho(UV:Z) \tag{3.3}$$

as $\alpha_2^2 \geq \alpha_3^2$, with equality if and only if

$$\alpha_1^2 = 0 \quad \text{or} \quad \alpha_2^2 = \alpha_3^2 \tag{3.4}$$

provided $\rho < 1$. Substituting (3.2) into (3.4) yields

$$\mu = 0 \quad \text{or} \quad \mathbf{corr}(Z,\ 1/Z) = \pm 1.$$

The correlation between Z and $1/Z$ is ± 1 if and only if $kZ + \ell = 1/Z$ for some constants k and ℓ. In this event $kZ^2 + \ell Z - 1 = 0$ and so Z then takes on only two values with probability one.

We summarize our results in the following:

THEOREM 3.1. *Let the random variables* X *and* Y *have the same mean* μ, *the same variance* σ^2, *and correlation coefficient* ρ. *Let the random variable* Z *be distributed independently of* X *and* Y *and suppose that*

$$U = b + aX/Z, \qquad V = d + cY/Z,$$

where $Z \neq 0$ *with probability one, and where* $a, b, c,$ *and* d *are constants with* $ac > 0$. *Then*

$$\rho(UV|Z) = \rho(UV:Z) = \rho \leq \rho(UV\cdot Z) \tag{3.5}$$

as given by (3.3), *with equality if and only if* $\mu = 0$ *or* Z *takes on only two values with probability one.*

Fleiss and Tanur (1971) assumed that both X and Z were uniformly distributed on the integers $1, 2, \ldots, 100$. When $\rho = 0$ the Yule-Pearson partial correlation coefficient $\rho(UV\cdot Z)$ can be made as large as 0.6630 in contrast to $\rho(UV|Z) = \rho(UV:Z) = \rho = 0$, cf. Fleiss and Tanur (1971, p.44). The difference between $\rho(UV\cdot Z)$ and $\rho(UV|Z) = \rho(UV:Z)$ can, however, be made even larger. We have

$$\rho(UV\cdot Z) = \rho(UV|Z) = (1 - \rho)\alpha_1^2(\alpha_2^2 - \alpha_3^2)/[1 + \alpha_1^2(\alpha_2^2 - \alpha_3^2)].$$

We will use:

LEMMA 3.2. *Let* X *and* Z *be uniformly distributed on the integers* $1, 2, \ldots, n$, *or continuously on the interval* $(1,n)$, *and let* $\alpha_1^2, \alpha_2^2, \alpha_3^2$ *be defined as in* (3.2). *Then as* $n \to \infty$ *it follows that* $\alpha_1^2 \to 3$, $\alpha_2^2 \to 1$, *and* $\alpha_3^2 \to 0$.

The proof is straightforward (cf. Lewis, 1980, Lemma 4.4) and, therefore, omitted.

It follows from (3.3) and Lemma 3.2 that as $n \to \infty$

$$\rho_{uv} \to \tfrac{1}{4}(\rho + 3), \quad \rho_{uz}^2 \to 0, \quad \rho(UV \cdot Z) \to \tfrac{1}{4}(\rho + 3)$$

and so

$$\rho(UV \cdot Z) - \rho(UV|Z) = \rho(UV \cdot Z) - \rho \to 3(1 - \rho)/4.$$

When $\rho = 0$ as in Fleiss and Tanur (1971), the Yule-Pearson partial correlation coefficient may be as large as 3/4 in contrast to the common value 0 of the conditional correlation coefficient and the Sverdrup partial correlation coefficient. The difference between $\rho(UV \cdot Z)$ and $\rho(UV|Z) = \rho$, when $\rho \neq 0$, is maximized, however, when $\rho = -1$, for then as $n \to \infty$

$$-1 = \rho = \rho(UV|Z) = \rho(UV:Z) < \rho(UV \cdot Z) = \tfrac{1}{2}\ .$$

When $\rho(UV|Z) = \rho > 0$, then $\rho(UV \cdot Z) > 0$ for all values of n since $\alpha_2^2 \geq \alpha_3^2$. If $\rho(UV|Z) = \rho < 0$, however, then $\rho(UV \cdot Z) > 0$ (and so $\rho(UV|Z)$ and $\rho(UV \cdot Z)$ have opposite signs) whenever

$$\beta = \alpha_1^2(\alpha_2^2 - \alpha_3^2) \geq 1$$

since $\rho \geq -1$. For the discrete uniform distribution this holds for all $n \geq 21$, while for the continuous uniform this holds for all $n \geq 29$ (cf. Lewis, 1980, Tables 4.2 and 4.3).

While β may exceed 1 for the uniform distribution, this does not appear to be possible for either the Poisson or the binomial distributions. The largest value we were able to find for the Poisson distribution with parameter λ was $\beta = 0.78364$, and then $\lambda = 6.8$. The largest value which we obtained for the binomial $B(n,p)$ distribution was $\beta = 0.76372$, and then $n = 64$ and $p = 0.1$; the binomial mean is then 6.4 (compared with the Poisson mean of 6.8).

Despite these findings with the Poisson and binomial distributions we have been able to construct a distribution where

$$\rho(UV \cdot Z) - \rho(UV|Z) > 3(1 - \rho)/4.$$

Let the random variables X and Z each take on three values: -1, $+1$ and h with probabilities 0.1, 0.1 and 0.8, respectively. Then as $h \to \infty$ we find that $\beta \to 4$. We obtain:

$$\alpha_1^2 = \frac{(0.64)h^2}{0.2 + (0.16)h^2}, \quad \alpha_2^2 = \frac{0.2 + 0.16/h^2}{0.2 + 0.8/h^2}, \quad \alpha_3^2 = \frac{0.1296}{0.168 + (0.032)h^2 + 0.16/h^2}\ .$$

As $h \to \infty$ it is clear that $\alpha_1^2 \to 4$, while $\alpha_2^2 \to 1$ and $\alpha_3^2 \to 0$. Thus $\beta \to 4$ and so $\rho(UV \cdot Z) - \rho(UV|Z) \to 4(1 - \rho)/5$. The convergence is quite fast. For $h = 10$, for example, we obtain

$$\alpha_1^2 = 3.9506, \quad \alpha_2^2 = 0.9692, \quad \alpha_3^2 = 0.0385, \quad \beta = 3.6768;$$

thus

$$\rho(UV \cdot Z) - \rho(UV|Z) = \frac{\beta}{1 + \beta}(1 - \rho) = 0.7862(1 - \rho) \to 1.5724,$$

with equality when $\rho = -1$. We summarize these results as

THEOREM 3.2. *Let the random variables* U, V, X, Y, *and* Z *be defined as in Theorem 3.1. Then*

$$\rho(UV|Z) = \rho(UV:Z) = \rho \le \rho(UV\cdot Z) = (\rho + \beta)/(1 + \beta),$$

where

$$\beta = \frac{E^2(X)\mathrm{var}(1/Z)}{\mathrm{var}(X)E(1/Z^2)}\,[1 - \mathrm{corr}^2(Z, 1/Z)]\ .$$

The difference

$$\rho(UV\cdot Z) - \rho(UV:Z) = \rho(UV\cdot Z) - \rho(UV|Z) = \beta(1 - \rho)/(1 + \beta)$$

is maximal when $\rho = -1$ *and* β *is maximized. Moreover,*

(1) *if* X *and* Z *are uniformly distributed on the integers* 1, 2,...,n *or on the interval* (1,n) *then* β *increases monotonically to* 3 *as* $n \to \infty$, *while*

(2) *if* X *and* Z *each take on three values:* -1, +1 *and* h *with probabilities* 0.1, 0.1 *and* 0.8, *respectively, then* β *increases monotonically to* 4 *as* $h \to \infty$.

We have not been able to find a distribution for X and Z for which $\beta > 4$.

ACKNOWLEDGEMENTS

This paper is based on part of the M.Sc. thesis by the first author, Lewis (1980). The research of the second author was supported in part, by the Natural Sciences and Engineering Research Council Canada under Grant No. A7274, and by the Gouvernement du Québec, Programme de formation de chercheurs et d'action concertée, subvention no. EQ-961.

4. REFERENCES

[1] Anderson, T.W., *An Introduction to Multivariate Statistical Analysis* (Wiley, New York, 1958).

[2] Fleiss, J.L. and Tanur, J.M., A note on the partial correlation coefficient, *Amer. Statist.* 25 (1971) 43-45.

[3] Gokhale, D.V., On partial and multiple correlation, *J. Indian Statist. Assoc.* 14 (1976) 17-21.

[4] Lawrance, A.J., On conditional and partial correlation, *Amer. Statist.* 30 (1976) 146-149.

[5] Lewis, Michael Charles, Inequalities and Equalities for Conditional and Partial Correlation Coefficients, M.Sc. Thesis, Dept. of Math., McGill University, Montréal.

[6] Lewis, Michael C. and Styan, George P.H., Further inequalities for conditional and partial correlation coefficients, To appear (1981).

[7] Pearson, Karl, Mathematical contributions to the theory of evolution - XI: On the influence of natural selection on the variability and correlation of organs, *Philos. Trans. Roy. Soc. London Ser. A* 200 (1902) 1-66.

[8] Sverdrup, Erling, *Laws and Chance Variations: Basic Concepts of Statistical Inference, Volume II: More Advanced Treatment* (North-Holland, Amsterdam, 1967).

[9] Yule, G. Udny, On the significance of Bravais' formulae for regression & c., in the case of skew correlation. *Proc. Roy. Soc. London* 60 (1897) 477-489.

PART II
PARAMETRIC INFERENCE AND GOODNESS-OF-FIT

STATISTICS AND RELATED TOPICS
M. Csörgö, D.A. Dawson, J.N.K. Rao, A.K.Md.E. Saleh (eds.)

QUANTILE PROCESSES AND SUMS OF WEIGHTED SPACINGS FOR COMPOSITE GOODNESS-OF-FIT

Miklós Csörgő[1] and Pál Révész

Department of Mathematics and Statistics
Carleton University
Ottawa, Ontario
Canada

Mathematical Institute of the
Hungarian Academy of Sciences
Budapest
Hungary

This paper is devoted to a historical review of recent developments in composite goodness-of-fit theory and to a discussion of some of our main new results. The proofs will appear in Csörgő-Révész (1980) in References.

1. INTRODUCTION AND LEGEND

Let $X_1, X_2, \ldots$ be a sequence of independent and identically distributed (i.i.d.) random variables (rv) with distribution function F. Let $F_n(x)$ denote the proportion of those X_i $(1 \leq i \leq n)$ which are less than or equal to a real number x ; i.e., F_n is the empirical distribution function of the random sample $X_1, \ldots, X_n$. The <u>empirical process</u> based on $X_1, \ldots, X_n$ is defined by

$$\alpha_n(x) = n^{\frac{1}{2}}(F_n(x) - F(x)), \quad -\infty < x < +\infty \tag{1.1}$$

and, as it is well known, its Kolmogorov-Smirnov functional $\sup_{-\infty<x<+\infty} |\alpha_n(x)|$ and its Cramér-von Mises functional $\int_{-\infty}^{+\infty} \alpha_n^2(x)\,dF(x)$ can be used to test the goodness-of-fit hypothesis that our random sample $X_1, \ldots, X_n$ has a completely specified distribution function. Most goodness-of-fit problems arising in practice, however, do not usually specify F completely, and, instead of one specified F , we are frequently given a whole parametric family of distribution functions $\{F(x;\theta);\ \theta \in \Theta \subseteq R^d\}$, where R^d is the d-dimensional Euclidean space $(d \geq 1)$. From a goodness-of-fit point of view the unknown parameters θ are a nuisance (nuisance parameters), which render most goodness-of-fit hypotheses composite ones. There are many ways of "getting rid of θ" so as to reduce composite goodness-of-fit hypotheses to simple ones. As far as the empirical process is concerned, one natural way of doing this is to "estimate out θ" by using some kind of a "good estimator" sequence $\{\hat{\theta}_n\}$, based on random samples $X_1, \ldots, X_n$ $(n = 1, 2, \ldots)$ on $F(x;\theta)$.

Concerning the classical Cramér-von Mises and Kolmogorov-Smirnov statistics, Darling (1955) and Kac, Kiefer and Wolfowitz (1955) investigated their asymptotic distributions when the unknown parameters of some <u>special</u> continuous distribution functions were to be estimated first. Durbin (1973a) considered the more global question of weak convergence of the empirical process under a given sequence of alternative hypotheses when parameters of <u>any</u> continuous distribution function $F(x;\theta)$ are estimated from the data. The estimators themselves were to satisfy certain maximum likelihood-like conditions. Durbin (1973a) showed that, for such a general class of estimators $\{\hat{\theta}_n\}$ the <u>estimated empirical process</u>

$$\hat{\alpha}_n(x) = n^{\frac{1}{2}}(F_n(x) - F(x;\hat{\theta}_n)) , \quad x \in R^1 , \tag{1.2}$$

converges weakly to a Gaussian process, whose mean and covariance functions he also gave. Neuhaus (1974, 1976a) studied the asymptotic properties of the Cramér-von Mises statistic when parameters are estimated and the asymptotic power properties of the same test under contiguous alternatives. Using the recently developed strong approximation methodology of Kiefer (1972), Csörgő-Révész (1975) and Komlós-Major-Tusnády (1975) for the empirical process α_n of (1.1), Burke et al (1979) obtained asymptotic in-probability and almost sure representations, in terms of Gaussian processes in both x and n, of the estimated empirical process $\hat{\alpha}_n$ of (1.2). The latter approach was earlier demonstrated in the preliminary drafts of Csörgő-Komlós-Major-Révész-Tusnády (1974) and Burke-Csörgő (1976) under substantially stronger regularity conditions. As to the type of estimation of the parameters $\theta \in R^d$ of $F(x;\theta)$ in these papers, the sequence of estimators $\{\hat{\theta}_n\}$ for θ are to satisfy maximum likelihood-like conditions à la Durbin (1973a).

Neuhaus (1976b) studied the weak convergence of the multivariate empirical process under contiguous alternatives when parameters are estimated. Using a recent almost sure approximation result of Philipp-Pinzur (1978) for the multivariate empirical process, Burke-Csörgő-Csörgő-Révész (1979) obtained an asymptotic in-probability representation of the estimated multivariate empirical process in terms of a Gaussian process in $x \in R^k$ and n, given again maximum likelihood-like estimators $\{\hat{\theta}_n\}$ for θ of $F(x;\theta)$ $(x \in R^k,\ k \geq 1;\ \theta \in R^d,\ d \geq 1)$.

A common feature of the results of Durbin (1973a) and those of the papers quoted afterwards in the above two paragraphs is that the resulting Gaussian processes which approximate the estimated empirical process depend, in general, not only on the form of the hypothetical distribution function but also on the true values of the unknown parameters of the latter. This of course creates difficulties when trying to construct the very tests of goodness-of-fit we were aiming at via these asymptotic results. The very nuisance parameters we were trying to "estimate out" generally reappear at the end (in the limit), resulting in a situation where the statistics themselves which we were to construct on the basis of the estimated empirical process $\hat{\alpha}_n$ of (1.2) become, given the form of F, computable, but the distributions of the corresponding functionals of the approximating Gaussian process do not. There have been a number of ways suggested to overcome this difficulty.

One of them is the so-called random substitution method suggested by Durbin (1961), initially for the finite-sample case. It is to transform tests of composite hypotheses into tests of simple hypotheses. This is done by replacing the estimator $\hat{\theta}_n$ of the unknown θ calculated from the sample by a corresponding "estimator" $\hat{\theta}_n^*$, external to the sample, of a known value of θ. It is shown (Durbin 1976) that when this method is used and, on the null hypothesis H_o, F is of the form $F(x;\theta) = F_o(\frac{x-\mu}{\sigma})$ $(\theta = (\mu,\sigma),\ \mu \in R^1,\ \sigma > 0)$ with a strictly positive density which is also differentiable in μ and σ, then, given a prescribed way of estimating the unknown θ from the sample and choosing a value of θ and "estimating" it independently of the sample, the limiting distribution of the thus "estimated" empirical process on the null hypothesis is the same as if the values of the nuisance parameters were known. At first sight this random substitution method seems to offer a relatively simple solution to the problem under discussion at least for families of distributions of the form $F(x;\mu,\sigma) = F_o(\frac{x-\mu}{\sigma})$. In spite of that, it has not found favour in practice. Repeating Durbin (1976, p.38) in terms of our notation, the reasons seem to be

(1) Practical workers do not like introducing into the analysis of real data the element of artificial randomisation involved in selecting the random vector $\hat{\theta}_n^*$ independently of the data.

(2) It is suspected that the use of randomisation must inevitably entail a loss of power relative to the corresponding test based on $\hat{\alpha}_n$.

(3) The computational labor required to obtain $\hat{\theta}_n^*$ and effect the required transformation is burdensome.

A much simpler way of constructing a form of the estimated empirical process $\hat{\alpha}_n$ which has the same limiting distribution as α_n under both null and alternative hypotheses is the so-called half-sample device. The purpose of the latter is the same as that of the asymptotic form of the random substitution method, namely to obtain a version of the estimated empirical process whose limiting distribution in the composite hypothesis case is the same as that of the usual form of the empirical process in the simple hypothesis case. It turns out that under general conditions this objective can be achieved simply by calculating the estimate of θ from a randomly chosen half of the given random sample instead of from the whole sample. This device was suggested by Durbin (1973b) following up an earlier related proposal by Rao (1972). For further details we refer to Section 4 of Durbin (1976) and to Theorem 4.2 of Burke-Csörgő-Csörgő-Révész (1979). The random substitution and half-sample methods are asymptotically equivalent in that on the null hypothesis their respective estimated empirical processes have the same limiting distribution as if the values of the nuisance parameters were known. Whence, while it is true that the half-sample device is intuitively more appealing than the random substitution method in that it does not entail going outside the set of observed values, the effect of the randomisation element required to select the half-sample is asymptotically equivalent to that involved in the use of the random substitution method. The half-sample device is computationally much easier to apply in practice. As to whether it should be recommended for practical use, Durbin (1976, p.40) writes: "Obviously, if the corresponding full-sample procedure is available and convenient this should be used in preference. If, however, appropriate full-sample procedures are either unavailable or inconvenient the half-sample device might provide a useful interim procedure".

As we have already noted above, the availability of appropriate full-sample procedures is generally hindered by dependence in the limit not only on the form of the hypothetical distribution function but also on the true value of the unknown parameters of the latter. An approximate solution to the latter problem was proposed by Csörgő-Komlós-Major-Révész-Tusnády (1974) along the following lines. Let $G(x,n;\theta_o)$ be a Gaussian process for which we have (cf. e.g., Theorem 3.1 in Burke et al. (1979))

$$\sup_{-\infty<x<+\infty} |\hat{\alpha}_n(x) - G(x,n;\theta_o)| \to 0 \tag{1.3}$$

in some manner (i.e., almost surely or in probability) as $n \to \infty$, where θ_o is the unknown true value of the parameter $\theta \in \Theta \subseteq R^d$. Let $\hat{G}(x,n) = G(x,n;\hat{\theta}_n)$. Then

$$\sup_{-\infty<x<+\infty} |\hat{\alpha}_n(x) - \hat{G}(x,n)| \to 0 \tag{1.4}$$

in the manner of (1.3). We note that (1.5) is true essentially under the same conditions which render (1.3) to be true. As a consequence of (1.4) we conclude that, e.g., $\sup_{-\infty<x<+\infty} |\hat{\alpha}_n(x)|$ has the same asymptotic distribution as that of $\sup_{-\infty<x<+\infty} |\hat{G}(x,n)|$ and the asymptotic distribution of the latter is, at least in principle, calculable. We should note, however, that this latter problem is still a very difficult one and that there are no workable general methods available so far. Durbin (1976, Section 6) developed an analogue of the reflection method appropriate in this context to the parameter unknown case.

However he writes (Durbin 1976, p.34): "This requires the use of a random boundary instead of the fixed boundary used for the usual Kolmogorov-Smirnov tests. Because of this, the practical value of the method is questionable owing to the amount of computing required to implement it."

So far we have seen that, while the available asymptotic results for the estimated empirical process $\hat{\alpha}_n$ are quite general and interesting enough from a probabilistic point of view, it has been somewhat difficult for this theory to find its way back to statistics which inspired it in the first place. The aim of this exposition is to call attention to further possibilities concerning this problem.

Before going into this we should also mention that in some important special cases the finite-sample, and hence also the asymptotic distribution of the full-sample estimated empirical process is independent of unknown parameters. For example the independence of $\hat{\alpha}_n$ of the unknown parameters μ and σ^2 of the normal distribution $N(\mu,\sigma^2)$ when μ and σ^2 are maximum likelihood estimated was noted in Kac, Kiefer and Wolfowitz (1956) who, utilizing this fact, have also produced the first tables for the Cramér-von Mises and Kolmogorov-Smirnov statistics, based on their above $\hat{\alpha}_n$. In the just mentioned situation and also in general when the finite-sample distribution of $\hat{\alpha}_n$ is independent of unknown parameters, it is possible to simulate the distribution of appropriate test statistics by Monte-Carlo methods and to prepare tables of significance points from the observed distributions. Such tables of Kolmogorov-Smirnov statistics for testing normality and exponentiality were prepared by Lilliefors (1967, 1969) and more extensively by Stephens (Table 54 of Pearson and Hartley (1972)). Durbin (1975) also studied the problem of finite-sample distributions of the Kolmogorov-Smirnov statistics for $\hat{\alpha}_n$ when the distribution of the latter is independent of both θ_o and the form of $\hat{\theta}_n$ for estimators like $\hat{\theta}_n = n^{-1} \sum_{j=1}^{n} h_j(X_j)$. He developed general techniques and applied them to constructing tables of percentage points for the one- and two-sided Kolmogorov-Smirnov statistics for the test of hypothesis that the data have come from an exponential distribution. In this connection he writes (Durbin 1976, p.41): "The exponential distribution is, of course, a particularly simple case and it must be acknowledged that to extend the application of these techniques to other distributions would be very onerous."

Shapiro and Wilk (1965), and Shapiro and Francia (1972) have proposed a test statistic for normality (parameters unknown) which is perhaps the most accepted and successful one so far. At the first sight their procedure has nothing to do with the estimated empirical process $\hat{\alpha}_n$. This however is not so, as we will see later on. First we mention that DeWet and Venter (1972) have shown that the Shapiro and Francia test for normality, which itself is a large sample version of the original Shapiro-Wilk procedure, is asymptotically equivalent to rejecting the null hypothesis that the data have come from a normal distribution for large values of the statistic

$$L_n' = \sum_{i=1}^{n} \{\frac{X_{i:n} - \bar{X}_n}{S_n} - \Phi^{-1}(\frac{i}{n+1})\}^2 , \tag{1.6}$$

where $X_{i:n}$ $(i=1,2,\ldots,n)$ are the order statistics of the random sample $X_1,\ldots,X_n$, $\bar{X}_n$ is the sample mean S_n^2 is the sample variance and $\Phi^{-1}(u) = \inf\{x : \Phi(x) \geq u\}$ is the inverse of the standard normal distribution function Φ. Assuming normality they show that

$$L_n = (L_n' - a_n) \xrightarrow{\mathcal{D}} \sum_{i=3}^{\infty} (Z_i^2 - 1)/i , \tag{1.7}$$

where $Z_1, Z_2, \ldots$ are independent standard normal rv, $a_n \sim EL_n'$ whose approximate values they also tabulate, and $\xrightarrow{\mathcal{D}}$ stands for convergence in distribution. DeWet and Venter (1972) also provide tables for their just mentioned asymptotic distribution of L_n, which itself can be viewed as the first large sample theory for the Shapiro-Wilk (1965) and Shapiro-Francia (1972) tests for normality.

For the sake of further explaining L_n above and also for later use, we define now the empirical quantile function Q_n by

$$Q_n(u) = \begin{cases} X_{k:n} & \text{if } \frac{k-1}{n+1} < u \le \frac{k}{n+1}, \quad k = 1,\ldots,n, \\ X_{n:n} & \text{if } \frac{n}{n+1} < u < 1, \end{cases} \tag{1.8}$$

the empirical quantile process q_n by

$$q_n(u) = n^{\frac{1}{2}}(Q_n(u) - F^{-1}(u)), \quad 0 < u < 1, \tag{1.9}$$

and, if the continuous distribution function F of the random sample $X_1,\ldots,X_n$ has a density function f, the standardized empirical quantile process ρ_n by

$$\rho_n(u) = n^{\frac{1}{2}} f(F^{-1}(u))(Q_n(u) - F^{-1}(u)), \quad 0 < u < 1, \tag{1.10}$$

where $F^{-1}(u) = \inf\{x : F(x) \geq u\}$, the inverse of F.

Our reason for calling ρ_n the standardized empirical quantile process stems from the fact that $\{\rho_n(u);\ 0 < u < 1\}$ for large n is like a Brownian bridge $\{B(u);\ 0 \leq u \leq 1\}$, i.e., a zero-mean Gaussian process with covariance function $EB(u_1)B(u_2) = \min(u_1,u_2) - u_1u_2$. Namely we have

Theorem 1 (Csörgő-Révész 1978). Let $X_1, X_2, \ldots$ be i.i.d. rv with a continuous distribution function F which is also twice differentiable on (a,b), where $-\infty \leq a = \sup\{x : F(x) = 0\}$, $+\infty \geq b = \inf\{x : F(x) = 1\}$ and $F' = f > 0$ on (a,b). Assume that for some $\gamma > 0$

$$\sup_{a<x<b} F(x)(1 - F(x)) \frac{|f'(x)|}{f^2(x)} \leq \gamma . \tag{1.11}$$

One can then define a Brownian bridge $\{B_n(u);\ 0 \leq u \leq 1\}$ for each n such that

$$\sup_{\delta_n \leq u \leq 1-\delta_n} |\rho_n(u) - B_n(u)| \overset{\text{a.s.}}{=} O(n^{-\frac{1}{2}} \log n), \tag{1.12}$$

where $\delta_n = 25n^{-1} \log\log n$.

If, in addition to (1.11), we also assume that

$$A = \lim_{x \downarrow a} f(x), \qquad B = \lim_{x \uparrow b} f(x)$$

with

$$\left.\begin{array}{l} \text{(i)} \quad \min(A,B) > 0 \\ \text{or} \\ \text{(ii)} \quad \text{if } A = 0\ (B = 0), \text{ then } f \text{ is nondecreasing (nonincreasing)} \\ \qquad \text{on an interval to the right of } a \text{ (to the left of } b) \end{array}\right\} \tag{1.13}$$

then

$$\sup_{0<u<1} |\rho_n(u) - B_n(u)|$$

$$\stackrel{a.s.}{=} O(n^{-\frac{1}{2}} \log n) \qquad \text{if} \quad \gamma < 2 \tag{1.14}$$

$$\stackrel{a.s.}{=} O(n^{-\frac{1}{2}}(\log\log n)(\log n)^{(1+\varepsilon)(\gamma-1)}) \quad \text{if} \quad \gamma \geq 2 \ .$$

Remark 1.1. We note that under condition (i) of (1.13) (which can, of course, occur only if $-\infty < a,b < \infty$) we have, independently of γ of (1.11), that

$$\sup_{0\leq u\leq 1} |\rho_n(u) - B_n(u)| \stackrel{a.s.}{=} O(n^{-\frac{1}{2}} \log n) \ , \tag{1.15}$$

i.e., in the latter case we have (1.12) immediately with $\sup_{0\leq u\leq 1}$ instead of $\sup_{\delta_n\leq u\leq 1-\delta_n}$ in its statement.

Remark 1.2. Theorem 1 above is essentially Theorem 6 in Csörgő-Révész (1978), only here condition (1.13) is somewhat more extended than our earlier similar one for Theorem 6 (cf. (3.4) in the just quoted paper). In part II of our present exposition we will extend our earlier proof of (1.14) in terms of (1.13) so that (1.15) will also follow.

Let now F be a continuous distribution function with two unknown parameters $\mu \in R^1$ and $\sigma > 0$ respectively and assume that F is the form $F(x;\mu,\sigma) = F_o(\frac{x-\mu}{\sigma})$, $x \in R^1$. Let $\mathcal{F}$ be the class of all continuous distribution functions of this latter form, i.e.,

$$\mathcal{F} = \{F(x;\mu,\sigma) : F(x;\mu,\sigma) = F_o(\tfrac{x-\mu}{\sigma});\ \mu \in R^1,\ \sigma > 0\} \ . \tag{1.16}$$

Given a random sample $X_1,\ldots,X_n$ from a distribution F, our task is to test the composite goodness-of-fit null hypothesis

$$H_o : F \in \mathcal{F} \quad \text{with } F_o \text{ specified.} \tag{1.17}$$

Let us assume that F_o of (1.16) has a density function f_o, and define the standardized empirical quantile process ρ_n of the family $\mathcal{F}$ by

$$\begin{aligned}
\rho_n(u;\mu,\sigma) &= n^{\frac{1}{2}} f_o(F_o^{-1}(u))((Q_n(u) - \mu)/\sigma - F_o^{-1}(u)) \\
&= n^{\frac{1}{2}} f_o(F_o^{-1}(u))(Q_n^o(u) - F_o^{-1}(u)) \\
&= f_o(F_o^{-1}(u))q_n^o(u) = \rho_n^o(u) \ , \quad 0 < u < 1,
\end{aligned} \tag{1.18}$$

where $q_n^o(u) = n^{\frac{1}{2}}(Q_n^o(u) - F_o^{-1}(u))$, $Q_n^o(u) = (Q_n(u) - \mu)/\sigma$ and $Q_n(u)$ is as in (1.8).

We go back now to L_n of (1.7) and consider first the result of DeWet and Venter (1972) for its variant

$$L_n^o = \left(\sum_{i=1}^n \{Y_{i:n} - \Phi^{-1}(\tfrac{i}{n+1})\}^2 - a_n^o\right) \xrightarrow{\mathcal{D}} \sum_{i=1}^\infty (Z_i^2 - 1)/i \ , \tag{1.19}$$

where $Y_{i:n} = (X_{i:n} - \mu)/\sigma$ $(i = 1,\ldots,n)$, $Z_1, Z_2, \ldots$ are independent standard normal rv and

$$a_n^o = n^{-1} \sum_{i=1}^{n} \frac{i}{n+1}(1 - \frac{i}{n+1})/\phi^2(\Phi^{-1}(\frac{i}{n+1})) \tag{1.20}$$

with $\phi(\cdot)$ being the standard normal density function.

The statistic L_n^o can be used as a goodness-of-fit test statistic for the simple statistical hypothesis that the distribution of the $Y_i = (X_i - \mu)/\sigma$, with μ and σ known, is the standard normal distribution function. Now the standardized empirical quantile process of the normal family of distribution functions $N = \{N(x;\mu,\sigma) = \Phi(\frac{x-\mu}{\sigma});\ \mu \in R^1,\ \sigma > 0\}$ with density $\phi = \Phi'$ is (cf. (1.18))

$$\rho_n^o(u) = n^{\frac{1}{2}}\phi(\Phi^{-1}(u))(Q_n^o(u) - \Phi^{-1}(u)) \ , \tag{1.21}$$

and hence, in terms of the latter, L_n^o can be written as

$$L_n^o = \sum_{i=1}^{n} \frac{(\rho_n^o(\frac{i}{n+1}))^2 - \frac{i}{n+1}(1-\frac{i}{n+1})}{\phi^2(\Phi^{-1}(\frac{i}{n+1}))} \frac{1}{n}$$

$$= \int_{-\infty}^{+\infty} \frac{(\rho_n^o(\frac{n}{n+1} F_n(x)))^2 - \frac{n}{n+1} F_n(x)(1-\frac{n}{n+1} F_n(x))}{\phi^2(\Phi^{-1}(\frac{n}{n+1} F_n(x))} dF_n(x) \ , \tag{1.22}$$

where F_n is the empirical distribution function of the Y_i $(i = 1,\ldots,n)$.

Viewing L_n^o of (1.19) this way we see that it is the Cramér-von Mises functional of the "inverse" of a weighted empirical process, namely that of the standardized empirical quantile process for the normal distribution centered at expectations and weighted with the function $1/\phi^2(\Phi^{-1}(\cdot))$. Also the way we have written L_n^o in (1.22) combined with Theorem 1 suggests that one should have

$$L_n^o \to \int_0^1 \frac{B^2(u) - u(1-u)}{\phi^2(\Phi^{-1}(u))} du \ . \tag{1.23}$$

Indeed, using (1.19) and the method of Csörgő-Révész (1979), the above argument can be made rigorous and then the result of DeWet and Venter (1972) as quoted in (1.19) gives the infinite series expansion of the integral rv of (1.23). The normalizing factor $u(1-u)/\phi^2(\Phi^{-1}(u)) = EB^2(u)/\phi^2(\Phi^{-1}(u))$, and hence also a_n^o in L_n^o , is there to make the integral in question exist. Moreover, we can now go back to L_n^o of (1.22) and replace Φ^{-1} resp. ϕ in its definition by F_o^{-1} resp. f_o of H_o of (1.17) and say that in its new form then, it becomes a goodness-of-fit statistic for testing the simple statistical hypothesis that the distribution of the $Y_i = (X_i - \mu)/\sigma$, with μ and σ known, is F_o. Also, for the asymptotic distribution of the thus modified L_n^o we will have (1.23) with $f_o^2(F_o^{-1}(u))$ instead of $\phi^2(\Phi^{-1}(u))$ in the integral and the latter will exist by (1.23) for any $f_o(F_o^{-1}(u))$ which for u near zero and near one behaves like $\phi(\Phi^{-1}(u))$. Naturally in the latter case (1.19) will not be true any more, due to the dependence of the integral of (1.23) on f_o and, whence, new expansion and tables are needed for each given f_o .

In order to be able to view L_n of (1.7) the way we did L_n^o in (1.22) and (1.23), via (1.18) and (1.21) we introduce the <u>estimated standardized empirical quantile process</u> $\hat{\rho}_n$ <u>for the normal family</u> $N = \{N(x;\mu,\sigma) = \Phi(\frac{x-\mu}{\sigma});\ \mu \in R^1,\ \sigma > 0\}$:

$$\hat{\rho}_n(u) = \rho_n(u;\bar{X}_n,S_n)$$
$$= n^{\frac{1}{2}}\phi(\Phi^{-1}(u))((Q_n(u)-\bar{X}_n)/S_n - \Phi^{-1}(u))\ ,\quad 0 < u < 1\ , \tag{1.24}$$

where $\bar{X}_n$ and S_n^2 are as in (1.6), $Q_n(u)$ is as in (1.8) and ϕ and Φ are as in (1.21). Then L_n of (1.7) can be written as

$$L_n = \sum_{i=1}^{n} \frac{(\hat{\rho}_n(\frac{i}{n+1}))^2 - E\,\hat{\rho}_n(\frac{i}{n+1})}{\phi^2(\Phi^{-1}(\frac{i}{n+1}))}\,\frac{1}{n}\ , \tag{1.25}$$

i.e., as the Cramér-von Mises functional of the "inverse" of a weighted estimated empirical process, namely that of the estimated standardized empirical quantile process for the normal family centered at expectations and weighted with the function $1/\phi^2(\Phi^{-1}(\cdot))$. Given then that the original Shapiro-Wilk (1965) test for normality and its Shapiro-Francia (1972) asymptotic variant are asymptotically equivalent to the DeWet-Venter (1972) procedure for testing for normality (cf. (1.6) and (1.7)), and since the latter can be written as in (1.25), our earlier contention that the Shapiro-Francia-Wilk procedure for testing for normality is also based on some kind of an estimated empirical process, namely on $\hat{\rho}_n$ of (1.24), is now borne out.

Using (1.7) and the method of Csörgő-Révész (1979), an analogue of (1.23) based on Theorem 1 can also be written down for L_n of (1.25). As to the form of the latter, it is not so easy now to guess it as it was guessing (1.23) itself from (1.22). First one shows that $\hat{\rho}_n$ of (1.24) can be written as a function of ρ_n^o of (1.21) (cf. (3.17) in Csörgő-Révész (1979)), bearing out the fact that the distribution of the former is independent of the unknown parameters μ and σ. Operating then on the thus represented $\hat{\rho}_n$ à la (1.25) and using Theorem 1, one gets an analogue of the integral of (1.23) for L_n which will exist by (1.7), with the latter being the infinite series expansion of the former. Moreover, just like in case of L_n^o, we can now go back to L_n of (1.25) and replace Φ^{-1} resp. ϕ in its definition by F_o^{-1} resp. f_o of H_o of (1.17) and say that in its new form then, it becomes a goodness-of-fit test statistic for testing the <u>composite</u> statistical hypothesis that the distribution of the X_i with μ and σ unknown, is $F \in \mathcal{F}$ with F_o specified. Also, for the asymptotic distribution of the thus modified L_n we will have the here not detailed analogue of (1.23) with $f_o^2(F_o^{-1}(u))$ instead of $\phi^2(\Phi^{-1}(u))$ in the appropriate integral, and the latter will exist by (1.7) for any $f_o(F_o^{-1}(u))$ which for u near zero and one behaves like $\phi(\Phi^{-1}(u))$. Naturally in the latter case (1.7) will not be true any more, due to the dependence of the just mentioned integral analogue of (1.23) on f_o and, whence, new expansion and tables will be needed for each given f_o.

The just described program was carried out for H_o of (1.17) in Csörgő-Révész (1979) so that F_o of the latter should not necessarily be the standard normal distribution. In our quoted paper we were also aiming at such statistics for the just mentioned composite hypothesis which would not require norming factors, like for example a_n^o and a_n of L_n^o and L_n respectively, for their limits to exist, and which would still have a tail-sensitivity, though less pronounced than that of L_n^o and L_n. Estimating μ resp. σ of the family $\mathcal{F}$ by $\bar{X}_n$ resp. S_n, we derived the asymptotic distribution of the following

Cramér-von Mises type statistics

$$M_n^o(\lambda) = \sum_{i=1}^{n}\{(\rho_n^o(\tfrac{i}{n+1}))^2/nf_o(F_o^{-1}(\tfrac{i}{n+1}))\}(F_o^{-1}(\tfrac{i}{n+1}))^{\lambda-1} \tag{1.26}$$

and

$$M_n(\lambda) = \sum_{i=1}^{n}\{(\hat\rho_n(\tfrac{i}{n+1}))^2/nf_o(F_o^{-1}(\tfrac{i}{n+1}))\}(F_o^{-1}(\tfrac{i}{n+1}))^{\lambda-1}\ , \tag{1.27}$$

where $\lambda \geq 1$ is a fixed integer, ρ_n^o is as in (1.18) and $\hat\rho_n$ is gained from the latter via replacing (estimating) μ by $\overline{X}_n$ and σ by S_n in it. For example it is easily guessed, but not so easily proved, that by Theorem 1 we should have

$$M_n^o(1) \xrightarrow{\mathcal{D}} \int_0^1 B^2(u)dF_o^{-1}(u)\ , \tag{1.28}$$

provided the slope of $F_o^{-1}(u)$ is not too steep as $u \uparrow 1$ and $u \downarrow 0$. The asymptotic distribution of $M_n(\lambda)$ is somewhat more complicated. We prove (cf. Theorem 3.1 in Csörgő-Révész (1979)) that $M_n(\lambda)$ has the same form of asymptotic representation for all F_o, provided μ resp. σ are estimated by $\overline{X}_n$ resp. S_n, and just like that of $M_n^o(\lambda)$, its asymptotic distribution depends on F_o but not on the unknown parameters μ and σ of F. We should also emphasize here that the nature of dependence of $M_n(\lambda)$ on F_o is also a function of the method of estimation of the unknown parameters μ and σ of the family $\mathcal{F}$. We (Csörgő-Révész 1979), as just mentioned here, have worked out the details with estimating μ by $\overline{X}_n$ and σ by S_n. The same kind of details can be worked out with general maximum likelihood estimators for the family $\mathcal{F}$, like, for example, those of our paper Burke et al (1979) replacing $\overline{X}_n$ and S_n in constructing $\hat\rho_n$. The resulting asymptotic distribution of the thus modified $M_n(\lambda)$ would again depend on F_o but not on the unknown parameters μ and σ of the family $\mathcal{F}$.

Summing up then, we (Csörgő-Révész 1979) demonstrated that via an estimated standardized empirical quantile process $\hat\rho_n$ for the family $\mathcal{F}$, the Shapiro-Francia-Wilk approach for testing for normality can be extended to testing also for the composite hypothesis H_o of (1.17). DeWet and Venter (1973) also treated the latter H_o with $\mu = 0$ and $\sigma > 0$. Along the lines of the original Shapiro-Wilk (1965) procedure, they proposed a test statistic consisting of the ratio of two asymptotically efficient estimates of σ and derived its asymptotic null distribution. Their procedure can be again expressed as the Cramér-von Mises functional of an estimated standardized empirical quantile process centered at expectations and weighted with a somewhat complicated (at least in the general case) weight function. DeWet and Venter (1973) worked out their asymptotics for the gamma family of order r (r is assumed to be known). For example when testing for the exponential family $F(x;\theta) = 1 - e^{-x/\theta}$ they arrive at the statistic

$$E_n = \sum_{i=1}^{n}\left(\frac{X_{i:n}}{\overline{X}_n} - \log\frac{1}{1-\frac{i}{n+1}}\right)^2 \frac{1}{\log\dfrac{1}{1-\frac{i}{n+1}}}\ , \tag{1.29}$$

and centering at expectations $e_n = EE_n$ they prove

$$(E_n - e_n) \xrightarrow{\mathcal{D}} \sum_{i=2}^{\infty} (Z_i^2 - 1)/i \tag{1.30}$$

where $Z_1, Z_2, \ldots$ are independent standard normal rv (compare to (1.7)). When testing for the same exponential family, we (Csörgő-Révész 1979) arrive at a similar statistic $M_n(\lambda)$ ($\lambda \geq 1$, a fixed integer) which looks like

$$M_n(\lambda) = \sum_{i=1}^{n} \left(\frac{X_{i:n}}{\bar{X}_n} - \log \frac{1}{1 - \frac{i}{n+1}} \right)^2 \left(1 - \frac{i}{n+1}\right) \left(\log \frac{1}{1 - \frac{i}{n+1}} \right)^{\lambda - 1} \tag{1.31}$$

and derive its asymptotic distribution (cf. Theorem 4.1 of our quoted paper) without any further norming factors required. As to what value of λ we should take, it would of course depend on the type of alternatives. Just as an omnibus test for exponentiality, in our case we would suggest $M_n(1)$. E_n gives more emphasis to deviations on the tails than $M_n(1)$. In fact it emphasises the tails so much that it would explode (i.e., E_n would not have an asymptotic null distribution) without subtracting $e_n = o(\log n)$ from it (the value of e_n are tabulated in DeWet and Venter (1973) for several values of n). While this phenomenon is certainly in favour of detecting deviations from exponentiality on the tails, it definitely works against detecting the same in the middle range. Indeed, large deviations from the null hypothesis in the middle range may very well end up being less emphasized by E_n than total agreement with H_o on the tails. Thus E_n would tend to accept H_o more frequently than it should against alternatives deviating from it in the middle range and, in the same time, totally agreeing with it on the tails. The omnibus nature of $M_n(1)$ would tend to handle such situations less pronouncedly on the tails and more emphatically in the middle range. The same can be said about $M_n(\lambda)$ of (1.27) in general, and also in particular when comparing it to the L_n procedure (cf. (1.7)) and to the Shapiro-Francia-Wilk procedure for testing for normality, i.e., when $F_o = \Phi$ in H_o of (1.17).

So far we have seen that a large number of composite goodness-of-fit procedures can be interpreted in terms of functionals of some kind of an estimated empirical or quantile process in general and that these procedures are quite successful in testing for H_o of (1.17) in particular. Naturally further studies of many kinds (tables, Monte Carlo and analytic power studies, etc.) are still very much needed in order to sort them out as far as their performance against different types of alternatives is concerned. Here we have only projected our own personal views on them.

In this exposition we are going to reexamine the problem of testing for H_o of (1.17) from a different angle. The latter will amount to saying that one can test for this composite goodness-of-fit hypothesis H_o with any specified F_o without estimating the unknown parameters of the family $\mathcal{F}$ at all, provided F_o satisfies the conditions of Theorem 1 with a continuously differentiable density f_o on (a,b). Our research in this direction was prompted by a number of remarks and a conjecture of Parzen (1979a,b) concerning H_o. Namely, in terms of our notation he suggested that, assuming H_o with $f_o = F_o'$, on the bases of a random sample $X_1, \ldots, X_n$ one should consider the stochastic process

$$\{p_n(u);\ 0 \leq u \leq 1,\ n = 1,2,\ldots\} \tag{1.32}$$

with

$$p_n(u) = \begin{cases} 0 & \text{if } 0 \le u < 2/n \\ n^{\frac{1}{2}}\left\{\dfrac{\sum_{j=1}^{[nu]-1} f_o(F_o^{-1}(\frac{j}{n+1}))(X_{j+1:n} - X_{j:n})}{\sum_{j=1}^{n-1} f_o(F_o^{-1}(\frac{j}{n+1}))(X_{j+1:n} - X_{j:n})} - u\right\} & \text{if } 2/n \le u \le 1 , \end{cases} \tag{1.33}$$

where $[x]$ stands for the integer part of the real number x.

Parzen (cf. Conjecture on page 110 in 1979a and the last paragraph on page 252 in 1979b) conjectured that on H_o of (1.17) $\{p_n(u);\ 0 \le u \le 1\}$ is asymptotically distributed as a Brownian bridge $\{B(u);\ 0 \le u \le 1\}$ (cf. our paragraph preceeding Theorem 1). Since $p_n(u)$ is independent of the unknown parameters μ and σ of the family $\mathcal{F}$ of (1.16), its Kolmogorov-Smirnov and Cramér-von Mises functionals can always be used in principle to test for H_o with any specified F_o, without first estimating μ and σ, and if Parzen's conjecture were to be correct, one would not even need any new tables for this procedure.

As we will see in Section 2 of this paper, Parzen's conjecture is indeed true for the uniform and exponential families of distributions. However the process $\{p_n(u);\ 0 \le u \le 1\}$ turns out to be asymptotically somewhat more complicated in general. Nevertheless it can be still viewed as the basis for many statistical investigations concerning the family $\mathcal{F}$ and whence we propose to study it in more detail in Section 2. In Section 3 we discuss the usual Kolmogorov-Smirnov and Cramér-von Mises goodness-of-fit statistics based on $p_n(u)$ and their consistency against all alternatives to H_o of (1.17). Section 4 is devoted to studying Weiss (1961, 1963)' approach to estimating the unknown scale parameter σ of the family $\mathcal{F}$.

2. DISCUSSION OF A NEW RESULT

Throughout this section and also in the sequel we will assume that $X_1,\dots,X_n$ form a random sample on a distribution with $F \in \mathcal{F}$ of (1.16). Then $(X_1 - \mu)/\sigma,\dots,(X_n - \mu)/\sigma$ are i.i.d. rv with distribution function $F_o(x)$. In terms of the order statistics $X_{1:n} \le \dots \le X_{n:n}$ of our random sample $X_1,\dots,X_n$ on $F \in \mathcal{F}$ we define

$$Y_j = (X_{j:n} - \mu)/\sigma , \quad j = 1,\dots,n . \tag{2.1}$$

Then the stochastic process $\{p_n(u);\ 0 \le u \le 1,\ n = 1,2,\dots\}$ of (1.32) can be written as

$$p_n(u) = \begin{cases} 0 & \text{if } 0 \le u < 2/n \\ n^{\frac{1}{2}}\left\{\dfrac{\sum_{j=1}^{[nu]-1} f_o(F_o^{-1}(\frac{j}{n+1}))(Y_{j+1} - Y_j)}{\sum_{j=1}^{n-1} f_o(F_o^{-1}(\frac{j}{n+1}))(Y_{j+1} - Y_j)} - u\right\} & \text{if } 2/n \le u \le 1 . \end{cases} \tag{2.2}$$

Now if F_o of H_o of (1.17) is the uniform-(0,1) distribution, i.e., if F of H_o is $F(x) = (x-A)/(B-A)$, $A \le x \le B$, with $F_o(y) = y$, $0 \le y \le 1$, then

$$p_n(u) = n^{\frac{1}{2}}\left(\frac{Y_{[nu]} - Y_1}{Y_n - Y_1} - u\right) ,$$

and applying Theorem 1 with $\rho_n(u)$ being the uniform quantile process $u_n(u) = n^{\frac{1}{2}}(Q_n(u) - u)$, where $Q_n(u)$ is now defined (cf. (1.8)) in terms of the uniform order statistics of (2.1), with the help of a little calculation one gets that for a suitable sequence of Brownian bridges $\sup_{0\le u\le 1} |p_n(u) - B_n(u)| \to 0$ in probability as $n \to \infty$. The latter statement implies, of course, that Parzen's conjecture is true for the family of uniform distributions. The same can be said about the case when F_o of H_o of (1.17) is the unit exponential distribution, i.e., if F of H_o is $F(x) = 1 - e^{-(x-A)/B}$, $x \ge A$, $(A \in R^1, B > 0)$, with $F_o(y) = 1 - e^{-y}$, $y \ge 0$. Namely then $f_o(F_o^{-1}(u)) = 1 - u$, $0 \le u \le 1$, and

$$p_n(u) = n^{\frac{1}{2}}\left\{\frac{\sum_{j=1}^{[nu]-1}\left(1 - \frac{j}{n+1}\right)(Y_{j+1} - Y_j)}{\sum_{j=1}^{n-1}\left(1 - \frac{j}{n+1}\right)(Y_{j+1} - Y_j)} - u\right\} . \qquad (2.3)$$

In order to see why the latter process should behave asymptotically like a Brownian bridge, we consider here an asymptotically equivalent form of p_n of (2.3), namely $\tilde{p}_n$ which is defined as

$$\tilde{p}_n(u) = n^{\frac{1}{2}}\left\{\frac{\sum_{j=1}^{[nu]-1}\left(1 - \frac{j}{n}\right)(Y_{j+1} - Y_j)}{\sum_{n=1}^{n-1}\left(1 - \frac{j}{n}\right)(Y_{j+1} - Y_j)} - u\right\}$$

$$= n^{\frac{1}{2}}\left\{\frac{\sum_{j=1}^{[nu]-1}(n - j)(Y_{j+1} - Y_j)}{\sum_{j=1}^{n-1}(n - j)(Y_{j+1} - Y_j)} - u\right\} . \qquad (2.4)$$

As to the asymptotic equivalence in distribution of p_n of (2.3) to $\tilde{p}_n$ of (2.4), it will follow from our more general Theorem 2 (cf. (2.12)). Now it is well known that on H_o the rv $\delta_j(n) = (n-j)(Y_{j+1} - Y_j)$ $j = 1,\ldots,n-1$, are themselves independent $F_o(y) = 1 - e^{-y}$ $(y \ge 0)$ exponential rv and whence (cf., e.g., Theorems 1 and 2 in Csörgő-Seshadri-Yalovsky 1975) for each $n \ge 4$ the joint distribution of $(Z_{r:n-2} = (\sum_{j=1}^{r}\delta_j(n))/(\sum_{j=1}^{n-1}\delta_j(n));\ 1 \le r \le n-2)$ is that of $(n-2)$ order statistics of $(n-2)$ independent uniform -(0,1) rv. Consequently $\tilde{p}_n(u)$ of (2.4) can be written as the uniform quantile process $u_n(u) = n^{\frac{1}{2}}(Q_n(u) - u)$ with $Q_n(u)$ now defined in terms of $(Z_{r:n-2};\ 1\le r\le n-2)$ and Theorem 1 implies that for a suitable sequence of Brownian bridges we have $\sup_{0\le u\le 1} |p_n(u) - B_n(u)| \to 0$ in probability as $n \to \infty$. Hence Parzen's conjecture is again true.

In order to be able to state our result for p_n of (1.33) or, equivalently, for that of (2.2) in general, we introduce the sequence of Gaussian processes $\{G_n(u);\ 0 \le u \le 1\}$ which is defined in terms of Brownian bridges $\{B_n(u);\ 0 \le u \le 1\}$ as follows:

$$\{G_n(u);\ 0 \le u \le 1\} = \{B_n(u)-I_n(u)+uI_n(1);\ 0 \le u \le 1\}\ , \tag{2.5}$$

where

$$I_n(u) = \int_0^u \{f_o'(F_o^{-1}(y))/f_o^2(F_o^{-1}(y))\}B_n(y)dy\ . \tag{2.6}$$

Our main theorem is

<u>Theorem 2</u>. Assume that for F_o of the family $\mathcal{F}$ of (1.16) conditions (1.11) and (1.13) of Theorem 1 hold true, and that

$$f_o' \text{ is continuous on } (a,b)\ . \tag{2.7}$$

Let $\{B_n(u);\ 0 \le u \le 1\}$ be that sequence of Brownian bridges for which we have (1.14) in terms of the standardized empirical quantile process ρ_n^o (cf. (1.18)) of the family $\mathcal{F}$. Define the sequence of Gaussian processes $\{G_n(u);\ 0 \le u \le 1\}$ of (2.5) via the latter sequence of Brownian bridges. Then, as $n \to \infty$,

$$\sup_{0<u<1} |p_n(u) - G_n(u)| \to 0 \text{ in probability.} \tag{2.8}$$

The proof of Theorem 2 is Theorem 1 based and somewhat lengthy. It will be given in Csörgő-Révész (1980).

Now define the Gaussian process $\{G(u);\ 0 \le u \le 1\}$ by

$$\{G(u);\ 0 \le u \le 1\} = \{B(u) - I(u) + uI(1);\ 0 \le u \le 1\}\ , \tag{2.9}$$

where $B(u)$ is a Brownian bridge and

$$I(u) = \int_0^u \{f_o'(F_o^{-1}(y))/f_o^2(F_o^{-1}(y))\}B(y)dy\ . \tag{2.10}$$

Then, in the light of (2.8) and the fact that $\{G(u);\ 0 \le u \le 1\} \overset{\mathcal{D}}{=} \{G_n(u);\ 0 \le u \le 1\}$ for each n, we also have that under the conditions of Theorem 2

$$h(p_n(\cdot)) \xrightarrow{\mathcal{D}} h(G(\cdot)) \quad \text{as } n \to \infty\ , \tag{2.11}$$

for every continuous functional h on the space of real valued functions on $[0,1]$ endowed with the supremum topology; i.e., the stochastic process $\{p_n(u);\ 0 \le u \le 1\}$ of (2.2) is, in general, asymptotically distributed as the Gaussian process $\{G(u);\ 0 \le u \le 1\}$ of (2.9). Here $\overset{\mathcal{D}}{=}$ stands for equality in distribution and $\xrightarrow{\mathcal{D}}$ denotes convergence in distribution.

Going back again to the uniform and exponential distributions, we get by inspecting $G(u)$ of (2.9) that $\{G(u);\ 0 \le u \le 1\} = \{B(u);\ 0 \le u \le 1\}$, a Brownian bridge if $f_o(y) = 1,\ 0 \le y \le 1$, and a somewhat tedious calculation yields that

$$EG(u_1)G(u_2) = \min(u_1,u_2) - u_1u_2,\ \text{if } f_o(y) = e^{-y}\ ,\ y \ge 0\ , \tag{2.12}$$

i.e., in the latter case our Gaussian process $\{G(u);\ 0 \le u \le 1\}$ is a Brownian bridge.

It can be also seen that the Gaussian process $\{G(u);\ 0 \le u \le 1\}$ is not always a Brownian bridge. The latter fact can be easily checked, for example, with $F_o(x) = x^2$, $0 \le x \le 1$. Indeed, in general, we believe that the following conjecture is true.

Conjecture. The Gaussian process $\{G(u);\ 0 \le u \le 1\}$ of (2.9) is a Brownian bridge only in the case of uniform and exponential densities, i.e., the only solutions of the covariance integral equation

$$EG(u_1)G(u_2) = \min(u_1,u_2) - u_1u_2$$

are the density functions $f_o(y) = 1$ $(0 \le y \le 1)$ and $f_o(y) = e^{-y}$ $(y \ge 0)$.

Some lengthy calculations easily yield of course the general form of $EG(u_1)G(u_2)$. Unfortunately the latter turns out to be quite complicated and does not appear to give an immediate handle to this problem.

In any case, in the light of Theorem 2, the stochastic process $p_n(u)$ of (1.33) can be viewed as a natural nuisance parameter free empirical process for all those members of the family of distribution functions $\mathcal{F}$ of (1.16) whose F_o satisfies the conditions of Theorem 2 and, given any specified one of these F_o (normal, exponential, Weibull, Cauchy, etc.), goodness-of-fit statistics for the composite hypothesis H_o of (1.17), like for example

$$t_1(n) = \sup_{0<u<1} |p_n(u)| \quad \text{and} \quad t_2(n) = \int_0^1 p_n^2(u)\,du\ , \tag{2.12}$$

can be based on it, without first estimating the parameters μ and σ of $\mathcal{F}$.

3. NUISANCE PARAMETER FREE GOODNESS-OF-FIT STATISTICS FOR THE SHIFT AND SCALE FAMILY

Let $t_1(n)$ and $t_2(n)$ be as in (2.12), and let $\{G(u);\ 0 \le u \le 1\}$ be as in (2.9). By Theorem 2 via (2.11) we get

Corollary 1. Assume that F_o of $\mathcal{F}$ of (1.16) satisfies the conditions of Theorem 2. Then, as $n \to \infty$, we have

$$t_1(n) \xrightarrow{\mathcal{D}} \sup_{0<u<1} |G(u)| \quad \text{and} \quad t_2(n) \xrightarrow{\mathcal{D}} \int_0^1 G^2(u)\,du\ . \tag{3.1}$$

As we have seen (cf. paragraph of (2.12)) this procedure does not require any new tables if F_o of H_o of (1.17) is $F_o(y) = y$, $0 \le y \le 1$, or $F_o(y) = 1 - e^{-y}$, $y \ge 0$, for then $t_1(n) \xrightarrow{\mathcal{D}} \sup_{0<u<1} |B(u)|$ and $t_2(n) \xrightarrow{\mathcal{D}} \int_0^1 B^2(u)du$ with $\{B(u);\ 0 \le u \le 1\}$ a Brownian bridge, and hence the readily available tables for the standard Kolmogorov-Smirnov and Cramér-von Mises statistics can be used.

In fact the performance of testing for exponentiality via a number of functionals of $p_n(u)$ of (2.4) has already been studied extensively by Csörgő, Seshadri and Yalovsky (1975) against several alternatives, where it was found that test procedures based on $\{Z_{r:n-2};\ 1 \le r \le n-2\}$ (cf. the paragraph following (2.4)) are preferable to the Lillifors (1969) test, that is to the Kolmogorov-Smirnov test when parameters are replaced by their estimates, at least in the case of the exponential family $F(x) = 1 - e^{-x/B}$ (cf. Table 1 in Csörgő, Seshadri and Yalovski 1975). The same was found concerning test procedures based on the above $\{Z_{r:n-2};\ 1 \le r \le n-2\}$ as compared to the Shapiro-Wilk (1972b) test for the exponential family $F(x) = 1 - e^{-(x-A)/B}$ (cf. Table 2 in Csörgő, Seshadri and Yalovski 1975). In particular we should also mention here in passing that, as far as testing for any of the just mentioned exponential families on the basis of the above $\{Z_{r:n-2};\ 1 \le r \le n-2\}$ is being concerned, we can also construct the following K. Pearson type exact χ^2 statistic:

$$\chi^2_{2(n-2)} = -2 \sum_{r=1}^{n-2} \log Z_{r:n-2}, \qquad (3.2)$$

where our notation is to indicate that on H_o the latter rv is a χ^2 rv with $2(n-2)$ degrees of freedom. A number of properties of the latter rv were mentioned in Csörgő and Seshadri (1970), where it was also suggested that perfect or near perfect fit to exponentiality would be evidenced by the computed value of $\chi^2_{2(n-2)}$ for any given random sample being near to its most probable theoretical value, i.e., to that of the mode which in this case is $2(n-2)-2 = 2n-6$. Evidence against exponentiality would be supported by much larger and much smaller values of the rv $\chi^2_{2(n-2)}$ than this theoretical mode value, and readily available exact χ^2 tables can be used to make this statement more precise for any fixed size type one error test for exponentiality. This K. Pearson type exact χ^2 procedure for testing exponentiality was found to be a good overall omnibus test against a number of alternatives when compared to other procedures, except against the log-normal one (cf. Tables 1 and 2 in Csörgő, Seshadri and Yalovski 1975).

As to the general case, i.e., when testing for H_o of (1.17) with any F_o, other than uniform or exponential, satisfying the conditions of our Theorem 2, new tables of significance points of $t_1(n)$ and $t_2(n)$, say, will be required for each specified F_o of interest, should our Conjecture of Section 2 turn out to be true. This situation is not new, of course, when testing for composite goodness-of-fit hypotheses. Indeed, as we have seen in Section 1, most procedures for the latter hypotheses were at best parameter free but certainly always F_o bound in the limit. For example, independently of whatever statistical merits $M_n(\lambda)$ of (1.27) might have vis-à-vis $t_1(n)$ and $t_2(n)$ of (2.12), or vice versa, it is certainly in favour of the latter two that they are applicable under less stringent conditions on F_o than the former (i.e., Theorem 2 of the present paper is proved under less stringent conditions on F_o than Theorem 3.1 in Csörgő-Révész 1979).

The results of (3.1) can, of course, be used at least in principle to evaluate the asymptotic distributions of the statistics $t_1(n)$ and $t_2(n)$ for any specified F_o of interest which satisfy the conditions of Theorem 2. Unfortunately they do not, however, give an immediate handle as to how to do this in general or, in particular, for any specified F_o. On the other hand, since p_n of (1.33) is independent of the unknown parameters of the family $\mathcal{F}$, one can certainly simulate the distributions of $t_1(n)$ and $t_2(n)$ for any specified F_o. For example, as far as the standard normal distribution is concerned, the books by Van der Waerden and Nievergelt (1956) and Hajek (1969) include extensive tables for the inverse of the unit normal distribution $F_o^{-1}(y) = \Phi^{-1}(y)$, which can be used to tabulate and study the performance of $t_1(n)$ and $t_2(n)$ by Monte Carlo methods when testing for normality.

As far as the general performance of $t_1(n)$ and $t_2(n)$ is concerned, we will prove in the second part of this exposition that tests based on them for H_o of (1.17) with any specified F_o satisfying the conditions of Theorem 2 are consistent against any alternative H_1 to H_o. Namely we will prove (cf. Csörgő-Révész 1980)

Corollary 2. Given any H_1 other than H_o of (1.17), then $t_1(n)$ and $t_2(n) \to \infty$ in probability as $n \to \infty$.

4. ON WEISS' ESTIMATE OF THE SCALE PARAMETER OF THE SHIFT AND SCALE FAMILY

In the second part of this paper we will prove

Corollary 3. Given that F_o of the family $\mathcal{F}$ of (1.16) satisfies the conditions of Theorem 2, there exists a sequence of Brownian bridges $\{B_n(u);\ 0 \le u \le 1\}$ such that

$$\sum_{j=1}^{n-1} f_o(F_o^{-1}(\tfrac{j}{n+1}))(Y_{j+1} - Y_j) = 1 - n^{-\frac{1}{2}} I_n(1) + o_P(n^{-\frac{1}{2}}) , \tag{4.1}$$

where the Y_j are as in (2.1), $I_n(\cdot)$ is as in (2.6) and $o_P(n^{-\frac{1}{2}})$ stands for a remainder term so that $n^{\frac{1}{2}} o_P(n^{-\frac{1}{2}}) \to 0$ in probability as $n \to \infty$.

From Corollary 3 and the definition of Y_j in (2.1) it follows immediately that we also have

Corollary 4. Given the conditions of Corollary 3,

$$\begin{aligned}\hat{\sigma}_n = \sum_{j=1}^{n-1} f_o(F_o^{-1}(\tfrac{j}{n+1}))(X_{j+1:n} - X_{j:n}) &= \sigma - n^{-\frac{1}{2}} \sigma I_n(1) + o_P(n^{-\frac{1}{2}}) \\ &= \sigma + O_P(n^{-\frac{1}{2}}) .\end{aligned} \tag{4.2}$$

Whence also

Corollary 5. Given the conditions of Corollary 3,

$$n^{\frac{1}{2}}(\hat{\sigma}_n - \sigma) \xrightarrow{\mathcal{D}} \sigma\, I(1) . \tag{4.3}$$

Proof. By (4.2) and the fact that, by (2.6) and (2.10), $I_n(1) \overset{\mathcal{D}}{=} I(1)$ for each n.

Remark 4.1. Corollary 4 states that the statistic $\hat{\sigma}_n = \sum_{j=1}^{n-1} f_o(F_o^{-1}(\frac{j}{n+1}))\cdot(X_{j+1:n} - X_{j:n})$ is a weakly consistent estimator of the scale parameter σ of the family $\mathcal{F}$ of (1.16), provided that F_o of the latter satisfies the conditions of Theorem 2. Such an estimator of σ of the family $\mathcal{F}$ was first proposed by Weiss (1961), who showed its weak consistency under somewhat stronger conditions on F_o than ours. Corollary 5 states that the rv $n^{\frac{1}{2}}(\hat{\sigma}_n - \sigma)$ converges in distribution to the normal rv $\sigma I(1)$ with mean zero and variance given by

$$2\sigma^2 \int_0^1 \left[(1-y)\, \frac{f_o(F_o^{-1}(y))}{f_o^2(F_o^{-1}(y))} \int_0^y x\, \frac{f_o'(F_o^{-1}(x))}{f_o^2(F_o^{-1}(x))}\, dx\right] dy . \tag{4.4}$$

The asymptotic distribution of a modified version of the rv $n^{\frac{1}{2}}(\hat{\sigma}_n - \sigma)$ was again first studied by Weiss (1963), who showed his version's asymptotic normality under somewhat stronger conditions on F_o than ours. In Csörgő-Révész (1980) we will prove

Lemma 1. Given that F_o of the family $\mathcal{F}$ of (1.16) satisfies the conditions of Theorem 2, there exists a sequence of Brownian bridges $\{B_n(u);\ 0 \le u \le 1\}$ such that

$$\sup_{0<u<1} \left| \sum_{j=1}^{[nu]-1} f_o(F_o^{-1}(\tfrac{j}{n+1}))(Y_{j+1}-Y_j)-(n^{-\frac{1}{2}}B_n(u)-n^{-\frac{1}{2}}I_n(u)+u) \right| = o_P(1) . \quad (4.5)$$

In the light of the latter one can give a complete description of the asymptotic behaviour of Weiss' somewhat more restricted modified version of our $\hat{\sigma}_n$ (cf. Weiss 1963) under the less stringent conditions of Theorem 2. We mention two examples of $\hat{\sigma}_n$ here. It follows from Corollary 5 and by (2.10) that in case of a random sample from the uniform-(A,B) distribution (i.e., when $F(x) = (X-A)/(B-A)$, $A \le x \le B$) $\hat{\sigma}_n = X_{n:n} - X_{1:n}$ is the maximum likelihood estimator of $\sigma = B-A$ and the normal rv $\sigma I(1)$ becomes degenerate with $n^{\frac{1}{2}}\{(X_{n:n} - X_{1:n}) - (A-B)\} \to 0$ in probability as $n \to \infty$. In case of a random sample from the exponential family $F(x) = 1 - e^{-(x-A)/B}$, $x \ge A$, $(A \in R^1,\ B > 0)$, Corollary 5 gives that $\hat{\sigma}_n = (n+1)^{-1}(\sum_{j=1}^{n} X_j + X_{n:n} - (n+1)X_{1:n})$ is essentially the maximum likelihood estimator of B and that, in this case, $n^{\frac{1}{2}}(\hat{\sigma}_n - B) \overset{\mathcal{D}}{\to} BI(1)$ with the latter normal rv now being a $N(0,B^2)$ rv. Further examples are given in Weiss (1961, 1963).

5. REFERENCES

[1] Burke, M.D., and Csörgő, M. (1976), "Weak approximations of the empirical process when parameters are estimated", in Empirical Distributions and Processes, ed. P. Gaenssler, and P. Révész, Lecture Notes in Mathematics 566, Berlin: Springer-Verlag, 1-16.

[2] Burke, M.D., Csörgő, M., Csörgő, S., and Révész, P. (1979), "Approximations of the Empirical Process when Parameters are Estimated," Annals of Probability, 9, 790-810.

[3] Csörgő, M., and Révész, P. (1975), "A New Method to Prove Strassen Type Laws of Invariance Principle. II.," Zeitschrift für Wahrscheinlichkeitstheorie und Verwandte Gebiete, 31, 261-269.

[4] Csörgő, M., and Révész, P. (1978), "Strong Approximations of the Quantile Process," Annals of Statistics, 6, 282-294.

[5] Csörgő, M., and Révész, P. (1979), "Quadratic...ity Tests", Carleton Mathematical Series No. 162.

[6] Csörgő, M., and Révész, P. (1980), "Quantile processes for Composite Goodness-of-fit." To appear.

[7] Csörgő, M., Komlós, J., Major, P., Révész, P., and Tusnády, G. (1974), "On the Empirical Process when Parameters are Estimated", in Transactions Seventh Prague Conference 1974, Prague: Academia 1977, 87-97.

[8] Csörgő, M., and Seshadri, V. (1970), "On the Problem of Replacing Composite Hypotheses by Equivalent Simple Ones", Review of the International Statistical Institute, 38, 351-368.

[9] Csörgő, M., Seshadri, V., and Yalovsky, M. (1975), "Applications of Characterizations in the Area of Goodness-of-Fit" in Statistical Distributions in Scientific Work, Vol. 2, eds. G.P. Patil, S. Kotz, and J.K. Ord, Dordrecht-Holland: D. Reidel Publishing Co., 79-90.

[10] Darling, D.A. (1955), "The Cramér-Smirnov Test in the Parametric Case," Annals of Mathematical Statistics, 26, 1-20.

[11] DeWet, T., and Venter, J.H. (1972), "Asymptotic Distributions of Certain Tests Criteria of Normality", South African Statistical Journal, 6, 135-149.

[12] DeWet, T., and Venter, J.H. (1973), "A Goodness of Fit Test for a Scale Parameter Family of Distributions", South African Statistical Journal, 7, 35-46.

[13] Durbin, J. (1961), "Some Methods of Constructing Exact Tests," Biometrika, 48, 41-55.

[14] Durbin, J. (1973a), "Weak Convergence of the Sample Distribution Function when Parameters are Estimated," Annals of Statistics, 1, 279-290.

[15] Durbin, J. (1973b), Distribution Theory for Tests Based on the Sample Distribution Function, Regional Conference Series in Applied Mathematics, 9, Philadelphia: Society for Industrial and Applied Mathematics.

[16] Durbin, J. (1975), "Kolmogorov-Smirnov Tests when Parameters are Estimated with Applications to Tests of Exponentiality and Tests on Spacings", Biometrika, 62, 5-22.

[17] Durbin, J. (1976), "Kolmogorov-Smirnov Tests when Parameters are Estimated", in Empirical Distributions and Processes, ed. P. Gaenssler, and P. Révész, Lecture Notes in Mathematics 566, Berlin: Springer-Verlag, 33-44.

[18] Hajek, Jaroslav (1969), A Course in Nonparametric Statistics, San Francisco: Holden-Day.

[19] Kac, M., Kiefer, J., and Wolfowitz, J. (1955), "On Tests of Normality and Other Tests of Goodness-of-Fit Based on Distance Methods," Annals of Mathematical Statistics, 26, 189-211.

[20] Kiefer, J. (1972), "Skorohod Embedding of Multivariate R.V.'s and the Sample D.F.", Zeitschrift für Wahrscheinlichkeitstheorie und Verwandte Gebiete, 24, 1-35.

[21] Komlós, J., Major, P., and Tusnády, G. (1975), "An Approximation of Partial Sums of Independent R.V.'s and the Sample D.F. I," Zeitschrift für Wahrscheinlichkeitstheorie und Verwandte Gebiete, 32, 111-131.

[22] Lilliefors, H.W. (1967), "On the Kolmogorov-Smirnov Test for Normality with Mean and Variance Unknown," Journal of the American Statistical Association, 62, 399-402.

[23] Lilliefors, H.W. (1969), "On the Kolmogorov-Smirnov Test for the Exponential Distribution with Mean Unknown," Journal of the American Statistical Association, 64, 387-389.

[24] Neuhaus, G. (1974), "Asymptotic Properties of the Cramér-von Mises Statistic when Parameters are Estimated," in Proceedings of the Prague Symposium on Asymptotic Statistics 1973, ed. J. Hajek, Prague: Charles University, 2, 257-297.

[25] Neuhaus, G. (1976a), "Asymptotic Power Properties of the Cramér-von Mises Test under Contiguous Alternatives," Journal of Multivariate Analysis, 6, 95-110.

[26] Neuhaus, G. (1976b), "Weak Convergence under Contiguous Alternatives of the Empirical Process when Parameters are Estimated: the D_k Approach," in Empirical Distributions and Processes, eds. P. Gaenssler, and P.Révész, Lecture Notes in Mathematics 566, Berlin: Springer-Verlag, 68-82.

[27] Parzen, Emanuel (1979a), "Nonparametric Statistical Date Modeling," Journal of the American Statistical Association, 74, 105-131.

[28] Parzen, Emanuel (1979b), "A Density-Quantile Function Perspective on Robust Estimation," in Robustness in Statistics, eds. Robert L. Launer, and Graham N. Wilkinson, New York: Academic Press, 237-258.

[29] Pearson, E.S., and Hartley, H.O. (1972), Biometrika Tables for Statisticians, 2, Cambridge University Press.

[30] Philipp, W. and Pinzur, I. (1978), "An Almost Sure Approximation for the Multivariate Empirical Process." To appear.

[31] Rao, K.C. (1972), "The Kolmogorov, Cramér-von Mises Chi-square Statistics for Goodness-of-Fit Tests in the Nonparametric Case," Abstract: IMS Bulletin, 1, 87.

[32] Shapiro, S.S., and Francia, R.S. (1972), "An Approximate Analysis of Variance Tests for Normality", Journal of the American Statistical Association, 67, 215-216.

[33] Shapiro, S.S., and Wilk, M. B. (1965), "An analysis of Variance Test for Normality (complete samples)," Biometrika, 52, 591-611.

[34] Shapiro, S.S., and Wilk, M.B. (1972b), "An Analysis of Variance Test for the Exponential Distribution", Technometrics, 14, 355-370.

[35] Van der Waerden, B.L., Nievergelt, E. (1956). Tables for Comparing Two Samples by X-test and Sign Test, Berlin: Springer-Verlag.

[36] Weiss, Lionel (1961), "On the Estimation of Scale Parameters," Naval Research Logistics Quarterly, 8, 245-256.

[37] Weiss, Lionel (1963), "On the Asymptotic Distribution of an Estimate of a Scale Parameter," Naval Research Logistics Quarterly, 10, 1-9.

[1] This research was supported by a NSERC Canada Grant and a Canada Council Killam Senior Research Fellowship at Carleton University, Ottawa.

STATISTICS AND RELATED TOPICS
M. Csörgö, D.A. Dawson, J.N.K. Rao, A.K.Md.E. Saleh (eds.)

ASYMPTOTIC DISTRIBUTIONS OF FUNCTIONS OF THE EIGENVALUES OF THE REAL AND COMPLEX NONCENTRAL WISHART MATRICES*

C. Fang and P. R. Krishnaiah

University of Pittsburgh
and
Carnegie-Mellon University

University of Pittsburgh

In this paper, the authors give asymptotic expressions for the joint distributions of the eigenvalues of the noncentral Wishart matrices in the real and complex cases. These expressions are derived by applying perturbation theory. Some applications of the above results are also discussed.

1. INTRODUCTION

The distributions of functions of the eigenvalues of the real and complex Wishart matrices are very useful in studying the structures of the covariance matrices of the real and complex multivariate normal distributions respectively and other problems. Krishnaiah and Lee (1977) derived the joint asymptotic distributions of the linear functions as well as the ratios of the linear functions of the roots of the central Wishart matrix when the population covariance matrix has simple roots. Fujikoshi (1978) derived an asymptotic expression for the distribution of a function of the roots of the central Wishart matrix when the population roots have multiplicity whereas Krishnaiah and Lee (1979) obtained corresponding expressions for the joint density of the functions of the roots. In this paper, we obtain asymptotic expressions for the joint densities of various functions of the noncentral real and complex Wishart matrices. These expressions are in terms of multivariate normal density and multivariate Hermite polynomials. Percentage points of some test statistics are computed by using the above asymptotic expressions and these percentage points are compared with the results obtained by simulation. Applications of the above results are also discussed in problems of studying the structure of interactions and mixtures of multivariate normal populations. Finally, the joint asymptotic distribution of the functions of the roots of the complex Wishart matrix is derived.

2. PERTURBATION TECHNIQUE

Let $\ell_1 \geq \ldots \geq \ell_p$ be the eigenvalues of a symmetric matrix T: $p \times p$, and $\lambda_1 \geq \ldots \geq \lambda_p$ are the eigenvalues of a symmetrix matrix V: $p \times p$

*This work is sponsored by the Air Force Office of Scientitic Research, Air Force Systems Command under Contract F49620-79-C-0161. Reproduction in whole or in part is permitted for any purpose of the United States Government.

where

$$T = V + \varepsilon V^{(1)} + \varepsilon^2 V^{(2)} + \ldots \tag{2.1}$$

Then, there exist orthogonal matrices G and Γ such that $T = GLG'$ and $V = \Gamma\Lambda\Gamma'$, where $L = \text{diag}(\ell_1, \ldots, \ell_p)$, $\Lambda = \text{diag}(\lambda_1, \ldots, \lambda_p)$. The columns of G and Γ consist of the eigenvectors of T and V respectively.

Lawley (1956), Mallows (1961), Izenman (1972) and Fujikoshi (1978) have approximated the eigenvalues and eigenvectors of T in various papers. These authors have either assumed that λ_i's do not have multiplicity or the approximations were established by tacitly assuming that the eigenvalues and eigenvectors admit series expansions in the infinitesimal parameter ε as follows:

$$\begin{aligned} \ell_j &= \lambda_j + \varepsilon\lambda_j^{(1)} + \varepsilon^2\lambda_j^{(2)} + \ldots \\ \underset{\sim}{G}_j &= \underset{\sim}{\Gamma}_j + \varepsilon\underset{\sim}{\Gamma}_j^{(1)} + \varepsilon^2\underset{\sim}{\Gamma}_j^{(2)} + \ldots \end{aligned} \tag{2.2}$$

and no attempts were made to prove that the series converge. A more insight treatment to settle this question of convergence is found in Kato (1976).

Now T and V are linear transformations which operate on the p-dimensional complex vector field C^p, ε is also complex, $\lambda_1 \geq \ldots \geq \lambda_p$ are the eigenvalues of V: $p \times p$ such that

$$\lambda_{q_1 + \ldots + q_{\alpha-1}+1} = \ldots = \lambda_{q_1 + \ldots + q_\alpha} = \theta_\alpha \tag{2.3}$$

for $\alpha = 1,2,\ldots, r$, $q_1 +\ldots+ q_r = p$, $q_o = 0$ and let $J_\alpha(\alpha=1,\ldots, r)$ denote the set of integers $q_1 +\ldots+ q_{\alpha-1}+1, \ldots, q_1 +\ldots+ q_\alpha$. We need the following lemma in the sequel.

Lemma 2.1. For Hermitian matrices T and V

$$T = V + \varepsilon V^{(1)} + \varepsilon^2 V^{(2)} + \ldots$$

and V is diagonalized as $V = \text{diag}(\lambda_1, \ldots, \lambda_p)$. Then the mean eigenvalue of T corresponding to θ_α which is the eigenvalue of V with multiplicity q_α, is

$$\bar{\ell}_\alpha = \theta_\alpha + \varepsilon\, \bar{\ell}_\alpha^{(1)} + \varepsilon^2\, \bar{\ell}_\alpha^{(2)} + \ldots \tag{2.4}$$

where

$$\begin{aligned} \bar{\ell}_\alpha^{(1)} &= \frac{1}{q_\alpha} \operatorname{tr} V_{\alpha\alpha}^{(1)} \\ \bar{\ell}_\alpha^{(2)} &= \frac{1}{q_\alpha} \operatorname{tr}[V_{\alpha\alpha}^{(2)} + \sum_{\beta\neq\alpha} \theta_{\alpha\beta}^{-1} V_{\alpha\beta}^{(1)} V_{\beta\alpha}^{(1)}] \\ \theta_{\alpha\beta} &= \theta_\alpha - \theta_\beta \end{aligned} \tag{2.5}$$

with

$$V^{(i)} = \begin{bmatrix} V_{11}^{(i)} & V_{12}^{(i)} & \cdots & V_{1r}^{(i)} \\ \vdots & \vdots & \cdots & \vdots \\ V_{r1}^{(i)} & V_{r2}^{(i)} & \cdots & V_{rr}^{(i)} \end{bmatrix}$$

and $V_{\alpha\beta}^{(i)}$ is of order $q_\alpha \times q_\beta$.

When $q_1 = \ldots = q_r = 1$, the above lemma was proved in Lawley (1956). When $q_\alpha \geq 1$ ($\alpha = 1, \ldots, r$), the lemma was given implicitly in Kato (1976). For $q_\alpha = 1$ the normalized eigenvector of linear transformation T corresponding to θ_α is $\underset{\sim}{G}_\alpha = (G_{1\alpha}, \ldots, G_{p\alpha})'$, with

$$G_{i\alpha} = \frac{a_{i\alpha}(\varepsilon)}{(\theta_\alpha - \lambda_i)} + \sum_{j\neq\alpha} \frac{a_{ij}(\varepsilon) a_{j\alpha}(\varepsilon)}{(\theta_\alpha - \lambda_i)(\theta_\alpha - \lambda_j)} - \frac{a_{i\alpha}(\varepsilon) a_{\alpha\alpha}(\varepsilon)}{(\theta_\alpha - \lambda_i)^2} + \ldots, \quad i \neq \alpha$$

$$G_{\alpha\alpha} = 1 - \frac{1}{2} \sum_{j\neq\alpha} \frac{a_{ij}(\varepsilon)\, a_{ji}^c(\varepsilon)}{(\theta_\alpha - \lambda_j)(\theta_\alpha - \lambda_j)^c} + \ldots \tag{2.6}$$

where

$$A(\varepsilon) = T - V = (a_{ij}(\varepsilon))$$

and z^c denote the complex conjugate of z. The series (2.4), (2.6) are convergent for

$$|\varepsilon| < \left(\frac{2c_1}{d} + c_2\right)^{-1} \tag{2.7}$$

where $c_1, c_2 \geq 0$ such that $||V^{(j)}|| \leq c_1 c_2^{j-1}$ for $j = 1, 2, \ldots$, and $d = \min(|\theta_\alpha - \theta_{\alpha-1}|, |\theta_\alpha - \theta_{\alpha+1}|)$.

3. ASYMPTOTIC JOINT DISTRIBUTION OF FUNCTIONS OF THE ROOTS OF NONCENTRAL WISHART MATRIX

Let the columns of X: p×n be distributed as multivariate normal with covariance matrix $\Sigma = (\sigma_{ij})$ and means given by $E(X) = U = (\underset{\sim}{\mu}_1, \ldots, \underset{\sim}{\mu}_n)$, where $\underset{\sim}{\mu}_j' = (\mu_{j1}, \ldots, \mu_{jp})$. Then, $S = XX' = (S_{ij})$ is distributed as the central or noncentral Wishart matrix $W_p(n, \Sigma, M)$ with n degrees of freedom according as $M = 0$ or $M \neq 0$ where $M = \sum_{j=1}^{n} \underset{\sim}{\mu}_j \underset{\sim}{\mu}_j' = n(\nu_{ij})$. Now, let $\ell_1 \geq \cdots \geq \ell_p$ denote the eigenvalues of S/n whereas $\lambda_1 \geq \cdots \geq \lambda_p$ denote the eigenvalues of $E(S/n) = \Sigma + M/n = \Lambda$. Without loss of generality, we assume that $\Lambda = \text{diag.}\ (\lambda_1, \ldots, \lambda_p)$. Also, let

$$\lambda_{q_1+\ldots+q_{\alpha-1}+1} = \cdots = \lambda_{q_1+\ldots+q_\alpha} = \theta_\alpha \tag{3.1}$$

for $\alpha = 1, 2, \ldots, r$, $q_1 + \ldots + q_r = p$, and $q_0 = 0$.

In this section, we derive the joint asymptotic distribution of $L_1,\cdots,L_k$ where $L_j = \sqrt{n}\,\{T_j(\ell_1,\ldots,\ell_p)-T_j(\lambda_1,\ldots,\lambda_p)\}$ and $T_j(\ell_1,\cdots,\ell_p)$ satisfy the following assumptions:

(i) $T_j(\ell_1,\cdots,\ell_p)$ is analytic about $\lambda_1,\cdots,\lambda_p$

(ii)
$$\left.\frac{\partial T_i(\ell_1,\cdots,\ell_p)}{\partial \ell_{j_1}}\right|_{\underset{\sim}{\ell}=\underset{\sim}{\lambda}} = c_{ij_1} = a_{i\alpha}$$

$$\left.\frac{\partial^2 T_i(\ell_1,\cdots,\ell_p)}{\partial \ell_{j_2}\;\partial \ell_{j_1}}\right|_{\underset{\sim}{\ell}=\underset{\sim}{\lambda}} = c_{ij_1j_2} = a_{i\alpha\beta}$$

$$\left.\frac{\partial^3 T_i(\ell_1,\cdots,\ell_p)}{\partial \ell_{j_3}\;\partial \ell_{j_2}\;\partial \ell_{j_1}}\right|_{\underset{\sim}{\ell}=\underset{\sim}{\lambda}} = c_{ij_1j_2j_3} = a_{i\alpha\beta\gamma} \tag{3.2}$$

for $j_1 \in J_\alpha$, $j_2 \in J_\beta$, $j_3 \in J_\gamma$, $\underset{\sim}{\lambda}' = (\lambda_1,\cdots,\lambda_p)$, $\underset{\sim}{\ell}' = (\ell_1,\cdots,\ell_p)$ and J_α denotes the set of integers $q_1+\cdots+q_{\alpha-1}+1,\cdots,q_1+\cdots+q_\alpha$ for $\alpha = 1,2,\cdots,r$.

Expanding $T_i(\ell_1,\cdots,\ell_p)$ as the Taylor series, we obtain

$$T_i(\ell_1,\cdots,\ell_p) = T_i(\lambda_1,\cdots,\lambda_p) + \sum_{\alpha=1}^{r} a_{i\alpha} \sum_{j_1\in J_\alpha} (\ell_{j_1} - \theta_\alpha)$$

$$+ \frac{1}{2}\sum_{\alpha=1}^{r}\sum_{\beta=1}^{r} a_{i\alpha\beta} \sum_{j_1\in J_\alpha}\sum_{j_2\in J_\beta} (\ell_{j_1}-\theta_\alpha)(\ell_{j_2}-\theta_\beta)$$

$$+ \frac{1}{6}\sum\sum\sum_{\alpha\beta\gamma} a_{i\alpha\beta\gamma} \sum_{j_1\in J_\alpha}\sum_{j_2\in J_\beta}\sum_{j_3\in J_\gamma} (\ell_{j_1}-\theta_\alpha)(\ell_{j_2}-\theta_\beta)(\ell_{j_3}-\theta_\gamma)$$

$$+ \text{terms of higher degree.} \tag{3.3}$$

Now, let

$$\frac{S}{n} = \Lambda + \frac{1}{\sqrt{n}}\, V \tag{3.4}$$

where

$$V = \sqrt{n}\,(\frac{S}{n} - \Lambda) = \begin{pmatrix} V_{11} & V_{12} & \cdots & V_{1r} \\ V_{21} & V_{22} & \cdots & V_{2r} \\ \vdots & \vdots & \cdots & \vdots \\ V_{r1} & V_{r2} & & V_{rr} \end{pmatrix}.$$

By using Lemma 2.1 and (3.4), we obtain

$$L_i = \sum_{\alpha=1}^{r} a_{i\alpha} \text{ tr } Z_\alpha^{(1)} + \frac{1}{\sqrt{n}} \{ \sum_{\alpha=1}^{r} a_{i\alpha} \text{ tr } Z_\alpha^{(2)} + \frac{1}{2} \sum_{\alpha=1}^{r} \sum_{\beta=1}^{r} a_{i\alpha\beta}$$

$$(\text{tr } Z_\alpha^{(1)})(\text{tr } Z_\beta^{(1)})\} + O(n^{-1}) \qquad (3.5)$$

where $Z_\alpha^{(1)} = V_{\alpha\alpha}$, $Z_\alpha^{(2)} = \sum_{\beta\neq\alpha} \theta_{\alpha\beta}^{-1} V_{\alpha\beta} V_{\beta\alpha}$ and $\theta_{\alpha\beta} = \theta_\alpha - \theta_\beta$.

Also,

$$\text{tr } Z_\alpha^{(1)} = \sqrt{n} \sum_{j_1 \varepsilon J_\alpha} \left(\frac{S_{j_1 j_1}}{n} - \lambda_{j_1} \right),$$

$$\text{tr } Z_\alpha^{(2)} = \frac{1}{n} \sum_{\beta\neq\alpha} \theta_{\alpha\beta}^{-1} \sum_{j_1 \varepsilon J_\alpha} \sum_{j_2 \varepsilon J_\beta} S_{j_1 j_2}^2 .$$

After some algebraic manipulations, we obtain the following expression for the joint characteristic function of $L_1, \ldots, L_k$:

$$\Psi(t_1, \cdots, t_k) = E\{\exp(i \sum_{j=1}^{k} t_j L_j)\}$$

$$= E[\exp(i \sum_{j=1}^{k} \sum_{\alpha=1}^{r} t_j a_{j\alpha} \text{ tr } Z_\alpha^{(1)}) \qquad (3.6)$$

$$\times\{1 + \frac{1}{\sqrt{n}} (i \sum_{j=1}^{k} \sum_{\alpha=1}^{r} t_j a_{j\alpha} \text{ tr } Z_\alpha^{(2)} + \frac{i}{2} \sum_{j=1}^{k} \sum_{\alpha=1}^{r} \sum_{\beta=1}^{r} t_j a_{j\alpha\beta}$$

$$\times \text{ tr } Z_\alpha^{(1)} \text{tr } Z_\beta^{(1)}) + O(n^{-1})\}]$$

$$= E_1(\underset{\sim}{t}) + E_2(\underset{\sim}{t}) + E_3(\underset{\sim}{t}) + O(n^{-1})$$

where $\underset{\sim}{t}' = (t_1, \ldots, t_k)$. In Eq. (3.6),

$$E_1(\underset{\sim}{t}) = E[\exp(i \sum_{j=1}^{k} \sum_{\alpha=1}^{r} t_j a_{j\alpha} \text{ tr } Z_\alpha^{(1)})]$$

$$= \exp(-i\sqrt{n} \text{ tr } B\Lambda) | I - 2iB\Sigma/\sqrt{n} |^{-n/2}$$

$$\times \exp\{i \text{ tr}[M(I - 2iB\Sigma/\sqrt{n})^{-1} B / \sqrt{n}]\} \qquad (3.7)$$

where $B = \sum_{j=1}^{k} t_j \text{ diag}(c_{j1}, \cdots, c_{jp})$. Also,

$$E_2(\underset{\sim}{t}) = E[\frac{i}{\sqrt{n}} \sum_{i_1=1}^{k} \sum_{\alpha=1}^{r} t_{i_1} a_{i_1\alpha} \text{tr } Z_\alpha^{(2)} \times \exp\{i \sum_{i_2=1}^{k} \sum_{\alpha_1=1}^{r} t_{i_2} a_{i_2\alpha_1} \times \text{tr } Z_{\alpha_1}^{(1)}\}]$$

$$= E_1(\underset{\sim}{t}) \frac{i}{\sqrt{n}} \sum_{i_1=1}^{k} \sum_{\alpha=1}^{r} \sum_{\beta\neq\alpha} \sum_{j_1 \varepsilon J_\alpha} \sum_{j_2 \varepsilon J_\beta} t_{i_1} a_{i_1\alpha} \theta_{\alpha\beta}^{-1}$$

$$\times \frac{1}{n}\{ \sum_{j=1}^{n} (\sigma^*_{j_1j_1} \sigma^*_{j_2j_2} + \sigma^{*2}_{j_1j_2} + \sigma^*_{j_1j_1} \xi^2_{jj_2} + 2\sigma^*_{j_1j_2} \xi_{jj_1} \xi_{jj_2}$$

$$+ \sigma^*_{j_2j_2} \xi^2_{jj_1})$$

$$+ [\sum_{j=1}^{n} (\sigma^*_{j_1j_2} + \xi_{jj_1} \xi_{jj_2})]^2\} \tag{3.8}$$

$$E_3(\underset{\sim}{t}) = E[\frac{i}{2\sqrt{n}} \sum_{i_1=1}^{k} \sum_{\alpha=1}^{r} \sum_{\beta=1}^{r} t_{i_1} a_{i_1\alpha\beta} (\text{tr } Z_\alpha^{(1)})(\text{tr } Z_\beta^{(1)})$$

$$\times \exp\{i \sum_{i_2=1}^{k} \sum_{\alpha_1=1}^{r} t_{i_2} a_{i_2\alpha_1} \text{ tr } Z_{\alpha_1}^{(1)}\}]$$

$$= E_1(\underset{\sim}{t}) \frac{i}{2\sqrt{n}} \sum_{i_1=1}^{k} \sum_{\alpha=1}^{r} \sum_{\beta=1}^{r} \sum_{j_1 \varepsilon J_\alpha} \sum_{j_2 \varepsilon J_\beta} t_{i_1} a_{i_1\alpha\beta}$$

$$\times \{\frac{1}{n} \sum_{j=1}^{n} (2\sigma^{*2}_{j_1j_2} + 4\sigma^*_{j_1j_2} \xi_{jj_1} \xi_{jj_2})$$

$$+ \frac{1}{n} \sum_{j=1}^{n} \sum_{m=1}^{n} (\sigma^*_{j_1j_1} + \xi^*_{jj_1})(\sigma^*_{j_2j_2} + \xi^2_{mj_2}) - \lambda_{j_1} \sum_{j=1}^{n} (\sigma^*_{j_2j_2} + \xi^2_{jj_2})$$

$$- \lambda_{j_2} \sum_{j=1}^{n} (\sigma^*_{j_1j_1} + \xi^2_{jj_1}) + n \lambda_{j_1} \lambda_{j_2}\} \tag{3.9}$$

where

$$\Sigma^* = \Sigma \left(I - \frac{2iB\Sigma}{\sqrt{n}} \right)^{-1} = (\sigma^*_{ij}) \tag{3.10}$$

$$\underset{\sim}{\xi}_j = \left(I - \frac{2i\Sigma B}{\sqrt{n}} \right)^{-1} \underset{\sim}{\mu}_j = (\xi_{j1}, \cdots, \xi_{jp})'$$

Using the following expansions

$$|I-A|^{-\beta} = \exp \beta \left(\sum_{j=1}^{\infty} \frac{\text{tr } A^j}{j} \right), (I-A)^{-1} = \sum_{j=0}^{\infty} A^j$$

in (3.7) and (3.10), we obtain

$$\sigma^*_{j_1j_2} = \sigma_{j_1j_2} + \frac{2i}{\sqrt{n}} \sum_{i_1=1}^{k} \sum_{j=1}^{p} t_{i_1} c_{i_1j} \sigma_{j_1j} \sigma_{j_2j} + O(n^{-1})$$

$$\xi_{jj_1} = \mu_{jj_1} + \frac{2i}{\sqrt{n}} \sum_{i_1=1}^{k} \sum_{m=1}^{p} t_{i_1} c_{i_1m} \sigma_{j_1m}\, \mu_{jm} + O(n^{-1})$$

Eq. (3.7), (3.8), (3.9) lead to

$$\Psi(\underset{\sim}{t}) = \exp(-\tfrac{1}{2}\underset{\sim}{t}'Q\underset{\sim}{t}) \times \{1 + \frac{1}{\sqrt{n}} \sum_{j=1}^{k} i\, t_j(d_1+d_2) + \frac{1}{\sqrt{n}} \sum_{i_1i_2i_3}^{k} (i^3 t_{i_1} t_{i_2} t_{i_3})(g_1+g_2+g_3)\} + O(n^{-1}) \tag{3.11}$$

where $Q = (Q_{i_1i_2})$, $Q_{i_1i_2} = 2\mathrm{tr}\, R^{(i_1)} R^{(i_2)} + 4\mathrm{tr}\, R^{(i_1)} \psi^{(i_2)}$, and Q is assumed to be nonsingular. Also,

$$d_1 = \sum_{\alpha=1}^{r} \sum_{\beta\neq\alpha}^{r} \sum_{j_1\varepsilon J_\alpha} \sum_{j_2\varepsilon J_\beta} a_{j\alpha}\theta^{-1}_{\alpha\beta} \; (\sigma_{j_1j_1}\sigma_{j_2j_2} + \sigma^2_{j_1j_2} + 2\sigma_{j_1j_2}\nu_{j_1j_2} + \sigma_{j_1j_1}\nu_{j_2j_2} + \sigma_{j_2j_2}\nu_{j_1j_1})$$

$$d_2 = \sum_{\alpha=1}^{r} \sum_{\beta=1}^{r} \sum_{j_1\varepsilon J_\alpha} \sum_{j_2\varepsilon J_\beta} a_{j\,\alpha\beta}(\sigma^2_{j_1j_2} + 2\sigma_{j_1j_2}\nu_{j_1j_2})$$

$$g_1 = \frac{4}{3}\mathrm{tr}\, R^{(i_1)} R^{(i_2)} R^{(i_3)} + 4\,\mathrm{tr}\, R^{(i_1)} R^{(i_2)} \psi^{(i_3)}$$

$$g_2 = 4 \sum_{\alpha=1}^{r} \sum_{\beta\neq\alpha} \sum_{j_1\varepsilon J_\alpha} \sum_{j_2\varepsilon J_\beta} a_{i_1\alpha}\, \theta^{-1}_{\alpha\beta}(\Xi^{(i_2)}_{j_1j_2} + T^{(i_2)}_{j_1j_2} + T^{(i_2)}_{j_2j_1}) \times (\Xi^{(i_3)}_{j_1j_2} + T^{(i_3)}_{j_1j_2} + T^{(i_3)}_{j_2j_1})$$

$$g_3 = 2 \sum_{\alpha=1}^{r} \sum_{\beta=1}^{r} \sum_{j_1\varepsilon J_\alpha} \sum_{j_2\varepsilon J_\beta} a_{i_1\alpha\beta}(\Xi^{(i_2)}_{j_1j_1} + 2T^{(i_2)}_{j_1j_1})(\Xi^{(i_3)}_{j_2j_2} + 2T^{(i_3)}_{j_2j_2}) \tag{3.12}$$

We define here $M/n = (\nu_{ij})$

$$C^{(i)} = \mathrm{diag}(c_{i1},\ldots, c_{ip})$$

$$R^{(i)} = C^{(i)}\Sigma,\quad \psi^{(i)} = C^{(i)}\frac{M}{n},\quad \Xi^{(i)} = \Sigma C^{(i)}\Sigma,\quad T^{(i)} = \frac{M}{n}C^{(i)}\Sigma \tag{3.13}$$

where A_{ij} denotes the (i,j)th element of matrix $A = (A_{ij})$.

Now inverting (3.11) we obtain the following expansion for density of $\underset{\sim}{L} = (L_1, \ldots, L_k)'$:

$$f(L_1, \ldots, L_k) = N(\underset{\sim}{L}, Q) \times$$

$$[1 + \frac{1}{\sqrt{n}} \sum_{j=1}^{k} H_j(L)(d_1 + d_2)$$

$$+ \frac{1}{\sqrt{n}} \sum_{i_1, i_2, i_3}^{k} H_{i_1 i_2 i_3}(\underset{\sim}{L})(g_1 + g_2 + g_3)] + O(n^{-1}) \quad (3.14)$$

where $$N(\underset{\sim}{L}, Q) = \frac{1}{(2\pi)^{k/2} |Q|^{1/2}} \exp(-\frac{1}{2} \underset{\sim}{L}' Q^{-1} \underset{\sim}{L})$$

$$H_{j_1, \ldots, j_s}(\underset{\sim}{L}) N(\underset{\sim}{L}, Q) = (-1)^s \frac{\partial^s}{\partial L_{j_1}, \ldots, \partial L_{j_s}} N(\underset{\sim}{L}, Q) \quad (3.15)$$

Now, let $T_i(\ell_1, \ldots, \ell_p) = \ell_i$. Then $L_i = \sqrt{n}(\ell_i - \lambda_i)$. Using Eq. (3.14), we obtain the following expressions for the joint density of the roots $\ell_1, \ldots, \ell_p$ when $\Sigma = \sigma^2 I$:

$$f(L_1, \ldots, L_p) = N(\underset{\sim}{L}, Q)\{1 + \frac{1}{\sqrt{n}} \sum_{i=1}^{p} H_i(\underset{\sim}{L}) \sum_{j \neq i} \theta_{ij}^{-1}(\sigma^2\lambda_i + \sigma^2\lambda_j - \sigma^4)$$

$$+ \frac{4\sigma^4}{\sqrt{n}} \sum_{i=1}^{p} H_{iii}(\underset{\sim}{L})(\lambda_i - \frac{2}{3}\sigma^2)\} + O(n^{-1})$$

$$(3.16)$$

where $Q = \mathrm{diag}(Q_1, \ldots, Q_p)$ and $Q_i = 2\sigma^2(2\lambda_i - \sigma^2)$. When $\Sigma = \sigma^2 I$ and $\lambda_1 > \ldots > \lambda_a = \lambda_{a+1} = \ldots \lambda_p = \sigma^2$, Carter and Srivastava (1979) obtained an alternative expression for the joint density of $\ell_1, \ldots, \ell_p$ by using a different method.

4. APPLICATIONS IN INVESTIGATION OF THE STRUCTURES OF INTERACTIONS

In this section, we discuss some applications of the results of Section 3 in studying the power functions of various tests for the hypothesis of no interaction in two-way classification model with one observation per cell.

Consider the model

$$x_{ij} = \mu + \alpha_i + \beta_j + \eta_{ij} + \varepsilon_{ij} \quad (4.1)$$

for $i = 1, \ldots, u$, $j = 1, 2, \ldots, s$. Here x_{ij} denotes the observation in i-th row and j-th column, μ is the general mean, α_i

denotes the effect due to i-th row, β_j denotes the effect due to j-th column and η_{ij} denotes the interaction of i-th row and j-th column. Also, we assume that ε_{ij}'s are distributed independently and normally with mean 0 and variance σ^2. Now, let

$d_{ij} = x_{ij} - \bar{x}_{i\cdot} - \bar{x}_{\cdot j} + \bar{x}_{\cdot\cdot}$, where $s\bar{x}_{i\cdot} = \sum_{j=1}^{s} x_{ij}$, $u\bar{x}_{\cdot j} = \sum_{i=1}^{u} x_{ij}$ and $us\bar{x}_{\cdot\cdot} = \sum_{i=1}^{u} \sum_{j=1}^{s} x_{ij}$. Also, let $D = (d_{ij})$, $X = (x_{ij})$, $W = C_u' X C_s C_s' X' C_u$ where C_u is chosen such that $C_u'C_u = I_{u-1}$ and $C_uC_u' = I_u - \frac{1}{u} J_u$ where J_u is the u×u matrix with all its elements equal to unity. The non-zero eigenvalues of DD' are the same (e.g., see Johnson and Graybill (1972)) as the nonzero eigenvalues of W. Also, the columns of $C_u'X$ are distributed independently as multivariate normal with mean vectors given by $E(C_u'X) = C_u'M_o$ and a common covariance matrix $\sigma^2 I_{u-1}$ where $M_o = (m_{ij})$ and $m_{ij} = \mu + \alpha_i + \beta_j + \eta_{ij}$. In addition,

$$E(W/(s-1)) = \sigma^2 I_{u-1} + \{C_u'M_oC_sC_s'M_oC_u/(s-1)\}$$
$$= \Sigma_o \tag{4.2}$$

and $C_u'M_oC_sC_s'M_oC_u = C_u'\eta\eta'C_u = \Omega$ where $\eta = (\eta_{ij})$. So, W is distributed as the noncentral Wishart matrix with (s-1) degrees of freedom and noncentrality matrix Ω. When $\eta=0$, W is distributed as the central Wishart matrix with (s-1) degrees of freedom.

Let $\ell_1 \geq \cdots \geq \ell_{u-1}$ be the eigenvalues of W/(s-1) and let $\lambda_1 \geq \cdots \geq \lambda_{u-1}$ be the nonzero roots of Σ_0. Then, the problem of testing the hypothesis $H : \Omega = 0$ is equivalent to testing the hypothesis that the eigenvalues of Σ_0 are equal.

Suppose $\eta = \gamma\, \underset{\sim}{\alpha}\underset{\sim}{\beta}'$ where $\underset{\sim}{\alpha}' = (\alpha_1, \ldots, \alpha_u)$ and $\underset{\sim}{\beta}' = (\beta_1, \ldots, \beta_s)$. Then $\Omega = \gamma^2 C_u'\underset{\sim}{\alpha}\underset{\sim}{\beta}'\underset{\sim}{\beta}\underset{\sim}{\alpha}'C_u$, and the nonzero root of Ω is $\gamma\underset{\sim}{\beta}'\underset{\sim}{\beta}\ \underset{\sim}{\alpha}'\underset{\sim}{\alpha}$. Next, we will assume that the rank of η is c. Then, using the well-known singular value decomposition of the matrix, we can write η as

$$\eta = \gamma_1 \underset{\sim}{w}_1 \underset{\sim}{v}_1' + \cdots + \gamma_c \underset{\sim}{w}_c \underset{\sim}{v}_c'. \tag{4.3}$$

The nonzero eigenvalues of $\eta\eta'$ are $\gamma_1^2, \ldots, \gamma_c^2$ and the associated eigenvectors are $\underset{\sim}{w}_1, \ldots, \underset{\sim}{w}_c$. The eigenvectors of $\eta'\eta$ corresponding to the eigenvalues $\gamma_1^2, \ldots, \gamma_c^2$ are $\underset{\sim}{v}_1, \ldots, \underset{\sim}{v}_c$. The nonzero eigenvalues of Ω are $\gamma_1^2, \ldots, \gamma_c^2$.

The problem of testing the hypothesis of no interaction in two-way classification with one observation per cell was studied by several authors (e.g., see Tukey (1949) and Williams (1952)). The statistic proposed by Tukey is given by $(\hat{\underset{\sim}{\alpha}}'\hat{\eta}\hat{\underset{\sim}{\beta}})^2/(\hat{\underset{\sim}{\alpha}}'\hat{\underset{\sim}{\alpha}})(\hat{\underset{\sim}{\beta}}'\hat{\underset{\sim}{\beta}})$ where $\hat{\underset{\sim}{\alpha}}' = (\hat{\alpha}_1,\ldots,\hat{\alpha}_u)$, $\hat{\underset{\sim}{\beta}}' = (\hat{\beta}_1,\ldots,\hat{\beta}_s)$, $\hat{\hat{\eta}} = (\hat{\eta}_{ij})$, $\hat{\alpha}_i = \bar{x}_{i.} - \bar{x}_{..}$, $\hat{\beta}_j = \bar{x}_{.j} - \bar{x}_{..}$ and $\hat{\eta}_{ij} = x_{ij} - \bar{x}_{i.} - \bar{x}_{.j} + \bar{x}_{..}$. Gollob (1968) and Mandel (1969) considered the problem of testing the hypotheses $\gamma_j = 0$ individually under the model (4.3) by using the statistics $F_j = \ell_j/(\ell_1+\ldots+\ell_{u-1})$. Gollob treated $\ell_1,\ldots,\ell_{u-1}$ as independent chi-square variables to get an approximation to the distribution of F_j. But ℓ_i's are neither independent nor distributed as chi-square variables. Corsten and Van Eijnsbergen (1972) showed that the likelihood ratio statistic for testing the hypothesis $\gamma_1 =\ldots= \gamma_c = 0$ under the model (4.3) is $(\ell_1+\ldots+\ell_c)/(\ell_{c+1}+\ldots+\ell_{u-1})$. When $c=1$, this statistic was derived independently by Johnson and Graybill (1972). Schuurmann, Krishnaiah and Chattopadhyay (1973) and Krishnaiah and Schuurmann (1974) discussed the problem of testing the hypotheses $\gamma_i = 0$ simultaneously by applying the simultaneous tests of Krishnaiah and Waikar (1971) for the equality of the eigenvalues of the covariance matrix of the multivariate normal population. Ghosh and Sharma (1963) studied the power function of Tukey's test for $\eta_{ij} = 0$ against the alternative that $\eta_{ij} = \gamma\,\alpha_i\,\beta_j$. Yochmowitz and Cornell (1978) derived the likelihood ratio test for $\gamma_1 =\ldots= \gamma_a = 0$ against the alternative that $\gamma_a \neq 0$ and $\gamma_{a+1} =\ldots= \gamma_c = 0$. We now compare the power functions of various procedures for testing the hypothesis of no interaction.

Let $T_1 = \ell_1/\ell_{u-1}$, $T_2 = (\mathrm{tr}\ W/(u-1))^{u-1}/|W|$, $T_3 = (u-c-1) \times \ell_1/(\ell_{c+1}+\ldots+\ell_{u-1})$, $T_4 = (u-c-1)(\ell_1+\ldots+\ell_c)/c(\ell_{c+1}+\ldots+\ell_{u-1})$. When σ^2 is unknown, we can use any of the above statistics for testing the hypothesis of no interaction.

If we use T_i, we accept or reject H_0 according as

$$T_i \lessgtr c_{\alpha i} \tag{4.4}$$

where

$$P[T_i \leq c_{\alpha i}|H_0] = (1-\alpha) . \tag{4.5}$$

The test statistic T_i is based upon the statistic considered by Krishnaiah and Waikar (1971) for testing the sphericity, whereas the test statistic T_2 is based upon the likelihood ratio test statistic for sphericity when the underlying population is multivariate normal. The statistic T_4 is the likelihood ratio test statistic (see Corsten and Van Eijnsbergen (1972)) for testing the hypothesis of no interaction in multiplicative components model (4.3).

Table 1 gives a comparison of the power functions of various procedures for testing the hypothesis of no interaction when σ^2 is unknown. The rows corresponding to S denote the simulated values. The multivariate normal deviates are generated by the IMSL subroutine GGNRM, and 10,000 trials are performed for each case, the 95% confidence limit for each value is then $1.96\{\hat{p}(1-\hat{p})/10{,}000\}^{\frac{1}{2}}$, where $\hat{p}$ is the actual value from the empirical trials. The rows corresponding to N denote the values corresponding to the first term in the asymptotic expansion (3.14). The rows corresponding to $N + O(n^{-\frac{1}{2}})$ give the values corresponding to the first two terms of the expansion (3.14).

The table reveals that results based on normal approximations are not sufficiently accurate for n as large as 100, while the asymptotic expression taking the term of order $n^{-\frac{1}{2}}$ achieves numerical accuracy for moderate sample sizes. This suggests that care should be given for the statistical inferences which are based on the normal approximations.

Next, consider the model (4.1) when σ^2 is known. In this case, we accept or reject the hypothesis $\gamma_1 = \ldots = \gamma_c = 0$ according as

$$\frac{\ell_1}{\sigma^2} \lessgtr d_\alpha \tag{4.6}$$

where

$$P[\frac{\ell_1}{\sigma^2} \leq d_\alpha | H] = (1-\alpha)\ . \tag{4.7}$$

When H is rejected, the hypothesis $\gamma_i = 0$ is accepted or rejected according as $(\ell_i/\sigma^2) \lessgtr d_\alpha$. When H is true, ℓ_1 is the largest eigenvalue of the central Wishart matrix. Exact distribution of this statistic is given in Krishnaiah and Chang (1971) and exact percentage points are given in Krishnaiah (1980). When H is not true, an asymptotic expression for the distribution of ℓ_1 can be obtained as a special case of (3.14) if γ_1 is different from $\gamma_2, \ldots, \gamma_c$.

When $\lambda_1 > \ldots > \lambda_{u-1} > 0$, Srivastava and Carter (1980) obtained asymptotic expression of $\log(\ell_1/\ell_1 + \ldots + \ell_{u-1})$ and $T_2^{1/(u-1)}$ by a different method. For a review of the literature on tests for no interaction in two way classification model with one observation per cell, the reader is referred to Krishnaiah and Yochmowitz (1980)

5. APPLICATIONS IN CLUSTER ANALYSIS

Let $\underset{\sim}{X}_1, \ldots, \underset{\sim}{X}_N$ be independent p-dimensional random variables. We consider using the sample covariance matrix

$$S = \sum_{i=1}^{N} (\underset{\sim}{X}_i - \bar{\underset{\sim}{X}})(\underset{\sim}{X}_i - \bar{\underset{\sim}{X}})'$$

TABLE 1

Comparison of the Power Functions of the Tests for no Interaction When σ^2 is Unknown

$p = 3,\ \alpha = 0.05$

n	$(\lambda_1,\lambda_2,\lambda_3)$	Type of Approximation	ℓ_1/ℓ_3	$(trW/3)^3/\lvert W\rvert$	$2\ell_1/(\ell_2+\ell_3)$	$(\ell_1+\ell_2)/2\ell_3$
		N	0.57	0.49		0.59
10	(12,6,1)	$N+0(n^{-\frac{1}{2}})$	0.83	0.78	–	0.83
		S	0.81	0.77		0.82
		N	0.57	0.63		0.44
10	(12,3,1)	$N+0(n^{-\frac{1}{2}})$	0.85	0.90	–	0.71
		S	0.82	0.87		0.73
10	(12,10,1)	N	0.57	0.55	–	0.72
		$N+0(n^{-\frac{1}{2}})$	0.88	0.83		0.95
10	(7,1,1)	N	–	–	0.79	–
		$N+0(n^{-\frac{1}{2}})$			0.95	
10	(12,12,1)	N	–	–	–	0.76
		$N+0(n^{-\frac{1}{2}})$				0.99
		N	0.51	0.54		0.68
25	(4,3.5,1)	$N+0(n^{-\frac{1}{2}})$	0.82	0.79	–	0.85
		S	0.79	0.77		0.84

TABLE 1 (continued)

n	$(\lambda_1, \lambda_2, \lambda_3)$	Type of Approximation	ℓ_1/ℓ_3	$(\mathrm{tr}W/3)^3/\lvert W\rvert$	$2\ell_1/(\ell_2+\ell_3)$	$(\ell_1+\ell_2)/2\ell_3$
25	(4,1,1)	N	-	-	0.89	-
		$N+O(n^{-\frac{1}{2}})$			0.98	
25	(4,4,1)	N				0.74
		$N+O(n^{-\frac{1}{2}})$	-	-	-	0.90
		S				0.89
50	(2.5,1.7,1)	N	0.46	0.42		0.45
		$N+O(n^{-\frac{1}{2}})$	0.66	0.68	-	0.62
		S	0.68	0.68	-	0.64
50	(2.5,1,1)	N			0.83	
		$N+O(n^{-\frac{1}{2}})$	-	-	0.92	-
		S			0.92	
50	(2.5,2.5,1)	N	-	-		0.72
		$N+O(n^{-\frac{1}{2}})$				0.86
100	(2,1.5,1)	N	0.57	0.54		0.57
		$N+O(n^{-\frac{1}{2}})$	0.75	0.77	-	0.72
		S	0.75	0.76		0.73

TABLE 1 (continued)

$p = 4, \quad \alpha = 0.05$

n	$(\lambda_1,\lambda_2,\lambda_3,\lambda_4)$	Type of Approximation	ℓ_1/ℓ_4	$(trW/4)^4/\lvert W\rvert$	$(\ell_1+\ell_2)/(\ell_3+\ell_4)$	$(\ell_1+\ell_2+\ell_3)/3\ell_4$
		N	0.45	0.36		0.59
100	(3,2.5,2,1)	$N+O(n^{-\frac{1}{2}})$	0.78	0.76		0.80
		S	0.79	0.76		0.81
100	(3,2.5,1,1)	N			0.90	
		$N+O(n^{-\frac{1}{2}})$			1.00	
100	(3,2.5,2.5,1)	N				0.70
		$N+O(n^{-\frac{1}{2}})$				0.88
100	(3,3,2,1)	N				0.70
		$N+O(n^{-\frac{1}{2}})$				0.88
		N				0.83
100	(3,3,3,1)	$N+O(n^{-\frac{1}{2}})$				0.98
		S				0.96
100	(3,3,1,1)	N			0.96	
		$N+O(n^{-\frac{1}{2}})$			1.00	

where $\bar{X} = N^{-1}(X_1+\ldots+X_N)$. We wish to test the hypothesis that X_i's come from a single multivariate normal population with covariance Σ against the hypothesis that they come from a mixture of $k \leq p$ such populations, differing in means. We assume $\Sigma = \sigma^2 I$. For k=2 the null hypothesis H_1 and the alternative hypothesis H_2 are given by

$$H_1: \quad X_i \sim N(\mu, \sigma^2 I)$$

$$H_2: \quad X_i \sim \pi N(\mu_1, \sigma^2 I) + (1-\pi) N(\mu_2, \sigma^2 I)$$

where π is the mixing probability. Under H_2 it is known (e.g. see Bryant (1975)) that

$$S \sim \sum_{j=0}^{N} \binom{N}{j} \pi^j (1-\pi)^{N-j} W_p(N-1, \sigma^2 I, M_j),$$

$$M_j = N^{-1} j(N-j)(\mu_1-\mu_2)(\mu_1-\mu_2)',$$

and M_j is of rank 1. Now, let $\Delta^2 = (\mu_1-\mu_2)'(\mu_1-\mu_2)/\sigma^2$ which is proportional to the nonzero root of $M_j/(N-1)$. When the null hypothesis is true, we know that

$$S \sim W_p(N-1, \sigma^2 I, 0).$$

Let the test statistics T_1 and T_2 be given as below:

$$T_1 = \ell_1/\sigma^2 \tag{5.1}$$

$$T_2 = \frac{(p-1)\ell_1}{\ell_2+\ldots+\ell_p} \tag{5.2}$$

where $\ell_1 \geq \ldots \geq \ell_p$ are the eigenvalues of $S/(N-1)$. If we use the statistics T_i to test H_1, then we accept or reject H_1 according as

$$T_i \lessgtr c_{\alpha i} \tag{5.3}$$

where

$$P\{T_i \leq c_{\alpha i} \mid H_1\} = (1-\alpha). \tag{5.4}$$

Let $f_j(\cdot)$ be the asymptotic density of a function of the eigenvalues of $S/(N-1)$ when j of the samples come from population 1. Under this condition $S \sim W_p(N-1, \sigma^2 I, M_j)$. So the unconditional asymptotic density function of the function of eigenvalue of S is

$$\sum_{j=0}^{N} \binom{N}{j} \pi^j (1-\pi)^{N-j} f_j(\cdot).$$

The following table gives a comparison of the asymptotic power value with the simulated value of tests of H_1 against H_2 for $p=4$, $\alpha=0.05$ and $\Delta = ||\underset{\sim}{\mu}_1-\underset{\sim}{\mu}_2||/\sigma$.

	Test		$\pi = .25$ $\Delta=1$	2	3	$\pi = .50$ $\Delta=1$	2	3
N=51	T_1		.02	.42	.95	.02	.63	1.00
		Simu.	.07	.45	.94	.07	.62	1.00
	T_2		.07	.68	.99	.11	.88	1.00
		Simu.	.11	.70	.98	.14	.86	1.00
N=101	T_1		.05	.76	1.00	.08	.96	1.00
		Simu.	.13	.78	1.00	.21	.96	1.00
	T_2		.13	.93	1.00	.22	1.00	1.00
		Simu.	.13	.93	1.00	.27	1.00	1.00

When $\Delta = 1$, the largest roots of $M_j/(N-1)$ are close to zero and the asymptotic expression does not give good approximation. Note, that if the means under H_2 are separated by more than two or three standard deviations, that is, for $\Delta = 2,3$, one may reasonably expect to detect the presence of two components, while if they are separated by less than two standard deviations the detection generally will not be good.

6. ASYMPTOTIC DISTRIBUTIONS OF FUNCTIONS OF THE ROOTS OF THE COMPLEX WISHART MATRIX

Let $Z = Z_1 + i\, Z_2$ be a $p\times n$ matrix and let the rows of $(Z_1' : Z_2')$ be distributed independently as multivariate normal with covariance matrix

$$\begin{pmatrix} \Sigma_1 & \Sigma_2 \\ -\Sigma_2 & \Sigma_1 \end{pmatrix}$$

and let the mean vector of j-th row of $(Z_1' : Z_2')$ be $\underset{\sim}{\mu}_j' = (\underset{\sim}{\mu}_j^{(1)'} : \underset{\sim}{\mu}_j^{(2)'})$. Also, let $\tilde{S} = Z\bar{Z}'$ where $\bar{Z}$ denotes the complex conjugate of Z. Then, the distribution of $\tilde{S}$ is known to be central or noncentral complex Wishart matrix with n degrees of freedom according as $\tilde{M} = 0$ or $\tilde{M} \neq 0$. where $M = \tilde{U}\bar{\tilde{U}}'$, $E(Z) = \tilde{U}$. The expected value of $\tilde{S}$ is given by

$$E(\tilde{S}) = 2n(\Sigma_1 - i\Sigma_2) + \tilde{M}$$

The matrix $\tilde{S}$ is Hermitian and the eigenvalues of $\tilde{S}/n$ are denoted by

$$\ell_1 \geq \ldots \geq \ell_p .$$

In the sequel, we assume that $E(\tilde{S}) = n$ diag. $(\lambda_1, \ldots, \lambda_p)$ and $\lambda_1 \geq \ldots \geq \lambda_p$. In addition, we assume that λ_i's have multiplicity as in (3.1). Now, let

$$L_j = \sqrt{n}\, \{T_j(\ell_1, \ldots, \ell_p) - T_j(\lambda_1, \ldots, \lambda_p)\} \tag{6.1}$$

for $j = 1, 2, \ldots, k$ and the function $T_j(\ell_1, \ldots, \ell_p)$ satisfy the assumptions (3.2) and (3.3). Then, following the same lines as in Section 3 for the real case, we obtain the following asymptotic expression for the joint density of $\underset{\sim}{L} = (L_1, L_2, \ldots, L_k)'$:

$$\begin{aligned} f(L_1, \ldots, L_k) = N(\underset{\sim}{L}, \tilde{Q}) \, [1 + \frac{1}{\sqrt{n}} \sum_{i=1}^{k} H_i(L)(\tilde{d}_1 + \tilde{d}_2) \\ + \frac{1}{\sqrt{n}} \sum_{i_1, i_2, i_3 = 1}^{k} H_{i_1 i_2 i_3}(\underset{\sim}{L})(\tilde{g}_1 + \tilde{g}_2 + \tilde{g}_3)] \\ + O(n^{-1}) \end{aligned} \tag{6.2}$$

where

$$\begin{aligned} \tilde{Q} = (\tilde{Q}_{i_1 i_2}), \tilde{Q}_{i_1 i_2} &= 4 \operatorname{tr} \tilde{R}_1^{(i_1)} \tilde{R}_1^{(i_2)} + 4 \operatorname{tr} \tilde{R}_2^{(i_1)} \tilde{R}_2^{(i_2)} \\ &+ 8 \operatorname{tr} \tilde{R}_1^{(i_1)} \tilde{\psi}^{(i_2)} \end{aligned}$$

and $\tilde{Q}$ is assumed to be nonsingular. Also,

$$\tilde{d}_1 = 4 \sum_{\alpha=1}^{r} \sum_{\beta \neq \alpha}^{r} \sum_{j_1 \in J_\alpha} \sum_{j_2 \in J_\beta} a_{i\alpha} \theta_{\alpha\beta}^{-1} (\tilde{\sigma}_{j_1 j_1}^{(1)} \tilde{\sigma}_{j_2 j_2}^{(1)} + \tilde{\sigma}_{j_1 j_1}^{(1)} \tilde{\nu}_{j_2 j_2} + \tilde{\sigma}_{j_2 j_2}^{(1)} \tilde{\nu}_{j_1 j_1}) \tag{6.3}$$

$$\tilde{d}_2 = 2 \sum_{\alpha=1}^{r} \sum_{\beta=1}^{r} \sum_{j_1 \in J_\alpha} \sum_{j_2 \in J_\beta} a_{i\alpha\beta} \, (\tilde{\sigma}_{j_1 j_2}^{(1)^2} + 2\tilde{\sigma}_{j_1 j_2}^{(1)} \tilde{\nu}_{j_1 j_2} - \tilde{\sigma}_{j_1 j_2}^{(2)^2}) \tag{6.4}$$

$$\begin{aligned} \tilde{g}_1 = \frac{8}{3} \operatorname{tr} \tilde{R}_1^{(i_1)} \tilde{R}_1^{(i_2)} \tilde{R}_1^{(i_3)} + 8 \operatorname{tr} \tilde{R}_1^{(i_1)} \tilde{R}_1^{(i_2)} \tilde{\psi}^{(i_3)} \\ + 8 \operatorname{tr} \tilde{R}_2^{(i_1)} \tilde{R}_2^{(i_2)} \tilde{R}_1^{(i_3)} - 8 \operatorname{tr} \tilde{R}_2^{(i_1)} \tilde{R}_2^{(i_2)} \tilde{\psi}^{(i_3)} \end{aligned} \tag{6.5}$$

$$\tilde{g}_2 = 16 \sum_{\alpha=1}^{r} \sum_{\beta\neq\alpha}^{r} \sum_{j_1\in J_\alpha} \sum_{j_2\in J_\beta} a_{i_1\alpha}\, \theta_{\alpha\beta}^{-1}[(\tilde{\Xi}_{j_1j_2}^{(i_2)} + \tilde{T}_{j_1j_2}^{(i_2)} + \tilde{T}_{j_2j_1}^{(i_2)} + \tilde{G}_{j_1j_2}^{(i_2)})$$

$$\times (\tilde{\Xi}_{j_1j_2}^{(i_3)} + \tilde{T}_{j_1j_2}^{(i_3)} + \tilde{T}_{j_2j_1}^{(i_3)} + \tilde{G}_{j_1j_2}^{(i_3)})$$

$$+ (\tilde{U}_{j_1j_2}^{(i_2)} - \tilde{U}_{j_2j_1}^{(i_2)})(\tilde{U}_{j_1j_2}^{(i_3)} - \tilde{U}_{j_2j_1}^{(i_3)})] \tag{6.6}$$

$$\tilde{g}_3 = 8 \sum_{\alpha=1}^{r} \sum_{\beta=1}^{r} \sum_{j_1\in J_\alpha} \sum_{j_2\in J_\beta} a_{i_1\alpha\beta}\, (\tilde{\Xi}_{j_1j_1}^{(i_2)} + 2\tilde{T}_{j_1j_1}^{(i_2)} + \tilde{G}_{j_1j_1}^{(i_2)})$$

$$\times (\tilde{\Xi}_{j_2j_2}^{(i_3)} + 2\tilde{T}_{j_2j_2}^{(i_3)} + \tilde{G}_{j_2j_2}^{(i_3)}) \tag{6.7}$$

where $C^{(i)} = \text{diag}\,(c_{i1},\ldots,c_{ip})$, $\Sigma_1 = (\tilde{\sigma}_{i_1i_2}^{(1)})$, $\Sigma_2 = (\tilde{\sigma}_{i_1i_2}^{(2)})$

$$\tilde{M} = \sum_{j=1}^{n} (\underset{\sim}{\mu}_j^{(1)} \underset{\sim}{\mu}_j^{(1)'} + \underset{\sim}{\mu}_j^{(2)} \underset{\sim}{\mu}_j^{(2)'})/2n = (\tilde{\nu}_{j_1j_2})$$

$$\tilde{R}_1^{(i)} = C^{(i)}\,\Sigma_1,\quad \tilde{R}_2^{(i)} = C^{(i)}\,\Sigma_2,\quad \tilde{\psi}^{(i)} = C^{(i)}\tilde{M}$$

$$\tilde{\Xi}^{(i)} = \Sigma_1\, C^{(i)}\,\Sigma_1,\quad \tilde{G}^{(i)} = \Sigma_2\, C^{(i)}\,\Sigma_2$$

$$\tilde{T}^{(i)} = \tilde{M}\, C^{(i)}\,\Sigma_1,\quad \tilde{U}^{(i)} = \tilde{M}\, C^{(i)}\,\Sigma_2$$

Krishnaiah and Lee (1977) derived the asymptotic joint distributions of the linear combinations and ratios of the linear combinations of the eigenvalues of the central complex Wishart matrix when the roots are simple. These results are special cases of the results in this section.

7. REFERENCES

[1] Bryant, P., On testing for clusters using the sample covariance, J. Multivariate Anal. 5 (1975) 96-105.

[2] Carter, E. M. and Srivastava, M. S., Asymptotic distributions of the latent roots of the non-central Wishart distribution and power of the L.R.T. for non-additivity, Technical Report No. 1, Department of Statistics, University of Toronto. (1979).

[3] Corsten, L. C. A. and Van Eijnsbergen, A. C., Multiplicative effects in two-way analysis of variance, Statistica Neerlandica 26 (1972) 61-68.

[4] Fujikoshi, Y., Asymptotic expansions for the distributions of some functions of the latent roots of matrices in three situations, J. Multivariate Anal. 8 (1978) 63-72.

[5] Ghosh, M. N. and Sharma, D., Power of Tukey's test for non-additivity, J. Roy. Statist. Soc. Ser. B 25 (1963) 213-219.

[6] Gollob, H. F., A statistical model which combines features of factor analysis and analysis of variance techniques, Psychometrika 33 (1968) 73-116.

[7] Izenman, A. J., Reduced-rank regression for the multivariate linear model, Ph.D. Thesis, Univ. of California, Berkeley (1972).

[8] Johnson, D. E. and Graybill, F. A., An analysis of a two-way model with interaction and no replication, J. Amer. Statist. Assoc. 67 (1972) 862-868.

[9] Kato, T., Perturbation Theory for Linear Operators (Springer-Verlag, New York, 1976).

[10] Krishnaiah, P. R. and Chang, T. C., On the exact distributions of the extreme roots of the Wishart and MANOVA matrices, J. Multivariate Anal. 1 (1971) 108-117.

[11] Krishnaiah, P. R. and Waikar, V. B., Simultaneous tests for equality of latent roots against certain alternatives - I, Ann. Inst. Statist. Math. 23 (1971) 451-468.

[12] Krishnaiah, P. R. and Schuurmann, F. J., On the evaluation of some distributions that arise in simultaneous tests for the equality of the latent roots of the covariance matrix, J. Multivariate Anal. 4 (1974) 265-282.

[13] Krishnaiah, P. R. and Lee, J. C., Inference on the eigenvalues of the covariance matrices of real and complex multivariate normal populations, in: Krishnaiah, P. R. (ed.), Multivariate Analysis-IV(North-Holland, Amsterdam, 1977).

[14] Krishnaiah, P. R. and Lee, J. C., On the asymptotic joint distribution of certain functions of the eigenvalues of four random matrices, J. Multivariate Anal. 9 (1979) 248-258.

[15] Krishnaiah, P. R., Computations of some multivariate distributions, in: Krishnaiah, P. R. (ed.), Handbook of Statistics 1 (North-Holland, Amsterdam, 1980).

[16] Krishnaiah, P. R. and Yochmowitz, M. G., Inference on the structure of interactions in two-way classification model, in: Krishnaiah, P. R. (ed.), Handbook of Statistics 1 (North-Holland, Amsterdam, 1980).

[17] Lawley, D. N., Tests of significance for the latent roots of covariance and correlation, Biometrika 43 (1956) 128-136.

[18] Mallows, C. L., Latent vectors of random symmetric matrices, Biometrika 48 (1961) 133-149.

[19] Mandel, J., Partitioning the interaction in analysis of variance, Journal of Research of the National Bureau of Standards B 73 B (1969) 309-328.

[20] Schuurmann, F. J., Krishnaiah, P. R. and Chattopadhyay, A. K., On the distributions of the ratios of the extreme roots to the trace of the Wishart matrix, J. Multivariate Anal. 3 (1973) 445-453.

[21] Srivastava, M. S. and Carter, E. M., Asymptotic distribution of latent roots and applications. To appear in the Proceedings of the Halifax Conference on Multivariate Statistical Analysis (1980).

[22] Tukey, J. W., One degree of freedom for non-additivity, Biometrika 5 (1949) 232-242.

[23] Williams, E. J., The interpretation of interactions in factorial experiments, Biometrika 39 (1952) 65-81.

[24] Yochmowitz, M. G. and Cornell, R. G., Stepwise tests for multiplicative components of interaction, Technometrics 20 (1978) 79-84.

STATISTICS AND RELATED TOPICS
M. Csörgö, D.A. Dawson, J.N.K. Rao, A.K.Md.E. Saleh (eds.)

ON EFFICIENT INFERENCE IN SYMMETRIC STABLE LAWS AND PROCESSES

Andrey Feuerverger and Philip McDunnough

Department of Statistics
University of Toronto

1. INTRODUCTION

Distributions which are limits (except for scaling and recentering) of sums of independent identically distributed variates are termed stable. When the variates have second moment the possible limits necessarily are Gaussian and this is the best known case of the central limit theorem. However the wider class of distributions that share in the central limiting feature coincides exactly with the stable laws. These laws, first obtained by P. Levy (1925, 1954) possess a natural interest for statistical applications and for robustness studies in particular. For applications to modelling telephone line noise, see Berger and Mandelbrot (1963), Stuck and Kleiner (1974). Applications to modelling price changes in various financial markets are given by Mandelbrot (1963), Fama (1965), Fielitz and Smith (1972), Samuelson (1975), and Leitch and Paulson (1975). The use of symmetric versus skewed stable distributions, for example, carries implications for investment strategy. Properties of the stable laws are discussed in Gnedenko and Kolmogorov (1954), Feller (1966), Lukacs (1970), and Holt and Crow (1973).

The stable distributions admit unimodal densities having all derivatives but in general these are available only as numerically awkward infinite series. This lack of a closed form for the density has impeded development of statistical methods for this distribution family though a number of ad hoc procedures have been developed. See for example Fama and Roll (1971), Press (1972), Paulson, Halcomb and Leitch (1975), Heathcote (1977), de Haan and Resnick (1980), Brockwell and Brown (1980).

In this paper, we also are concerned with inference for stable distributions, however our interest centers exclusively on procedures which are asymptotically efficient, or at least on procedures whose asymptotic efficiency can be made arbitrarily high. The first indication of efficient inference for the stable laws was given by DuMouchel (1973) who showed that the MLE's were consistent, asymptotically normal and followed the well-known theory for maximum likelihood inference. The maximum likelihood procedure was implemented by DuMouchel (1971, 1975) and subsequently in unpublished independent work by the authors. Because maximum likelihood is so technically cumbersome, alternative asymptotically efficient techniques remain of considerable interest. Since the stable characteristic functions are readily available it is natural to ask if inference can be based directly on these and whether or not efficient procedures exist. This question was studied in Feuerverger and McDunnough (1980, 1979) and has an affirmative answer. The results of these two papers show that under very general conditions statistical procedures based on the empirical characteristic function (e.c.f.) may be used for a wide class of statistical problems and that suitable ecf-based procedures have arbitrarily high asymptotic efficiency. To explore the applicability of these ideas to the problem of inference for the stable laws is one purpose of the present paper.

The outline of our paper is as follows. In §2 we present a new continuous reparametrization of the stable laws and a slight, but useful extension of a result due to Zolotarev. Certain essential properties of the stable laws are reviewed. In §3-4 we discuss certain numerical aspects of maximum likelihood for these distributions and we present some Monte Carlo results for the symmetric case. In §5 we discuss the ecf and methods of efficient inference in the Fourier domain. The application to stable laws is considered in §6 and some numerical results pertaining to grid selection are obtained. The various procedures discussed extend easily to discrete time linear stable processes and in §7 we provide a brief Monte Carlo study for the stationary AR(1) case. Some unusual results are noted. Our numerical work is confined throughout to the symmetric case; the methods however are entirely general.

2. SOME PROPERTIES OF STABLE DISTRIBUTIONS

The stable distributions are defined through

$$\log \phi(t) = \begin{cases} -|t|^{\alpha}\left\{1 + i\,\beta\,\mathrm{sgn}(t)\tan\left(\frac{\pi\alpha}{2}\right)\right\}, & \alpha \neq 1 \\ -|t|\left\{1 + i\,\frac{2}{\pi}\,\beta\,\mathrm{sgn}(t)\log|t|\right\}, & \alpha = 1 \end{cases} \tag{2.1}$$

where $\phi(t) = E(\exp(itX))$ is the characteristic function, and $0 < \alpha \leq 2$, $-1 \leq \beta \leq 1$ are shape and skewness parameters. For $\alpha \neq 1$ an alternative representation is sometimes used:

$$\log \phi(t) = -|t|^{\alpha} \exp\{-i\tfrac{\pi}{2}\beta'\,\mathrm{sgn}(t)\cdot K(\alpha)\} \tag{2.2}$$

where $K(\alpha) = 1 - |1-\alpha|$ and $-1 \leq \beta \leq 1$. The two representations disagree on scaling as well as skewness. The relation between β and β' is given in Lukacs (1970, pp. 136-8).

For applications involving the full nonsymmetric class the discontinuity at $\alpha = 1$ is troublesome. We shall show that this discontinuity may be removed. To do so we first write the $\alpha \neq 1$ term of (2.1) in the form

$$-|t|^{\alpha} - i\,\beta\,t|t|^{\alpha-1}\tan\frac{\pi\alpha}{2}\ . \tag{2.3}$$

If we now shift the mean by an amount $\beta\tan\frac{\pi\alpha}{2}$ we obtain

$$-|t|^{\alpha} - i\,\beta\,t(|t|^{\alpha-1} - 1)\tan\frac{\pi\alpha}{2} \tag{2.4}$$

or

$$-|t|^{\alpha} - i\,\beta^{*}h(t,\alpha) \tag{2.5}$$

where

$$h(t,\alpha) = \frac{t(|t|^{\alpha-1} - 1)}{\alpha-1} \tag{2.6}$$

and

$$\beta^{*} = \beta(\alpha-1)\tan\frac{\pi\alpha}{2}\ . \tag{2.7}$$

Some analysis now establishes that the function $h(t,\alpha)$ is continuous on R^2 and can be defined by continuity as $t\,\ell n|t|$ when $\alpha = 1$.

That the discontinuity at $\alpha = 1$ can be removed by reparametrization is known; see Chambers, Mallows and Stuck (1977) and DuMouchel (1971). The approach given here seems more direct, however, and results in a parameter domain whose shape (roughly elliptical but having corners at $\alpha = 0,2$) is more consistent with the known behaviour of the densities near $\alpha = 0,2$ for varying skewness. The parametrizations β, β' and β^{*} are in 1-1 correspondence through (2.7) and

$$\beta^* = (1-\alpha)\tan\frac{\pi K(\alpha)\beta'}{2} \quad . \tag{2.8}$$

By verifying that the variation of the difference between characteristic functions within a shrinking neighbourhood in $(\alpha,\beta^*,\mu,\sigma)$ approaches zero we may prove:

Theorem 2.1. Let $p_{\alpha\beta^*}(x)$ be the density corresponding to (2.5). Then the family $\left\{\frac{1}{\sigma}p_{\alpha\beta^*}\left(\frac{x-\mu}{\sigma}\right)\right\}$ varies continuously in the sup-norm over the domain of its $(\alpha,\beta^*,\mu,\sigma)$ parametrization.

The Bergstrom-Feller expansions for the stable densities may be written as follows: if $0<\alpha<1$, then

$$f_{\alpha\beta'}(x) = \frac{1}{\pi x}\sum_{k=1}^{\infty}\frac{(-1)^{k-1}\Gamma(\alpha k+1)}{k!}\sin\left[\frac{\pi k\alpha}{2}\{\beta'+\operatorname{sgn}x\}\right]\cdot|x|^{-k\alpha} \tag{2.9}$$

and if $1<\alpha\le 2$, then

$$f_{\alpha\beta'}(x) = \frac{1}{\pi x}\sum_{k=1}^{\infty}\frac{(-1)^{k-1}\Gamma\left(\frac{k}{\alpha}+1\right)}{k!}\sin\left[\frac{\pi k}{2}\left\{\left(\frac{2-\alpha}{\alpha}\right)\beta'+\operatorname{sgn}x\right\}\right]\cdot|x|^{k} \quad . \tag{2.10}$$

For the asymptotic character of these series and for the case $\alpha=1$ we refer to Lukacs (1970), §5.8 and 5.9. We remark that from a statistical viewpoint expansions for $\log f$ would be of greater interest.

Several useful observations about (2.9) and (2.10) do not seem to have been made previously. First note that (2.10) converges also if $\alpha>2$ and secondly that both series remain convergent for arbitrary $-\infty<\beta'<\infty$. The series for $0<\alpha<1$ and $1<\alpha<\infty$ (and β' arbitrary) are closely related and we have the following slightly generalized form of a result due to Zolotarev: For $0<\alpha<\infty$, but $\alpha\ne\frac{1}{2},1$ we have

$$f_{\alpha\beta'}(x) = x^{-1}|x|^{-\alpha}f_{\alpha^{-1},\beta''}(|x|^{-\alpha}) \tag{2.11}$$

where

$$\beta'' = \begin{cases}\dfrac{\alpha\beta'+(\alpha-1)\operatorname{sgn}x}{2\alpha-1} & \text{if } 0<\alpha<1\\[2ex] (2-\alpha)\beta'+(\alpha-1)\operatorname{sgn}x & \text{if } 1<\alpha<\infty \quad .\end{cases}$$

The failure of (2.11) at $\alpha=\frac{1}{2}$ is, curiously, of an inessential kind. For if we reparametrize in (2.10) by replacing $\frac{2-\alpha}{\alpha}\beta'$ by β''' then the expansion (2.9) for some $0<\alpha<1$ and some β' will be related to the expansion (2.10) for α^{-1} and $\beta'''=\alpha\beta'+(\alpha-1)\operatorname{sgn}x$. This relation does not fail at $\alpha=\frac{1}{2}$. The significance of (2.11) is that numerical evaluation for any $0<\alpha<1$ (except $\alpha=\frac{1}{2}$) may be replaced by evaluation at α^{-1}; for small α discrete Fourier transformation is very difficult.

3. MAXIMUM LIKELIHOOD BY INVERSION

Our approach to maximum likelihood estimation for the stable laws is an application of the fast Fourier transform (FFT) algorithm (Cooley and Tukey, 1965). Suppose ϕ is an integrable characteristic function; then the evaluation of the integral

$$f(x) = \int_{-\infty}^{\infty}\phi(t)\,e^{-itx}\,dt \tag{3.1}$$

by means of the FFT effectively restricts x to values on an equispaced grid such as 0 , $\pm \Delta x$, $\pm 2\Delta x$,... . If the FFT is based on N points the available range for the density will be $\pm N \cdot \Delta x/2$ with one end-point missing; the corresponding spacing for the frequency variable will then be $\Delta t = 2\pi/N \cdot \Delta x$ and the range will be, not $\pm \pi/\Delta x$, but rather $\pm 2\pi/\Delta x$ with endpoints excluded. The range for t is doubled in this way due to the fact that $\phi(-t) = \overline{\phi(t)}$ so that we have the identity

$$\sum_{n=-N+1}^{N-1} \phi(\Delta t \cdot n) e^{-i\lambda n} = 2 \operatorname{Re}\left\{ \sum_{n=0}^{N-1} \phi_0 (\Delta t \cdot n) e^{-i\lambda n} \right\}$$

where ϕ_0 is identical to ϕ except that $\phi_0(0) = \tfrac{1}{2}$

Now, given a complex sequence $X(0), X(1), \ldots, X(N-1)$, the FFT algorithm produces the sequence $\sum_{n=0}^{N-1} X(n)\, e^{-i\lambda n}$ for $\lambda = \frac{2\pi s}{N}$, $s = 0, 1, \ldots, (N-1)$. Therefore if the FFT is applied to the sequence

$$\tfrac{1}{2} \,,\ \phi(\Delta t) \,,\ \phi(2\Delta t) \,,\ldots,\ \phi\big((N-1) \cdot \Delta t\big) \quad ,$$

and if the real parts of the resulting sequence are multiplied by $2/N \cdot \Delta x$ we obtain - except for the effects of truncation and discretization of the integral - values of the density

$$f(0),\ f(\Delta x),\ f(2\Delta x), \ldots,\ f(-2\Delta x),\ f(-\Delta x) \quad .$$

Note the circular format with values for f on the negative axis occuring at the end of the sequence.

The effect of truncation (e.g. Brillinger, 1975, Ch. 3) is that we obtain a convoluted form of the transform required. One possibility would be to use a tapering function; the one due to Bohman (1960) seems especially appropriate. DuMouchel (1971, p. 35) gives a better resolution and shows how the truncation effect may be eliminated using a "wrapped summation" method. In our work we used a 10% cosine taper (Tukey, 1967) and found this gave satisfactory results provided α was not less than about 0.5 .

The effect of discretization (e.g. Brillinger, 1975, §5.11) is that we obtain an aliased version $\sum_{j=-\infty}^{\infty} f(x + j N \Delta x)$ of the density. Since $N \Delta x$ typically is not small, de-aliasing could be achieved using the asymptotic expansion for $|x| \to \infty$; in fact this same expansion is required also for the tails where the inversion is inaccurate numerically. DuMouchel (1971) replaced the Fourier integral by $N/2$ intervals and quadratically interpolated $\phi(t)$ in each interval (Filon's method). Our approach to de-aliasing was based on the fortuitously discovered observation - which is supported by simple numerical arguments - that within a suitable range the aliasing error essentially is constant and thus may be determined immediately by subtracting the known exact expression $f_{\alpha,\beta'}(0) = \pi^{-1}\Gamma(1 + \alpha^{-1})$ for the density at $x = 0$ from the corresponding FFT determined value. Further details about this as well as other aspects of this paper may be found in an unpublished technical report by the authors.

4. MAXIMUM LIKELIHOOD SIMULATION STUDY

Using the methods in §3 a maximum likelihood procedure was developed in FORTRAN on the University of Toronto IBM 360/170. Versions for both the symmetric and non-symmetric laws were produced but only the symmetric case was subjected to simulation. A noteworthy feature of the programs, particularly for comparison with the ecf procedures of §5-6, is their essential technical complexity and the considerable programming effort required. On this point, see also DuMouchel (1971, 1974).

The stable variates required for the simulation study were generated using the algorithm RSTAB (Chambers, Mallows and Stuck, 1976) corrected for an error noticed in the published function D2: the fourth DATA line should read "&,.18001 33704 07390 023 D3" (cf. approximation 1801 in Hart et. al., 1968). The FFT subroutine used was the November 1967, Bell Laboratories, Murray Hill version of AR1DFT written by W.M. Gentleman and G. Sande. The FFT was used to obtain the standardized density at the current estimate $\hat{\alpha}$ as well as at $\hat{\alpha} \pm \Delta\alpha$ where $\Delta\alpha = 0.025$. The data was then subjected to standardization at the current estimates $\hat{\mu}$ and $\hat{\sigma}$ and as well at $\hat{\mu} \pm \Delta\mu$, $\hat{\sigma} \pm \Delta\sigma$ where $\Delta\mu = .05$ and $\Delta\sigma = .05$. In this manner, the likelihood was calculated on the subset of the $3 \times 3 \times 3$ grid of parameter points needed in order to perform a Newton-Raphson procedure. We used $\Delta x = 0.1$ and $N = 1024$ and found single precision adequate. For $|x| \le 7.5$ values for the density were obtained using quadratic interpolation on the FFT values which then were corrected for aliasing as described in §3. For $|x| > 7.5$ values of the density were determined using three terms of the asymptotic series. These values and rules were determined after numerical experimentation.

The simulation study spanned the values $\alpha = 0.7\ (.1)\ 1.9$ and sample sizes $N = 50, 100, 200$ of standard variates, with each sample being used once only. (Note N now is no longer the FFT length.) The initial estimates were taken to be the actual true values and five full iterations of the Newton-Raphson procedure were carried out for sample sizes $N = 50, 100$, and four iterations for $N = 200$. For $N = 100$ and 200, fifty trials $(n = 50)$ were conducted and for $N = 50$ we conducted $n = 100$ trials.

Table 4.1 summarizes the results of this Monte Carlo. For the three parameters in each cell we give the sample average and the sample standard deviation of the estimates resulting from the n trials. We also give the value n for the number of trials; whenever n differs from the value declared at the top of the table, this indicates failures of the MLE procedure to terminate normally. This could occur if the latest update exceeded the boundaries of the parameter space or if an unacceptable level of numerical instability was detected. The results of table 4.1 are in good agreement with the asymptotic calculations given in DuMouchel (1975). A detailed discussion appears in the technical report mentioned above.

5. FOURIER METHODS FOR INFERENCE

Suppose $X_1, X_2, \ldots, X_n$ are iid variates with density in $\{f_\theta(x)\}$ where θ is a real univariate parameter. The equation of maximum likelihood may be written in the form

$$\int_{-\infty}^{\infty} \frac{\partial \log f_\theta(x)}{\partial \theta}\, d\big(F_n(x) - F_\theta(x)\big) = 0 \tag{5.1}$$

where F_θ is the cdf of f_θ and $F_n(x)$ is the empirical cdf. Define now the following transformed quantities: the characteristic function

$$c_\theta(t) = \int e^{itx}\, d\,F_\theta(x) \; ; \tag{5.2}$$

the empirical characteristic function (ecf)

TABLE 4.1
Simulation Results for Maximum Likelihood Estimation for Symmetric Stable Laws

		N = 50 n = 100 (# iter = 5)		N = 100 n = 50 (# iter = 5)		N = 200 n = 50 (# iter = 4)	
α = 1.9	μ	-	-	.030	.119	-.035	.122
	α	-	-	-	-	1.870	.063
	σ	-	-	-	-	.987	.062
	n	26		23		30	
α = 1.8	μ	-	-	-.046	.137	-.023	.103
	α	-	-	1.780	.098	1.758	.094
	σ	-	-	.987	.087	.987	.064
	n	53		36		41	
α = 1.7	μ	.018	.247	.048	.133	.021	.097
	α	-	-	1.629	.124	1.689	.097
	σ	-	-	.985	.094	.993	.077
	n	68		43		45	
α = 1.6	μ	.008	.174	.002	.128	.017	.101
	α	1.582	.173	1.593	.160	1.608	.119
	σ	.997	.159	1.005	.088	.995	.066
	n	90		46		50	
α = 1.5	μ	.019	.233	.034	.113	-.037	.116
	α	1.492	.162	1.474	.154	1.503	.107
	σ	.989	.152	.983	.136	.989	.069
	n	90		50		50	
α = 1.4	μ	.036	.221	-.002	.177	.012	.110
	α	1.434	.183	1.391	.149	1.399	.092
	σ	1.028	.162	.959	.101	1.006	.056
	n	93		50		50	
α = 1.3	μ	-.024	.223	-.005	.152	.011	.112
	α	1.302	.194	1.320	.137	1.324	.106
	σ	.970	.163	.983	.123	1.105	.094
	n	93		50		50	
α = 1.2	μ	.020	.205	-.054	.168	.001	.123
	α	1.229	.187	1.191	.139	1.209	.092
	σ	1.012	.179	1.010	.132	1.003	.094
	n	93		50		50	
α = 1.1	μ	-.022	.189	.031	.163	.021	.093
	α	1.146	.139	1.088	.109	1.107	.096
	σ	1.039	.169	.995	.136	1.005	.097
	n	88		49		50	
α = 1.0	μ	-.013	.198	-.005	.135	-.007	.101
	α	1.061	.158	1.003	.123	.997	.077
	σ	.989	.188	1.022	.161	1.000	.108
	n	89		50		50	
α = .9	μ	-.018	.187	-.032	.149	-.014	.084
	α	.958	.140	.911	.102	.927	.081
	σ	.996	.203	1.015	.141	1.033	.104
	n	82		50		50	
α = .8	μ	-	-	-.038	.146	.007	.096
	α	-	-	.808	.084	.804	.052
	σ	-	-	.992	.158	1.028	.136
	n	72		46		50	
α = .7	μ	-	-	.031	.122	.015	.083
	α	-	-	.740	.059	.717	.043
	σ	-	-	1.026	.168	1.025	.111
	n	39		32		42	

$$c_n(t) = \int e^{itx}\, d\,F_n(x) = \frac{1}{n}\sum_{j=1}^{n} e^{itx_j} \tag{5.3}$$

and the inverse transform of the score

$$w_\theta(t) = \frac{1}{2\pi}\int_{-\infty}^{\infty} \frac{\partial \log f_\theta(x)}{\partial\theta}\, e^{-itx}\,dx \ . \tag{5.4}$$

Our starting point is to note that under very general conditions, we may apply the Parseval theorem to (5.1) to obtain

$$\int_{-\infty}^{\infty} w_\theta(t)\,\big(c_n(t) - c_\theta(t)\big)\,dt = 0 \ . \tag{5.5}$$

This is the Fourier domain version of the likelihood equation. Note that $w_\theta(t)$ must usually be regarded as a generalized function. (The multiparameter extension is straightforward.) This result at once suggests that the empirical characteristic function may have valuable applications; procedures based on the ecf and their asymptotic efficiency were explored in Feuerverger and McDunnough (1980, 1979).

Consider the process $c_n(t)$. We are indebted to R.A. Mureika for pointing out the following result which is due to an anonymous referee and which generalizes a result of Feuerverger and Mureika (1977):

Theorem 5.1: If F is any distribution function and $\log T_n = o(n)$ then

$$P\{\lim_{n\to\infty}\ \sup_{|t|\le T_n} |c_n(t) - c(t)| = 0\} = 1 \ .$$

Proof. For fixed $\varepsilon > 0$ let A be such that $F(-A) + 1 - F(A) \le \varepsilon$, and replace $c_n(t) - c(t)$ by

$$\gamma(t,X_1,\ldots,X_n) \equiv \frac{1}{n}\sum_{|X_j|\le A} e^{itx_j} - \int_{[-A,A]} e^{itx}\,d\,F(x) \ .$$

Clearly

$$|\gamma(t,X_1,\ldots,X_n) - \gamma(t',X_1,\ldots,X_n)| \le |t_1-t_2|\,A \ .$$

Thus, it suffices to show that

$$\lim_{n\to\infty}\ \sup_{|k|\le \frac{AT_n}{\varepsilon}} \left|\gamma\left(\frac{k\varepsilon}{A}, X_1,\ldots,X_n\right)\right| = 1 \quad \text{w.p.1.}$$

But

$$P\left\{\sup_{|k|\le \frac{AT_n}{\varepsilon}} \left|\gamma\left(\frac{k\varepsilon}{A}, X_1,\ldots,X_n\right)\right| \ge \varepsilon\right\}$$

$$\le \frac{AT_n}{\varepsilon}\ \sup_t P\{|\gamma(t,X_1,\ldots,X_n)| \ge \varepsilon\} \ .$$

The result follows using standard exponential bounds: the latter probability decreases exponentially, because $n \cdot \gamma$ is the sum of n iid bounded random variables with mean zero. □

Define $Y_n(t) = \sqrt{n}\big(c_n(t) - c(t)\big)$. As $c_n(t)$ is a sum of iid bounded processes we have at once $EY_n(t) = 0$ and $\operatorname{cov}(Y_n(s), Y_n(t)) = EY_n(s)\overline{Y_n(t)} = c(s-t) - c(s)\overline{c(t)}$ and the covariance structure of the real and imaginary parts:

$$\begin{aligned}
\operatorname{Cov}(\operatorname{Re} Y_n(s), \operatorname{Re} Y_n(t)) &= \tfrac{1}{2}[\operatorname{Re} c(s-t) + \operatorname{Re} c(s+t)] - \operatorname{Re} c(s) \operatorname{Re} c(t) \\
\operatorname{Cov}(\operatorname{Re} Y_n(s), \operatorname{Im} Y_n(t)) &= \tfrac{1}{2}[\operatorname{Im} c(s-t) + \operatorname{Im} c(s+t)] - \operatorname{Re} c(s) \operatorname{Im} c(t) \\
\operatorname{Cov}(\operatorname{Im} Y_n(s), \operatorname{Im} Y_n(t)) &= \tfrac{1}{2}[\operatorname{Re} c(s-t) - \operatorname{Re} c(s+t)] - \operatorname{Im} c(s) \operatorname{Im} c(t) \quad .
\end{aligned} \tag{5.6}$$

Let $Y(t)$ be a zero mean complex Gaussian process having covariance structure identical to Y_n. By the central limit theorem Y_n converges in distribution to Y at finite numbers of points. Feuerverger and Mureika (1977) prove the weak convergence of $Y_n(t)$ to $Y(t)$ in any finite interval provided that $E|X|^{1+\delta} < \infty$. Csorgo (1980) shows that the moment condition is not easily removed and gives a general treatment of convergence questions. Necessary and sufficient conditions for the weak convergence are given by Marcus (1980). Feuerverger and McDunnough (e.g. lemma 2.1 of 1980) show that weak convergence of the ecf process is not critical for many statistical purposes since one can exploit the essentially simple stochastic structure of $c_n(t)$ to obtain needed results. A quadratic version of the quoted lemma 2.1 may be proved upon evaluating the limiting cumulants:

<u>Lemma 5.2</u> Let $A(t_1,t_2)$ be a function having bounded variation on R^2. Then

$$n \cdot \iint (c_n(t_1) - c(t_1))(c_n(t_2) - c(t_2)) \, A(dt_1,dt_2) \xrightarrow{\mathcal{D}} \iint Y(t_1)\, Y(t_2)\, A(dt_1,dt_2) \quad .$$

The result holds also if matching factors in the integrands are conjugated.

Turning to inference, a comprehensive discussion of asymptotically efficient or arbitrarily highly efficient procedures based on the ecf is given in Feuerverger and McDunnough (1980, 1979). Here we emphasize only two of these - the harmonic-regression procedure, and the k-L procedure, both of "discrete type". The k-L procedure is so-called because it is of likelihood type, and based on a fixed number k of ecf points. Let $0 < t_1 < \ldots < t_k$ be this fixed grid. Define $\underset{\sim}{z}_0 = (\operatorname{Re} c(t_1), \ldots, \operatorname{Re} c(t_k), \operatorname{Im} c(t_1), \ldots, \operatorname{Im} c(t_k))'$ and let $\underset{\sim}{z}_n$ be its empirical counterpart. Letting $n^{-1} \Sigma$ be the covariance matrix of $\underset{\sim}{z}_n$, the entries of Σ will be given by (5.6). The k-L procedure estimates a vector parameter $\underset{\sim}{\theta}$ by maximizing the asymptotic normal form of the log-likelihood of $\underset{\sim}{z}_n$. This may be taken either as

$$-\tfrac{1}{2} \log \det \Sigma - \tfrac{n}{2}(\underset{\sim}{z}_n - \underset{\sim}{z}_\theta)' \Sigma^{-1} (\underset{\sim}{z}_n - \underset{\sim}{z}_\theta)$$

or as just the second term of this expression. Under very general conditions, the asymptotic efficiency of the k-L procedure can be made arbitrarily close to the Cramer-Rao bound by selecting the grid $\{t_j\}$ to be sufficiently fine and extended. The harmonic-regression procedure is equivalent, asymptotically, to the k-L procedure and involves finding $\underset{\sim}{\theta}$ by fitting $\underset{\sim}{z}_\theta$ to $\underset{\sim}{z}_n$ using nonlinear least squares and any consistent estimate of the asymptotically optimal weights. This is carried out using a first order expansion for $\underset{\sim}{z}_\theta$; to preserve the asymptotic properties, a single iteration starting from any consistent estimates suffices. The ease with which this procedure may be implemented contrasts sharply with the methods of §3.

The harmonic regression procedure was implemented for the parametrization

$\underset{\sim}{\theta} = (\mu,\sigma,\alpha)$ for the symmetric stable laws: $c_\theta(t) = e^{i\mu t} \cdot e^{-|\sigma t|^\alpha}$. We used centered variates $\hat{X}_j = (X_j - \hat{\mu})/\hat{\sigma}$ where $\hat{\mu}, \hat{\sigma}$ were current estimates: for symmetric families this gives a convenient block-diagonal structure for $\underset{\sim}{\Sigma}$. In particular the 2k×3 regression may then be separated into a $k \times 2$ and a $k \times 1$ regression - the former involving only α and σ, and the latter only μ. The procedure was tested extensively and proved to be well behaved, however, because the question of optimal gridpoints is not resolved we did not carry out a Monte Carlo study for this procedure.

6. FOURIER GRIDPOINTS FOR THE SYMMETRIC STABLE LAWS

For fixed k the optimal $\{t_j\}$ depends on the unknown $\theta = (\mu,\sigma,\alpha)$. Since iteration takes place at standardized variates, we may presume without loss of generality that $\mu = 0$, $\sigma = 1$; the dependence on α is therefore the one of greatest interest. In general, for fixed k we propose to use those points which minimize the asymptotic variances (at the current estimates). For arbitrary spacing, the asymptotic covariance of the estimators is $\underset{\sim}{G}' \underset{\sim}{\Sigma}^{-1} \underset{\sim}{G}$ where $\underset{\sim}{G} = \left.\left(\frac{\partial \underset{\sim}{z}_\theta}{\partial\mu}, \frac{\partial \underset{\sim}{z}_\theta}{\partial\sigma}, \frac{\partial \underset{\sim}{z}_\theta}{\partial\alpha}\right)\right|_{\theta = (0,1,\alpha)}$. Actually the optimal $\{t_j\}$ depends on which parameter variance, or which joint-criterion we wish to minimize (the determinant of the covariance matrix being one possibility). Now the updating algorithm is such that the μ adjustment depends only on the imaginary ecf, while the σ, α adjustments depend only on the real part. (Note that the increased computational burden in using distinct grids for the real and imaginary components is very slight. The components for the imaginary parts covariances in (5.6) will now differ from those for the real parts.) The important tradeoff therefore takes place on the real axis between σ and α.

The question of optimal gridpoints requires numerical treatment. For the case of uniform spacing we refer to Feuerverger and McDunnough (1979), especially table 1. According to these results, α is the parameter least amenable to uniform spacing, particularly for small α, and is hence the parameter of greatest interest here. Adjustments were made to the programme which calculates the asymptotic variances. By an iterative procedure of arbitrary starts, sequential optimization using steepest ascent on lattices, and further checking, grid-points were obtained which appear to be optimal (asymptotically) for the estimation of α. These were obtained for $k = 2, 3, 4$ and 5 with $\alpha = 1.0\ (.1)\ 1.9$ and are given in table 6.1.

Table 6.2 compares the asymptotic values of $n \cdot \mathrm{var}(\hat{\alpha})$ for α-optimal spacing of $k = 2, 3, 4$ and 5 points. The Cramer-Rao bound value ranges shown are determined from DuMouchel (1975). It may be noted that the change from optimal *uniform* spacing to optimal spacing involves a sharp improvement in the asymptotic efficiencies. As before, however, for fixed k, efficiencies are seen to decrease with α; in particular, for $k = 5$ the efficiency is seen to drop below 90% for $\alpha < 1.1$. We may remark that while the *optimal* spacings become costlier and more difficult to determine as k increases, the results of table 6.1 provide a useful guide in determining good spacings for larger k. In practice there are no special difficulties in using values of $k = 10$ or even 20. For further results and discussion (and an indication of the intrinsic complexity of the k-dimensional surfaces optimized here) we refer the reader to our technical report.

7. A SIMULATION STUDY FOR AR(1) STABLE PROCESSES

The closure under convolutions property of the stable distributions provides a

α	k = 2		k = 3			k = 4				k = 5				
	t_1	t_2	t_1	t_2	t_3	t_1	t_2	t_3	t_4	t_1	t_2	t_3	t_4	t_5
1.0	.09	1.83	.03	.14	1.73	.03	.16	1.48	2.43	.016	.07	.20	1.4	2.4
1.1	.11	1.70	.04	.17	1.62	.04	.18	1.39	2.22	.022	.09	.23	1.3	2.1
1.2	.13	1.59	.05	.19	1.53	.05	.21	1.30	2.03	.026	.10	.25	1.3	2.0
1.3	.15	1.50	.06	.22	1.44	.07	.24	1.23	1.87	.032	.12	.28	1.2	1.8
1.4	.18	1.41	.07	.24	1.36	.08	.26	1.17	1.72	.042	.14	.31	1.1	1.7
1.5	.20	1.33	.09	.27	1.29	.048	.15	.31	1.28	.050	.16	.33	1.1	1.6
1.6	.22	1.26	.10	.29	1.22	.058	.17	.33	1.20	.062	.18	.36	1.0	1.4
1.7	.24	1.17	.12	.32	1.14	.070	.19	.36	1.12	.074	.20	.37	1.0	1.3
1.8	.25	1.09	.13	.33	1.05	.082	.21	.38	1.04	.082	.21	.39	0.9	1.1
1.9	.27	0.95	.15	.35	0.92	.100	.23	.40	.90	.096	.22	.38	1.0	1.6

TABLE 6.1 α-optimal spacings for k = 2 to k = 5

TABLE 6.2

Asymptotic values of $N \cdot VAR(\hat{\alpha})$
for α-optimal spacings

α	k = 2	k = 3	k = 4	k = 5
1.0	1.8327	1.5890	1.4637	1.393
1.1	2.0993	1.8426	1.7207	1.648
1.2	2.3432	2.0826	1.9710	1.898
1.3	2.5478	2.2924	2.1974	2.127
1.4	2.6918	2.4524	2.3774	2.314
1.5	2.7508	2.5363	2.4831	2.427
1.6	2.6942	2.5120	2.4691	2.434
1.7	2.4839	2.3402	2.3079	2.290
1.8	2.0698	1.9682	1.9464	1.940
1.9	1.3721	1.3152	1.3038	1.284

class of highly tractable linear stationary processes. With an MLE algorithm already available, it is not very difficult to carry out maximum likelihood estimation for stable autoregressive processes. We give here the results of a brief simulation study for the AR(1) case.

Suppose $X(t)$ is a discrete stationary process satisfying

$$\{X(t) - \mu\} = a\{X(t-1) - \mu\} + e(t)$$

where the $e(t)$ are iid variates of the form σS_α where S_α is the standardized symmetric stable law. Then

$$\{X(t) - \mu\} = \sum_{j=0}^{\infty} a^j e(t-j) \sim \sigma(1 - |a|^\alpha)^{-\frac{1}{\alpha}} S_\alpha \ .$$

The joint density for a length T of the stationary process

$$f(x_1, x_2, \ldots, x_T) = f(x_1) \sum_{j=2}^{T} f(x_j | x_{j-1})$$

$$= \frac{1}{\sigma^*} f_\alpha \left(\frac{x_1 - \mu}{\sigma^*}\right) \frac{1}{\sigma^{T-1}} \prod_{j=2}^{T} f_\alpha \left(\frac{(x_j - \mu) - a(x_{j-1} - \mu)}{\sigma}\right)$$

where $\sigma^* = \sigma(1 - |a|^\alpha)^{-1/\alpha}$ and where f_α is the density of the standard stable variate. We may therefore easily carry out the full (unconditional) MLE procedure for such AR processes by adapting the methods of §3-4.

Tables 7.1 and 7.2 summarize the results of a limited simulation study for the cases $\alpha = 1.9, 1.8, 1.6, 1.0$ and $a = 0, .25, .5$ for series lengths of $T = 200$ and $T = 500$. The data series used were constructed with $\mu = 0$, $\sigma = 1$ and initial estimates were taken at the true values and followed by five Newton-Raphson iterations. Convergence was, in general, extremely rapid. We carried out $n = 25$ trials for each cell; values $n < 25$ indicate that some trials did not terminate normally (see §4). As before, the tables show the mean and standard deviation of the n trials for each parameter.

Of particular interest, though not apparent in these tables, is an instability in the $Var(\hat{a})$ values produced by the MLE procedure, with many of these values being exceedingly small. This phenomenon occurs because the accuracy of estimates for AR coefficients clearly is conditional on whether or not there are some extreme outliers present to help us. This is a numerical confirmation of the limiting infinite Fisher information per observation for this parameter: see Hannan and Kanter (1977) and Kanter and Steiger (1974). A related observation is made by Cox (1966).

TABLE 7.1
Simulation Results for Symmetric Stable Time Series of Length T = 200 with 25 Trials per Cell and 5 Iterations per Trial

		a = 0		a = .25		a = .5	
$\alpha = 1.9$	μ	.013	.109	-.054	.166	-.017	.153
	α	1.877	.051	1.853	.055	1.859	.064
	σ	.984	.049	1.003	.045	.981	.046
	a	.024	.042	.248	.066	.499	.037
	n	16		18		18	
$\alpha = 1.8$	μ	.032	.090	-.005	.105	.011	.082
	α	1.792	.132	1.815	.082	1.826	.115
	σ	.998	.057	.989	.067	.990	.058
	a	.000	.052	.237	.072	.486	.041
	n	25		25		25	
$\alpha = 1.6$	μ	-.021	.129	.007	.131	.039	.275
	α	1.633	.109	1.609	.107	1.644	.112
	σ	.996	.048	1.014	.061	1.005	.061
	a	.014	.049	.256	.043	.499	.044
	n	25		25		25	
$\alpha = 1.0$	μ	-.032	.109	-.048	.133	.023	.190
	α	1.007	.080	.986	.084	1.008	.065
	σ	.985	.082	.989	.087	1.005	.102
	a	-.003	.009	.251	.008	.501	.007
	n	23		22		25	

TABLE 7.2
Simulation Results for Symmetric Stable Time Series of Length T = 500 with 25 Trials per Cell and 5 Iterations per Trial

		a = 0		a = .25		a = .5	
$\alpha = 1.9$	μ	.024	.074	.001	.094	-.011	.140
	α	1.895	.039	1.879	.048	1.887	.056
	σ	1.012	.044	.994	.031	.998	.049
	a	.013	.053	.247	.044	.504	.032
	n	19		18		20	
$\alpha = 1.8$	μ	.023	.173	.023	.080	.036	.127
	α	1.783	.063	1.807	.074	1.824	.069
	σ	.999	.033	1.003	.049	.984	.037
	a	.005	.029	.253	.032	.507	.035
	n	25		25		25	
$\alpha = 1.6$	μ	.016	.052	.007	.092	.002	.166
	α	1.612	.071	1.598	.061	1.604	.062
	σ	.991	.039	1.001	.056	1.005	.038
	a	.000	.023	.254	.031	.494	.019
	n	25		25		24	
$\alpha = 1.0$	μ	-.006	.072	.017	.092	-.030	.146
	α	1.013	.052	1.010	.046	1.006	.032
	σ	1.013	.063	1.014	.045	.980	.057
	a	.000	.005	.249	.006	.501	.003
	n	25		25		25	

Acknowledgments
The research of both authors is supported by NSERC. The manuscript was typed by Margot Burrows.

8. REFERENCES

[1] Berger, J.M. and Mandelbrot, B., A new model for error clustering in telephone circuits, I.B.M. Journal of Research and Development 7 (1963) 224-236.

[2] Bohman, H., Approximate Fourier analysis of distribution functions, Ark. Mat. 4 (1960) 99-157.

[3] Brillinger, D.R., Time Series: Data analysis and theory (Holt, Rinehart and Winston, New York, 1975).

[4] Brockwell, P.J. and Brown, B.M., Efficient estimation for the positive stable laws. Submitted for publication.

[5] Chambers, J.M., Mallows, C.L. and Stuck, B.W., A method for simulating stable random variables, J. Amer. Statist. Assoc. 71 (1976) 340-344.

[6] Cooley, J.W. and Tukey, J.W., An algorithm for the machine calculation of complex Fourier series, Math. Comp. 19 (1965) 297-301.

[7] Cox, D.R., The null distribution of the first serial correlation coefficient, Biometrika 53 (1966) 623-626.

[8] Csorgo, S., Limit behaviour of the empirical characteristic function, Annals Probab., to appear.

[9] de Haan, L. and Resnick, S.I., A simple asymptotic estimate for the index of a stable distribution, J. Royal Statist. Soc. (B) 42 (1980) 83-87.

[10] DuMouchel, W.H., Stable distributions in statistical inference: 2 - Information from stably distributed samples, J. Amer. Statist. Assoc. 70 (1975) 386-393.

[11] DuMouchel, W.H., Stable distributions in statistical inference: 3 - Estimation of the parameter of a stable distribution by the method of maximum likelihood. Unpublished manuscript. Presented to NBER-NSF Conference on Bayesian Econometrics, Ann Arbor, Michigan, April 1974.

[12] DuMouchel, W.H., On the asymptotic normality of maximum-likelihood estimates when sampling from a stable distribution, Annals of Statist. 1 (1973) 948-957.

[13] DuMouchel, W.H., Stable distributions in statistical inference, Ph.D. dissertation, Yale University.

[14] Fama, E.F. and Roll, R., Parameter estimates for symmetric stable distributions, J. Amer. Statist. Assoc. 66 (1971) 331-338.

[15] Fama, E.F. and Roll, R., Some properties of symmetric stable distributions, J. Amer. Statist. Assoc. 63 (1968) 817-836.

[16] Fama, E.F., The behaviour of stock market prices, J. Bus. 38 (1965) 34-105.

[17] Feller, W., An Introduction to Probability Theory and its Application, Vol. 2 (Wiley, New York, 1966).

[18] Feuerverger, A. and McDunnough, P., On some Fourier methods for inference, submitted to J. Amer. Statist. Assoc., 1979.

[19] Feuerverger, A. and McDunnough, P., On the efficiency of empirical characteristic function procedures, J. Royal Statist. Soc., Part 3 (1980) to appear.

[20] Feuerverger, A. and Mureika, R.A., The empirical characteristic function and its applications, Annals. Statist. 5 (1977) 88-97.

[21] Fielitz, B.D. and Smith, E.W., Asymmetric stable distributions of stock price changes, J. Amer. Statist. Assoc. 67 (1972) 813-814.

[22] Gnedenko, B.V. and Kolmogorov, A.N., Limit Distributions for Sums of Independent Random Variables, (trans. by K.L. Chung)(Addison-Wesley, Massachusetts, 1954).

[23] Hannan, E.J. and Kanter, M., Autoregressive processes with infinite variance, J. Appl. Prob. 14 (1977) 411-415.

[24] Hart, J.F. et al, Computer Approximations (Wiley, New York, 1968).

[25] Heathcote, C.R., The integrated squared error estimation of parameters, Biometrika 64 (1977) 255-264.

[26] Holt, H.R. and Crow, E.L., Tables and graphs of the stable probability density functions, J. Res. of the Nat. Bur. of Standards 77B (1973) 143-198.

[27] Kanter, M. and Steiger, W.L., Regression and autogression with infinite variance, Adv. Appl. Prob. 6, (1974) 768-783.

[28] Leitch, R.A. and Paulson, A.S., Estimation of stable law parameters: stock, price behaviour application, J. Amer. Statist. Assoc. 70 690-697.

[29] Lévy, P., Théorie de l'addition des Variables Aléatoires, 2nd ed., (Gauthier-Villars, Paris, 1954).
[30] Lévy, P., Calcul des Probabilités, (Gauthier-Villars, Paris, 1925).
[31] Lukacs, E., Characteristic Functions, (Hafner, Connecticut, 1970).
[32] Mandelbrot, B., The variation of certain speculative prices, J. Bus. of the Univ. of Chicago 26 (1963), 394-419.
[33] Mandelbrot, B., The Pareto-Levy law and the distribution of income, Int. Econ. Rev. 1 (1960), 79-106.
[34] Marcus, M.B., Weak convergence of the empirical characteristic function, Annals Probab., to appear.
[35] Paulson, A.S., Halcomb, E.W. and Leitch, R.A., The estimation of the parameters of the stable laws, Biometrika 62 (1975), 163-170.
[36] Press, S.J., Estimation in univariate and multivariate stable distributions, J. Amer. Statist. Assoc. 67 (1972), 842-846.
[37] Samuelson, P., The mathematics of speculative price, S.I.A.M. Review 15, No. 1 (1973), 1-41.
[38] Stuck, B.W. and Kleiner, B., A statistical analysis of telephone noise, Bell System Technical Journal 53, No. 7 (1974), 1263-1312.
[39] Tukey, J.W., An introduction to the calculations of numerical spectrum analysis, in Advanced Seminar on Spectral Analysis of Time Series, ed. B. Harris, (Wiley, New York, 1967, 25-46).

STATISTICS AND RELATED TOPICS
M. Csörgö, D.A. Dawson, J.N.K. Rao, A.K.Md.E. Saleh (eds.)

SOME COMMENTS ON THE MINIMUM MEAN SQUARE ERROR AS A CRITERION OF ESTIMATION

C. Radhakrishna Rao

Department of Mathematics and Statistics
University of Pittsburgh
Pittsburgh, Pennsylvania U.S.A.

It is shown that estimators obtained by MMSE (minimizing the mean square error) may not have optimum properties with respect to other criteria such as PN (probability of nearness to the true value in the sense of Pitman) or PC (probability of concentration around the true value). In particular, a detailed study is made of estimators obtained by shrinking the minimum variance unbiased estimators to reduce the MSE. It is suggested that because of mathematical convenience and some intuitive considerations, MMSE could be used as a primitive postulate to derive estimators, but their acceptability should be judged on more intrinsic criteria such as PN and PC.

1. INTRODUCTION

The concept of minimum mean square error (MMSE) as a criterion of estimation is attributed to Gauss and figures prominently in the discussion of problems of statistical estimation. No doubt, the criterion is a valid one if the problem of estimation is considered in a decision theoretic frame work with the loss function specified as the square of the error in an estimator. Otherwise, the criterion is arbitrary as Gauss himself has observed in a paper presented to the Royal Society of Göttingen in 1809:

> "From the value of the integral $\int_{-\infty}^{\infty} x\phi(x)dx$, i.e., the average value of x (defined as deviation in the estimator from the true value of the parameter) we learn the existence or non-existence of a constant error as well as the value of this error; similarly, the integral $\int_{-\infty}^{\infty} x^2\phi(x)dx$, i.e., the average value of x^2 , seems very suitable for defining and measuring, in a general way, the uncertainty of a system of observations. ... If one objects that this convention is arbitrary and does not appear necessary, we readily agree. The question which concerns us here has something vague about it from its very nature, and cannot be made really precise except by some principle which is arbitrary to a certain degree. ... It is clear to begin with that the loss should not be proportional to the error committed, for under this hypothesis, since a positive error would be considered as a loss, a negative error would be considered as a gain; the magnitude of a loss ought, on the contrary, to be evaluated by a function of the error whose value is always positive. Among the infinite number of functions satisfying this condition, it seems natural to choose the simplest, which is, without doubt, the square of the error, and in this way we are led to the principle proposed above".

Karlin (1958) expresses the same opinion:

> "The justification for the quadratic loss as a measure of the discrepancy of an estimate derives from the following two characteristics: (i) in the case where $a(x)$ represents an unbiased estimate of $h(\omega)$, MSE may

be interpreted as the variance of $a(x)$ and, of course, fluctuations as measured by the variance is very traditional in the domain of classical estimation; (ii) from a technical and mathematical viewpoint square error lends itself most easily to manipulation and computations".

Thus, the criterion of MMSE is used not because of its practical relevance in a given problem but for its simplicity and mathematical convenience. We may, therefore, accept MMSE as a primitive postulate providing a rule of estimation like other methods such as maximum likelihood, minimum chi-square, etc., and examine the properties of estimators so obtained in terms of other criteria.

The present study is limited to the examination of estimators obtained by "shrinking" unbiased estimators with a view to decrease the MSE. We compare the shrunken estimator with the unbiased estimator in terms of its bias (B), mean absolute error (MAE), mean square error (MSE), mean quartic error (MQE), and more intrinsic properties like the probability of nearness to the true value (PN) due to Pitman (1937), and probability of concentration in intervals round the true value (PC).

In the discussion on a recent paper by Berkson (1980), the author (Rao, 1980) has pointed out some anomalies that may result in accepting MMSE as a criterion of estimation. Examples were given of estimators which have a smaller MSE but perform poorly in terms of more intrinsic criteria such as PN and PC when compared to other estimators. Such anomalies are expected since the quadratic loss function places undue emphasis on large deviations which may occur with small probability, and minimizing MSE may insure against large errors in an estimator occurring more frequently rather than providing greater concentration of an estimator in neighborhoods of the true value. A more detailed study of such situations is made in the present paper.

2. ESTIMATION OF A SINGLE PARAMETER

Let X be an unbiased estimator of a parameter θ with $V(X) = \sigma^2$. It is well known that with respect to a quadratic loss function, cX is an admissible estimator of θ if $0 < c \leq 1$ (see Rao, 1976b for instance). The MSE of cX is

$$E(cX - \theta)^2 = \sigma^2[c^2 + (1-c)^2\delta^2] \leq E(X - \theta)^2 \tag{2.1}$$

iff $\delta^2 \leq (1+c)/(1-c)$ where $\delta = \theta/\sigma$. Thus, if we have some knowledge of δ, we can make an appropriate choice of c to ensure the inequality in (2.1). The minimum of $E(cX-\theta)^2$ is attained at $c = \delta^2/(1+\delta^2)$, and if it is known that the true δ is near about δ_o, we may try the estimator

$$X_o = \frac{\delta_o^2}{1+\delta_o^2} X \tag{2.2}$$

which has the property

$$E_2 = [E(X_o - \theta)^2/E(X-\theta)^2]^{\frac{1}{2}} \leq 1 \quad \text{if} \quad |\delta| \leq (2\delta_o^2 + 1)^{\frac{1}{2}} . \tag{2.3}$$

But the property (2.3) does not ensure that

$$PN = Pr(|X_o - \theta| < |X - \theta|) \geq 0.5 \tag{2.4}$$

for the same range of δ. Table 1 gives the approximate values of δ below which $PN \geq 0.5$ and $E_2 \leq 1$ for different values of the shrinkage factor $c = \delta_o^2/(1+\delta_o^2)$ and the associated values of δ_o, assuming that X is normally distributed.

TABLE 1

Values of $|\delta|$ below which $E_2 \leq 1$ and $PN \geq .05$ for different shrinkage factors

c	0	.1	.2	.3	.4	.5	.6	.7	.8	.9
δ_o	0	.33	.50	.65	.82	1.00	1.22	1.53	2.00	3.00
$E_2 \leq 1$	1.0	1.2	1.3	1.4	1.6	1.8	2.0	2.5	3.0	4.5
$PN \geq 0.5$	0.7	0.8	0.8	0.9	1.0	1.1	1.2	1.4	1.6	1.8

Table 1 shows that the range of δ for which (2.4) holds is much smaller than that for (2.3) to hold. It is also interesting to note that the optimum choice of c corresponding to a given δ_o for reducing the MSE does not ensure that $PN \geq 0.5$ even for $\delta = \delta_o$ unless δ_o is below 1.2 (approximately). Thus, shrinking an unbiased estimator is useful only when the true value of the parameter under estimation is smaller than about 1.2 times the standard error of estimation.

If σ^2, the variance of the estimator X, is unknown, but an estimator s^2 of σ^2 is available, we can define an empirical version of (2.2)

$$X_e = \frac{(X/s)^2}{(1+(X/s)^2} X \qquad (2.5)$$

and study its performance. The MSE of X_e compared to that of X has been extensively studied by Thompson (1968) under various distributional assumptions on X. We shall examine other properties of (2.5) assuming that X is normally distributed and σ^2 is known. As shown by Thompson, the conclusions are not likely to be different when σ^2 is used instead of s^2 in (2.5) even for small values of f, the degrees of freedom on which σ^2 is estimated.

Table 2 gives the values of

$$B = \sigma^{-1}E(X_e - \theta), \qquad PN = Pr(|X_e - \theta| \leq |X - \theta|),$$

$$E_1 = \sigma^{-1}\sqrt{\pi/2}\, E|X_e - \theta|, \quad E_2 = \sigma^{-1}[E(X_e - \theta)^2]^{\frac{1}{2}}, \quad E_4 = \sigma^{-1}[E(X_e - \theta)^4/3]^{\frac{1}{4}},$$

obtained by simulation. It is seen that the empirically shrunken estimator X_e is better than the unbiased estimator X only when $\delta \leq 1.4$ (approximately), i.e., when the standard error of the estimator of a parameter is more than 70% of the value of the parameter. But a serious drawback of the estimator (2.5) may be the large negative bias it has unless δ is very small or very large.

TABLE 2

Values of E_1, E_2, E_3, E_4, $E = \Sigma E_i$, PN and B for the estimator $X^3/(s^2 + X^2)$ for different values of $\delta = \theta/\sigma$.

δ	B	E_1	E_2	E_4	E	PN
0.0	- .005	.549	.702	.813	.688	1.000
0.5	- .171	.757	.764	.818	.780	.706
1.0	- .290	.946	.893	.856	.898	.565
1.4	- .344	1.047	.993	.919	.986	.504
1.5	- .352	1.066	1.015	.938	1.006	.487
2.0	- .367	1.124	1.097	1.031	1.084	.444
2.5	- .350	1.136	1.131	1.094	1.120	.436
3.0	- .317	1.124	1.133	1.121	1.126	.437
3.5	- .283	1.105	1.118	1.118	1.114	.444
4.0	- .252	1.086	1.100	1.102	1.096	.453
8.0	- .130	1.026	1.037	1.035	1.033	.476
10.0	- .105	1.018	1.028	1.026	1.024	.480
20.0	- .055	1.007	1.017	1.015	1.013	.490
100.0	- .015	1.003	1.013	1.012	1.009	.495

3. ESTIMATION OF VARIANCE

If S^2 denotes the corrected sum of squares of n i.i.d. observations from $N(\mu,\sigma^2)$, it is well known that $s^2 = S^2/(n-1)$ is the minimum variance unbiased estimator of σ^2. But $s_2^2 = S^2/(n+1)$ has smaller MSE than s^2 uniformly for all σ^2 and all n, so that s^2 is inadmissible as an estimator of σ^2 with respect to the MSE criterion. How does s_2^2 compare with s^2 with respect to other criteria? Table 3 gives the values of the following for different degrees of freedom (n-1):

$$B = E(s_2^2 - \sigma^2)/\sigma^2 ,$$

$$PN = Pr(|s_2^2 - \sigma^2| \le |s^2 - \sigma^2|)$$

$$PC = Pr(-\log a \le \log s^2 - \log \sigma^2 \le \log a),$$

$$PC_2 = Pr(-\log a \le \log s_2^2 - \log \sigma^2 \le \log a),$$

$$E_2 = [E(s_2^2 - \sigma^2)^2/E(s^2 - \sigma^2)^2]^{\frac{1}{2}} .$$

It is seen that PN, the probability that s_2^2 is closer to σ^2 than s^2, is less than 0.5 uniformly for all σ^2 and for all n although E_2 is uniformly less than unity for all σ^2 and for all n. Similarly, $\log s^2$ has a greater concentration probability in any symmetrical interval around $\log \sigma^2$ than $\log s_2^2$ uniformly for all σ^2 and all n. Thus shrinking the unbiased estimator s^2 has resulted in a smaller MSE but has not brought the estimator closer to the true value of σ^2 in any sense. The unbiased estimator s^2 seems to have better intrinsic properties than s_2^2.

It may be noted that the optimum shrinkage of s^2 depends on the loss function chosen. If instead of the MSE, we choose the MQE = $E(cs^2 - \sigma^2)^4$ as the loss, then the optimum c is a solution of the cubic equation

$$(1 + \frac{6}{n-1})(1 + \frac{4}{n-1})(1 + \frac{2}{n-1})c^3 - 3(1 + \frac{4}{n-1})(1 + \frac{2}{n-1})c^2 + 3(1 + \frac{2}{n-1})c - 1 = 0. \quad (3.1)$$

The estimator so obtained is denoted by s_4^2.

TABLE 3

Values of E_2, B, PN, PC and PC_2 for different degrees of freedom (DF)

D.F. (n-1)	E_2	B*	PN	PC (first row) and PC_2 (second row) a = 1.5	a = 2.0	a = 2.5	a = 3
1	.577	.677	.221	.193 .123	.322 .206	.413 .267	.480 .315
2	.707	.500	.264	.290 .213	.471 .349	.588 .442	.667 .511
3	.774	.400	.290	.360 .285	.571 .457	.695 .566	.772 .643
4	.816	.333	.308	.416 .345	.644 .540	.768 .658	.838 .734
5	.845	.286	.323	.463 .395	.701 .608	.821 .727	.883 .801
6	.866	.250	.334	.503 .440	.747 .663	.859 .780	.913 .849
7	.882	.222	.346	.539 .479	.784 .709	.888 .822	.935 .885
8	.894	.200	.352	.570 .514	.815 .747	.911 .855	.951 .911
9	.904	.182	.359	.599 .545	.840 .780	.928 .882	.963 .932
10	.912	.167	.365	.624 .574	.862 .808	.942 .903	.972 .947
20	.953	.091	.400	.793 .762	.963 .945	.992 .987	.998 .996
40	.976	.048	.428	.926 .912	.996 .994	.999+ .999+	.999+ .999+

* The shrinkage factor is (1-B) where $B = \text{Bias}/\sigma^2$

On the other hand, the optimum c which minimizes the MAE = $E|cs^2 - \sigma^2|$ is the one which minimizes the function

$$(c-1) + 2\, G_{n-1}(\frac{n-1}{c}) - 2\, c\, G_{n+1}(\frac{n-1}{c}) \quad (3.2)$$

where

$$G_k(a) = \frac{1}{2^{k/2}\Gamma(\frac{k}{2})} \int_0^a e^{-t/2} t^{\frac{k}{2}-1} dt .$$

The estimator so obtained may be denoted by s_1^2 .

Table 4 gives the values of $E_i = [E(|s_i^2 - \sigma^2|^i)/E(|s^2 - \sigma^2|^i)]^{1/i}$, $B_i = E(s_i^2 - \sigma^2)$, $PN = Pr(|s_i^2 - \sigma^2| \leq |s^2 - \sigma^2|)$ and PC - PC_i where

$$PC = Pr(- \log a \leq \log s^2 - \log \sigma^2 \leq \log a)$$

$$PC_i = Pr(- \log a \leq \log s_i^2 - \log \sigma^2 \leq \log a)$$

for $i = 1$ and 4. It is seen that s^2 performs better than s_1^2 and s_4^2 in terms of PN and PC. Among the estimators s_1^2, s_2^2 and s_4^2, s_1^2 appears to be better than s_2^2 and s_4^2. The results are not unexpected since the distribution of s^2 is skew on the right and minimization of an expression of the type $E(cs^2 - \sigma^2)^m$ pulls the estimator away from σ^2 in the region around and below the modal value of s^2.

It is not clear why in statistical literature much emphasis is laid on the estimation of σ^2 and not on σ although in practice the latter should be the parameter of direct interest. Unfortunately, none of the properties such as unbiasedness and MMSE are preserved under transformations of estimators and parameters. For instance, the minimum variance unbiased estimator of σ is

$$s^* = (\frac{n-1}{2})^{\frac{1}{2}} \frac{\Gamma(\frac{n-1}{2})}{\Gamma(\frac{n}{2})} s = ts \tag{3.3}$$

which is different from s while the MMSE of σ is

$$s_2^* = (\frac{n-1}{2})^{-\frac{1}{2}} \frac{\Gamma(\frac{n}{2})}{\Gamma(\frac{n-1}{2})} s \tag{3.4}$$

which is different from s_2. Now

$$E(s^* - \sigma)^2 = \sigma^2(t^2 - 1) > 2\sigma^2(1 - \frac{1}{t}) = E(s - \sigma)^2 \tag{3.5}$$

so that s has a smaller MSE than s^* as an estimation of σ .

We shall compare the relative performances of s and s^* as estimators of σ and of s^2 and $(s^*)^2$ as estimators of σ^2. Table 5 gives the values of the following for different degrees of freedom:

$$E_2 = [E(s - \sigma)^2/E(s^* - \sigma)^2]^{\frac{1}{2}} ,$$

$$PN_1 = Pr(|s^* - \sigma| \leq |s - \sigma|) ,$$

$$PN_2 = Pr(|(s^*)^2 - \sigma^2| \leq |s^2 - \sigma^2|) ,$$

$$PC = Pr(-\log a \leq \log s^2 - \log \sigma^2 \leq \log a) ,$$

$$PC_2 = Pr(-\log a \leq \log (s^*)^2 - \log \sigma^2 \leq \log a).$$

TABLE 4

Values of B_i, E_i, PN_i and $PC-PC_i$ for i = 1,4 for different degrees of freedom

	s_1^2							s_4^2						
				PC - PC_1							PC - PC_4			
DF	E_1	B_1^*	PN_1	a=1.5	a=2	a=2.5	a=3	E_4	B_4^*	PN_2	a=1.5	a=2	a=2.5	a=3
1	.78	.577	.24	.04	.07	.10	.11	.32	.764	.20	.11	.18	.22	.25
2	.85	.404	.29	.04	.07	.09	.10	.46	.619	.23	.14	.21	.24	.25
3	.89	.311	.31	.04	.07	.08	.08	.55	.521	.25	.14	.20	.22	.22
4	.91	.252	.33	.04	.06	.07	.07	.61	.450	.27	.14	.19	.20	.18
5	.93	.212	.35	.04	.05	.06	.05	.66	.396	.28	.13	.17	.17	.14
6	.94	.183	.36	.03	.05	.05	.04	.70	.354	.29	.13	.16	.14	.12
7	.94	.161	.37	.03	.04	.04	.03	.73	.320	.30	.13	.15	.12	.09
8	.96	.144	.38	.03	.04	.03	.02	.75	.292	.31	.12	.13	.10	.07
9	.96	.130	.38	.03	.04	.03	.02	.77	.268	.32	.11	.12	.09	.06
10	.96	.118	.39	.03	.04	.02	.01	.79	.248	.33	.11	.11	.07	.05
20	.98	.063	.42	.02	.03	.00	.00	.88	.142	.37	.07	.04	.01	.00
40	.99	.032	.44	.01	.00	.00	.00	.93	.077	.40	.03	.00	.00	.00

* The optimum shrinkage factor if 1-B where $B = \text{Bias}/\sigma^2$

TABLE 5

Values of E_2, PN_1, PN_2, PC and PC_2 for various degrees of freedom

D.F.	E_2	PN_1	PN_2	PC	PC_2	PC	PC_2	PC	PC_2	PC	PC_2
				a = 1.5		a = 2		a = 2.5		a = 3.0	
1	.841	.625	.622	.193	.186	.322	.313	.413	.407	.480	.478
2	.912	.586	.585	.290	.284	.471	.467	.588	.590	.667	.675
3	.940	.570	.569	.360	.356	.571	.570	.695	.702	.772	.784
4	.954	.560	.559	.416	.413	.644	.646	.768	.776	.838	.850
5	.963	.553	.553	.463	.461	.701	.705	.821	.829	.883	.893
6	.969	.549	.548	.503	.502	.747	.751	.859	.868	.913	.923
7	.973	.545	.545	.539	.538	.784	.789	.888	.896	.935	.943
8	.976	.542	.542	.570	.570	.815	.820	.911	.918	.951	.957
9	.979	.539	.539	.599	.599	.840	.845	.928	.934	.963	.968
10	.982	.537	.537	.624	.625	.862	.867	.942	.947	.972	.976
20	.991	.526	.526	.793	.794	.963	.966	.992	.993	.998	.998
40	.995	.519	.519	.926	.927	.996	.997	.999+	.999+	.999+	.999+

It is seen that although $E_2 \le 1$ uniformly for all σ and DF so that s has a smaller MSE than s^* as an estimator of σ, PN_1 is uniformly above 0.5 so that s^* is nearer to σ more often than s. What is more interesting is that PN_2 is also uniformly above 0.5 indicating that $(s^*)^2$ is nearer to σ^2 more often than s^2. Further, $\log (s^*)^2$ has greater concentration around $\log \sigma^2$ than $\log s^2$ around $\log \sigma^2$ if the DF is not small and the interval chosen is not short. It appears that the biased estimator $(s^*)^2$ of σ^2 has better properties in terms of PN and PC than s^2, although highly inadmissible with respect to MSE.

Since the cube root of χ^2 has an approximate normal distribution, it may be of interest to compare the expressions

$$PC = \Pr[1 - \delta < (\frac{s^2}{\sigma^2})^{\frac{1}{3}} < 1 + \delta] \quad (3.6)$$

$$PC_2 = \Pr[1 - \delta < (\frac{s_2^2}{\sigma^2})^{\frac{1}{3}} < 1 + \delta] . \quad (3.7)$$

The following table gives the values of PC and PC_2 for $\delta = 0.1$ and 0.5 for different degrees of freedom.

	$\delta = 0.1$		$\delta = 0.5$	
D.F.	PC	PC_2	PC	PC_2
1	.145	.093	.657	.538
5	.354	.306	.982	.972
10	.491	.456	.999	.999
20	.654	.632	.999+	.999+

It is seen that s^2 is better than s_2^2 in terms of the criterion of concentration as defined by (3.6) and (3.7).

4. DIRECT OR INVERSE REGRESSION

Consider a pair of random variables (θ, Y) such that

$$Y = \theta + \varepsilon, \quad E(\varepsilon) = 0, \quad \text{cov}(\theta,\varepsilon) = 0, \quad V(\varepsilon) = \sigma_o^2 . \quad (4.1)$$

In practice θ stands for the true value of a quantity (such as the cholesterol level of a blood sample) and Y is a measurement of θ subject to error. Only Y is observable and not θ, in which case the problem is one of estimating or predicting θ given Y.

From (4.1), the regression of Y on θ is θ itself so that the inverse regression estimate of θ is Y which is also an unbiased estimator of θ. On the other hand, if the mean (μ) and variance (σ_θ^2) of the unconditional distribution of θ are known, then the regression of θ on Y is

$$\hat{\theta} = \mu + \frac{\sigma_\theta^2}{\sigma_\theta^2 + \sigma_o^2} (Y - \mu) \quad (4.2)$$

which provides a direct regression estimate of θ. In practice, the estimation procedure (4.2) can be implemented by estimating μ, σ_θ^2 and σ_o^2 from past data on Y

(cholesterol determinations) on a large number of individuals (see Rao, 1973, p.337), and updating the estimates as more data accumulate. The estimator $\hat{\theta}$ can be identified as the Bayes estimator using a quadratic loss function and a relevant prior distribution for θ.

Suppose that an individual's blood sample has been referred to a clinic for the determination of cholesterol and the clinic reports the measurement as Y. What should we record as the estimate of blood cholesterol for the individual, the unbiased estimator Y or the Bayes estimator $\hat{\theta}$ of (4.2) using a relevant estimated prior distribution? There has been considerable controversy on this subject, in a slightly different context, in the calibration problem (see Berkson, 1969; Halperin, 1970; Krutchkoff, 1967, 1969, 1971 and Williams, 1969). We shall examine this problem in the set up of (4.1) assuming that the parameters μ, σ_θ^2 of the prior distribution and the variance σ_o^2 of the error of measurement are known. Now

$$E(\hat{\theta} - \theta)^2 = \frac{\sigma_\theta^2 \sigma_o^2}{\sigma_\theta^2 + \sigma_o^2} \leq \sigma_o^2 = E(Y - \theta)^2 \tag{4.3}$$

and the strict inequality holds if $\sigma_o \neq 0$, so that the mean square error of prediction is smaller for $\hat{\theta}$. Does this mean that $\hat{\theta}$ is closer to θ than Y in some sense? To examine this question we have to consider the distributions of Y and $\hat{\theta}$ for given θ.

The MSE's and Y and $\hat{\theta}$ for given θ are

$$E[(Y - \theta)^2|\theta] = \sigma_o^2 \tag{4.4}$$

$$E[(\hat{\theta} - \theta)^2|\theta] = \sigma_o^2\delta^2(\delta^2 + \lambda^2)/(1 + \delta^2)^2 \tag{4.5}$$

where $(\theta - \mu)/\sigma_\theta = \lambda$ and $\delta = \sigma_\theta/\sigma_o$. From (4.4) and (4.5),

$$E[(\hat{\theta} - \theta)^2|\theta] \leq E[(Y - \theta)^2|\theta] \tag{4.6}$$

iff $\lambda^2 \leq (1 + 2\delta^2)/\delta^2$. Then the efficiency of $\hat{\theta}$ compared to Y with respect to MSE depends on the magnitude of the deviation of the true value of θ from the a priori mean. If the deviation is large, $\hat{\theta}$ is less efficient than Y.

The estimator Y is unbiased while the bias in $\hat{\theta}$ is

$$E[(\hat{\theta} - \theta)|\theta] = -\lambda\sigma_\theta\sigma_o^2/(\sigma_o^2 + \sigma_\theta^2) \tag{4.7}$$

so that large values of θ are under-estimated and small values are over-estimated.

Table 6 gives for different combinations of δ and λ the values of

$$E_2 = [E\{(\hat{\theta} - \theta)^2|\theta\}/E(Y - \theta)^2]^{\frac{1}{2}},$$

$$PN = Pr(|\hat{\theta} - \theta| < |Y - \theta|),$$

where the regions for which (i) $E_2 < 1$, $PN > 0.5$, (ii) $E_2 < 1$, $PN < 0.5$ and (iii) $E_n > 1$, $PN < 0.5$ are marked. It is seen that $\hat{\theta}$ performs better than Y when the error of measurement is large and the true value is near the mean of the a priori distribution. But if precise estimation of large deviations from the a priori mean is more important (as it should be in a problem like the estimation of blood cholesterol), Y should be preferred to $\hat{\theta}$.

TABLE 6

Values of E_2 (first entry) and PN (second entry) for different combinations of λ and δ

δ \ λ	0.5	1.0	1.5	2.0	2.5	3.0
0.5	.283 .835	.447 .679	.632 .535	.825 .411	1.020 .308	1.217 .226
1.0	.559 .742	.707 .528	.901 .375	1.118 .275	1.346 .209	1.581 .160
1.5	.729 .673	.832 .460	.979 .353	1.154 .294	1.346 .248	1.548 .207
2.0	.825 .615	.894 .435	1.000 .371	1.131 .329	1.281 .290	1.442 .253
2.5	.879 .569	.928 .433	1.005 .391	1.104 .356	1.219 .322	1.346 .290
3.0	.912 .535	.949 .439	1.006 .407	1.082 .376	1.171 .346	1.273 .318
5.0	.966 .487	.980 .461	1.004 .442	1.035 .422	1.075 .403	1.121 .384
10.0	.991 .490	.995 .480	1.001 .470	1.010 .461	1.020 .451	1.033 .441
15.0	.996 .493	.997 .487	1.000 .480	1.004 .474	1.009 .467	1.015 .460

5. SIMULTANEOUS ESTIMATION OF TWO PARAMETERS

Let $X_1 \sim N(\theta_1,\sigma^2)$, $X_2 \sim N(\theta_2,\sigma^2)$ and $fs^2 \sim \sigma^2\chi^2(f)$ be independent random variables, and consider the following estimators of θ_1, θ_2:

$$t_1 = \frac{X_1 + X_2}{2} + c\,\frac{X_1 - X_2}{2} \tag{5.1}$$

$$t_2 = \frac{X_1 + X_2}{2} + c\,\frac{X_2 - X_1}{2} \tag{5.2}$$

as alternatives to the unbiased estimators X_1 and X_2 . Then

$$E(t_i - \theta_i)^2 = \sigma^2\left[\frac{1+c^2}{2} + \frac{(1-c)^2\delta^2}{4}\right], \quad i = 1,2. \tag{5.3}$$

and the expected compound quadratic loss (ECQL) is

$$E\sum_1^2 (t_i - \theta_i)^2 = \sigma^2\left[1 + c^2 + \frac{(1-c)^2\delta^2}{2}\right] \tag{5.4}$$

where $\delta = (\theta_1 - \theta_2)/\sigma$. The expression (5.4) attains the minimum when $c = \delta^2/(2+\delta^2)$. Since δ^2 is not known, we may consider the empirical versions of (5.1) and (5.2).

$$t_1^{(e)} = \frac{X_1 + X_2}{2} + \frac{(X_1 - X_2)^2/s^2}{2 + (X_1 - X_2)^2/s^2} \frac{X_1 - X_2}{2}$$

$$t_2^{(e)} = \frac{X_1 + X_2}{2} + \frac{(X_1 - X_2)^2/s^2}{2 + (X_1 - X_2)^2/s^2} \frac{X_2 - X_1}{2} .$$

We shall compare $t_1^{(e)}$ and $t_2^{(e)}$ with X_1 and X_2, assuming that σ^2 is known, with respect to the following criteria:

$$B_1 = \sigma^{-1}(t_1^{(e)} - \theta_1), \quad B_2 = \sigma^{-1}(t_2^{(e)} - \theta_2) ,$$

$$PN = \frac{1}{2}\left[Pr(|t_1^{(e)} - \theta_1| \le |X_1 - \theta_1|) + Pr(|t_2^{(e)} - \theta_2| \le |X_2 - \theta_2|)\right]$$

$$E = (E_1 + E_2 + E_4)/3 ,$$

where

$$E_1 = \sigma^{-1} \sqrt{\pi/2}\left(\frac{1}{2}E\,|t_1^{(e)} - \theta_1| + \frac{1}{2}E\,|t_2^{(e)} - \theta_2|\right) ,$$

$$E_2 = \sigma^{-1}\left[\frac{1}{2}E\,(t_1^{(e)} - \theta_1)^2 + \frac{1}{2}E\,(t_2^{(e)} - \theta_2)^2\right]^{\frac{1}{2}} ,$$

$$E_4 = \sigma^{-1}\left[\frac{1}{6}E\,(t_1^{(e)} - \theta_1)^4 + \frac{1}{6}E\,(t_2^{(e)} - \theta_2)^4\right]^{\frac{1}{4}} .$$

Table 7 gives the values of E_1, E_2, E_4, E, PN and B_1, B_2 based on a simulation study using 1000 samples, for various values of $\delta = (\theta_1 - \theta_2)/\sigma$. It is seen that simultaneous estimation of θ_1, θ_2 by $t_1^{(e)}$ and $t_2^{(e)}$ has some advantage over X_1 and X_2 when $\delta \le 2$ (approximately), i.e., when the parameters under estimation do not differ by more than twice the standard error of the estimator of a single parameter.

The above analysis raises a slightly different issue concerning the detection and elimination of spurious observations in estimating a location parameter. Suppose we have N observations from $N(\mu,\sigma^2)$ with mean $\bar{X}$ and M spurious observations from $N(\nu,\sigma^2)$ with mean $\bar{Y}$. Let us ignore the fact that $\bar{Y}$ is a contaminating observation and estimate μ by $\hat{\mu} = (N\bar{X} + M\bar{Y})/(N+M)$. Then, denoting $\nu - \mu = \delta\sigma$,

$$E(\hat{\mu} - \mu)^2 = \frac{\sigma^2}{N+M}\left(1 + \frac{M^2\delta^2}{N+M}\right) < \frac{\sigma^2}{N} = V(\bar{X})$$

if $\delta^2 < M^{-1} + N^{-1}$ which is always true when $\delta \le 1$ and $M = 1$ whatever N may be. Thus, under the MMSE criterion, it pays to include a spurious observation coming from a population whose mean may differ by as much as one standard deviation from the parameter under estimation!

TABLE 7

Values of E_1, E_2, E_4, E, B_1, B_2 and PN for $t_1^{(e)}$ and $t_2^{(e)}$ for different values of $\delta = (\theta_1 - \theta_2)/\sigma$

δ	B_1	B_2	PN	E_1	E_2	E_4	E
0	-.027	.005	.702	.854	.864	.872	.863
.5	.052	-.132	.674	.871	.871	.878	.873
1.0	.121	-.164	.635	.894	.887	.883	.888
1.5	.216	-.145	.567	.962	.956	.954	.957
2.0	.257	-.259	.517	1.013	.998	.976	.996
2.5	.269	-.277	.485	1.041	1.029	1.000	1.023
3.0	.199	-.243	.475	1.046	1.045	1.041	1.044
3.5	.279	-.234	.455	1.081	1.064	1.037	1.061
4.0	.229	-.251	.442	1.087	1.083	1.080	1.083
5.0	.188	-.132	.454	1.037	1.034	1.023	1.031
6.0	.207	-.181	.465	1.027	1.029	1.025	1.027
7.0	.197	-.118	.468	1.042	1.040	1.049	1.044
8.0	.085	-.102	.491	1.022	1.018	1.010	1.017
9.0	.070	-.108	.490	1.015	1.024	1.029	1.023
10.0	.097	-.105	.485	1.006	1.009	1.025	1.013

6. ESTIMATION OF SEVERAL PARAMETERS

Let $X_i \sim N(\theta_i, \sigma^2)$, $i = 1,\ldots,p$ and $fs^2 \sim \sigma^2\chi^2(f)$ be independent random variables, where $(\theta_1,\ldots,\theta_p) = \theta'$ is a <u>fixed</u> vector parameter. James and Stein (1961) have found the remarkable result that when $p \geq 3$ there exist statistics

$$T_i = T_i(X_1,\ldots,X_p, s^2), \quad i = 1,\ldots,p \tag{6.1}$$

such that

$$E[\Sigma(T_i - \theta_i)^2] < E[\Sigma(X_i - \theta_i)^2] \tag{6.2}$$

uniformly for all θ_i, which implies that $X' = (X_1,\ldots,X_p)$ as an estimator of θ is inadmissible with respect to the CQL (compound quadratic loss) function. The result (6.2) gives the impression that we stand to gain by answering several problems, possibly unrelated, simultaneously. It is well known that there do not exist statistics t_i alternative to X_i such that

$$E(t_i - \theta_i)^2 < E(X_i - \theta_i)^2, \quad i = 1,\ldots,p \tag{6.3}$$

uniformly for all θ_i, so that the overall reduction in the ECQL possibly takes place by an increase in the MSE for some parameters and decrease to a larger extent for the others. To examine this phenomenon in some detail, we shall consider a number of alternative joint estimators of $\theta_1,\ldots,\theta_p$ of the type suggested by James and Stein and study the performance of individual estimators.

Specifically, we consider the following types of estimators of $\theta_1,\ldots,\theta_p$:

$$T_{1i} = b\,X_i\,, \quad i = 1,\ldots,p, \tag{6.4}$$

$$T_{2i} = a + b(X_i - a), \quad i = 1,\ldots,p, \tag{6.5}$$

$$T_{3i} = a + b_i(X_i - a), \quad i = 1,\ldots,p, \tag{6.6}$$

which may be represented by T_1, T_2 and T_3 in vector notation.

Now

$$E[\Sigma(T_{1i} - \theta_i)^2] = pb^2\sigma^2 + (1-b)^2\Sigma\theta_i^2 \tag{6.7}$$

which attains the minimum value at $b = \nu^2/(1+\nu^2)$ where $\nu^2 = \Sigma\theta_i^2/p\sigma^2$. If ν is known, then the optimum estimator of the type (6.4) is

$$T_1 = (1 - \frac{1}{1+\nu^2})\,\underline{X} \tag{6.8}$$

and the ECQL is

$$E[\Sigma(T_{1i} - \theta_i)^2] = p\sigma^2\,\frac{\nu^2}{1+\nu^2} \geq p\sigma^2 = E[\Sigma(X_i - \theta_i)^2]\,. \tag{6.9}$$

The MSE for an individual estimator is

$$E(T_{1i} - \theta_i)^2 = \sigma^2(\nu^4 + \nu_i^2)/(1 + \nu^2)^2 \tag{6.10}$$

where $\nu_i = \theta_i/\sigma$. The expression exceeds the MSE of X_i if $\nu_i^2 > 2\nu^2 + 1$ indicating the possibility that in joint estimation of the T_1-type, the larger parameters are less efficiently estimated than the smaller ones.

If ν^2 is not known, we may estimate $1/(1 + \nu^2)$ by $cs^2/\Sigma X_i^2$, where c is a suitable constant, and obtain an empirical version of (6.8)

$$T_1^{(e)} = (1 - \frac{cs^2}{\Sigma X_i^2})\,X\,. \tag{6.11}$$

The best choice of c obtained by minimizing the ECQL of (6.11) is $f(p-2)/(f+2)$ if $p \geq 3$, which leads to the James-Stein estimator

$$T_1^{(e)} = \left(1 - \frac{f(p-2)}{f+2}\,\frac{s^2}{\Sigma X_i^2}\right) X\,. \tag{6.12}$$

James and Stein have shown that for $p \geq 3$

$$E[(T_1^{(e)} - \theta)'(T_1^{(e)} - \theta)] = p\sigma^2 - \frac{f(p-2)^2\sigma^2}{f+2} E[(2K_1 + p - 2)^{-1}] \tag{6.13}$$

where K_1 is a Poisson variable with parameter $\nu^2/2$. The expression (6.13) is smaller than $p\sigma^2$ for all ν. Ullah (1974) computed the bias and MSE of individual estimators:

$$\sigma^{-1}E(T_{1i}^{(e)} - \theta_i) = -\frac{(p-2)f}{f+2}\nu_i \; E[(2K_1 + p)^{-1}] \tag{6.14}$$

$$\sigma^{-2}E(T_{1i}^{(e)} - \theta_i)^2 = 1 - \frac{f(p-2)^2}{2(f+2)}\Big[(1+g)E\{(2K_1 + p - 2)^{-1}\} - (1 + \frac{pg}{p-2})E\{(2K_1 + p)^{-1}\}\Big] \tag{6.15}$$

where

$$g = [(p+2)\theta_i^2 - 2\Sigma\theta_i^2]/\Sigma\theta_i^2 \; .$$

Similarly, the best estimator of the type (6.5) is

$$T_{2i} = \bar{\theta} + \left(1 - \frac{1}{1+\eta^2}\right)(X_i - \bar{\theta}) \; , \quad i = 1,\ldots,p \tag{6.16}$$

where $\bar{\theta} = (\theta_1 + \ldots + \theta_p)/p$ and $\eta^2 = \Sigma(\theta_i - \bar{\theta})^2/p\sigma^2$. The ECQL for T_2 as in (6.16) is

$$p\sigma^2\left(\frac{\eta^2}{1+\eta^2}\right) \le p\sigma^2\left(\frac{\nu^2}{1+\nu^2}\right) \le p\sigma_o^2 \tag{6.17}$$

so that if $\bar{\theta}$ and η^2 are known further improvement in ECQL over X and T_1 is possible. The MSE for an individual estimator is

$$E(T_{2i} - \theta_i)^2 = \sigma^2(\eta^4 + \eta_i^2)/(1 + \eta^2)^2 \tag{6.18}$$

where $\eta_i^2 = (\theta_i - \bar{\theta})^2/\sigma^2$. The expression (6.18) exceeds σ^2, the MSE of X_i, if $\eta_i^2 > 2\eta^2 + 1$, indicating the possibility that in joint estimation of the type T_2, the extreme parameters are less efficiently estimated and the middle parameters more efficiently than the corresponding unbiased estimators.

As in the case of T_1, we can estimate $\bar{\theta}$ and $1/(1+\eta^2)$ by $\bar{X}$ and $(p-3)fs^2/(f+2)[\Sigma(X_i - \bar{X})^2]$ respectively if $p \geq 4$ and obtain an empirical version of T_2

$$T_{2i}^{(e)} = \bar{X} + \left(1 - \frac{f(p-3)}{(f+2)}\frac{s^2}{\Sigma(X_i - \bar{X})^2}\right)(X_i - \bar{X}) \; , \tag{6.19}$$

$$i = 1,\ldots,p \; (\geq 4).$$

It can be shown (Efron and Morris, 1971, 1972a, Rao, 1976a) that

$$E\{(T_2^{(e)} - \theta)'(T_2^{(e)} - \theta)\} = p\sigma^2 - \frac{f(p-3)^2\sigma^2}{f+2} E\{(2K_2 + p - 3)^{-1}\} \tag{6.20}$$

where K_2 is a Poisson variable with parameter $\eta^2/2$. The expression (6.20) is less than $p\sigma_o^2$ so that $T_2^{(e)}$ is uniformly better than X with respect to ECQL. Rao and Shinozaki (1978) have shown that for individual estimators

$$E(T_{2i} - \theta_i) = \frac{-f(p-3)}{f+2}(\theta_i - \bar{\theta})E\{(2K_2 + p - 1)^{-1}\} \tag{6.22}$$

$$E(T_{2i} - \theta)^2 = \sigma^2 - \frac{f(p-3)^2\sigma^2}{2(f+2)}$$

$$\times \left[(c+d)E\{(2K_2 + p-3)^{-1}\} - \frac{c(p-3)+d(p-1)}{(p-3)} E\{(2K_2 + p-1)^{-1}\}\right]$$

where

$$c = (p-1)/p, \; d = \{(p+1)(\theta_i - \bar{\theta})^2 - \frac{2(p-1)}{p} \Sigma(\theta_i - \bar{\theta})^2\}/\Sigma(\theta_i - \bar{\theta})^2.$$

Finally, the best estimator of the type (6.6) is

$$T_{3i} = \bar{\theta} + \frac{(\theta_i - \bar{\theta})^2}{(\theta_i - \bar{\theta})^2 + \sigma_o^2}(X_i - \bar{\theta}), \quad i = 1,\ldots,p \tag{6.23}$$

and its empirical version is

$$T_{3i}^{(e)} = \bar{X} + \frac{(X_i - \bar{X})^2}{(X_i - \bar{X})^2 + s^2}(X_i - \bar{X}), \quad i = 1,\ldots,p . \tag{6.24}$$

It is difficult to compute the ECQL of $\underline{T}_3^{(e)}$ or the MSE of $T_{3i}^{(e)}$.

The relative performance of the estimators $T_1^{(e)}$ and $T_2^{(e)}$ and the ranges of parametric values for which the individual $T_1^{(e)}$ and $T_2^{(e)}$ - estimators are better than the X-estimators are examined in Rao and Shinozaki (1978). Table 8 contains the results of simulation studies based on thousand samples for the estimation of four parameters $\theta_1 = a\sigma$, $\theta_2 = (a+d)\sigma$, $\theta_3 = (a+2d)\sigma$ and $\theta_4 = (a+3d)\sigma$ for various combinations of a and d, assuming σ to be known, i.e., by replacing $fs^2/(f+2)$ by σ^2 in the formulae (6.12) and (6.19) for $T_2^{(e)}$ and $T_3^{(e)}$. The broad conclusions remain the same if the random variable s^2 is used provided f, the degrees of freedom, is not small.

For each combination of $\bar{\theta}/\sigma$ and d, Table 8 gives the values of E, PN and B in the first row for $T_1^{(e)}$, in the second row for $T_2^{(e)}$ and in the third row for $T_3^{(e)}$, where for any statistic

$$E = [E(t_i - \theta_i)^2/E(X_i - \theta_i)^2]^{1/2},$$

$$PN = \Pr(|t_i - \theta_i| \le |X_i - \theta_i|),$$

$$B = E(t_i - \theta_i).$$

TABLE 8

Values of E_2, PN and B for estimators of the type T_1, T_2 and T_3 for various combinations of $\bar{\theta}/\sigma$ and d

$\frac{\bar{\theta}}{\sigma}$	d		E2				PN				Bias			
			θ_1	θ_2	θ_3	θ_4	θ_1	θ_2	θ_3	θ_4	θ_1	θ_2	θ_3	θ_4
1.25	0.5	T_1	.835	.884	.955	1.069	.770	.589	.479	.440	.094	-.183	-.297	-.410
		T_2	.927	.863	.864	.950	.634	.766	.751	.620	.197	.079	-.067	-.209
		T_3	.761	.614	.613	.771	.692	.727	.723	.692	.246	-.092	.105	.232
2.00	1.0	T_1	.919	.947	.985	.953	.793	.521	.446	.453	.041	-.123	-.215	-.304
		T_2	1.004	.912	.914	1.010	.500	.739	.721	.510	.217	.078	-.074	-.220
		T_3	.814	.688	.693	.816	.652	.712	.704	.645	.028	-.015	.019	-.033
2.75	1.5	T_1	.950	.972	.993	1.030	.800	.478	.456	.466	-.022	-.088	-.157	-.230
		T_2	1.015	.949	.950	1.016	.476	.696	.662	.480	.175	.061	-.059	-.177
		T_3	.901	.793	.802	.899	.575	.626	.612	.570	.160	.005	-.006	-.159
3.50	2.0	T_1	.972	.983	.995	1.019	.912	.462	.465	.476	-.013	-.068	-.127	-.183
		T_2	1.011	.969	.970	1.012	.482	.646	.597	.482	.139	.048	-.047	-.140
		T_3	.967	.888	.894	.964	.521	.486	.477	.524	.224	.008	-.011	-.220
7.00	4.0	T_1	.993	.997	.999	1.004	.651	.464	.481	.488	-.057	-.036	-.065	-.093
		T_2	1.003	.991	.992	1.003	.501	.536	.500	.492	.073	.025	-.024	-.073
		T_3	1.054	1.035	1.035	1.056	.465	.199	.201	.469	.229	.000	-.001	-.226

On the basis of previous investigations by Efron and Morris (1971, 1972a,b, 1973a,b), Rao (1975a, 1975b, 1975c, 1977) and Rao and Shinozaki (1978) and the present simulation studies, the following broad conclusions emerge.

(a) There is some advantage in using $T_2^{(e)}$ and $T_3^{(e)}$ when the range of parameter values is small, and $T_1^{(e)}$ when both the range and values of the parameters are small compared to the standard error of the unbiased estimators.

(b) When the range of parameters is large both $T_1^{(e)}$ and $T_2^{(e)}$ tend to have the same properties as X. But the performance of $T_3^{(e)}$ tends to be erratic.

(c) When the range of parameters is moderate, $T_1^{(e)}$ gives higher precision for the parameters with smaller values at the expense of lower precision for higher values. According to the PN criterion, only the smallest of the four parameters is better estimated than the corresponding unbiased estimator. In the case of $T_2^{(e)}$ and $T_3^{(e)}$, extreme values of the parameters suffer at the expense of increased precision for the middle values.

(d) One drawback in using estimators $T_1^{(e)}$, $T_2^{(e)}$ and $T_3^{(e)}$ in preference to X, is the bias in these estimators. The bias is of a substantial magnitude for the higher values in the case of $T_1^{(e)}$ and for the extreme values of the parameters in the case of $T_2^{(e)}$ and $T_3^{(e)}$. There are situations where bias in the estimators may have serious consequences such as the following.

Suppose there are four regions and periodical estimates of a certain characteristic are needed for sharing some resources in proportion to the values of the characteristic of the four regions. If each time, estimates of the type $T_1^{(e)}$, $T_2^{(e)}$ and $T_3^{(e)}$ are used, some regions stand to gain at the expense of the others in the long run (Rao, 1977, 1979).

In some situations, the individual parameters may not be of direct interest but certain linear combinations may be important. If $c'\theta = c_1\theta_1 + \ldots + c_4\theta_4$ is a linear combination to be estimated, should one estimate it by $c'X$ or $c'T_1^{(e)}$ or $c'T_2^{(e)}$ or $c'T_3^{(e)}$? Naturally, the answer depends on the vector c (Rao, 1975b, 1975c), and the optimal properties of $T_1^{(e)}$, $T_2^{(e)}$ and $T_3^{(e)}$ with respect to a single criterion like the CQL do not insure their efficiency in different ways they may be used for practical purposes. If multiple uses are intended, the best plan is to place on record X as the estimator of θ (together with an estimate of σ^2) leaving it to the user to make any optimal adjustments in X depending on particular problems under study.

Note 1. It may be of historical interest to note that estimators of the type T_2 have been constructed under more general conditions, in multivariate analysis, for purposes of genetic selection by Smith (1936), Hazel (1943) and Rao (1953) based on an idea suggested by Fisher. The problem was as follows. Let $(\underline{\theta}, \underline{y}_1, \ldots, \underline{y}_m)$ be $(m+1)$ vector variables representing the unobservable genetic values $\underline{\theta}$ and repeated independent phenotypic vector measurements $\underline{y}_1, \ldots, \underline{y}_m$ on an individual. The variables are related by the model

$$\underline{y}_i = \underline{\theta} + \underline{\varepsilon}_i \,, \quad i = 1, \ldots, m \,. \tag{6.24}$$

The genetic worth of an individual is measured by a linear function $\underline{g}'\underline{\theta}$. Suppose that we have observed p individuals from a population, with phenotypic measurements

$$(\underline{y}_{1j},\ldots,\underline{y}_{mj})\ ,\quad j=1,\ldots,p\ . \tag{6.25}$$

What is the best way estimating the genetic worths $\underline{g}'\underline{\theta}_1,\ldots,\underline{g}'\underline{\theta}_p$ of these individuals for purposes of ranking and selecting a given proportion of the individuals with the largest genetic values? If $\bar{\underline{y}}_j$ represents the mean of the measurements (6.25) for individual j, then $\underline{g}'\bar{\underline{y}}_j$ is an unbiased estimator of $\underline{g}'\underline{\theta}_j$. However, Fisher suggested the regression of $\underline{g}'\underline{\theta}_j$ on $\bar{\underline{y}}_j$ as the appropriate selection index, which involves the knowledge of $\underline{\mu}=E(\underline{\theta})$, $\Gamma=\mathrm{cov}(\theta,\theta)$, $\Delta=\mathrm{cov}(\bar{\varepsilon},\bar{\varepsilon})$, where $\bar{\underline{\varepsilon}}=m^{-1}\Sigma\underline{\varepsilon}_i$. The regression estimator of $\underline{g}'\underline{\theta}_j$ is $\underline{g}'\hat{\underline{\theta}}_j$ where

$$\hat{\theta}_j=\mu+[I-\Delta(\Gamma+\Delta)^{-1}](\bar{\underline{y}}_j-\underline{\mu})\ . \tag{6.26}$$

By multivariate analysis of variance and covariance of the data (6.25), we obtain dispersion matrices B and W as between and within individuals with degrees of freedom $(p-1)$ and $f=m(p-1)$ respectively, which supply estimates $B/(p-1)$ of $(\Gamma+\Delta)$ and W/mf of Δ. Then an empirical version of (6.26) is

$$\hat{\underline{\theta}}_j=\bar{\underline{y}}+[I-\frac{p-1}{mf}WB^{-1}](\bar{\underline{y}}_j-\bar{\underline{y}}) \tag{6.27}$$

where $p\bar{\underline{y}}=\Sigma\bar{\underline{y}}_j$. The details leading to the formula (6.27) are given in Rao (1953), pp.237-8). When all the variables are one dimensional, (6.27) is the same as $T_{2j}^{(e)}$ of (6.19) except that the multiplying factor $(p-1)/f$ is replaced by $(p-3)(f+2)$. In the 1953 paper, Rao also considered some distributional problems for testing hypotheses concerning the rank of the Γ matrix and the efficiency of the regression estimator.

It should be noted that the regression estimators (or empirical Bayes estimators) $\underline{g}'\underline{\theta}_j$ are appropriate in the problem of selection where the <u>total genetic worth</u> of the selected subset of individuals has to be maximized. In such a case, it is well known that the best ordering of the observed individuals is achieved by using, as the selection index, the regression of genetic worth on phenotypic measurements (see Cochran, 1951 and Henderson, 1963). But the regression estimator may not be appropriate if the genetic worth of each individual has to be assessed for other purposes which may demand equal precision for the individual estimators.

<u>Note 2</u>. In his presidential address delivered to the Royal Statistical Society, Finney (1974) suggested that the problem of simultaneous estimation may be approached through the principle of maximum likelihood, thus avoiding the use of the arbitrary compound quadratic loss function. Let $X_i \sim N(\theta_i,\phi)$, $i=1,\ldots,p$ be p independent observations. If the θ_i's arise as a random sample from $N(\mu,\tau)$, then the log likelihood is, apart from a constant,

$$L=-\frac{\Sigma(X_i-\theta_i)^2}{2\phi}-\frac{\Sigma(\theta_i-\mu)^2}{2\tau}\ . \tag{6.28}$$

Finney maximizes L with respect to μ and θ_i and obtains the estimates

$$\hat{\theta}_i=\bar{X}+(1-\frac{\phi}{\tau+\phi})(X_i-\bar{X})\ . \tag{6.29}$$

It is not known whether the maximum likelihood principle applies in situations such as (6.28) where the likelihood is a function of both the unknown parameters and unobserved random variables. Finney says that an unbiased estimate of $1/(\tau+\phi)$

is $(p-3)/\Sigma(X_i - \bar{X})^2$, so that when τ is not known, the estimator of θ_i is

$$\hat{\theta}_i = \bar{X} + \left(1 - \frac{(p-3)\phi}{\Sigma(X_i - \bar{X})^2}\right)(X_i - \bar{X}) \tag{6.30}$$

which is the same as the expression given by Lindley (1962) using Bayes theorem and quadratic loss function.

However, if τ is unknown, the appropriate log likelihood is proportional to

$$-\frac{p}{2}\log\tau - \frac{\Sigma(X_i - \theta_i)^2}{2\phi} - \frac{\Sigma(\theta_i - \mu)^2}{2\tau} \quad . \tag{6.31}$$

The expression (6.31) can be made arbitrarily large by choosing $\theta_i = \mu = \bar{X}$ for all i and letting $\tau \to 0$. Thus, the m.l. estimator is $\theta_i = \bar{X}$ for all i! Such anomalies do occur when unobserved random variables are included as unknowns and a "full likelihood" function such as (6.31) is considered for drawing inference.

ACKNOWLEDGEMENT

I wish to thank Robert Boudreau for the simulation studies reported in the various tables. This work is sponsored by the Air Force Office of Scientific Research under Contract F49620-79-0161.

7. REFERENCES

[1] Berkson, J., Estimation of a linear function for a calibration line: consideration of a recent proposal, Technometrics 11 (1969) 649-660.

[2] Berkson, J., Minimum chi-square, not maximum likelihood, Ann. Statist. 8 (1980) 457-469.

[3] Cochran, W.G., Improvement by means of selection, Proc. Second Berkeley Symp. Math. Statist. Prob. (1951) 449-470.

[4] Efron, B. and Morris, C., Limiting the risk of Bayes and empirical Bayes estimators, I. The Bayes case, J. Amer. Statist. Assoc., 66 (1971) 807-815.

[5] Efron, B. and Morris, C., Empirical Bayes on vector observations: an extension of Stein's method, Biometrika 59 (1972) 335-347.

[6] Efron, B. and Morris, C., Limiting the risk of Bayes and empirical Bayes estimators, Part II. The empirical Bayes case, J. Amer. Statist. Assoc. 67 (1972) 130-139.

[7] Efron, B. and Morris, C., Stein's estimation rule and its competitors -- an empirical Bayes approach, J. Amer. Statist. Assoc. 68 (1973) 117-130.

[8] Efron, B. and Morris, C., Combining possibly related estimation problems. (With discussion.) J. Roy. Statist. Soc. Ser. B 35 (1973) 379-421.

[9] Finney, D.J., Problems, data, and inference. (With discussion.) J. Roy. Statist. Soc. Ser A 137 (1974) 1-22.

[10] Halperin, M., On inverse estimation in linear regression, Technometrics 12 (1970) 727-736.

[11] Hazel, L.N., The genetic basis for constructing selection indices, Genetics 28 (1943) 476-490.

[12] Henderson, C.R., Selection index and expected genetic advance, Statistical Genetics and Plant Breeding NAS-NRC, 982 (1963) 141-163.

[13] James, W. and Stein, C., Estimation with quadratic loss, Proc. Fourth Berkeley Symp. Math. Statist. Prob. 1 (1961) 362-379.

[14] Karlin, S., Admissibility for estimation with quadratic loss, Ann. Math. Statist. 29 (1958) 406-436.

[15] Krutchkoff, R.G., Classical and inverse regression methods of calibration, Technometrics 9 (1967) 425-439.

[16] Krutchkoff, R.G., Classical and inverse regression methods of calibration in extrapolation, Technometrics 11 (1969) 605-608.

[17] Krutchkoff, R.G., The calibration problem and closeness, J. Statist. Comput. Simul. 1 (1971) 87-95.

[18] Lindley, D.V., Discussion on Professor Stein's paper "Confidence sets for the mean of a multivariate normal distribution", J. Roy. Statist. Soc. Ser. B 24 (1962) 285-287.

[19] Pitman, E.J.G., The "closest" estimates of statistical parameters, Proc. Cambridge Philos. Soc. 33 (1937) 212-222.

[20] Rao, C.R., Discriminant functions for genetic differentiation and selection, Sankhyā 12 (1953) 229-246.

[21] Rao, C.R., Linear Statistical Inference and Its Applications, Second Edition (Wiley, New York, 1973).

[22] Rao, C.R., Simultaneous estimation of parameters in different linear models and applications to biometric problems, Biometrics 31 (1975) 545-554.

[23] Rao, C.R., Some thoughts on regression and prediction, Sankhyā Ser. C 37 (1975) 102-120.

[24] Rao, C.R., Some problems of sample surveys, Suppl. Adv. Prob. 7 (1975) 50-61.

[25] Rao, C.R., Characterization of prior distributions and solution to a compound decision problem, Ann. Statist. 4 (1976) 823-835.

[26] Rao, C.R., Estimation of parameters in a linear model, Ann. Statist. 4 (1976) 1023-1037.

[27] Rao, C.R., Simultaneous estimation of parameters -- a compound decision problem, Decision Theory and Related Topics, S.S. Gupta and D.S. Moore, eds. (Academic Press, New York, 1977), 327-350.

[28] Rao, C.R., Presidential address, 42nd Session of the International Statistical Institute, Manila, Philippines (1979).

[29] Rao, C.R., Discussion of [2], Ann. Statist. 8 (1980) 482-485.

[30] Rao, C.R. and Shinozaki, N., Precision of individual estimators in simultaneous estimation of parameters, Biometrika 65 (1978) 23-30.

[31] Smith, H.F., A discriminant function for plant selection, Ann. Eugenics 7 (1936) 240-250 and 8 (1937) 219.

[32] Thompson, J.R., Some shrinkage techniques for estimating the mean, J. Amer. Statist. Assoc. 63 (1968) 113-122.

[33] Ullah, A., On the sampling distribution of improved estimators for coefficients in linear regression, J. Econometrics 2 (1974) 143-150.

[34] Williams, E.J., A note on regression methods in calibration, Technometrics 11 (1969) 189-192.

8. APPENDIX

In Section 3 of the paper it was shown that $s^2 = S^2/(n-1)$, the unbiased estimator of σ^2, is better than $s_2^2 = S^2/(n+1)$ if we use the PN criterion of Pitman. This can be theoretically demonstrated by using the fact that in a chi-square distribution, the mean exceeds the median, which is easily established.

We now raise the question about the best choice of the divisor of S^2 for estimating σ^2 by the PN criterion. Since the PN criterion is based on a comparison of two estimators, we shall determine the divisor d such that

$$\Pr\left(\left|\frac{S^2}{d} - \sigma^2\right| < \left|\frac{S^2}{ad} - \sigma^2\right|\right) > 0.5$$

for any a. It is easily seen that the left hand side expression is the same as

$$\Pr(S^2 \leq 2ad/(1+a)) \quad \text{if} \quad a > 1$$

$$\Pr(S^2 \geq 2ad/(1+a)) \quad \text{if} \quad a < 1 .$$

These values exceed 0.5 if d is chosen as the median of the chisquare distribution on (n-1) degrees of freedom. With such a choice of d, S^2/d is better than $S^2/(n-1)$ by the PN criterion. The estimator S^2/d was first suggested by Pitman (1937).

STATISTICS AND RELATED TOPICS
M. Csörgö, D.A. Dawson, J.N.K. Rao, A.K.Md.E. Saleh (eds.)

ESTIMATING QUANTILES OF EXPONENTIAL DISTRIBUTION

A. K. Md. Ehsanes Saleh*
Carleton University, Ottawa

ABLUE of the quantiles x_ξ $(0 < \xi < 1)$ is given based on $k(\leq n)$ selected order statistics for the two parameter exponential distribution. The coefficients of linear combination and the variance expressions are given along with the ARE results relative to the usual nonparametric estimator and the complete sample BLUE or M.L.E.

1. INTRODUCTION

Let $x_1, x_2, \ldots x_n$ be a random sample of size n from a continuous distribution $F(x)$. Recently, there has been considerable interest in the problem of estimating the quantiles, x_ξ $(0 < \xi < 1)$. The initial attempts towards estimating x_ξ are the nonparametric techniques such as using sign test for testing the quantiles to obtain the sample quantile $x_{[n\xi]+1}$ as the estimator of x_ξ. Wilks (1963) presents the nonparametric confidence interval for x_ξ based on two order statistics. Parzen (1978) presents nonparametric estimation of quantile-function and density-quantile function. Reiss (1980) suggests using quasi-quantiles and studies their properties. Point estimation of x_ξ is rather sketchy. Recent attempts are due to Kubat (1975), Robertson (1977) and Weissman (1978). The object of this paper is to propose ABLUE (asymptotically best linear unbiased estimate) of x_ξ for the two-parameter exponential distribution based on k $(\leq n)$ selected order statistics and discuss the ARE (asymptotic relative efficiency) relative to the complete sample BLUE and sample quantile $x_{[n\xi]+1}$. Recently, Ali et al (1980) proposed ABLUE of x_ξ based on two selected order statistics but their method cannot be generalized to k-selected order statistics. We adopt the method proposed in Saleh and Ali (1966) and modify it to obtain the generalization.

2. ESTIMATION OF QUANTILES OF EXPONENTIAL DISTRIBUTION BASED ON k ARBITRARY ORDER STATISTICS

Let $x_{(1)} \leq x_{(2)} \leq \ldots \leq x_{(n)}$ be the order statistics from a sample of size n from the exponential distribution

$$F(x) = 1 - \exp\left\{-\frac{x-\mu}{\sigma}\right\},\ x \geq \mu,\ \sigma > 0\ . \tag{2.1}$$

Consider the k $(\leq n)$ order statistics $x_{(n_1)}, \ldots, x_{(n_k)}$ where $n_1, \ldots, n_k$ are the respective ranks satisfying the inequality $1 \leq n_1 < \ldots < n_k \leq n$. The integers $n_1, n_2, \ldots, n_k$ are determined by the k fixed spacings $p_1, p_2, \ldots p_k$ which satisfy the inequality $0 < p_1 < \ldots < p_k < 1$ and $n_i = [np_i] + 1$, $(i=1,2,\ldots,k)$. $[np_i]$ is the Euler notation denoting the largest integer contained in np_i. Denote $p_o = 0$ and $p_{k+1} = 1$. Then, it is well-known that $(x_{(n_1)}, \ldots, x_{(n_k)})$ has a k-dimensional normal distribution with mean-vector $(\mu + \sigma u_1, \ldots, \mu + \sigma u_k)$ and dispersion matrix $W = \frac{\sigma^2}{n}((w_{ij}))$ where $w_{ii} = w_{ij} = w_{ji} = (e^{u_i} - 1)$ $(i < j)$;

$i,j = 1,2,\ldots,k$ and $u_i = \ell n(1-p_i)^{-1}$, $(i=1,\ldots,k)$ are the quantiles from the standardized esponential distribution $1-e^{-z}$, corresponding to $p_1,\ldots,p_k$ respectively. Also, it is well-known from Saleh and Ali (1966) that the ABLUE of μ and σ based on $x_{(n_1)},\ldots,x_{(n_k)}$ are

$$\hat{\mu} = x_{(n_1)} - \hat{\sigma} u_1 \tag{2.2}$$

$$\hat{\sigma} = \sum_{i=1}^{k} b_i \, x_{(n_i)} \tag{2.3}$$

where

$$b_1 = -\frac{(u_2 - u_1)}{(e^{u_2} - e^{u_1})L} \tag{2.4}$$

$$b_i = \frac{1}{L}\left\{\frac{u_i - u_{i-1}}{e^{u_i} - e^{u_{i-1}}} - \frac{u_{i+1} - u_i}{e^{u_{i+1}} - e^{u_i}}\right\} \tag{2.5}$$

$i = 2,\ldots,k-1$

$$b_k = \frac{1}{L}\left(\frac{u_k - u_{k-1}}{e^{u_k} - e^{u_{k-1}}}\right) \tag{2.6}$$

$$L = \sum_{i=2}^{k} \frac{(u_i - u_{i-1})^2}{e^{u_i} - e^{u_{i-1}}}. \tag{2.7}$$

The variances and covariance of the estimates of μ and σ are

$$V(\hat{\mu}) = \frac{\sigma^2}{n}\left\{\frac{u_1^2}{L} + (e^{u_1} - 1)\right\} \tag{2.8}$$

$$V(\hat{\sigma}) = \frac{\sigma^2}{nL}, \quad \text{Cov}(\hat{\mu},\hat{\sigma}) = -\frac{\sigma^2}{n}\frac{u_1}{L}. \tag{2.9}$$

The estimate of x_ξ is then given by

$$\hat{x}_\xi = \hat{\mu} + \hat{\sigma} z_\xi = x_{(n_1)} - \hat{\sigma}(u_1 - z_\xi) \tag{2.10}$$

$$= \{1 - b_1(u_1 - z_\xi)\} x_{(n_1)} - \sum_{i=2}^{k} b_i (u_1 - z_\xi) x_{(n_i)} \tag{2.11}$$

where $\Sigma_{i=1}^{k} b_i = 0$ and z_ξ is the ξ-quantile from $1-e^{-z}$. The variance of $\hat{x}_\xi$ is given by

$$V(\hat{x}_\xi) = \frac{\sigma^2}{n}\left\{\frac{(z_\xi - u_1)^2}{L} + (e^{u_1} - 1)\right\}. \tag{2.12}$$

It is seen that the coefficients of $x_{(n_1)},\ldots,x_{(n_k)}$ add to 1, which is the condition for unbiasedness of $\hat{x}_\xi$.

3. ESTIMATION OF QUANTILES BASED ON k SELECTED ORDER STATISTICS

In this section, we propose two estimates of x_ξ based on k selected order statistics. Firstly, we consider the results of Saleh and Ali (1966) for the estimation of μ and σ based on k optimum order statistics. They are:

$$\hat{\mu} = x_{(1)} - \hat{\sigma}\,\ell n \frac{2n-1}{2n+1} \tag{3.1}$$

$$\hat{\sigma} = b_0^o x_{(1)} + \sum_{i=2}^{k-1} b_i^o x_{(n_i^o)} \tag{3.2}$$

where the coefficients $b_0^o, b_1^o, \ldots, b_{k-1}^o$ are available in Sarhan and Greenberg (1962) Table II.D.1 with the optimum spacings $\lambda_1^o, \ldots, \lambda_{k-1}^o$ for $k=2,\ldots,16$ to determine $n_1^o, \ldots, n_{k-1}^o$ by the relation

$$p_{i+1}^o = \{2 + (2n-1)\lambda_i^o\}/(2n+1) \ . \tag{3.3}$$

We shall refer to the table of Sarhan and Greenberg (1962) as S-G table. Thus, using the estimates of μ and σ we have the estimate of x_ξ as

$$\hat{x}_\xi = \{1 - b_0^o(\ell n \frac{2n-1}{2n+1} - z_\xi)\}x_{(1)} - \sum_{i=1}^{k-1} b_i^o(\ell n \frac{2n-1}{2n+1} - z_\xi)x_{(n_i^o)} \tag{3.4}$$

for all $\xi \in (0,1)$ and $n_i^o = [np_{i+1}^o] + 1$, $(i = 1,2,\ldots,k-1)$. The variance of this estimate is

$$\hat{V}(x_\xi) = \frac{\sigma^2}{n}\left\{\frac{2n-1}{2n+1} \frac{(z_\xi - \ell n \frac{2n-1}{2n+1})^2}{Q_{k-1}^o} - \frac{2}{2n+1}\right\} \tag{3.5}$$

where Q_{k-1}^o is the ARE of $\hat{\sigma}$ based on $k-1$ selected order statistics relative to the complete sample estimate of σ. The full sample BLUE is given by

$$\bar{x}_\xi = \frac{n}{n-1}(1 - z_\xi)x_{(1)} - \frac{1}{n-1}(1 - nz_\xi)\bar{x} \tag{3.6}$$

whose variance is given by

$$V(\bar{x}_\xi) = \frac{\sigma^2}{n(n-1)}(1 + 2z_\xi + nz_\xi^2) \ . \tag{3.7}$$

Hence,

$$\text{ARE}(\hat{x}_\xi : \bar{x}_\xi) = Q_{k-1}^o \quad \text{for all} \quad \xi \in (0,1). \tag{3.8}$$

We know that the nonparametric estimate of x_ξ is $x_{([n\xi]+1)} = \dot{x}_\xi$ with variance $\frac{\sigma^2}{n}\xi(1-\xi)^{-1}$. Therefore the ARE of $\hat{x}_\xi$ relative to $\dot{x}_\xi$ is given by

$$\text{ARE}(\hat{x}_\xi : \dot{x}_\xi) = \frac{\xi Q_{k-1}^o}{(1-\xi)\,\ell n^2(1-\xi)} > 1 \quad \text{for all} \quad \xi \in (0,1). \tag{3.9}$$

Hence, $\hat{x}_\xi$ is superior than $\dot{x}_\xi$ for the exponential case.

In the derivation of ABLUE of x_ξ we have used the optimum order statistics of (μ,σ) and not of $\mu + \sigma z_\xi$ $(= x_\xi)$. Therefore, we consider the second estimator of x_ξ based on k suitable order statistics by minimizing the variance of $\hat{\mu} + \hat{\sigma} z_\xi$ based on k arbitrary order statistics. The variance of such an estimator is

$$V(\hat{\hat{x}}_\xi) = \frac{\sigma^2}{n}\left\{\frac{(z_\xi - u_1)^2}{L} + (e^{u_1} - 1)\right\} . \tag{3.10}$$

We minimise this variance for suitable choices of $u_1, u_2, \ldots, u_k$ which will determine the estimate $\hat{\hat{x}}_\xi$ of x_ξ. An alternative form of $V(\hat{\hat{x}}_\xi)$ is given by

$$V(\hat{\hat{x}}_\xi) = \frac{\sigma^2}{n}\left\{\frac{t_1^2 e^{z_\xi - t_1}}{Q_{k-1}} + (e^{z_\xi - t_1} - 1)\right\} \tag{3.11}$$

where $t_i = z_\xi - u_i$ $(i = 1,2,\ldots,k)$ and Q_{k-1} is defined by

$$Q_{k-1} = \sum_{i=1}^{k-1} \frac{(t_i^* - t_{i-1}^*)^2}{e^{t_i^*} - e^{t_{i-1}^*}} \quad \text{with } t_o^* = 0 \text{ and } t_i^* = t_{i+1} - t_1, \; i=1,2,\ldots,k-1 . \tag{3.12}$$

In order to minimize (3.11) we first maximize Q_{k-1} as in Saleh and Ali (1966) with maximum value Q_{k-1}^o which are tabulated in SG(1962) with corresponding values of $t_1^*, \ldots, t_{k-1}^*$. Next, we minimize

$$\frac{t_1^2 e^{z_\xi - t_1}}{Q_{k-1}^o} + (e^{z_\xi - t_1} - 1) \tag{3.13}$$

over the region $t_1 \leq z_\xi$. By differentiating one finds that (3.13) is decreasing for $t_1 < z_{\xi_Q}$ where $z_{\xi_Q} = 1 - \sqrt{1 - Q_{k-1}^o}$ and for $t_1 > 2 - z_{\xi_Q}$ and increasing in between. Now, we must have $t_1 \leq z_\xi$, so if $z_\xi \geq z_{\xi_Q}$, we get

$$(z_{\xi_Q}^2 / Q_{k-1}^o + 1)\exp(z_\xi - z_{\xi_Q}) - 1 \tag{3.14}$$

at $t_1 = z_{\xi_Q}$ and z_ξ^2 / Q_{k-1}^o at $t_1 = z_\xi$. These two values are equal at $z_\xi = z_{\xi_Q}$ and $z_\xi = z^*_{\xi_Q^*}$, where $z^*_{\xi_Q^*}$ is to be determined iteratively solving the equation

$$z_\xi = \ell n \frac{(1 + z_\xi^2 / Q_{k-1}^o)}{(1 + z_{\xi_Q}^2 / Q_{k-1}^o)} + z_{\xi_Q} \tag{3.14a}$$

as a function of z_ξ. Hence, we get the minimum at $z_\xi = z_{\xi_Q}$ and at $z_\xi = z^*_{\xi_Q^*}$ with $Q_{k-1} = Q_{k-1}^o$. Thus, if $z_{\xi_Q} \leq z_\xi \leq z^*_{\xi_Q^*}$, we have the solutions for the u's as follows:

$$\left.\begin{aligned} u_1^o &= z_\xi - z_{\xi_Q} \\ u_{i+1}^o &= z_\xi + t_i^o - z_{\xi_Q} \end{aligned}\right\} \tag{3.15}$$

for $i = 1,\ldots,k-1$. The corresponding spacings of the selected order statistics are

$$p_1^o = 1 - (1-\xi)\exp(z_{\xi_Q}) \tag{3.17}$$

$$p_{i+1}^o = 1 - (1-\xi)(1-\lambda_i^o)\exp(z_{\xi_Q}) \tag{3.18}$$

$i = 1,\ldots,k-1$. Thus, for $\xi_Q \leq \xi \leq \xi_Q^*$ (where $\xi_Q = 1 - e^{-z^*_{\xi^*_Q}}$), the ABLUE of x_ξ is given by

$$\hat{\hat{x}}_\xi = c_0^o x_{([np_1^o]+1)} + \sum_{i=1}^{k-1} c_i^o x_{([np_{i+1}^o]+1)}, \tag{3.19}$$

where $c_0^o = 1 + b_0^o z_{\xi_Q}$, $c_i^o = b_i^o z_{\xi_Q}$, $i = 1,2,\ldots,k-1$. (3.20)

The coefficients are tabulated in SG(1962).

For the case $z_\xi < z_{\xi_Q}$ or $z_\xi > z^*_{\xi^*_Q}$, the minimum occurs at $t_1 = z_\xi$ since we must have $t_1 \leq z_\xi$ and (3.13) is decreasing over the intervals $(0,z_{\xi_Q}) \cup (z^*_{\xi^*_Q},\infty)$. In this case u_1 must be near zero and by the reasoning of Saleh and Ali (1966) we must choose

$$u_1^o = \ln\frac{2n-1}{2n+1} \quad \text{and} \quad u_i^o = t_i^o, \quad i = 2,\ldots,k. \tag{3.21}$$

Thus, for $\xi \in (0, \xi_Q) \cup (\xi_Q^*, 1)$, the ABLUE of x_ξ is given by the expressions (3.4). Hence, we write the estimate of x_ξ for all cases as

$$\hat{\hat{x}}_\xi = \left\{1 - b_0^o\left(\ln\frac{2n-1}{2n+1} - z_\xi\right)\right\}x_{(1)} - \sum_{i=1}^{k-1} b_i^o\left(\ln\frac{2n-1}{2n+1} - z_\xi\right)x_{(n_i^o)}$$

$$\text{for } \xi \in (0, \xi_Q) \cup (\xi_Q^*, 1)$$

$$= c_0^o x_{([np_1^o]+1)} + \sum_{i=1}^{k-1} c_i^o x_{([np_{i+1}^o]+1)} \quad \text{for } \xi \in [\xi_Q, \xi_Q^*]. \tag{3.22}$$

It may be noted that the estimate x_ξ depends on where ξ is located on (0,1).

4. ARE OF $\hat{\hat{x}}_\xi$ RELATIVE TO $\bar{x}_\xi$ AND $\dot{x}_\xi$

The ARE of $\hat{\hat{x}}_\xi$ relative to $\bar{x}_\xi$ is given by

$$\text{ARE}(\hat{\hat{x}}_\xi:\bar{x}_\xi) = \begin{cases} Q_{k-1}^o & \text{if } \xi \in (0, \xi_Q) \cup (\xi_Q^*, 1) \\ z_{\xi_Q}^2\left[\exp(z_\xi - z_{\xi_Q})\left(1 + \dfrac{z^2_{\xi_Q}}{Q_{k-1}^o}\right) - 1\right]^{-1} & \text{if } \xi \in [\xi_Q, \xi_Q^*] \end{cases} \tag{3.23}$$

Similarly, the ARE of $\hat{\hat{x}}_\xi$ relative to $\dot{x}_\xi$ is given by

$$\mathrm{ARE}(x_\xi : \dot{x}_\xi) = \begin{cases} \xi(1-\xi)^{-1}Q^o_{k-1} & \text{if } \xi \in (0,\xi_Q) \cup (\xi^*_Q, 1) \\ \xi(1-\xi)^{-1}[(1-\xi)^{-1}\exp(-z_{\xi_Q})(1+\frac{z^2_{\xi_Q}}{Q^o_{k-1}}-1]^{-1} & \text{if } \xi \in [\xi_Q,\xi^*_Q]. \end{cases} \quad (3.24)$$

It may be noted that for $k = 2$, the results agree with Ali et al (1980). Some tabular values of $\mathrm{ARE}(\hat{\hat{x}}_\xi : \bar{x}_\xi)$ are given for $k = 2(2)14$ and $\xi = .60(.10).90$ with values of ξ^*_Q.

Table 1. $\mathrm{ARE}(\hat{\hat{x}}_\xi:\bar{x}_\xi)$ when $\xi \in [\xi_Q,\xi^*_Q]$

k	2	4	6	8	10	12	14
ξ_Q	.3339	.4882	.5375	.5617	.5759	.5854	.5922
ξ ξ^*_Q	.9296	.8292	.7779	.7475	.7230	.7135	.7030
.60	.7705	.9095	.9512	.9697	.9794	.9848	.9882
.70	.8115	.9275	.9606	.9752	.9826	.9869	.9896
.80	.8147	.9108					
.90	.7205	N o t	A d m i s s i b l e				
Q^o_{k-1}	.6475	.8910	.9476	.9693	.9798	.9857	.9894

Note: ξ_Q and ξ^*_Q are the bounds of ξ for each k values. Q^o_{k-1} are the maximum values of Q_{k-1} for each k and also represents the ARE value of $\hat{\hat{x}}_\xi$ when $\xi \in (0, \xi_Q) \cup (\xi^*_Q,1)$. Note that all ARE values are greater than Q^o_{k-1}.

Example. Suppose we wish to estimate $x_{.70}$ using 4 selected order statistics in a sample of size $n = 50$. Since $.70 \in [.4882, .8292]$ we compute the optimum spacings as $p^o_1 = .3994$, $p^o_2 = .7175$, $p^o_3 = .8979$ and $p^o_4 = .9792$. The ranks are $n^o_1 = 20$, $n^o_2 = 36$, $n^o_3 = 45$ and $n^o_4 = 49$. The coefficients of these order statistics are $c^o_1 = .4966$, $c^o_2 = .2998$, $c^o_3 = .1518$ and $c^o_4 = .0519$. The ARE of this estimate is .9275.

5. REFERENCES

[1] Ali, M. M., Umbach, Dale and Hassnein, K. (1980). Estimation of quantiles of Exponential and Double-Exponential distributions based on two order statistics. Ball State Tech. Reports. Dept. of Math. Sciences.

[2] Kubat, P. (1975). Simplified estimation procedure based on order statistics. D.Sc. Thesis, Technion - Israel Institute of Technology.

[3] Parzen, E. (1979). Nonparametric Statistical Data Analysis. JASA, 74, 105-120.

[4] Reiss, R-D. (1980). Estimation of quantiles in certain nonparametric models. Ann. Statist. 8, 87-105.

[5] Robertson, C.A. (1966). Estimation of quantiles of exponential distribution with minimum error in predicted distribution function. JASA 72, 162-164.

[6] Saleh, A.K.M.E. and Ali, M.M. (1966). Asymptotic optimum quantiles for the estimation of the parameters of the negative exponential distribution. Ann. Math. Statist. 37, 143-151.

[7] Sarhan, A.E. and Greenberg, B.G. (1962). Contributions to Order Statistics. (John Wiley & Sons Inc.)

[8] Weissman, I. (1978). Estimation of parameters and large quantiles based on k largest observations. JASA 73, 812-815.

[9] Wilks, S.S. (1962). Mathematical Statistics. (John Wiley & Sons Inc.)

* This research is part of the activity supported by the NSERC A3088.

PART III
BIOSTATISTICS

STATISTICS AND RELATED TOPICS
M. Csörgö, D.A. Dawson, J.N.K. Rao, A.K.Md.E. Saleh (eds.)

SOME ASPECTS OF THE ANALYSIS OF EVOKED RESPONSE EXPERIMENTS

David R. Brillinger[1]

Department of Statistics
University of California
Berkeley, California
U.S.A.

Evoked response experiments provide an important class of situations in which the basic responses recorded are curves. In this paper a variety of modifications, to the usual statistical procedures, are proposed for handling such data. In particular analogs of the mean, the general linear model, robust/resistant estimates, experimental designs, analysis of variance and parametric models are investigated.

1. INTRODUCTION

A traditional, (dating back to Caton in 1875), means of studying the nervous system involves applying sensory stimuli to a subject and examining the ongoing electroencephalogram (EEG) for an evoked response (ER) . The stimulus may be auditory, visual, olfactory (an odour), somatosensory (an electric shock) or gustatory (a taste) in character. Generally the stimulus is applied for a time interval that is brief in comparison to the duration of the response. The response, if one occurs, takes place with a small delay (latency) and perhaps lasts half a second.

A general description of the evoked response technique may be found in Regan (1975). He lists as principal applications: (1) revelation of specific brain activities, (2) provision of an objective indicator of sensory function and (3) distinction of organic disorders from psychogenic ones. The technique is fast and provides an effectively risk free means of testing hearing, vision and spinal cord function that may be applied even to infants.

One specific example of the use of the procedure is related in Bergamini et al (1967). Siamese twins were joined in such a way that it was not possible to determine, by traditional means, if the peripheral nervous pathways were dependent. Before operating to separate the twins it was desired to examine for their independence. Ongoing EEG's were recorded for each twin. A series of experiments were carried out in which the twins legs were stimulated, by electrical shocks, in turn. When a leg of one twin was stimulated, EEG activity was noted only for that twin. On the basis of this information the twins were separated - successfully.

A second example of the use of the ER technique is provided by hearing exams for newborn infants, (including sleeping infants). Ongoing EEG's are recorded. These are examined for responses after loud clicks are made near the infant. Rapin and Graziani (1967) present examples of average evoked responses for an infant with hearing difficulties wearing and not-wearing a hearing aid. It was found that the aid had an objectively measurable effect.

The first and most basic obstacle to making use of the ER procedure is that of seeing the ER's in an EEG. In almost all circumstances the ER's are much smaller than the level of the continuing noise. Dawson (1951) demonstrated that one way to surmount this difficulty was to apply the stimulus periodically at well spaced time points and then to average together the EEG values that occur at the same time lag after the application of the stimulus. Specifically if $Y(t)$ is the observed series and if the stimulus is applied at the times $j\sigma$, $j = 1,\ldots,M$ one computes

$$\overline{Y}(u) = \frac{1}{M} \sum_{j=1}^{M} Y(u + j\sigma) \tag{1}$$

$0 \lesssim u \lesssim U$. This statistic is referred to as an average evoked response (AER) . The interval width σ is to be taken large enough that neighboring ER's do not interfere with each other.

In fact it turns out that Laplace (1825) had earlier suggested the consideration of sums of values of a series at fixed time lags relative to the times of certain external events. Namely, in Volume 5 of his Traité de Mécanique Celeste he summed the difference between morning and evening low and high tidal heights at lags of -1, 0, 1, 2, 3, 4 days relative to the times of equinoctial syzygies. Another early example of the use of the statistic (1) is provided by the table of Buys-Ballot (1847). Buys-Ballot's concern was the detection of an effect of period σ, so that the values (1) were computed for different σ's as well as different u's . Yet another early example of a related procedure was mentioned by Professor F. N. David . In the late 1800's Galton superposed photographic negatives of faces of criminals, in a search for common features. Doing this may be viewed as analagous to superposing the separate curves $Y(u + j\sigma)$, $0 \lesssim u \lesssim U$, of (1) and performing the averaging mentally. The radar memory tube that proved so important in World War II, see Watson-Watt (1946), provides an electronic realization of Galton's procedure.

Other diverse applications of the statistic (1) include: (i) the stacking technique that exploration seismologists employ in combining the seismic traces obtained at nearby locations after letting off a series of shots, (see for example Waters (1978)), (ii)the examination of average hourly rainfall curves by Neyman (1977) in a search for an effect due to cloud seeding and (iii) the aligned activity records of animals prepared by biologists in a search for circadian rhythms (see for example Figure 4.4.1 in Pavlidis (1973)).

The computing of AER's has now become routine with fairly fleeible special purpose computers available for the analysis including the ARC (average response computer) and the CAT (computer of average transients). In a sense these computers may be viewed as providing the $\overline{x}$ button of hand-held calculators for data consisting of curves.

General references to the use and interpretation of ER's include: Donchin and Lindsley (1969), Shagass (1972), John (1977), Thatcher and John (1977), Callaway et al (1978) . References going into statistical concerns in some detail are Glaser and Ruchkin (1976) and Freeman (1980).

2. SOME FORMALIZATION

In order to proceed to an investigation of the statistical properties of AER's and related quantities it is necessary to set down some notation and assumptions. It will be assumed that the experiment is not evolving in time, that the noise processes present are stationary stochastic processes and that functional transformations are time invariant.

The times of application of stimuli will be denoted by σ_j , $j = 0, \pm 1, \pm 2, \ldots$ $M(\cdot)$ will denote the corresponding point process with the number of σ_j satisfying $0 < \sigma_j \leq t$ denoted by $M(t)$. The response observed at time t will be denoted by $Y(t)$. The time period of observation will be taken as the interval $(0,T]$. Further, for convenience, we will set $M = M(T)$. The AER

$$\overline{Y}(u) = \frac{1}{M} \sum_{j=1}^{M} Y(u + \sigma_j) = \frac{1}{M} \int_0^T Y(u + t)\, dM(t) \qquad (2)$$

may now be seen to be able to be viewed as an estimate of

$$E\{Y(u + t) \mid \text{stimulus at time } t\} \quad . \qquad (3)$$

As such it is not a system invariant, but has the distribution of the σ_j melded into it. Suppose one defines

$$E\{Y(t) \mid \text{single stimulus at } t-u\} = \mu + a(t-u) \qquad (4)$$

with $a(t)$ tending to 0 as t tends to ∞ . This function $a(\cdot)$ may be viewed as a system invariant. Supposing that stimuli are applied sufficiently apart in time that their effects do not overlap, then from (4)

$$E\{Y(t) \mid M(u),\, u \leq t\} = \mu + \sum_{\sigma_j \leq t} a(t - \sigma_j)$$

$$= \mu + \int_{-\infty}^{t} a(t - u)\, dM(u) \quad . \qquad (5)$$

This last expression leads to a consideration of the following model for the response series when stimuli of intensity I are applied at times σ_j .

<u>Model 1</u>. Suppose $M(t)$ is a step function jumping by 1 at each σ_j . Suppose μ is a constant and that $a(\cdot)$ is a fixed function vanishing for $t < 0$. Suppose that $\varepsilon(\cdot)$ is a stationary noise process. The response series is given by

$$Y(t) = \mu + I \int_{-\infty}^{\infty} a(t-u)\, dM(u) + \varepsilon(t) \quad . \qquad (6)$$

There have been some investigations that suggest that Model 1 is reasonable in certain practical situations. Suppose that the function $a(\cdot)$ vanishes outside the interval $[0,V]$. Suppose that the stimuli are applied farther than V time units apart, then from (6),

$$\overline{Y}(u) = \mu + I\, a(u) + \overline{\varepsilon}(u) \qquad (7)$$

$0 \leq u \leq V$. Under these conditions the AER is seen to be estimating the function $a(\cdot)$ directly. The experiments of Biedenbach and Freeman (1965) may now be seen as examining the linearity in I and the superposability of the σ_j effects required in Model 1 . They were concerned with responses evoked in the prepyriform cortex of the cat by olfactory tract stimulation. In some of their experiments stimuli of various intensities were applied at well separated times. The AER's appeared to be linear in the intensity I provided I was above a threshold, but not overly large. In others of their experiments the stimuli times were paired, $\sigma_{2j} - \sigma_{2j-1} = d$, for various d . The AER's obtained were compared with the result of superposing at lag d the AER's obtained from well-separated times. It was found that superposability was a reasonable assumption provided that d was not too small . (The first AER's here were computed assuming the stimulus to be the pair of pulses d units apart.)

3. INVESTIGATION OF THE AER FOR MODEL 1

Under regularity conditions large sample approximations may be derived for the mean, variance and distribution of the AER in the case that Model 1 holds. From expression (6) it is apparent that $\bar{Y}(u)$ is made up of a fixed part, $E\,\bar{Y}(u)$, and a stochastic part, $\bar{\varepsilon}(u)$. These two parts may be discussed separatly. The investigation may be carried through under the conditions of Brillinger (1973). Suppose I = 1.

These conditions include: (i) $|M(t) - M(s)| \leq A + B|t-s|$ for some finite A and B, (ii) $\hat{p} = M(T)/T$ tends to p as T tends to ∞ and (iii) for $u > 0$

$$\hat{P}(u) = \int_0^{T-u} (M(u+v) - M(v))dM(v)/T \tag{8}$$

tends to P(u) as T tends to ∞, for almost all u. These conditions require that the stimulus be applied at fixed, but stationarly distributed, time points. Now one has

$$\begin{aligned} E\,\bar{Y}(u) &= \int_0^T (\mu + \int a(u+v-w)\,dM(w))dM(v)\ /M(T) \\ &= \mu + \int a(u-v)\ d\hat{P}(v)\ /\hat{p} \\ &\rightarrow \mu + \int a(u-v)\ dP(v)\ /p \end{aligned} \tag{9}$$

as T tends to ∞.

Expression (9) is not generally the desired $\mu + a(u)$. The distribution of the times of application of the stimulus has been convolved in. Some particular cases include:

a) Purely periodic. In this case, $\sigma_j = j\sigma$ and so $p = 1/\sigma$,

$$\frac{dP(v)}{dv} = \sum_{j=-\infty}^{\infty} \delta(v - j\sigma)/\sigma$$

with $\delta(\cdot)$ the Dirac delta function, giving

$$E\,\bar{Y}(u) = \mu + \sum_j a(u - j\sigma) \quad . \tag{10}$$

This reduces to $\mu + a(u)$, for $0 \leq u \leq V$, in the case that a(u) vanishes for u outside the interval $[0,V]$ and that $\sigma > V$.

b) Poisson process rate p. Suppose that the times of stimulation are those of a Poisson process, then

$$\frac{dP(v)}{dv} = \delta(v)p + p^2$$

giving

$$E\,\bar{Y}(u) = \mu + a(u) + p\int a(v)dv \quad . \tag{11}$$

This is the desired function up to the level value. The clear advantage of this stimulation procedure is that no restrictions are placed on the support of the function $a(\cdot)$ and the rate of application of the stimulus.

c) Stationary mixing point process. Suppose that $M(\cdot)$ is taken to be a realization of a stochastic point process with rate p and autointensity function p(t), then

$$\frac{dP(v)}{dv} = \delta(v)p + p(v)$$

giving

$$E\,\overline{Y}(u) = \mu + a(u) + \int p(u-v)a(v)dv/p \quad . \tag{12}$$

A deconvolution is seen to be required before arriving at the desired $a(\cdot)$.

The variability of the estimate $\overline{Y}(u)$ is that of the stochastic part $\overline{\varepsilon}(u)$. Suppose that $c(t)$ denotes the autocovariance function and $f(\lambda)$ the power spectrum of the series $\varepsilon(t)$. Then

$$\operatorname{var} \overline{Y}(u) = \operatorname{var} \overline{\varepsilon}(u)$$

$$\sim \int c(v)\, dP(v) / Tp^2 \tag{13}$$

and the estimate is asymptotically normal, as $T \to \infty$, (using the results of Brillinger (1975).)

In the case of the purely periodic example above expression (13) becomes

$$\operatorname{var} \overline{Y}(u) \doteq \sum_k f\left(\frac{2\pi k}{\sigma}\right) / 2\pi T \quad . \tag{14}$$

The AER will have inflated variance when the spectrum of $\varepsilon(\cdot)$ has a peak at frequency $2\pi/\sigma$. Now EEG spectra do have peaks, for example corresponding to alpha rhythm. The experimenter needs to be careful not to chose a stimulus interval corresponding to such a peak.

The problem of variance estimation will be returned to later in the paper. Traditional procedures include splitting the data stretch into a number of equal disjoint segments and viewing the AER's of the segments as independent estimates and the ($\pm$) method based on the difference between the successive responses of pairs of responses (see Schimmel (1967).)

4. USES OF MODEL 1

An advantage resulting from having set down Model 1 is that a variety of hypotheses of scientific interest may be examined in a formal fashion. These include:

i) Is there an evoked response ? This is one of the questions examined in the Rapin and Graziani (1967) paper mentioned earlier. They were concerned with whether or not infants were hearing certain loud clicks. The formal hypothesis here is; is $a(t) = 0$ for all t ?

ii) Do two individuals have the same evoked response ? Lewis et al (1972) were concerned with the degree of similarity of evoked responses for monozygotic twins, dizygotic twins and non-twins. Supposing separate experiments are carried out to estimate $a_1(\cdot)$ and $a_2(\cdot)$ for two individuals. The formal hypothesis may be written; is $a_1(t) = a_2(t)$ for all t ?

iii) Are the evoked responses of an individual with respect to two different stimuli the same ? McCormack (1977) measured the visual responses evoked in an individual when different patterns were presented. This hypothesis may be formalized as in ii) .

iv) Are the responses evoked at two symmetrically related locations on the skull the same ? John (1977) is concerned with this question in looking for learning disabilities. In this situation the response recorded is a vector of curves. The component curves may each be modelled as in expression (6) . The possibility that the noise processes are correlated now has to be addressed.

v) If stimulus A results in a response and stimulus B results in another response, is the result of simultaneously applying stimuli A and B the sum of those two responses ? Diamond (1964) is concerned with this question in the case that A is a flash of red light and B a flash of blue light. To examine this question the model (6) must be expanded to include point processes corresponding to each type of stimulus.

vi) Are the effects of the individual stimulus applications superposable or are there interactions (nonlinearities) ? Biedenbach and Freeman (1965) examined this hypothesis in one situation. They found interaction effects in the case that the σ_j were close together.

5. A FORMAL (LINEAR) APPROACH

Suppose that r separate stimuli are available. Let the counting function $M_k(t)$ provide the application times of the k-th of these stimuli. Collect the $M_k(t)$ into an r vector, $\mathbf{M}(t)$. Consider the model

$$\mathbf{Y}(t) = \boldsymbol{\mu} + \int \mathbf{a}(t-u)\, d\mathbf{M}(u) + \boldsymbol{\varepsilon}(t) \tag{15}$$

with $\mathbf{Y}(t)$ a vector of s response series, $\boldsymbol{\mu}$ an s vector, $\mathbf{a}(t)$ an $s \times r$ matrix (with the entry in row j and column k providing the effect of the k-th stimulus on the j-th series), and with $\boldsymbol{\varepsilon}(t)$ a vector of s noise series.

Given data $\mathbf{Y}(t)$, $\mathbf{M}(t)$ $0 \leq t \leq T$, with T sufficiently large, the parameters of the model (15) may be estimated and hypotheses concerning them examined. In this connection it is easier to proceed in the frequency domain. To this end define

$$\mathbf{A}(\lambda) = \int_{-\infty}^{\infty} \exp(-i\lambda t)\, \mathbf{a}(t)\, dt \tag{16}$$

$$\mathbf{d}_Y^T(\lambda) = \int_0^T \exp(-i\lambda t)\, \mathbf{Y}(t)\, dt \tag{17}$$

$$\mathbf{d}_M^T(\lambda) = \int_0^T \exp(-i\lambda t)\, d\mathbf{M}(t) \quad . \tag{18}$$

The transfer function $\mathbf{A}(\lambda)$ may now be estimated by

$$\hat{\mathbf{A}}(\lambda) = \left(\sum_j \mathbf{d}_Y^T\left(\frac{2\pi j}{T}\right) \overline{\mathbf{d}_M^T\left(\frac{2\pi j}{T}\right)}^{\tau} \right) \left(\sum_j \mathbf{d}_M^T\left(\frac{2\pi j}{T}\right) \overline{\mathbf{d}_M^T\left(\frac{2\pi j}{T}\right)}^{\tau} \right)^{-1} \tag{19}$$

with the sums in (19) over n distinct frequencies $2\pi j/T$ near λ .

Suppose next that, among other things, the noise process $\boldsymbol{\varepsilon}(\cdot)$ is stationary, has mean $\mathbf{0}$, spectral density matrix $\mathbf{f}(\lambda)$ and satisfies, for example, the mixing assumption of Brillinger (1974). Then, for $\lambda \neq 0, \pi$ the estimate (19) may be shown to be asymptotically complex normal with mean $\mathbf{A}(\lambda)$ and with vec $\hat{\mathbf{A}}(\lambda)$ having covariance matrix

$$2\pi T\, \mathbf{f}(\lambda) \otimes \left(\sum_j \mathbf{d}_M^T\left(\frac{2\pi j}{T}\right) \overline{\mathbf{d}_M^T\left(\frac{2\pi j}{T}\right)}^{\tau} \right)^{-1} \quad . \tag{20}$$

Further estimates at distinct frequencies are asymptotically independent.

Various hypotheses of scientific interest were indicated in the previous section. In the cases i) – v) these may be written out as hypotheses concerning the entries of the matrix $\mathbf{A}(\lambda)$. With the approximate distribution for $\hat{\mathbf{A}}(\lambda)$ indicated above, a general linear hypothesis concerning the matrix $\mathbf{A}(\lambda)$ may be examined in an analysis of variance fashion.

If desired, the function $\mathbf{a}(u)$ may be estimated by an expression of the form

$$\hat{\mathbf{a}}(u) = Q^{-1} \sum_{q=0}^{Q-1} \exp\left(\frac{2\pi uq}{Q}\right) \hat{\mathbf{A}}\left(\frac{2\pi q}{Q}\right) \quad . \tag{21}$$

Advantages of setting down the model (15) are now seen to include: 1) different sorts of stimuli (including steady ones) may be handled with no greater difficulty than single ones, 2) randomization analyses (where the stimulus applied is selected randomly from those available) may be developed, 3) experimental designs (eg. cross-over, systematic, rotation, repeated measurement) may be incorporated, 4) analysis of variation may be formalized, 5) best linear unbiased estimates may be constructed, approximately. (This approach was introduced in Brillinger (1978).)

6. AN ALTERNATIVE VIEWPOINT

Consider once again the single stimulus, single response series case. Suppose that the function $a(u)$ vanishes outside the interval $[0,V]$ and that the times of application of the stimulus are more than V units apart. Suppose, further, that the autocovariance function of the noise series is essentially zero after lag V . Then the linear model may be written

$$\begin{aligned} Y_j(u) &= Y(u+\sigma_j) \\ &= \mu + a(u) + \varepsilon_j(u) \end{aligned} \tag{22}$$

for $0 \leq u \leq V$, $j = 1,\ldots,M(T)$, with the noise process $\varepsilon_j(u) = Y(u+\sigma_j)$ uncorelated with $\varepsilon_k(v)$, $j \neq k$. The AER $\bar{Y}(u)$ is simply the mean of the separate responses. The model (22) is now seen to be the linear model of multivariate analysis and the model of growth curves with replicated observations.

Adopting this viewpoint means then that one can take over the whole apparatus and procedures from those fields. For example, a situation where a number of distinct stimuli are applied may be modelled by

$$Y_{ij}(u) = \mu_{ij} + a(u) + b_i(u) + \varepsilon_{ij}(u) \tag{23}$$

with i indexing the various stimuli, j indexing the replicates, $a(\cdot)$ denoting an overall effect and $b_i(\cdot)$ providing the effect due to the i-th stimulus. The results that have been developed for MANOVA of complex experimental designs may be taken over directly.

Brillinger (1980) is a review paper on the analysis of variance of curves in the case that the noise process, ($\varepsilon_j(\cdot)$ in (22)), is stationary.

7. SUPERPOSABILITY

The next two sections examine certain effects resulting from a departure from the assumptions of Model 1 and (15). These models were motivated by a consideration of expression (3). In the superposable case it turned out to be enough to discuss the expected value (3) in the case of a single stimulus time, namely to consider expression (4).

It seems natural to move on to a consideration of characteristics like

$$E\{Y(u+t) \mid \text{stimulus at times } t,\ t-v\} \quad . \tag{24}$$

This latter may be estimated by an expression of the form

$$(2\beta T)^{-1} \sum_{j\neq k} Y(u+\sigma_j)\,\{\,|\sigma_j-\sigma_k-v| < \beta\,\} \quad . \tag{25}$$

with β small and $\{E\}$ here defined to be 1 if the event E is true and 0 otherwise.

This statistic crops up, for example, in a consideration of;
Model 2. Suppose M(t) is a step function jumping by 1 at each σ_j. Suppose that μ is a constant and that $a(\cdot)$, $b(\cdot,\cdot)$ are fixed functions. Suppose that $\varepsilon(\cdot)$ is a stationary noise process. The response series is given by

$$Y(t) = \mu + \int_{-\infty}^{\infty} a(t-u)\, dM(u) + \iint_{u \neq v} b(t-u,t-v)\, dM(u)\, dM(v) + \varepsilon(t) \quad . \tag{26}$$

This model allows an interaction between pairs of stimuli effects. If the stimulus process consists of an impulse at times 0 and d alone then (26) gives

$$Y(u) = \mu + a(u) + a(u-d) + b(u,u-d) + \varepsilon(t) \tag{27}$$

for $u \geqslant d$. In the causal case $a(u)$, $b(u,v)$ will vanish for $u < 0$, $v < 0$ and (27) will hold for all u.

It is clear from expression (27) may be examined via two-pulse experiments. The computations are also direct in the case that the times of stimulus application are those of a Poisson process, (see Krausz (1975).)

8. ROBUST/RESISTANT ESTIMATES

These days research workers are very much schooled in the sensitivity of the mean to outliers. Now large transients, that are artifacts, occur commonly in EEG's - movement, eye blink, EMG (muscle electromyograph), barbituate spikes, mains pulses all occur. These all affect the AER.

To deal with this problem some researchers have computed the median response at each lag, see for example Figure 2.22 in Rosenblith (1962). Some statistical properties of this estimate have been derived, see Section 4.7 in Glaser and Ruchkin (1976). However, being nonlinear the median computation can exhibit spurious harmonics of the noise components and behave in nonelementary fashions, a case in point is illustrated in Figure 1 of Ruchkin and Walter (1975). It is clear that various trimmed means could be used in place of the median.

The median and the trimmed mean just mentioned operate separately at each time point. It seems worthwhile to develop an estimate that weights, (possibly rejects), whole ER curves differentially.

Suppose that the notation of the Alternate Viewpoint section is adopted. Let $\hat{\theta}(u)$ denote the estimate, about to be constructed, at lag u. Let $\hat{\rho}$ denote an estimate of scale and $\|Y - \theta\|$ a measure of distance, for example,

$$\|Y - \theta\|^2 = \int_0^V (Y(u) - \theta(u))^2 du \quad . \tag{28}$$

Set

$$\hat{\theta}(u) = \sum_j W_j Y_j(u) \Big/ \sum_j W_j \tag{29}$$

with

$$W_j = W(\|Y_j - \hat{\theta}\| / \hat{\rho}) \tag{30}$$

$W(\cdot)$ being a non-negative weight function. (An example will be given shortly.)

Both sides of equation (29) involve the desired estimate $\hat{\theta}(\cdot)$. In practice an iterative procedure will be set up, based on an initial estimate, (perhaps the AER), with iterations carried out until the current estimate is changing little. (See below.)

As a specific example one can consider the trimmed mean

$$\hat{\theta}(u) = \Sigma' \, Y_j(u) \, / \, \beta M \tag{31}$$

with Σ' denoting summation over the βM smallest $\| Y_j - \hat{\theta} \|$.

By analogy with equations (2.11), (2.12) of Huber (1977) one is led to consider parameters satisfying the relationships

$$E\{(Y(u) - \theta(u)) \, W(\| Y(u) - \theta(u) \| / \rho)\} = 0 \tag{32}$$

$$E\{U(\| Y - \theta \| / \rho) \, \| Y - \theta \|^2 / \rho^2 - V(\| Y - \theta \| / \rho)\} = 0 \tag{33}$$

The choice

$$\begin{aligned} W(u) &= (1 - u^2)^2 & & 0 \leq u \leq 1 \\ &= 0 & & \text{otherwise} \end{aligned} \tag{34}$$

gives the biweight. The choice

$$\begin{aligned} U(u) &= a^2 & & 0 \leq u \leq a \\ &= u^2 & & a \leq u \leq b \\ &= b^2 & & b \leq u \end{aligned} \tag{35}$$

and $V(u) = 1$ seems useful in practice.

If all of the data is available in a computer at the same time then the estimate (29) may be computed in an iterative fashion as follows;

$$\hat{\theta}_{k+1}(u) = \sum_j W_{j,k} \, Y_j(u) \, / \, \sum_j W_{j,k} \tag{36}$$

$$\hat{\rho}^2_{k+1} = \sum_j U_{j,k} \, \| Y_j - \hat{\theta}_k \|^2 / \sum_j V_{j,k} \tag{37}$$

$$W_{j,k} = W(\| Y_j - \hat{\theta}_k \| / \hat{\rho}_k) \tag{38}$$

with $U_{j,k}$ and $V_{j,k}$ defined by expressions similar to (38).

9. A RECURSIVE PROCEDURE

The estimate introduced in the previous section has the disadvantage of requiring the experimenter to retain all the data. A recursive procedure requiring only the most recent estimate and the just-collected response will now be set down. The procedure is motivated by stochastic approximation and the known-scale location estimate of Martin and Masreliez (1975). Let $\hat{\theta}_j$ denote the estimate based on $Y_1, \cdots, Y_j$. Then set

$$\hat{\theta}_{j+1}(u) = \hat{\theta}_j(u) - \frac{L}{j} W(\| Y_{j+1} - \hat{\theta}_j \| / \hat{\rho}_j) \, (Y_{j+1}(u) - \hat{\theta}_j(u)) \tag{39}$$

$$\hat{\rho}_{j+1} = \hat{\rho}_j - \frac{L}{j}\left(U(\| Y_{j+1} - \hat{\theta}_j \| / \hat{\rho}_j) \| Y_{j+1} - \hat{\theta}_j \|^2 / \hat{\rho}_j^2 - V(\| Y_{j+1} - \hat{\theta}_j \| / \hat{\rho}_j)\right) \tag{40}$$

for some constants L .

The estimate of (39) will not be the same as (29), even when all the data has been collected; however the two estimates will converge to the same value, (given by (32)) as T tends to ∞ .

In the case of the 100β per cent trimmed mean the recursive equations become, for the choice $L = 1/\beta$ in the first

$$\hat{\theta}_{j+1}(u) = \hat{\theta}_j(u) - \frac{1}{\beta j}(Y_{j+1}(u) - \hat{\theta}_j(u)) \quad \text{if } \| Y_{j+1} - \hat{\theta}_j \| \leq \hat{\rho}_j \tag{41}$$

$$= \hat{\theta}_j(u) \quad \text{otherwise}$$

$$\hat{\rho}_{j+1} = \hat{\rho}_j - \frac{L}{j}\left(\frac{1}{\beta} - 1\right) \quad \text{if } \| Y_{j+1} - \hat{\theta}_j \| \leq \hat{\rho}_j \tag{42}$$

$$= \hat{\rho}_j + \frac{L}{\beta j} \quad \text{otherwise}$$

The effect of employing the estimate (41) is to exclude from the averaging any (complete) evoked response that deviates substantially from the bulk of the responses. The average response computers mentioned earlier typically have circuits to detect signals above an arbitrarily established amplitude at some lag u . This procedure has the disadvantage of not rejecting responses that are not quite abnormal at any u, but that over all lags are quite abnormal. The estimates (29) and (39) are of multivariate nature.

10. RESISTANCE (FREQUENCY DOMAIN)

The above discussion refers to time domain procedures. As the example of Ruchkin and Walter (1975) shows, such procedures can have undesired effects on quasi-sinusoidal signals or noise. Some frequency-based procedures are now presented for constructing resistant estimates - estimates not substantially affected by frequency domain abnormalities.

Suppose that the various ER's are separated in time so that the data may be described by expression (22) . In the frequency domain this leads to

$$d^V_{Y_j}(\lambda) = \int_0^V e^{-i\lambda u} Y_j(u)\, du \tag{43}$$

$$\doteq A(\lambda) + d^V_{\varepsilon_j}(\lambda) \quad \lambda \neq 0 \tag{44}$$

$$= a_j + ib_j = C_j \exp\{i\delta_j\} \tag{45}$$

$j = 1,\ldots,M$. In expression (45) here, dependence on λ has been suppressed in the notation. Forming the AER corresponds to averaging the a_j, b_j with respect to j .

One means of detecting abnormal values is to plot, say for $\lambda = 2\pi v/V$, $v = 0,1,\ldots, V/2$, the points (a_j,b_j), $j = 1,\ldots,M$. The point corresponding to the AER will sit in the middle of a cloud of points.

It is now apparent that one might proceed by applying scalar procedures to the individual components of (a_j,b_j) or (C_j,δ_j) or $(\log C_j,\delta_j)$. Alternatively, taking note of the Tukey (1980) procedure for a single series one might; (a) apply scalar resistance methods to δ_j , (or $\exp\{i\delta_j\}$) with result ϕ_j , (or $d_j\exp\{i\phi_j\}$); (b) apply scalar methods to $\log C_j$ with result $\log C_j^*$; (c) take the average of the $C_j^*\exp\{i\phi_j\}$, (or of the $C_j^* d_j \exp\{i\phi_j\}$) as the estimate at frequency λ .

11. A PARTIALLY PARAMETRIC MODEL

Suppose that the model (22) is replaced by

$$Y_j(u) = \mu_j + I_j S(u + \gamma_j) + \varepsilon_j(u) \tag{46}$$

for $0 \leq u \leq V$ and $j = 1,\ldots,M$. That is, it is allowed that the response not occur always at the same time lag (latency) after the application of the stimulus and it is allowed that the intensity of the response is not always the same. One hence has the problem of estimating the γ_j and I_j as well as the desired signal $S(\cdot)$.

These parameters may be estimated via a frequency domain procedure. To this end set

$$Y_{jv} = d_{Y_j}^V\left(\frac{2\pi v}{V}\right) , \quad S_v = d_S^V\left(\frac{2\pi v}{V}\right) \tag{47}$$

with a similar definition of ε_{jv} . Then expression (46) yields

$$Y_{jv} = S_v I_j \exp\{i\gamma_j 2\pi v/V\} + \varepsilon_{jv} \tag{48}$$

$v \neq 0$. Now the ε_{jv} may often be treated as independent complex normal variates with mean 0 and variance $2\pi V f(2\pi v/V)$. This last suggests setting down the (approximate) negative log-likelihood

$$\sum_j \sum_v \left(\log f\left(\frac{2\pi v}{V}\right) + |Y_{jv} - S_v I_j \exp\{i\gamma_j 2\pi v/V\}|^2 / 2\pi V f\left(\frac{2\pi v}{V}\right) \right) , \tag{49}$$

and the estimation of the unknown parameters by the minimization of (49) . In order that the model be identifiable constraints, such as,

$$\sum_j \gamma_j = 0 , \quad \sum_j I_j^2 = 1 \tag{50}$$

will have to be introduced. If an iterative procedure is employed in this minimization, then initial values for I_j, γ_j might be obtained by crosscorrelating $Y_j(u)$ with one of the resistant estimates indicating earlier.

In the case that the noise spectrum $f(\cdot)$ is constant the Woody (1967) adaptive filter may be seen to be one means of seeking the minimum of expression (49). The above formulation is seen to provide a maximum likelihood interpretation of Woody's procedure and an extension of it to handle autocorrelated noise.

12. A FULLY PARAMETRIC MODEL

Freeman (1975) has made substantial use of the following parametric model,

$$\begin{aligned} a(u|\theta) &= \sum_k \alpha_k \exp\{-\beta_k u\}\cos(\gamma_k u + \delta_k) && \text{for } u \geq u_0 \\ &= 0 && \text{for } u < u_0 \end{aligned} \tag{51}$$

with θ denoting the parameters . The model may be fit to each individual response or to the AER itself. If u_0 is known to sufficient accuracy that only $Y_j(u)$ values with $u > u_0$ are employed, then the fitting is simplified. The fitting may be carried through in either the time domain or the frequency domain. In the case that the error series $\varepsilon(\cdot)$ is stationary it seems simplest to work in the frequency domain. This is done in Bolt and Brillinger (1979) where the computations required are laid out and the asymptotic distributions of the estimates developed.

Freeman (1975) makes use of the model (51) as it represents the response for linear differential equations. In one case involving two damped sine waves, he views the larger wave as representing intracortical negative feedback and the smaller as representing other feedback loops.

13. REFERENCES

[1] Bergamini, L., Bergamasco, B., Fra, L., Gandiglio, G. and Mutani, R., Diagnostic recording of somato-sensory cortical evoked potentials in man, Electroenceph. clin. Neurophysiol. 22 (1967) 260-262.

[2] Biedenbach, M.A. and Freeman, W.J., Linear domain of potentials for the prepyriform cortex with respect to stimulus parameters, Experimental Neurology 11 (1965) 400-417.

[3] Bolt, B.A. and Brillinger, D.R., Estimation of uncertainties in eigenspectral estimates from decaying geophysical time series, Geophys. J. r. astr. Soc. 59 (1979) 595-603.

[4] Brillinger, D.R., Estimation of the mean of a stationary time series by sampling, J. Appl. Prob. 10 (1973) 419-431.

[5] Brillinger, D.R., The Fourier analysis of stationary processes, Proc. IEEE 62 (1974) 1628-1643.

[6] Brillinger, D. R., A note on the estimation of evoked response, Biol. Cybernetics 31 (1978) 141-144.

[7] Brillinger, D. R., Analysis of variance and problems under time series models, in Krishnaiah, P.R. (ed.), Handbook of Statistics, Vol. 1 (North-Holland, Amsterdam, 1980).

[8] Buys-Ballot, C.H.D., Les Changements Periodiques de Temperature, Utrect (1847).

[9] Callaway, E., Tueting, P. and Koslow, S.H., Event-Related Brain Potentials in Man (Academic, New York, 1978).

[10] Dawson, G.D., A summation technique for detecting small signals in a large irregular background, J. Physiol. 115 (1951) 2.

[11] Diamond, S. P., Input-output relations, Ann. New York Acad. Sci. 112 (1964) 160-171.

[12] Donchin, E. and Lindsley, D. B., Average Evoked Potentials (NASA, Washington, 1969).

[13] Freeman, W.J., Mass Action in the Nervous System (Academic, New York, 1975).

[14] Freeman, W.J., Measurement of cortical evoked potentials by decomposition of their wave forms, J. Cybernetics and Inf. Sci. in press (1980).

[15] Glaser, E.M. and Ruchkin, D.S., Principles of Neurobiological Signal Analysis (Academic, New York, 1976).

[16] Huber, P.J., Robust covariances, in: Gupta, S.S. and Moore, D.S. (eds.) Statistical Decision Theory and Related Topics II (Academic, New York, 1977).

[17] John, E.R., Neurometrics (Wiley, New York, 1977).

[18] Krausz, H.I., Identification of nonlinear systems using random impulse train inputs, Biol. Cybernetics 19 (1975) 217-230.

[19] Laplace, M., Traite de Mecanique Celeste 5 (Bachelier, Paris, 1825).

[20] Lewis, E.G., Dustman, R.E. and Beck, E.C., Evoked response similarity in monozygotic, dizygotic and unrelated individuals: a comparative study, Electroenceph. clin. Neurophysiol. 23 (1972) 309-316.

[21] Martin, R.D. and Masreliez, C.J., Robust estimation via stochastic approximation, IEEE Trans. Inf. Theory 21 (1975) 263-271.

[22] McCormack, G.L., The Albino Visual Pathway Defect and Human Strabismus, Ph.D. Thesis, School of Optometry, University of California, Berkeley (1977).

[23] Neitzel, E.B., Seismic reflection records obtained by dropping a weight, Geophysics 23 (1958) 58-80.

[24] Neyman, J., A statistician's view of weather modification technology (a review), Proc. Nat. Acad. Sci. USA 74 (1977) 4714-4721.

[25] Pavlidis, T., Biological Oscillators: Their Mathematical Analysis (Academic, New York, 1973).

[26] Rapin, I. and Graziani, L.J., Auditory-evoked responses in mormal, brain-damaged and deaf infants, Neurology 17 (1967) 881-891.

[27] Regan, D., Electrical responses evoked from the human brain, Sci. Amer. 242(1979) 134-146.

[28] Rosenblith, W.A., Processing Neuroelectric Data (MIT, Cambridge, 1962).

[29] Ruchkin, D.S. and Walter, D.O., A shortcoming of the median evoked response, IEEE Trans. Biomed. Eng. 22 (1975) 245.

[30] Schimmel, H., The (±) reference: accuracy of estimated mean components in average response studies, Science 157 (1967) 92-93.

[31] Shagass, C., Evoked Brain Potentials in Psychiatry (Plenum, New York,1972).

[32] Thatcher, R.W. and John, E.R., Foundations of Cognitive Processes (Wiley, New York, 1977).

[33] Tukey, J.W., Can we predict where "time series" should go next?, in: Brillinger, D.R. and Tiao, G.C. (eds.), Directions in Time Series (Institute of Mathematical Statistics, Hayward, 1980).

[34] Waters, K.H., Reflaction Seismology (Wiley, New York, 1978).

[35] Watson-Watt, R., The evolution of radar location, J. Inst. Elec. Eng. 93 (1946) 6-20.

[36] Woody, C.D., Characterization of an adaptive filter for the analysis of variable latency neuroelectric signals, Med. and Biol. Eng. 5 (1967) 539-553.

-0-0-0-

[1]Prepared with the support of the National Science Foundation Grant MCS-772986 .

STATISTICS AND RELATED TOPICS
M. Csörgö, D.A. Dawson, J.N.K. Rao, A.K.Md.E. Saleh (eds.)

MEAN RESIDUAL LIFE

W. J. Hall and Jon A. Wellner

Statistics Department
University of Rochester
Rochester, New York
U.S.A.

The mean residual life function e at age x is defined to be the expected remaining life given survival to age x; it is a function of interest in actuarial studies, survivorship analysis, and reliability. Here we characterize those functions which can arise as mean residual life functions, present and study an "inversion formula" which expresses the survival function in terms of e, and collect a variety of facts about e and other residual moments: inequalities for e, new characterizations of the exponential distribution, inequalities for coefficients of variation, and limiting behavior of e at 'great age'. We also discuss applications to parametric modelling.

1. INTRODUCTION

Let X be a non-negative random variable with right continuous distribution function (df) F, and survival function $\overline{F} = 1 - F$, on $\mathbb{R}^+$ and suppose that $F(0) = 0$ and $\mu \equiv E(X) = \int_0^\infty x dF(x) = \int_0^\infty \overline{F}(x)dx < \infty$; write $T \equiv T_F \equiv \inf\{x: F(x) = 1\} \leq \infty$. The <u>mean residual life</u> (MRL) function or remaining life expectancy function at age x is defined as

$$e(x) = e_F(x) \equiv E(X-x|X>x) = \int_x^\infty \overline{F}dI/\overline{F}(x), \qquad \text{for } x \geq 0, \tag{1.1}$$

and $e(x) \equiv 0$ whenever $\overline{F}(x) = 0$. We use I to denote the identity function and Lebesgue measure on $\mathbb{R}^+$.

The discretized version of the MRL function e has had considerable use in life table analysis (see e.g. Chiang, 1968, pages 189 and 213-214; Gross and Clark, 1975, page 25ff), and estimation of $e = e_F$ on the basis of samples from F has

---

<u>AMS 1970 Subject classifications</u>: Primary 62P05, 62N05, 62E10; Secondary 62G99

<u>Key words and phrases</u>: life expectancy, characterizations, residual variance, coefficient of variation, new better than used in expectation, renewal theory, regular variation, inequalities.

Research of the second author supported by National Science Foundation Grant MCS 77-02255.

recently been considered by Yang (1978) and Hall and Wellner (1980). Here we are concerned with the behavior of e as a function (of x and F). For example, what functions e can arise as MRL functions? (What is the image or range E of the set F of all df's with finite mean under the function $F \to e_F$?) Does the mean residual life function e_F determine the df F completely? (Is the function $F \to e_F$ one-to-one?) These questions are answered in Section 3, along with a brief review of the related literature.

The remainder of the paper presents a number of related properties of e, and of other residual moments, and applications thereof. Among them are elementary inequalities for e (Section 2); some characterizations of the exponential distribution, some inequalities for coefficients of variation, and Pyke's variance formula (Section 4); characterization of linear segments in e, decomposition of e, and relations with renewal theory (Section 5); and limiting properties of e 'at great age' (Section 6). Use of some of the results of Sections 5 and 6 in parametric modelling appears in Section 7.

2. BOUNDS FOR MRL

The following elementary inequalities yield bounds for the MRL function e_F: Since $e_F(x) + x = E(X \mid X > x)$, we have $\{e_F(x) + x\}\overline{F}(x) = E(X\cdot 1_{(X > x)}) = \mu - E(X\cdot 1_{(X \le x)})$. But $E(X\cdot 1_{(X > x)}) \le T_F\overline{F}(x)$, $\le \mu$, and $\le (EX^r)^{1/r}\overline{F}(x)^{1-(1/r)}$ for $r > 1$, the last by Hölder's inequality. Similarly, $E(X\cdot 1_{(X \le x)}) \le xF(x)$ and also $\le (EX^r)^{1/r}F(x)^{1-(1/r)}$ for $r > 1$. These five inequalities, together with conditions for equality, yield (a) - (e) below; (f) - intuitively trivial as is (a) - follows from (d).

Proposition 1. If F is non-degenerate with mean μ and $\nu_r \equiv EX^r \le \infty$,

(a) $e_F(x) \le (T - x)^+$ for all x, with equality if and only if $F(x) = F(T-)$ or 1;

(b) $e_F(x) \le (\mu/\overline{F}(x)) - x$ for all x with equality if and only if $F(x) = 0$;

(c) $e_F(x) < (\nu_r/\overline{F}(x))^{1/r} - x$ for all x and any $r > 1$;

(d) $e_F(x) \ge (\mu - x)^+/\overline{F}(x)$ for $x < T$ with equality if and only if $F(x) = 0$;

(e) $e_F(x) > \{\mu - F(x)(\nu_r/F(x))^{1/r}\}/\overline{F}(x) - x$ for $x < T$ and any $r > 1$;

(f) $e_F(x) \ge (\mu - x)^+$ for all x, with equality if and only if $F(x) = 0$ or 1.

If F is degenerate at μ, $e_F(x) = (\mu - x)^+$ for all x.

3. CHARACTERIZATIONS OF MRL; THE INVERSION FORMULA

We first present some properties of the MRL function e_F; some are apparently trivial but are included to provide the basis for a characterization theorem.

<u>Proposition 2</u>. (Properties of MRL). (a) e is non-negative and right-continuous, and $e(0) = \mu > 0$; (b) $v \equiv e + I$ is non-decreasing; (c) e has left limits everywhere in $(0,\infty)$ and has positive increments at discontinuities, if any; (d) $e(x-) > 0$ for $x \varepsilon (0,T)$; if $T < \infty$, $e(T-) = 0$ and e is continuous at T; (e) $\bar{F}(x) = \{e(0)/e(x)\}\exp\{-\int_0^x (1/e)dI\}$ for all $x < T$; (f) $\int_0^x (1/e)dI \to \infty$ as $x \to T$.

<u>Proof</u>. The continuity of the numerator in (1.1) and the right continuity and positivity of the denominator establishes much of (a). To prove (b), consider $t > 0$ and $x + t < T$; then $v(x+t) - v(x) \geq \{(\int_{x+t}^{\infty} \bar{F}dI - \int_x^{\infty} \bar{F}dI/\bar{F}(x)\} + t =$ $\{-\int_x^{x+t} \bar{F}dI/\bar{F}(x)\} + t \geq 0$. If $x < T \leq x + t$, this remains true, and if $T \leq x$ (b) is trivial. Now (c) follows from (b).

For $x < T$, $e(x-) \geq \int_{x-}^{T} \bar{F}dI > 0$. If $x < T < \infty$, $v(x) \leq v(T) = T$ so $e(x) \leq T - x$ and $e(T-) = 0$. Hence e is continuous at T (when $T < \infty$), and (d) is proved. To prove (e), write $k(x) \equiv \int_x^{\infty} \bar{F}dI = \bar{F}(x)e(x)$; then

$$\int_0^x (1/e)dI = -\int_0^x d(\log k(t)) = -\log\{k(x)/k(0)\} = -\log\{\bar{F}(x)e(x)/e(0)\}, \tag{3.1}$$

from which (e) follows after negative exponentiation. Now $k(x) \to 0$ as $x \to T$, so the right side in (3.1) $\to \infty$, proving (f). □

The function $v(x)$ in (b) is the <u>mean age at death conditional on survival to age x</u>; it is intuitive that it should be monotone. It increases from μ at $x = 0$ to T at $x = T$ $(\leq \infty)$, and is flat only on intervals having zero probability. The formula (e) shows that $\bar{F}$ may be recovered from e_F, and hence a one-to-one correspondence exists between survival functions (with $\mu \to \infty$) and MRL functions. This formula has been discovered, rediscovered, and generalized repeatedly: it was alluded to by Cox (1962, Exercise 1, page 128); given explicitly and proved very simply by Meilijson (1972); also presented in Swartz (1973), Laurent (1974), and Galambos and Kotz (1978); "extended" to expectations of the form $E\{h(X)|X>x\}$, h a fixed function, X with df F absolutely continuous, by a series of authors including Hamdan (1972), Gupta (1975), and Shanbag and Rao (1975) (see the discussion on pages 21 and 32 of Galambos and Kotz in this regard); and extended to df's on $(-\infty,\infty)$ in Kotz and Shanbag (1980). We demonstrate a variety of uses of the inversion formula (e) in the next two sections.

The property (f) describes a limitation on how fast e can grow; thus, for each $k \geq 1$, $e(x) \sim c\, x(\log x) \cdots (\log_{k-1} x)(\log_k x)^{1+\varepsilon}$ is seen to be possible for $\varepsilon = 0$, impossible for $\varepsilon > 0$. Behavior of e_F 'at great age' is pursued in Section 6.

We now proceed to find a list of characteristic properties, in that any function e satisfying them will be a MRL function for some survival function $\bar{F}$, namely

$\overline{F}$ defined by the inversion formula (e). Since property (c) above follows directly from (b), it need not be explicitly required; likewise, if $e(T) = 0$ for some $T < \infty$, then (b) implies $\int_0^T (1/e)dI = \infty$ (since $e(x) \leq T - x$) and so (f) need only be required when e is strictly positive everywhere. We thus are led to:

Characterization Theorem. Suppose $e: \mathbb{R}^+ \to \mathbb{R}^+$ satisfies (a) e is right-continuous and $e(0) > 0$; (b) $v \equiv e + I$ is non-decreasing; (c) if $e(x-) = 0$ for some $x = x_o$, then $e(x) = 0$ on $[x_o, \infty)$; (d) if $e(x-) > 0$ for all x, then $\int_0^\infty (1/e)dI = \infty$.
Let $T \equiv \inf\{x: e(x-) = 0\} \leq \infty$, and define $\overline{F}$ by (e) for $x < T$ and $\overline{F}(x) \equiv 0$ for $x \geq T$. Then $F \equiv 1 - \overline{F}$ is a df on $\mathbb{R}^+$ with $F(0) = 0$, $T_F = T$, finite mean $\mu_F = e(0)$ and MRL function $e_F = e$.

For related results, see Theorem 2.1 of Bhattacharjee (1980) (who has a more complex list of characteristic properties) and Proposition 2 of Kotz and Shanbag (1980) (more general and more complex).

Counterexamples can be constructed to demonstrate that none of (a) - (d) can be omitted. In particular, if $e(x) = 1 + x^2$, $x \geq 0$, then $\int_0^\infty (1/e)dI = \arctan(x) \to \pi/2 < \infty$ as $x \to \infty$. According to Proposition 2(f), e cannot be a MRL function; nevertheless, the inversion formula (e) may be used to define a df F whose MRL function e_F turns out to be $(1 + x^2)\{1 - \exp(\arctan(x) - \frac{\pi}{2})\} \neq e(x)$.

Proof. We need to prove: (a') $\overline{F}$ is non-negative, right-continuous, with $\overline{F}(0) = 1$; (b') $\overline{F}$ is non-increasing; (c') $\overline{F} > 0$ for $x < T$, and if $T = \infty$, $\overline{F}(x) \to 0$ as $x \to \infty$; and (d') $\mu_F < \infty$ and $e_F = e$.

Now (a') follows from (a) and (e). To prove (b'), consider $0 < t \leq t + x < T$; then $\overline{F}(x+t)/\overline{F}(x) = \{e(x)/e(x+t)\}\exp\{-\int_x^{x+t}(1/e)dI\}$. But

$$\begin{aligned}\int_x^{x+t}(1/e)dI &= \int_x^{x+t}(v(u) - u)^{-1}du \\ &\geq \int_x^{x+t}(v(x+t) - u)^{-1}du \quad \text{by (b)} \\ &\geq \log\{[e(x+t) + t]/e(x+t)\}\end{aligned} \tag{3.2}$$

and hence

$$\int_x^{x+t}(1/e)dI \geq \log\{e(x)/e(x+t)\}, \tag{3.3}$$

again by (b). Therefore $\overline{F}(x+t)/\overline{F}(x) \leq \{e(x)/e(x+t)\}\exp\{-\log[e(x)/e(x+t)]\} = 1$, proving (b') for $x < T$. The case of $x \geq T$ is trivial.

For $x < T$, there exists $\varepsilon > 0$ for which $\inf\{e(t): 0<t<x\} \geq \varepsilon$; hence $\int_0^x (1/e)dI < \infty$ and $\overline{F}(x) > 0$. Now consider (3.2) with $x = 0$ and $t < T = \infty$: $\int_0^t (1/e)dI > \log\{[e(t) + t]/e(t)\}$ so that $\overline{F}(t) \leq \{e(0)/e(t)\} \cdot \{e(t)/[e(t) + t]\}$ which decreases to 0 as $t \to \infty$ by (b). Note that this last inequality

$\overline{F}(x) \leq e(0)/v(x)$ is equivalent to $e(x) \leq (e(0)/\overline{F}(x)) - x$, stated in Proposition 1(b) for a MRL function $e = e_F$.

To prove (d'), we first show that e has property (f) (already assumed in (d) if $T = \infty$): simply apply (3.3) with $x = 0$ and let $t \to T$. Now note from (e) that $-\log\{\overline{F}(x)e(x)/e(0)\} = \int_0^x (1/e)dI$ for $x < T$ (and hence $\overline{F}e \downarrow 0$). The right side has derivative $1/e(x)$. Hence the left side is differentiable, and equating derivatives yields $\overline{F}(x) = -(d/dx)\{\overline{F}(x)e(x)\}$, and therefore

$$\int_x^T \overline{F}dI = -\overline{F}e\Big|_x^T = \overline{F}(x)e(x) \qquad \text{for all } x < T. \tag{3.4}$$

In particular, $\mu_F \equiv \int_0^\infty \overline{F}dI = \int_0^T \overline{F}dI < \infty$, and dividing (3.4) by $\overline{F}(x)$ yields $e_F = e$. □

How irregular or ill-behaved can e_F be? It inherits continuity and differentiability properties from F at all points except $T = T_F$; and although $e + I$ is monotone, e may oscillate with $0 \leq \liminf_{x\to\infty} e(x) < \limsup_{x\to\infty} e(x) \leq \infty$, with one or both equalities holding. For further discussion see Section 5.

4. RESIDUAL MOMENT FORMULAS AND SOME CHARACTERIZATIONS

Introduce the notation

$$\overline{F}^{(r)}(x) \equiv \int_x^\infty \overline{F}^{(r-1)}dI \ (\leq \infty), \qquad \text{for } r = 1,2,\ldots \tag{4.0}$$

where $\overline{F}^{(0)} \equiv \overline{F}$, and $\nu_r \equiv E_F X^r$ ($\leq \infty$). When normalized, these are the survival functions corresponding to the df's $F_{(r)}$ of Smith (1959, page 6). We find by successive integration by parts that

$$\nu_r = r!\ \overline{F}^{(r)}(0) \qquad \text{for } r = 1,2,\ldots \tag{4.1}$$

with $\overline{F}^{(r)}$ finite if and only if ν_r is finite. Hence $(d/dx)\overline{F}^{(r)} = -\overline{F}^{(r-1)}$.

Now introduce the <u>residual life distribution at age a</u>:

$$\overline{F}_a(x) = P(X > a + x \mid X > a) = \overline{F}(a + x)/\overline{F}(a)$$

for $a < T$, and let X_a represent a random variable with df F_a. Appending the subscript 'a' on previous symbols we have $\overline{F}_a^{(r)}(x) = \int_x^\infty \overline{F}_a^{(r-1)}dI$ which equals $\overline{F}^{(r)}(a + x)/\overline{F}(a)$ by induction on r. Hence (4.1) yields:

<u>Proposition 3</u>. For $r = 0,1,\ldots$ and $a < T$,

$$\nu_{r,a} = r!\ \overline{F}^{(r)}(a)/\overline{F}(a). \tag{4.2}$$

In particular,

$$\mu_a \equiv \nu_{1,a} = e(a) = \overline{F}^{(1)}(a)/\overline{F}(a), \tag{4.3}$$

$$\nu_{2,a} = 2\overline{F}^{(2)}(a)/\overline{F}(a), \tag{4.4}$$

$$\sigma_F^2(a) \equiv \sigma_a^2 = \nu_{2,a} - \mu_a^2 = \mu_a\{2\frac{\overline{F}^{(2)}(a)}{\overline{F}^{(1)}(a)} - \mu_a\}. \tag{4.5}$$

In the following pages we have used the notation $k \equiv \overline{F}^{(1)}$, $K \equiv \overline{F}^{(2)}$, so $K' = -k$, $k' = -\overline{F}$, etc.

Now for some characterizations of the exponential distribution. It is elementary (e.g. from the inversion formula (e)) that a constant e_F characterizes the exponential distribution. It likewise follows from the differential equation (4.2) with $r = 1$: $\mu = -k(x)/k'(x)$ for all x. A constant residual life moment of any order does likewise; see Theorem 2.3.2, page 33 of Galambos and Kotz (1978) and the accompanying discussion:

Proposition 4. Suppose r is a positive integer. Then $\nu_{r,a} = \nu\ (> 0)$ for all $a \in \mathbb{R}^+$ if and only if F is exponential.

Proof. This follows directly by expressing (4.2) as a differential equation, recalling that $\overline{F}(x) = (-1)^r(d^r/dx^r)\overline{F}^{(r)}(x)$, namely $\nu_r = r!\ (-1)^r g(x)/g^{(r)}(x)$ where $g = \overline{F}^{(r)}$ and $g^{(r)}$ is the rth derivative of g. □

If we only ask that e_F be constant a.e.(F), then other distributions are possible. Specifically, the geometric distribution on a lattice in $(0,\infty)$ has e_F constant on the lattice but has slope -1 off the lattice. (This is a characterization, among distributions with positive probability everywhere on the lattice.) But other e's (and hence F's) may be constructed: take e constant except on an interval $[a,b)$ where it is continuous with slope -1; then F is exponential, except flat on the interval and with a mass point at the right end of the interval; additional mass points may be inserted inside the interval. Other discrete distributions may also be constructed, non-lattice or lattice with 'holes'.

Also, the exponential distribution has a constant residual variance. Does this uniquely characterize the exponential law? Not quite, since it may be verified that a geometric distribution on a lattice has a constant residual variance. (The residual distribution remains geometric, but on a 'shifted lattice' which of course does not affect the variance.) We prove the following characterization and return to this question at the end of this section:

Proposition 5. If F is strictly increasing on $\mathbb{R}^+$ and $\sigma_F^2(x) = \sigma^2$ for all x, then F is exponential with mean $\mu = \sigma$.

Proof. Equation (4.5) yields

$$r(x) \equiv (\sigma^2(x)/e(x)) + e(x) = 2\ K(x)/k(x). \tag{4.6}$$

Since k has derivative $-\overline{F}$ and K has derivative $-k$, the right side has derivative $-2 + 2\overline{F}(x)K(x)/k^2(x) = -2 + r(x)/e(x)$ (using (4.6)) $= -s(x)$ where $s(x) \equiv 1 - \sigma^2/e^2(x)$. Hence from (4.6) r also has derivative $-s$.

Since e is right-continuous, we find, for $\delta > 0$,

$$\frac{r(x+\delta) - r(x)}{\delta} = \frac{e(x+\delta) - e(x)}{\delta}\{1 - \frac{\sigma^2}{e(x+\delta)e(x)}\}$$

$$= \frac{e(x+\delta) - e(x)}{\delta}\{s(x) + o(1)\} \quad \text{as} \quad \delta \to 0.$$

Since the left side has limit $-s(x)$, e has a right-derivative $e'(x)$ and $\{e'(x) + 1\}s(x) = 0$.

Now F strictly increasing may be shown to imply that $v \equiv e + I$ is strictly increasing, from which it follows that $e'(x) + 1$ is positive. Hence $s(x) = 0$ or $e(x) = \sigma$. □

We now go to the residual coefficient of variation $\gamma_a = \sigma_a/\mu_a$. It is identically unity for an exponential distribution, and this again is a characterization; this and some other related characterizations are given in

<u>Proposition 6</u>. $\sigma_F(x) = \gamma e_F(x)$ for all x in $\mathbb{R}^+$ and some $\gamma > 0$ if and only if $e_F(x) = (\mu + cx)^+$ with $c = (\gamma^2 - 1)/(\gamma^2 + 1)$, and hence F is Pareto (if $\gamma > 1$), exponential ($\gamma = 1$), rescaled beta(α,β) with $\alpha = 1$ $(0 < \gamma < 1)$, or degenerate $(\gamma = 0)$, respectively.

Hence a constant residual coefficient of variation characterizes the distributions listed, up to a scale factor.

<u>Proof</u>. Verification that $\gamma_F(x)$ is constant for each listed distribution is elementary, using (4.5).

That $e(x) = (\mu + cx)^+$ occurs if and only if F is of one of the given types follows from the inversion formula (e). For such an e_F, the corresponding $\sigma_F^2(x)$ may be found from (4.5) after substituting the inversion formula as noted in Section 5, and hence the constancy of $\gamma_F(x)$ established.

For the converse, consider $x < T$, replace the left side of (4.5) with $\gamma^2 e^2(x)$ to obtain $\beta e(x) = K(x)/k(x)$ where $\beta = (1/2)(\gamma^2 + 1)$. Differentiating both sides leads to $\beta e'(x) = -1 + \beta$, and hence to $e(x) = \mu + cx$ for $x < T$, and by continuity $e(x) = (\mu + cx)^+$ for all x. □

Several classes of distributions are defined in terms of the MRL: NBUE $(e_F(x) \le e_F(0) = \mu$ for all $x)$, NWUE $(e_F(x) \ge \mu$ for all $x)$, IMRL $(e_F \uparrow)$, and DMRL $(e_F \downarrow)$. These are larger than the classes IFRA and DFRA respectively (if $\mu < \infty$).

Watson and Wells (1961) show that γ_F is at most, or at least, unity in the IFR and DFR classes. Barlow and Proschan (1975), page 117, extended these inequalities to the IFRA and DFRA classes. We show that the same inequalities hold for the even larger NBUE and NWUE classes. (See Bryson and Siddiqui (1969) for the relationships of these classes; and Haines and Singpurwalla (1974) and Klefsjo (1979) for other classes and related material.)

Proposition 7. Suppose that F is in the NWUE [NBUE, resp.] class. Then $\gamma_F \geq 1$ [≤ 1 resp.], and the exponential distribution is the unique member of these classes with $\gamma = 1$. (If $\mu = \infty$ or $\mu < \infty = \sigma$, $\gamma \equiv \infty$.)

Proof. Assume F is NWUE. Then $e(x) \geq \mu$ for all $x \in \mathbb{R}^+$ and hence $k(x) \geq \mu\overline{F}(x)$. Integrating this inequality (0 to ∞) and using (4.1), we find $EX^2 \geq 2\mu^2$ or $\sigma^2 \geq \mu^2$. Equality holds only if $e(x) = \mu$ for all x. Similarly for F NBUE. []

An application in renewal theory is that $M(t) - (t/\mu)$ (M the renewal function) has a positive limit ($\gamma_F - 1$) as $t \to \infty$ whenever F is NBUE (F is also required to be non-arithmetic); see Karlin and Taylor (1975), page 195.

Proposition 7 can be extended to residual coefficients of variation: Define the class NWUE(a) as $\{F\colon e_F(a+x) \geq e_F(a)$ for all $x \in \mathbb{R}^+\}$ and similarly NBUE(a); hence F is in NWUE(a) if F_a is in NWUE, and F is in NWUE(a) for every $a \geq 0$ if and only if F is in IMRL. From Proposition 7, $F \in$ NWUE(a) implies $\sigma_a \geq \mu_a$, with equality if and only if F_a is exponential:

Proposition 8. If F is IMRL [DMRL, resp.], then $\gamma_a \equiv \sigma_a/\mu_a \geq 1$ [≤ 1, resp.] for every $a \in \mathbb{R}^+$.

It is easily seen that $\gamma_a \geq 1$ [≤ 1, resp.] does not imply that e is non-decreasing [non-increasing, resp.]; in fact, if e is such that $\sigma_0^2 = \infty$, then $\sigma_a^2 = \infty$ and $\gamma_a = \infty$ for all $a \geq 0$, but e need not be monotone. Or, take e arbitrary continuous with $e + I$ increasing on $[0,x_0]$, $e(x) = e(x_0) + 2(x - x_0)$ on $[x_0,\infty)$. Yet another counterexample is the geometric distribution: $\gamma(x) \geq 1$ for all x, but e decreases off the integers.

We close this section with an extension of 'Pyke's formula for the variance'--a curious formula relating σ_F^2 to e_F--and comment again on distributions with constant residual variance. The continuous case version of this formula appears in Pyke (1965, page 422).

Proposition 9. $\sigma_F^2 = E_F e(X)e(X-)$, $\sigma_F^2(x) = E_F[e(X)e(X-) \mid X > x]$.

Proof. It may be verified (by integrating over the continuity set for F and each of the discontinuity points) that $d(1/\overline{F}(x)) = [\overline{F}(x)\overline{F}(x-)]^{-1}dF(x)$ a.e.(F). Writing $Ee(X)e(X-)$ as a triple integral, applying Fubini and the formula just given, yields $\iint[\overline{F}(s \vee t) - \overline{F}(s)\overline{F}(t)]dsdt$, which equals σ^2 since $\iint\overline{F}(s \vee t)dsdt = \iiint_{s\vee t}^{\infty} dF(u)dsdt = \int u^2 dF(u)$ by Fubini. This proves the formula for σ^2; the residual variance formula holds similarly. []

Whether $E_F e(X)^2$, or even its finiteness, is related to σ_F^2 when F is not continuous is not known.

The formulas in Proposition 9 enable characterization of distributions with constant residual variance, alternative to Proposition 5. Specifically, it follows readily that $\sigma_F^2(x) = \sigma_F^2$ for all x, or a.e.(F), if e(x)e(x-) is constant a.e.(F). Among continuous F's, the exponentials are the only such distributions; this is consistent with Proposition 5. But discontinuous distributions are possible--e.g. the geometric distribution has this property as already noted; for it, e(x)e(x-) is easily seen to be constant on the lattice.

Other discrete distributions with constant residual variance can be constructed by modifying a geometric distribution--removing one mass point, or translating a mass point, and adjusting subsequent masses appropriately to preserve the "$e(x)e(x-) = \sigma^2$ a.e." property.

To construct a mixed distribution with constant residual variance, start with an exponential with $e(x) \equiv \mu$, say. Choose an interval $I = (a,b)$ with length $< \mu$ and set $e(x) = \mu - (x-a)$ on $[a,c)$ and $= \mu + b - c - (x-c)$ on $[c,b]$, with c in I so chosen that $e(c)e(c-) = \mu^2$ -- i.e., by solving the quadratic equation $(\mu+b-c)\cdot(\mu-c+a) = \mu^2$ for c. The resulting e is a MRL function, of a distribution F with $e(x)e(x-) = \mu^2$ a.e. This F is exponential on $[0,a)$ and on $[b,\infty)$, has a mass point at c, and (a,c) and (c,b) are null intervals. To have constant residual variance, F <u>must</u> be exponential on any interval on which it is strictly increasing.

5. APPLICATIONS OF THE INVERSION FORMULA

Recall the inversion formula from (e):

$$\bar{F}(x) = \{e(0)/e(x)\}\exp\{-\textstyle\int_0^x (1/e)dI\} \qquad \text{for } x < T \qquad (5.1)$$

or $k(x) = \mu\cdot\exp\{-\int_0^x(1/e)dI\}$. Applying (5.1) to the residual survival function $\bar{F}_a$ yields

$$\bar{F}_a(x) = \{e(a)/e(a+x)\}\exp\{-\textstyle\int_a^{a+x}(1/e)dI\} \qquad \text{for } x < T - a. \qquad (5.2)$$

Use of these formulas (4.1) - (4.5) is sometimes convenient. Thus from (4.1), $\nu_2 = 2\mu\int_0^\infty \exp\{-\int_0^x(1/e)dI\}dx$ and similarly for $\nu_2(a)$; the first of these yields an alternative easy proof of Proposition 7, and both are useful when e has a convenient form, as noted already in the proof of Proposition 6.

We now apply the inversion formula in the form (5.2) to infer properties of $\bar{F}_a$ on an interval $[0,b]$ from properties of e_F on $J = [a,a+b]$, and conversely. Since $\bar{F}(a+x) = \bar{F}(a)\bar{F}_a(x)$, we equivalently relate properties of $\bar{F}$ and e_F on J. Specifically, the next proposition characterizes MRL's containing linear segments.

<u>Proposition 10</u>. Let $J \equiv [a,a+b]$ (or $[a,\infty)$), $a \geq 0$, $a + b \leq T_F$. If

(1) $$e_F(x) = \lambda - cx \text{ on } J \qquad (-\infty < c \leq 1),$$

then

$$\text{(2)} \qquad \overline{F}(x) = \begin{cases} \overline{F}(a)\{(\lambda - cx)/(\lambda - ca)\}^{(1/c)-1} & \text{on } J \text{ if } c \neq 0 \\ \overline{F}(a)\exp(-(x-a)/\lambda) & \text{on } J \text{ if } c \neq 0 \end{cases}$$

and, trivially,

$$\text{(3)} \qquad \lambda = e_F(a+b) + ca + cb \qquad (\text{if } b < \infty).$$

Conversely, if (2) and (3) hold for some c, or if (2) holds and $\overline{F}(a+b) = 0$, then e_F is linear (if $c \neq 0$) or constant (if $c = 0$) on J; specifically, (1) holds.

Remarks. The case $c = 1$ corresponds to $P_F(J) = 0$; the case $0 < c < 1$ corresponds to F beta $(1,(1/c)-1)$ on J; $c = 0$ to F exponential on J; and $c < 0$ to F Pareto on J. The converse *without* (3) is not true (if $\overline{F}(a+b) > 0$), as will become apparent in the proof; this is not surprising: for (1) to hold at $x = a+b$ imposes a condition on the (mean of the) residual distribution beyond $a+b$, and (2) makes no imposition on this residual distribution.

Proof. We prove the case with $c = 0$; the other case is similar when $a+b < T_F$, and easier when $a+b = T_F$.

From the inversion formula, we have for $x \in J$, $\overline{F}(x)/\overline{F}(a) = \exp(-(x-a)/\lambda)$ which is (2). Conversely, assume (2) (and $b \leq \infty$); then for $x \in J$

$$\begin{aligned} \int_x^\infty \overline{F}dI &= \int_x^{a+b} \overline{F}dI + \int_{a+b}^\infty \overline{F}dI \\ &= \overline{F}(a)\int_x^{a+b}\exp\{-(t-a)/\lambda\}dt + \overline{F}(a+b)e_F(a+b) \\ &= \overline{F}(a)\ \lambda \exp\{-(x-a)/\lambda\} - \overline{F}(a)\ \lambda\exp\{-b/\lambda\} \\ &\qquad + \overline{F}(a)\exp\{-b/\lambda\}e_F(a+b). \end{aligned}$$

Therefore on J $e_F(x) = \lambda - \exp\{x-a)/\lambda\}\exp\{-b/\lambda\}\{\lambda - e_F(a+b)\}$ and (1) follows if and only if (3) is assumed or $b = \infty$. $\square$

Example: If $e_F(x) = \mu - c(x \wedge a)$ for some $a > 0$, then (from (2), or directly from the inversion formula)

$$\overline{F}(x) = \{1 - (c/\mu)(x \wedge a)\}^{(1/c)-1}\exp\{-(x-a)^+/(\mu - ca)\} \text{ on } \mathbb{R}^+. \qquad (5.3)$$

Conversely, if $\overline{F}(x) = \{1 - (c/\mu)(x \wedge a)\}^{(1/c)-1}\exp\{-(x-a)^+/(\mu-ca)\}$ on $\mathbb{R}^+$, then $e_F(x)$ is constant for $x \geq a$; also, $e_F(x)$ is linear on $[0,a]$ if and only if $\mu = e_F(a) + ca$, i.e., $\mu = \lambda + ca$. Also see Section 7.

For related results, see Proposition 9 of Kotz and Shanbag (1980) and the references therein. The relationship of e_F and $\overline{F}$ on $[a,\infty)$, for large a, is pursued similarly in Section 6.

We now discuss decomposition of e and F. From the definition of e, or from the inversion formula, and with $v = e + I$, we find

$$dv = (e/\overline{F})dF \qquad \text{or} \qquad dF = (\overline{F}/e)dv$$

and $dv = de + dI$ (treating e as v-I, the difference between two increasing functions). Hence, except for a possible mass point at T_F (if $< \infty$), where e and $\overline{F}$ vanish, absolutely continuous, singular and discrete components occur together or in neither of F and v. However, an absolutely continuous component in v need not correspond to one in e, and conversely; for if e (or F) is discrete or singular, v and hence F (or e) has an absolutely continuous component.

We thus conclude: If F has a discrete [singular, resp.] component then so does e, except a discontinuity in F at T_F does not lead to a discontinuity in e, and conversely. Either e or F alone can have an absolutely continuous component.

Brown (1980, page 238) noted that if F is in the IMRL class, then it must have an absolutely continuous component and raised the question whether a singular component is possible. The formula $dF = (\overline{F}/e)(de + dI)$, when e is increasing, provides an alternative proof of Brown's claim. Moreover, both discrete and singular components are seen to be possible: let e_1 be 1+G for a discrete df G, and $e_2 = (2-H)^{-1}$ for a singular df H, and let $e = e_1 + e_2$. Then F corresponding to e is in the IMRL class and has discrete, singular, and absolutely continuous components.

A final application of the inversion formula, in renewal theory, is as follows: When watching a cumulative sum of iid rv's until 'just before', and 'just after', it crosses a level t, δ_t (= t - 'the sum before crossing') is defined as the <u>current life at time t</u>, and γ_t (= 'the sum after crossing' - t) is defined as the <u>excess life at time t</u>. It is well-known (e.g. Karlin and Taylor (1975), page 193) that

$$\lim_{t\to\infty} P(\delta_t \geq y, \gamma_t \geq x) = \int_{x+y}^{\infty} \overline{F}dI/\mu = \overline{G}(x+y)$$

for all $x,y \geq 0$, where $G(x) \equiv \mu^{-1}\int_0^x \overline{F}dI$; this is the df $F_{(1)}$ of Smith (1959)- see (4.0) in Section 4. Alternative expressions for $\overline{G}(x)$ are $k_F(x)/\mu \equiv \overline{F}(x)e_F(x)/\mu$ and $\exp\{-\int_0^x(1/e)dI\}$, the latter from the inversion formula. It follows immediately that e_F is related to the hazard function of G as noted by Meilijson (1972) and by Bhattacharjee (1980). The df G has a monotone density $g = \overline{F}/\mu$ and its hazard function $\lambda_G \equiv g/\overline{G}$ equals $1/e_F$. Conversely, if g is an arbitrary non-increasing density on $(0,\infty)$ (right-continuous without loss of generality) with hazard function λ_G, then $\overline{F}(x) \equiv g(x)/g(0)$ defines a df F with $e_F = 1/\lambda_G$. Meilijson used these facts to give a very simple proof of the inversion formula (e).

6. MRL 'AT GREAT AGE'

Now suppose that $\overline{F}(x) > 0$ for all $x \in \mathbb{R}^+$ as well as $\mu = EX < \infty$. Recall that

$\overline{F}$ is said to be regularly-varying (at infinity) with exponent $-\gamma$, and we write $\overline{F} \in R_{-\gamma}$, if $\overline{F}(tx)/\overline{F}(t) \to x^{-\gamma}$ for all $x \geq 0$ as $t \to \infty$; $\overline{F}$ is regularly-varying with exponent $-\infty$, written $\overline{F} \in R_{-\infty}$, if $\overline{F}(tx)/\overline{F}(t) \to \infty, 1, 0$ according as $x <,=,> 1$ respectively; and that $\overline{F}$ satisfies the weak law of large numbers, written $\overline{F} \in$ WLLN, if $\overline{F}(t+x)/\overline{F}(t) \to 0$ for all $x > 0$. The following proposition is simply a restatement of various theorems concerning functions of regular variation -- which are conveniently stated in de Haan (1975).

Proposition 11. If $\overline{F}(x) > 0$ for all $x \in \mathbb{R}^+$ and $\mu = EX < \infty$ then, as $x \to \infty$,

(a) $e(x)/x \to \infty$ if $\overline{F} \in R_{-1}$,

(b) $e(x)/x \to c \in (0,\infty)$ if and only if $\overline{F} \in R_{-1-(1/c)}$,

(c) $e(x)/x \to 0$ if and only if $\overline{F} \in R_{-\infty}$,

(d) $e(x) \to 0$ if and only if $\overline{F} \in$ WLLN .

Now, in addition, suppose that $f = F'$ exists for large x and define $\lambda \equiv f/\overline{F}$, the hazard function. Then, as $x \to \infty$,

(e) $e(x)/x \to 0$ if $x\lambda(x) \to \infty$, and

(f) $e(x) \to 0$ if $\lambda(x) \to \infty$.

Finally, make the further assumption that f is non-increasing; then, as $x \to \infty$,

(g) $e(x)/x \to 0$ if and only if $x\lambda(x) \to \infty$, and

(h) $e(x) \to 0$ if and only if $\lambda(x) \to \infty$.

Proof. All of (a) - (h) are simply restatements of results in de Haan (1975): (a) and (b) are in Theorem 1.2.1, page 15; (c) is in Theorems 1.3.1, page 26, and 2.9.3, page 116; (d) is Theorem 2.9.3, page 119; (e) and (g) are given in Theorem 2.9.2, page 118; and (f) and (h) are given in Theorem 2.9.4, page 120. □

It would be interesting to know other sufficient conditions for limsup $e(x)/x$ to be finite; this would imply for example that $E_F e(X)^2 \leq E_F X^2$ (see Proposition 9 above and the remarks after Condition 1a in Hall and Wellner (1980)).

For best results concerning the residual distribution at great age, see Balkema and de Haan (1974).

The principal shortcoming of Proposition 11 is that many quite different MRL functions satisfy (c), (e), and (g): $e(x)/x \to 0$ is a relatively crude description of the behavior of e for large x. The next proposition presents an attempt to 'separate out' or distinguish a more complete range of possible behavior of e, and correspondingly of $\overline{F}$, at infinity.

Proposition 12. Suppose that (i) $e_F \sim e_G$ and (ii) $-\log(e_G\overline{G}) = O(-\log\overline{G})$. Then $-\log(\overline{F}) \sim -\log(\overline{G})$, or equivalently, $\overline{F}(x) = \overline{G}(x)^{1+o(1)}$.

A simple sufficient condition for (ii) is $\liminf_{x\to\infty} e_G(x) > 0$; another (see the Lemma below) is: e_G has a derivative e_G' with a limit at ∞.

Proof. Let $\eta(x) = \{e_G(x)/e_F(x)\} - 1 = o(1)$ by (i). Then the inversion formula yields

$$r(x) \equiv \frac{\overline{F}(x)}{\overline{G}(x)} = \frac{\mu_F}{\mu_G}\{1 + \eta(x)\}\exp\{-\int_0^x (\eta/e_G)dI\} .$$

But $|\int_0^x(\eta/e_G)dI| \le |\int_0^t(\eta/e_G)dI| + \delta\int_t^x(1/e_G)dI$ if $|\eta(x)| < \delta$ for all $x > t = t_\delta$ (for large x) and the right side $= A(t) + \delta\int_0^x(1/e_G)dI$. Now, by the inversion formula and (ii), $\int_0^x(1/e_G)dI = -\log\{e_G(x)\overline{G}(x)/\mu_G\} = O(-\log\overline{G})$ and hence $\int_0^x(\eta/e_G)dI = o(-\log\overline{G}(x))$. Then $\log(r(x)) = O(1) - \int_0^x(\eta/e_G)dI = o(-\log\overline{G}(x))$ or $-\log\overline{F} = -\log\overline{G} + o(-\log\overline{G})$. □

Lemma. Suppose F (or e_F) is absolutely continuous and e_F' has a limit ($\le \infty$) at infinity. Then $\lim e_F\lambda_F \ge 1$ and $\lim\{-\log(e_F\overline{F})/-\log\overline{F}\} \le 1$ where $\lambda_F \equiv f/\overline{F}$.

Proof. $e'(x) = -1 + \lambda(x)e(x)$ and $\lim e' \ge 0$ (since $\lim e' < 0$ implies $\lim e < 0$). Therefore $\lim(e\lambda) \ge 1$ and $\lim(1/e) \le 1$. But, by L'Hôpital's rule,

$$\lim[\log(e\overline{F})]/[\log(\overline{F})] = \lim[\log\int_x^\infty \overline{F}dI]/[\log(\overline{F}(x))] = \lim(1/e\lambda). \; \square$$

The following Corollary is an immediate consequence of Proposition 12. Note that $e_F(x)/x \to 0$ in cases (a), (b), and (e), but the related $-\log\overline{F}$'s are asymptotic to quite different $-\log\overline{G}$'s.

Corollary.

(a) If $e_F(x) \to c$ as $x \to \infty$, then $\overline{F}(x) = \overline{G}(x)^{1+o(1)}$ where $\overline{G}(x) - e^{-x/c}$.

(b) If $e_F(x) \sim cx^{1-\theta}$ for some $c \in (0,\infty)$, $\theta > 0$, then $\overline{F}(x) = \overline{G}(x)^{1+o(1)}$ where $\overline{G}(x) = \exp\{-\alpha x^\theta\}$, $a > 0$, a Weibull survival function.

(c) If $e_F(x) \sim cx$ for some $c \in (0,\infty)$, then $\overline{F}(x) = \overline{G}(x)^{1+o(1)}$ where $\overline{G}(x) = (1 + bx)^{-1/b}$, $b = c(c + 1)^{-1}$, a Pareto survival function.

(d) If $e_F(x) \sim c\,x\log(x)$ for some $c \in (0,\infty)$, then $\overline{F}(x) = \overline{G}(x)^{1+o(1)}$ where $\overline{G}(x) = x^{-1}\{\log(ex)\}^{-\tau}$, $\tau = 1+(1/c)$.

(e) If $e_F(x) \sim \sigma^2x/\log(x)$, $\sigma^2 > 0$, then $\overline{F} = \overline{G}^{1+o(1)}$ where $\overline{G}(x) = P(\exp(\sigma Z + \mu) > x)$, $Z \sim N(0,1)$, a lognormal survival function.

(See Watson and Wells (1961) page 289 in regard to (e).) Note that in (a) we could also have taken G to be a Gamma$(\beta,1/c)$ df for some $\beta > 0$ (i.e. $\overline{G}(x) = \int_x^\infty c^{-\beta}\Gamma(\beta)^{-1}t^{\beta-1}e^{-t/c}dt$); all of these gamma df's have $e_G(x) \to c$ as $x \to \infty$.

7. USE OF MRL IN MODELLING

Various authors have (at least implicitly) suggested that knowledge of the characteristic forms of MRL functions may be useful in modelling (e.g. Bhattacharjee (1980), Bryson and Siddiqui (1969), Chhikara and Folks (1977), Laurent (1974), Muth (1977), and Watson and Wells (1961)). We add a few comments supporting this view.

Specifically, if the empirical MRL function has some apparently linear segments, Proposition 10 characterizes the survival function on these segments (see example below). Also, the behavior of the empirical MRL 'at great age' may suggest a corresponding tail behavior of the underlying survival function, asserted in the Corollary of the previous section.

Of course, histograms, density estimates, empirical survival functions, total time on test plots, and empirical (cumulative) hazard functions may likewise be used to advantage in parametric modelling. We only suggest that the empirical MRL function is a useful addition to this arsenal -- one which identifies certain kinds of behavior more readily than others. For example, a flat (or linear) tail on the MRL suggests a gamma (or Pareto or beta) tail on F, features not so readily determined from other empirical plots.

As an example, consider the survival times of 72 guinea pigs injected with tubercle bacilli (Bjerkedal, 1960, regimen 4.3), illustrated in Figure 1 of Hall and Wellner (1980). The empirical plot of the MRL is not too different from that in the example of Section 5 above, namely, two line segments, the latter one horizontal: $e(x) = \mu - c(x \wedge a)$. This suggests the parametric model of (5.3) for the survival function: beta followed by exponential. By maximum likelihood methods, we fit the parameters as $\mu = 176.3$, $c = 0.8278$, and $a = 91.9$, yielding an asymptote of 100.2. By contrast, other plots (not shown) do not so clearly suggest a parametric model. This model suggests that an abrupt change in the mechanism of mortality occurs after an 'incubation period' of about 92 hours.

Examining Bjerkedal's Regimen 6.6 data somewhat analogously (see Figure 2 in Hall and Wellner (1980) and note from the text that the MRL curves downward eventually) the MRL plot suggests a linear segment followed by a function curving upward, flattening, and then proceeding downward. This can be fit by piecing together two beta distributions, consistent again with an abrupt change after an initial period.

Finally, study of the MRL plot of survival times (from date of diagnosis) of 43 leukemia patients, presented by Bryson and Siddiqui (1969), suggests a linear MRL tail from 1000 to 2500 days with slope $-\frac{1}{2}$; then, according to Proposition 10, this is consistent with a beta(1,1) = uniform distribution on this interval.

8. REFERENCES

[1] Balkema, A. A. and de Haan, L. (1974). Residual life time at great age. Ann. Probability 2 792-904.

[2] Barlow, Richard E. and Proschan, Frank (1975). Statistical Theory of Reliability and Life Testing: Probability Models. Holt Rinehard and Winston, New York.

[3] Bjerkedal, Tor (1960). Acquisition of resistance in guinea pigs infected with different doses of virulent tubercle bacilli. Amer. Jour. Hygiene 72 130-148.

[4] Bhattacharjee, Manish C. (1980). The class of mean residual lives and some consequences. Research Report. Indian Institute of Management, Calcutta.

[5] Brown, Mark (1980). Bounds, inequalities, and monotonicity properties of some specialized renewal processes. Ann. Probability 8 227-240.

[6] Bryson, Maurice C. and Siddiqui, M. M. (1969). Some criteria for aging. J. Amer. Statist. Assoc. 64 1472-1483.

[7] Chhikara, R. S. and Folks, J. L. (1977). The inverse Gaussian distribution as a lifetime model. Technometrics 19 461-468.

[8] Chiang, Chin Long (1968). Introduction to Stochastic Processes in Biostatistics. Wiley, New York.

[9] Cox, D. R. (1962). Renewal Theory. Methuen, London.

[10] de Haan, L. (1975). On regular variation and its application to the weak convergence of sample extremes.. Mathematical Centre Tract. 32, Mathematisch Centrum, Amsterdam.

[11] Galambos, Janos and Kotz, Samuel (1978). Characterizations of Probability Distributions. Lecture Notes in Mathematics #675. Springer-Verlag, New York.

[12] Gross, Alan J. and Clark, Virginia A. (1975). Survival Distributions: Reliability Applications in the Biomedical Sciences. Wiley, New York.

[13] Gupta, Ramesh C. (1975). On characterization of distributions by conditional expectations. Communications in Statist. 4 99-103.

[14] Haines, Andrew L. and Singpurwalla, Nozer D. (1974). Some contributions to the stochastic characterization of wear. Reliability and Biometry, Statistical Analysis of Lifelength, ed. Frank Proschan and R. Serfling, 47-80. SIAM, Philadelphia.

[15] Hall, W. J. and Wellner, Jon A. (1980). Estimation of mean residual life. J. Appl. Probability. To appear.

[16] Hamdan, M. A. (1972). On a characterization by conditional expectations. Technometrics 14 497-499.

[17] Karlin, Samuel and Taylor, Howard M. (1975). A First Course in Stochastic Processes, Second Edition. Academic Press, New York.

[18] Klefsjo, Bengt (1979). Some tests against aging based on the total time on test transform. Statistical Research Report No. 1979-4, University of Umea, Sweden.

[19] Klefsjo, Bengt (1979a). The class harmonic new better than used in expectation - some properties and tests. Statistical Research Report No. 1979-9, University of Umea, Sweden.

[20] Kotz, Samuel and Shanbag, D. N. (1980). Some new approaches to probability distributions. Adv. Appl. Prob. 12 (to appear).

[21] Laurent, Andre G. (1974). On characterization of some distributions by truncation properties. J. Amer. Statist. Assoc. 69 823-827.

[22] Meilijson, Isaac (1972). Limiting properties of the mean residual life function. Ann. Math. Statist. 43 354-357.

[23] Muth, Eginhard J. (1977) Reliability models with positive memory derived from the mean residual life function. The Theory and Applications of Reliability, Vol. II, Tsokos, C. P. and Shimi, I. N., editors, 401-435. Academic Press, New York.

[24] Pyke, R. (1965). Spacings. J. Roy. Statist. Soc. B 27 395-449.

[25] Shanbag, D. N. and Rao, Bhaskara M. (1975). A note on characterizations of probability distributions based on conditional expected values. Sankhyā Ser. A 37 297-300.

[26] Smith, Walter L. (1959). On the cumulants of renewal processes. Biometrika 46 1-29.

[27] Swartz, G. Boyd (1973). The mean residual life function. IEEE Transactions on Reliability R-22 108-109.

[28] Watson, G. S. and Wells, W. R. (1961). On the possibility of improving the mean useful life of items by eliminating those with short lives. Technometrics 3 281-298.

[29] Yang, Grace L. (1978). Estimation of a biometric function. Ann. Statist. 6 112-116.

STATISTICS AND RELATED TOPICS
M. Csörgö, D.A. Dawson, J.N.K. Rao, A.K.Md.E. Saleh (eds.)

THE PRODUCT FORM OF THE HAZARD RATE MODEL IN CARCINOGENIC TESTING

H. O. Hartley, H. D. Tolley and R. L. Sielken, Jr.

Department of Mathematics, Duke University, Durham, North Carolina, USA
Battelle Northwest, Richland, Washington, USA
Institute of Statistics, Texas A&M University, College Station, Texas, USA

The product form of the hazard rate is a general model for carcinogenic risk encompassing both the dose level and the time to response. Two broad biological foundations for the model are given. The first is based on a dynamic or compartment-analytic background in which each "compartment" involves a large number of "cells" and certain restrictions are imposed on the transfer of the carcinogen from compartment to compartment. The second is based on the administration of the carcinogen leading to a series of "attacks" on target areas with the time frequency of these attacks related to the experimental protocol.

1. *Introduction*

The product form of the hazard rate is a general model for carcinogenic experiments in which the "time-to-tumor", t, is observed. It yields the probability that an animal exposed to a "dose x" does not develop a tumor before time T in the form

$$Q(T, x) = \exp\{-g(x)H(T) + \phi(x)\} . \tag{1}$$

Here $g(x)$ is a positive convex function of the dose x and $H(T)$ a positive function of the time-to-tumor limit T. The ratio $\phi(x)/g(x)$ can, in certain cases, be equated to $H(T_o(x))$ where $T_o(x)$ is the minimum exposure time at which a tumor may be observed. However, $\phi(x)$ may represent other biological phenomena. Many applications of (1) set $\phi(x) \equiv 0$ and for this important special case we prove an interesting property of (1) in Appendix 2. Differentiating the logarithm of $Q(T, x)$ with regard to T yields

$$-\frac{\partial \log Q(T, x)}{\partial T} = \frac{p(T, x)}{Q(T, x)} = g(x)h(T) \tag{2}$$

where $p(T, x)$ is the ordinate of the time-to-tumor distribution given by

$$p(T, x) = -\frac{\partial Q(T, x)}{\partial T} \tag{3}$$

and

$$h(T) = \frac{dH(T)}{dT} . \tag{4}$$

Equation (2) implies that the so-called hazard rate (that is, the age specific tumor incidence) is a product of a function of dose ($g(x)$) and a function of the time ($h(T)$).

Special cases of the model (1) have been used by a number of authors. Perhaps the first instance of its use is the so-called multistage model by Armitage and Doll (1954), (1957) and (1961). Other instances are Peto and Lee (1973),

Crump et al (1976, 1977), Guess et al (1976, 1977, 1978), Crump (1979), and Hartley and Sielken (1977). The latter two papers are perhaps the most general applications of (1), although the inclusion of the term $\phi(x)$ can only be found in Hartley and Sielken (1978).

Most of the above papers are concerned with the so-called estimation of "virtually safe doses", that is, those (extremely small) doses $x^*(T)$ for which

$$Q(T, 0) - Q(T, x^*(T)) = \pi$$

or

$$Q(T, x^*(T))/Q(T, 0) = \varepsilon = 1 - \frac{\pi}{Q(T, 0)} . \qquad (5)$$

This application of (1) is, as might be expected, extremely sensitive to the assumptions made about the slope of g(x) at x = 0. This slope has, of course, the same sign as

$$-\frac{\partial Q}{\partial x}(T, x)\Bigg|_{x=0} = \frac{\partial P}{\partial x}(T, x)\Bigg|_{x=0}$$

that is, the slope of the risk curve at zero dose. Most of the recent controversies about the use of (1) for the estimation of virtually safe doses have been concerned with the question as to whether or not this slope should be assumed to be = 0 or > 0 or $\geq$ 0. Peto (1974, 1976) has strenously maintained that the slope is > 0. Cornfield and Mantel (1977) have taken the opposite view; namely, that the slope should be assumed to be 0 while Hartley and Sielken avoid the issue by assuming that it is $\geq$ 0 and let the experimental data settle the issue in every specific case. Of course, Hartley and Sielken pay a price for this in that their confidence limits for the virtually safe dose are often lower than those computed by fixing the slope (a priori) at 0. (More recently Cornfield (1978) has also recommended a data based decision using the Gamma function incidence curve.) Experimental evidence is available which shows that for some experimental protocols a 0 slope is clearly indicated and for others that a > 0 slope arises. Still others do not permit a clear cut decision. It is generally agreed that ideally issues of this kind should be decided by model validation based on biological reasoning. We have been painfully aware of the lack of knowledge of the biological mechanism which should be invoked concerning such issues, particularly since the mechanism may vary with each potential carcinogen. Perhaps the only biologically based model leading to a special case of (1) is the Armitage-Doll multistage model and this leads to a positive slope. However, the authors of this model have never claimed universal applicability for their model.

In this paper we do not, of course, settle the above issue. However, we provide two "biological underpinnings" for the model (1) which may be helpful in settling such problems in the future. We develop two "generic" biological models both leading to (1). Both "generic" models comprise a large number of more specific "deep" biological models as special cases by interpreting the "generic" concepts in terms of the specific biological mechanism applicable to a particular experimental protocol. The first generic model is based on a dynamic or compartment-analytic background in which each "compartment" involves a large number of "cells" but certain restrictions are imposed on the "transfer mechanism" of the "carcinogen" from "compartment" to "compartment". In the second model the administration of the "carcinogen" leads to a "series" of "attacks" on "target areas" and the "time frequency" of these "attacks" can be related to the experimental protocol. Some general considerations leading to these models are given in Appendix 1.

2. *The Dynamic, Compartment-Analytic Model*

2.1 Definitions and Assumptions

Let us assume that the organism has k compartments of interest. A compartment can be either a tissue site (a single tissue or collection of tissues) or a state in a sequential process. Not all compartments may actually be of interest as far as tumor growth is concerned. For example, intermediate states of a chemical agent or compartments of digestion are of interest only implicitly as necessary precursor states prior to carcinogenic exposure of a tissue site of interest. Such situations are easily accounted for in our model by setting certain parameters to zero, and some such modifications to the original model are given below. Each compartment is assumed to involve essentially an infinite number of homogeneous cells. For each such cell in the j-th compartment ($j = 1, \ldots, k$) we define the following:

$g(x)$ = effective body dose for administered dose x,
$\theta_j g(x)$ = effective dose to cells of compartment j,
$M_j(t)$ = concentration of carcinogen in a cell of compartment j at time t,
$P_j(t, x)$ = cumulative distribution function for time to tumor in compartment j for dose x,
$Q_j(t, x) = 1 - P_j(t, x)$, and
$F_j(t, x)$ = age specific tumor incidence rate or hazard rate at dose x for compartment j
$= -\frac{d}{dt} \log Q_j(t, x)$.

Also for the set of k compartments define

A = $k \times k$ matrix of kinetic constants for transition from compartment to compartment (all off-diagonal elements of A are assumed ≥ 0),
$D = g(x) \cdot (\theta_1, \theta_2, \ldots, \theta_k)^T$, and
$M(t) = (M_1(t), M_2(t), \ldots, M_k(t))^T$.

We list the following necessary assumptions:

Assumption I: The probability that any particularly cell is transformed is independent of the probability of any other cell being transformed in the same or different compartments. (When we say that a cell is transformed, we mean that it changes from a healthy cell to a cancerous cell.)

Assumption II: The probability of a cell transformation in compartment j during the interval $(t, t + \Delta t)$ given compartment j is tumor free at t, is proportional to the concentration $M_j(t)$. Explicitly the probability, $p_j(t)$, is given as

$$\Delta t\, p_j(t) = \gamma_j(t)\, M_j(t)\, \Delta t + o(\Delta t). \quad (6)$$

Assumption III: Prior to exposure, no carcinogen exists in any tissue; i.e., $M(0) = 0$.

Assumption IV: $g(x)$ is non-negative for $x \geq 0$.

Assumption V: The transfer of carcinogen from compartment to compartment follows a first order kinetic model.

At this point a few remarks are in order regarding the assumptions.

Remark I: It is well known that the cell wall of cancerous cells is different than normal cells. Among the causes for this are an increase in the freedom of agglutinating receptor sites and a change in glucose regulating cell membrane proteins. Additionally, cancerous cells excrete a proteolytic enzyme, called plasmin, which tends to digest cell walls. It seems that the major effects of these changes are on the transformed cells, giving them clonal growth advantage over other cells and not increasing the probability of transformation of neighboring cells (see Watson (1976)). Therefore Assumption I, although not strictly correct, is probably acceptable.

Remark II: Assumption II is a common assumption in kinetic situations. Discussion of this assumption is given by Gehring and Young (1977).

Remark III: The initial condition of no carcinogenic agent in the organism at time $t = 0$ does not necessarily imply that there is no background cancer since this can be invoked by postulating that $g(0) > 0$ in the relation between the administered dose, x, and the effective body dose $g(x)$.

Remark IV: Assumption IV implies that any dose is bad. The notion that a little carcinogen is "good for you" is dismissed here. This assumption will be made more strict in the following.

2.2 The Model

As implied in the definition of A and in Assumption V, we assume that the transport of carcinogen to cells of each compartment follows a first order kinetic equation. To determine the amount of carcinogen, $M_j(t)$, in tissue j at time t we have the differential equation

$$M'(t) = A\,M(t) + D\,. \tag{7}$$

Differentiating both sides with respect to t gives

$$M''(t) = A\,M'(t); \tag{8}$$

from whence comes the solution

$$M'(t) = \sum_{i=1}^{k} C_i \delta_i \exp(\lambda_i t), \tag{9}$$

where λ_i are the eigenvalues of A, the C_i are the standardized eigenvectors of A, and the δ_i are scalar scale factors of C_i determined by the initial conditions. Hence

$$\lambda_i C_i = A\,C_i. \tag{10}$$

Integrating (3) we get

$$M(t) = \sum_i C_i \frac{\delta_i}{\lambda_i} \exp(\lambda_i t) + b, \tag{11}$$

where b is a k-vector of constants. Substituting (9) and (11) into (7) we get

$$\sum_i \exp(\lambda_i t)\, C_i \delta_i = \sum_i \exp(\lambda_i t) \frac{\delta_i}{\lambda_i} A\,C_i + Ab + D\,. \tag{12}$$

Applying (10), we have

$$b = -A^{-1} D\,. \tag{13}$$

Using the condition $M(0) = 0$, we have

$$\sum_i C_i \frac{\delta_i}{\lambda_i} - A^{-1}D = 0\,. \tag{14}$$

Regarding the $\frac{\delta_i}{\lambda_i}$ as "unknowns" in a system of k linear equations with matrix C, whose column vectors are C_i, we can solve (14) to get

$$\left(\frac{\delta_1}{\lambda_1}, \ldots, \frac{\delta_k}{\lambda_k}\right)' = C^{-1}A^{-1}D\,. \tag{15}$$

Using the result of equation (15) in (11) we get for $M(t, x) \equiv M(t)$

$$M(t, x) = g(x)\left\{\sum_{i=1}^{k}\sum_{j=1}^{k}\theta_j f_{ij}\exp(\lambda_i t)\,C_i - A^{-1}\theta\right\} \quad (16)$$

where f_{ij} is the i, j-th element of $C^{-1}A^{-1}$ and $\theta^T = (\theta_1, \theta_2, \ldots, \theta_k)$. We note here that the term in brackets is a function of time only and not of dose x. For notational convenience the bracketed term in (16) will be denoted $r(t) = (r_1(t), r_2(t), \ldots, r_k(t))^T$. It has been proven by Hearon (1964, p. 48) that the condition

$$a_{ij} \geq 0 \text{ for all } i \neq j \quad (17)$$

is both necessary and sufficient for $M(t) \geq 0$ or, in our notation, for all $r_j(t)$ to be ≥ 0. Our model assumption of (17) representing non-negative transfer coefficients is, therefore, equivalent to the obviously reasonable feature that we never have a negative amount of carcinogen in any tissue.

The special case of a dose input into only one compartment is represented by $\theta_i > 0$, $\theta_j = 0$, $j \neq i$. A special case of this situation (chains of compartments) was considered by R. Whitmore (1978) who also showed that in this case M(t) factored into a function of the dose multiplied by a function of time.

Using Assumption II, the probability of a cell in compartment j being transformed into a cancerous state in the interval t to t + Δt given that compartment j is tumor free at t is given as

$$\Delta t\, F_j(t, x) = g(x)\,\gamma_j(t)\,r_j(t)\,\Delta t + o(\Delta t)\,. \quad (18)$$

Thus the time-to-tumor distribution is implied by

$$Q_j(t, x) = \exp\{-g(x)\int_0^t h_j(\tau)\,d\tau + z_j(x)\} \quad (19)$$

where $h_j(t) = r_j(t)\,\gamma_j(t)$ and $z_j(x)$ is a constant of integration. Since $\gamma_j(t) \geq 0$ and $r_j(t) \geq 0$ for all positive t, we have $H_j(T) = \int_0^T h_j(\tau)d\tau \geq 0$, for all $T \geq 0$.

If our Assumption III implies that $M_j(t) = 0$ (and hence $h_j(t) = 0$) for the "pre-experimental" period $t \leq 0$, then clearly $z_j(x) = 0$ since $Q_j(t, x)$ must be 1 for $t < 0$. If, however, we make the assumption that there is a "pre-experimental period" $-T \leq t \leq 0$ (starting with $M(-T) = 0$) during which the dynamic process has been running with a constant background rate of $D = g(0)\theta$, then clearly from the above argument $M_j(0) = g(0)\, r_j(T)$. If these values are used as initial values for the experimental period then it is easy to show that

$$M_j(t) = g(x)\,r_j(t) + s_j(t) \quad (20)$$

where

$$s_j(t) = \sum_i C_{ij}\exp(\lambda_i t)\sum_p C_{ip}^{-1} g(0)\,r_j(T)\,. \quad (21)$$

This leads to an additive t-dependent function in the hazard rate which will now be of the form

$$\frac{-\partial \log Q_j(t, x)}{\partial t} = g(x)\,h_j(t) + \ell_j(t)\,. \quad (22)$$

This represents a modification of (1) which can be easily treated by the maximum likelihood procedure of Hartley and Sielken.

Note here that there are two processes involved. The first is the deterministic transmission of the carcinogen into cells of each compartment. This process is governed by the differential equation given in (7). The second process is the probabilistic transformation of the cell into the cancerous state implied by equation (18). This transformation does not remove the cell from participation in the first process. There is reason to believe that the rate a cell being transformed will take up the carcinogen is different than a normal cell (see Remark I). However, this rate change applies to the transformed cells only and in general not to the whole compartment. Therefore, the kinetic process continues uneffected by the initiation or completion of transformation of some cells in the individual compartments.

2.3 Reaction Chains

Many believe that the production of a cancer cell is a sequential process. For example, the actual administered chemicals in a dosage may have to go through a chain of reactions to become carcinogenic (see e.g. Gehring and Young (1977)). Alternatively the dose may effect a constituent of the cell, for example one of the genes, which in time effects a series of evolutionary steps until the complete cell is ready for transformation (see e.g., Farber (1973)). Either the chemical chain reaction or cellular evolution can be represented as in Figure I. In the first case, however, the precursor substrates are chemical components of a reaction while in the second case the evolutionary precursors are transformed cytologic or nucleic material.

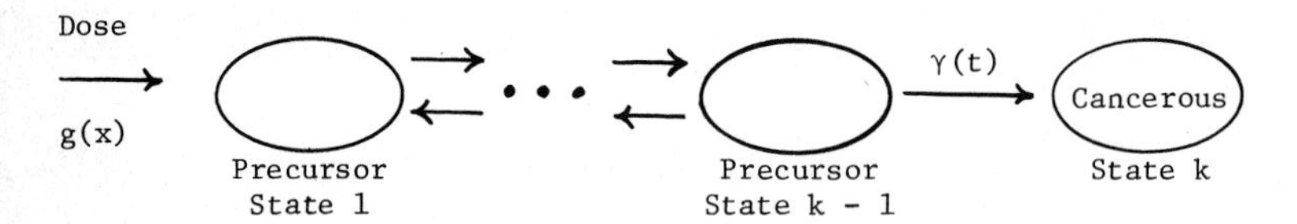

Figure I

It is not hard to see that mathematically, the foregoing model holds with the simplification that, if there is only one tumor site of interest, then $\theta_1 = 1$ and $\theta_j = 0$, $j \neq 1$. Also many of the entries in A are set to zero. Deviations from this model, such as multiple tumor sites or types of cancer as depicted, for example, in Figure II are also possible. Clearly many different variations on the reaction chain model exist. In any of these models the probability of no tumor is in the form given in equation (19).

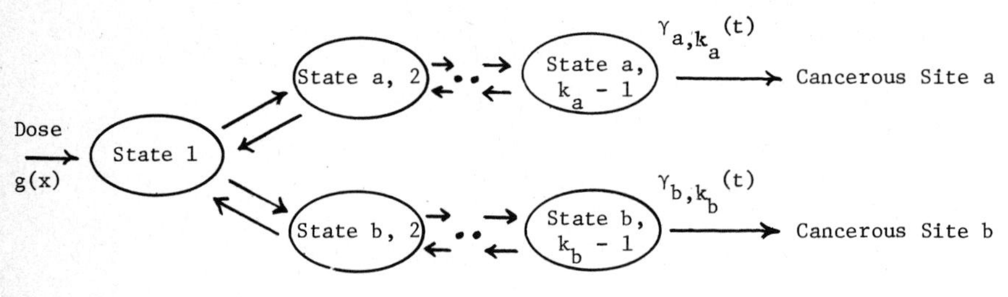

Figure II

2.4 Stochastic Transition

It is important to note that in the above models the source of stochasticity is in the final transition to the cancerous state. All steps leading up to this step are deterministic first order kinetic reactions. Increasingly stochastic models are being proposed in which the transition from state to state is represented by some probability formulation rather than simply a deterministic model.

Perhaps the most well known cancer model of this type is the multi-hit or multi-stage model of Armitage and Doll (1954) (see also Doll (1971)). In this stochastic formulation the precursor states usually represent more of a cell condition rather than a concentration of a carcinogen. In many situations, such as the multi-hit model, the transition of a cell from one precursor state to another and finally to the cancerous state is modeled as a random outcome. The probability of transition from state i to state j during the interval t to t + Δt is assumed to be of the form $\mu_{ij}(t)\Delta t + o(\Delta t)$ for any cell in the compartment. In the preceding section we have conglomerated all of these states by assuming a general transition from a deterministic state to a cancer state can be modeled by (6).

If each cell behaves independently of other cells while undergoing a transformation, then the probability of no tumor at time T in compartment k is given as follows (see Chiang (1968)):

$$Q(T, x) = \int_0^T \int_0^{t_{s-1}} \cdots \int_0^{t_2} \prod_{i=1}^{s-1} \mu_{i-1,\, i}(t_i) \exp\left(- \int_{t_{i-1}}^{t_i} \mu_{i-1,\, i}(\tau)d\tau\right) \exp\left(- \int_{t_{s-1}}^{T} \mu_{s-1,\, s}(\tau)d\tau\right) dt_1 \ldots dt_{s-1}, \tag{23}$$

where this equation assumes that transformation is equivalent to s sequential irreversible transitions. It is clear from equation (23) that if the transition intensity functions $\mu_{ij}(t)$ are of the form $\mu_{ij}(t) = M_k(t, x) \cdot \mu_{ij}^*(t)$, where $\mu_{ij}^*(t)$ is independent of dose, then Q(T, x) does <u>not</u> satisfy equation (19). In other words, for equation (19) to hold, the probability of transition from one state to another cannot be proportional to dose or carcinogenic concentration unless s = 1. Therefore, our restriction given by Assumption II is more inhibitive than one might expect.

2.5 Relationship to Product Form of Hazard Rate

In the preceding sections we have shown that a relatively simple set of assumptions concerning carcinogenesis leads to a biological model where the probability of a cell in compartment j, say, not being transformed into a cancerous cell by time T when constantly exposed to a dose x has the form

$$Q_j(T, x) = \exp\{-g(x)\, H_j(T) + z_j(x)\} \tag{24}$$

where $H_j(T) = \int_0^T r_j(t)\gamma_j(t)dt$ is a non-negative nondecreasing function of T. Furthermore, if the transformation of cells to a cancerous state in each of the k compartments are independent, the probability, Q(T, x), of having no cancer in <u>any</u> of the k compartments by time T at dose x is given by

$$Q(T, x) = \exp\{-g(x) \sum_{j=1}^{k} H_j(T) + \sum_{j=1}^{k} z_j(x)\} \equiv \exp\{-g(x)\, H(T) + z(x)\} . \tag{25}$$

In 1977 Hartley and Sielken suggested that the relationship among carcinogenesis, time, and dose could be analyzed using the model

$$Q(T, x) = \exp\{-g^*(x)\, H^*(T)\} \tag{26}$$

where the so-called "product form of the hazard rate" implied that $g^*(x)$ was a function of the dose x but not time T and $H^*(T)$ was a function of T but not x. In 1978 they extended their analytic techniques to include models of the form

$$Q(T, x) = \exp\{-g^*(x)\, H^*(T) + z^*(x)\} \tag{27}$$

where $z^*(x)$ was a function of x but not T.

The kinetic model of this section covers a great variety of interactions between "compartments" representing tissue compounds whose cells contain continuous time dependent concentrations $M_j(t)$ of the carcinogen. On the other hand the model is rather restrictive with regard to the experimental protocol for the administration of the carcinogen. Specifically, it requires input rates that are constant in time both during the pre-experimental period and also constant in time (but dose dependent) during the experimental period. It also requires that the time of onset of the concentration does not depend on the dose.

In our second model (see Section 3) we reverse these model characteristics. We shall provide for a rather general time dependence of the administration of the carcinogen but make a very restrictive assumption with regard to any transfer of the carcinogen from tissue to tissue or cell to cell.

3. *The "Multiple Attack" Model*

3.1 Definitions and Assumptions

We first describe this model in generic terms and then discuss only a limited number of specific biological interpretations. It is assumed that the "target tissue" is composed of an essentially infinite number of cells. It is further assumed that the administration of the carcinogen consists of discrete "packages" which may "attack" precisely one of the cells. A cell, if attacked by a "package", will have a probability $p(x) = 1 - q(x)$ to reach the cancer "initiation stage" and a probability $q(x)$ to "repair" any damage caused by the attack before there is a chance that it will suffer a second attack. (If the probability that an individual cell undergoes more than one attack is essentially zero, then the concept of a "repair" is not necessary, and $q(x)$ is simply the probability of not reaching the cancer "initiation stage".) The basic probabilities $p(x)$ and $q(x)$ depend on the "dose" x, which is the "concentration" of the carcinogen in the package. All the damage to the cell required to cause initiation is done at time t of the "attack". However, the initiation stage is not reached until time $t + \tau$ where τ is a constant time. This probability model for a single cell breakdown may, for example, consist of a single stage multi-hit process or may involve several stages. As soon as at least <u>one</u> cell reaches the initiation stage, a tumor growth is initiated in the tissue with a tumor apparent at time $T = t + \tau + \theta$ where θ is a constant time. Note that neither τ or θ depend on the dose x. However, they may be made to depend on "age" t.

The experimental protocol generates a "frequency of attacks" where $h(t)$ denotes the number of attacks between t and $t + dt$. This frequency distribution $h(t)$ may also depend on age dependent changes in the physiology of the animal.

If we assume that the initiation process is completely independent from cell to cell, then the probability of the tissue to be tumor free at time T is given by

$$Q(x, T) = q(x)^{H(T-\tau-\theta)-\psi(x)} \tag{28}$$

where

$$H(t) = \int_0^t h(u)\,du \tag{29}$$

and

$$\psi(x) = \int_{o}^{\omega(x)} h(u)du \tag{30}$$

where $0 \leq t \leq \omega(x)$ is a possibly dose dependent initial screening period. Introducing

$$g(x) = -\log q(x), \ \phi(x) = \frac{\psi(x)}{g(x)}, \tag{31}$$

we can write (28) in the form

$$Q(x, T) = \exp\{-g(x)H(T - \tau - \theta) + \phi(x)\} \tag{32}$$

which is of the form (1). (If the "frequency of the attacks" is itself dose dependent, say $h^*(t) = s(x)\,h(t)$, then the general form (32) still applies with $g(x)$ arising as $g(x) = -s(x) \log q(x)$.)

3.2 The Interpretation of g(x)

From equation (31) we see that the assumption (made by Hartley and Sielken) that $d^2g/dx^2 > 0$ is equivalent to

$$\frac{d}{dx} \frac{f(x)}{q(x)} \geq 0 \tag{33}$$

where $f(x) = -\frac{dq(x)}{dx}$ is the ordinate of the within cell dose risk distribution.

Thus, this assumption is seen to be identical with assuming that the probability tail area $q(x)$ has an increasing hazard rate. It can be shown that this condition is satisfied for the multi-hit tail area

$$q(x) = \{(k - 1)!\}^{-1} \int_{\lambda x+c}^{\infty} e^{-u} u^{k-1} du . \tag{34}$$

Indeed it is a reasonable assumption to make that an incremental dose δx when applied to a high basic dose level x will result in a larger number of initiated cells counted as a proportion of uninitiated cells.

Apart from the convexity no constraints are imposed on $g(x) = -\log q(x)$ and it is expected that the polynomial form used by Hartley and Sielken will approximate a great variety of models for the within cell hazard distribution. This would require a more comprehensive study.

3.3 The Interpretation of $H(T - \tau - \theta)$

Equation (29) defines H(T) as the cumulative integral (or sum) of the frequency distribution of "attacks". This frequency distribution is expected to depend on (a) the experimental protocol for dose administration and (b) distortions in the time distribution of (a) introduced through animal physiology, particularly through the growth of the animal. It is of interest to note that high exponent Weibull forms $H(t) = \gamma t^k$ (k = 5, 6, 7) can probably be explained more adequately by (a) and (b) than by a multi-stage cancer development. A particular case of interest arises when a large dose x is administered at one particular time. Then (a) would bring about a single step function for H(t) but (b) would presumably smooth out this step function. Nevertheless, it may be necessary to modify the polynomial form of H(t) used by Hartley and Sielken.

3.4 The Interpretation of $\phi(x)$

It has already been remarked that the additive dose dependent term $\psi(x) = g(x)\phi(x)$ may arise from an initial screening period during which the attacks are screened off from the target tissue. In a similar manner we could account for a repair mechanism which would increase Q(x, T) by a dose dependent proportion

$e^{\delta\psi(x)}$. However, such repair mechanisms can usually be accounted for through a modification of q(x).

3.5 Shortcomings

There are a number of shortcomings of which perhaps the most serious one is that the model does not allow for animal to animal variation in carcinogenic susceptibility. However, that phenomenon could be accounted for by introducing mixtures of the distribution given by (1).

3.6 Special Biological Interpretations of Generic Concepts

The restrictive features of this model are the complete absence of any interaction between cells in the single target tissue. Moreover, the concept of discrete "attacks" by "packages" limits the application of the model.

Examples may arise in the case of an impact of bursts of carcinogenic radiation on skin or other target tissue. Repeated discrete applications of a carcinogenic ointment or tincture onto the skin may follow this model. Food additives whose carcinogenic ingredients impinge directly onto the cells in the wall of the digestive system are another possibility. This subsection requires a more detailed study.

Acknowledgements

We would like to acknowledge with sincere appreciation, the inputs made by Dr. James H. Matis in discussions. Dr. H. Dennis Tolley's research has been supported by HEW Grant No. 1 R23 ES02067-01.

This collaborative effort was stimulated by H. O. Hartley prior to his death in December, 1980.

4. *References*

[1] Armitage, P. and Doll, R., The age distribution of cancer and a multistage theory of carcinogenesis, *British Journal of Cancer 8* (1954) 1-12.

[2] Armitage, P. and Doll, R., A two-stage theory of carcinogenesis in relation to the age distribution of human cancer, *British Journal of Cancer 11* (1957) 161-169.

[3] Armitage, P. and Doll, R., Stochastic models for carcinogenesis, *Fourth Berkeley Symposium on Mathematical Statistics and Probability*, University of California Press, Berkeley, California, (1961) 19-38.

[4] Cairns, John, The cancer problem, *Scientific American 233* (1975) 64-78.

[5] Chiang, C. L., *Introduction to stochastic processes in biostatistics* (John Wiley and Sons, Inc., New York).

[6] Cornfield, J., Chapter 11: Quantitative risk assessment. *Proposed System for Food Safety Assessment*, Report of the Scientific Committee of the Food Safety Council, submitted September 11, 1978.

[7] Cornfield, J. and Mantel, N., Discussion of "Estimation of 'safe doses' in carcinogenic experiments", *Biometrics 33* (1977) 21-24.

[8] Crump, K. S., Dose response problems in carcinogenesis, *Biometrics 35* (1979) 157-167.

[9] Crump, K. S., Guess, H. A., and Deal, K. L., Confidence intervals and tests

of hypothesis concerning dose-response relations inferred from animal carcinogenicity data, *Biometrics* 33 (1977) 437-451.

[10] Crump, K. S., Hoel, D. G., Langley, C. H. and Peto, R., Fundamental carcinogenic processes and their implications for low dose risk assessment, *Cancer Research* 36 (1976) 2973-2979.

[11] Doll, R., The age distribution of cancer: implication for models of carcinogenesis, *Journal of the Royal Statistical Society*, A 134 (1971) 133-166.

[12] Farber, Emmanuel, Carcinogenesis-cellular evolution as a unifying thread: presidential address, *Cancer Research* 33 (1973) 2537-2550.

[13] Gehring, P. J. and Young, J. D., Application of pharmacokinetic principles in practice, *Proceedings of the First International Congress on Toxicology* (1977) 119-142.

[14] Guess, H. and Crump, K., Low dose extrapolation of data from animal carcinogenicity experiments -- analysis of a new statistical technique, *Mathematical Biosciences* 32 (1976) 15-36.

[15] Guess, H. and Crump, K., Maximum likelihood estimation of dose-response functions subject to absolutely monotonic constraints, *Annals of Statistics* 6 (1978) 101-111.

[16] Guess, H. A., Crump, K. S., and Peto, R., Uncertainty estimates for low-dose-rate extrapolations of animal carcinogenicity data, *Cancer Research* 37 (1977) 3475-3483.

[17] Hartley, H. O. and Sielken, R. L., Estimation of 'safe doses' in carcinogenic experiments, *Biometrics* 33 (1977) 1-20.

[18] Hartley, H. O. and Sielken, R. L., Development of statistical methodology for risk estimation, Report to National Center for Toxicological Research (1978).

[19] Hearon, J. Z., Theorems on linear systems, *Annals of New York Academy of Sciences* 108 (1963) 36-68.

[20] Mantel, N., Discussion of presentation by Hartley and Sielken on estimation of 'safe doses' in carcinogenic experiments, *Journal of Environmental Pathology and Toxicology* 1 (1977) 267-269.

[21] Marcelletti, John and Funmanski, Philip, Infection of macrophages with Friend virus: relationship to the spontaneous regression of viral erythroleukemia, *Cell* 16 (1979) 649-659.

[22] Marx, Jean L., RNA tumor viruses: getting a handle on transformation, *Science* 199 (1978) 161-164.

[23] Peto, R., Time to occurrence models. Presented at a "Workshop meeting" sponsored by National Institute of Environmental Health Sciences (NIH), Wrightsville Beach, North Carolina. (October 1974).

[24] Peto, R., Presentation at NIEHS Conference, Pinehurst, North Carolina. (March 1976).

[25] Peto, R. and Lee, P., Weibull distributions for continuous carcinogenesis experiments, *Biometrics* 29 (1973) 457-470.

[26] Watson, James D., *Molecular biology of the gene*, W. A. Benjamin Inc., Menlo

Park, California.

[27] Whitmore, R. W., Compartmental modeling and analysis for carcinogenic experiments, Ph.D. dissertation, Texas A&M University, College Station, Texas (1978).

5. APPENDIX 1

General Biological Considerations Leading to the Formulation of the Models of Sections 2 and 3

The transformation of a healthy cell to that of a cancerous cell is almost universally felt to be due to a somatic change (change in genetic make-up of cells not destined to become sex cells). In fact, Watson (1976) says, "There...seems little doubt that much if not most carcinogenesis is the result of changes in DNA". However, how the genetic constituents are altered to transform healthy cells to morphologically distinguishable cancer cells is uncertain. Two major theses are now prominent. Since oncologists agree that cancer is really a collection of many diseases, both of these, and possibly other theories, can be correct for different cancers.

The best known theory of what could cause a somatic change giving rise to cancer is mutation. Critical mutations in DNA or RNA can be demonstrated in the lab by exposure of the cells to any of a variety of mutagens, such as certain hydrocarbons. Also ionizing radiation and visable light radiation have been shown to induce mutation. Additionally, weaker mutagens can be changed, once inside the cell, to highly mutagenic agents. For example, nitrates can be changed into highly mutagenic nitrosamines. Although most mutations may lead to death or inefficiency in the cell, a small percentage of mutated cells may become cancerous.

A second, and increasingly more popular theory of carcinogenesis is the cancer virus notation. A virus in the cell may be capable of one of four mechanistic behaviors. First is that it may disappear. Secondly, it may be lytic to the cell by proliferating to the point of killing the host cell. Thirdly, it may be unable to proliferate but may add its own genetic material to that of the cell. On occasion this last step gives rise to transformation of the healthy cell into a cancerous cell. Several viruses have been demonstrated to have such onocogenic potential, for example, SV40 (a polyoma type DNA virus), Rous Sarcoma, and RNA virus. Action of a virus may be immediate upon cell entry such as in SV40. Alternatively, it may be after mutation of the virus itself such as in the case of Herpes simplex (see Watson (1976)). Fourthly, it may be dormant in the cell until activated by some mechanism as, for example, in the Friend virus (see Marcelletti and Funmanski (1979)).

Under either of the above theories the mechanism of activating cell transformation, i.e., the method of initiating a mutation or activating a virus, can be dichotomized into two major classifications. The first mechanism is a moderate, relatively continuous exposure of the cell. Carcinogenic additives which are constantly metabolized would be one such example of continuous exposure. The second mechanism consists of a sequence of one or more shocks or assaults. High radiation doses or severe pollution for a short period of time would be examples of this type.

Characteristically, the difference in the two mechanisms is that in the first stress is the accumulation of constant exposure while in the second there are only one or a few stressful exposures. These considerations lead to the models described respectively in Sections 2 and 3.

6. APPENDIX 2

A Universal Property of the Product Form of the Hazard Rate

Many toxicological studies are concerned with the incidence $P(T, x)$ of a quantal response through an exposure of test animals to a "dose x" for an "exposure time T". More specifically, a frequently occuring problem is to determine the "incremental incidence rate" over the spontaneous background rate $P(T, 0)$ which results from the application of very small (residual) doses. It is well known that this incremental incidence rate varies with the exposure time T. For certain regulatory purposes it is of interest to know for which exposure time $T = T^*$ (say) the incremental risk is at a maximum, given a residual dose x. We now show that, if (1) (with $\phi(x) = 0$) is accepted, then T^* is essentially independent of the residual dose x. Hence in the context of animal tests with potential carcinogens, this exposure time T^* will result in the smallest safe dose for a specified incremental risk.

More specifically we show below that this critical exposure time T^* is the one which generates a spontaneous incidence rate of $100\,(1 - \frac{1}{e}) \doteq 63\%$. Since most spontaneous rate curves never reach 63% over the life-span of the animals, the maximum incremental risk will then occur at the maximum length of exposure.

To prove the above we compare the incidence rate for a given dose x with the spontaneous rate at $x = 0$ and write the difference as

$$\begin{aligned} \Delta P(T, x) &= P(T, x) - P(T, 0) \\ &= \exp\{-g(0)H(T)\} - \exp\{-g(x)H(T)\} \end{aligned} \tag{35}$$

and its derivative with regard to T as

$$\frac{\partial \Delta P(T, x)}{\partial T} = h(T)[g(x)\exp\{-g(x)H(T)\} - g(0)\exp\{-g(0)H(T)\}] \, . \tag{36}$$

Using a Taylor expansion with exact first order remainder term, we have

$$\frac{\partial \Delta P(T, x)}{\partial T} = (g(x) - g(0))h(T)\exp\{-\bar{g}H(T)\}[1-\bar{g}H(T)] \tag{37}$$

where $g(0) < \bar{g} < g(x)$.

The quantity on the right-hand side of (37) is positive as long as $\bar{g}H(T) < 1$ or $\exp\{-\bar{g}H(T)\} > 1/e$ or $1 - \exp\{-\bar{g}H(T)\} < 1 - 1/e$. Since the above equation is to be used for residual doses x only, $\bar{g}$ is close to $g(0)$ so that we can replace the above condition by $1 - \{\exp{-g(0)H(T)}\} = P(T, 0) < 1 - 1/e$. This means that the incremental risk $\Delta P(T, x)$ will grow with increasing T as long as the spontaneous rate for period T is below $100(1 - 1/e) = 63\%$.

7. APPENDIX 3

The Slope of $g(x)$ at Dose $x = 0$

In this appendix we attempt an interpretation of the slope of $g(x)$ at zero dose $(x = 0)$ in terms of the biological concepts introduced through our models.

We first consider the compartment-analytic model of Section 2. The function $g(x)$ in (19) was defined as the functional relation between the "effective body dose" and the administered dose x". Assume now that a "background dose, d" is equivalent to an "administered dose" of $x_0 = \theta(d)$ which is additive to the actually administered dose x. Denote by $X = x + \theta(d)$ the "total dose". If we now denote by $\gamma(X)$ the effective body dose which is equivalent to X, then we clearly have

$$g(x) = \gamma(x + \theta(d))$$

and (38)

$$g'(0) = \gamma'(\theta(d)) \ .$$

The following are reasonable assumptions concerning the functional relations $\gamma(X)$ and $\theta(d)$:

(a) $\gamma(0) = \theta(0) = 0$,
(b) $\gamma(X) > 0$ for $X > 0$, $\theta(d) > 0$ for $d > 0$, (39)
(c) $\gamma'(X) \geq 0$, $\theta'(d) \geq 0$.

We now consider two situations; viz, (i) a positive background $d > 0$, and (ii) a zero background $d = 0$.

(i) With $d > 0$ it follows from (b) that $g(0) = \gamma(\theta(d)) > 0$, and from (c) that $g'(0) = \gamma'(\theta(d)) \geq 0$. In fact $g'(0) = \gamma'(\theta(d)) > 0$ if $\gamma'(X) > 0$ for all positive X. This means that if there is a positive background, the slope $g'(0)$ is > 0 unless the function $\gamma(X)$ has a stationary point at argument $X = \theta(d)$ precisely. If this is considered unlikely, then a positive background will normally be associated with a positive slope $g'(0)$ at $x = 0$. The above argument is in agreement with a contention by R. Peto (1976).

(ii) The situation for a zero background $d = 0$ is less definite for it may well be that either $\gamma'(0) = g'(0) = 0$ or $\gamma'(0) = g'(0) > 0$.

With the compartment-analytic model, therefore, the decision as to whether the assumption $g'(0) = 0$ may be made depends in the first place on whether the background is positive or zero. If it is positive, then $g'(0) = 0$ is hardly justified. Of course, since the knowledge that $g'(0) > 0$ does not convey anything about the magnitude of $g'(0)$, the estimation technique will have to employ the inequality $g'(0) \geq 0$. However, if $d = 0$, then the decision will depend on the deterministic biological mechanism $\gamma(X)$ at $X = 0$. A superficial perusal of the literature of dose response curves indicates that both $\gamma'(0) = 0$ and $\gamma'(0) > 0$ may occur.

Next we turn to the "multiple attack" model. Here the interpretation of $g(x)$ involves both deterministic and stochastic concepts. If we retain the deterministic concepts of total dose $X = x + \theta(d)$ and effective body dose $\gamma(X)$, then equation (31) can be written in the form

$$g(x) = -\log q[x + \theta(d)] \ . \tag{40}$$

It follows that

$$g(0) = -\log q[\gamma(\theta(d))] \ . \tag{41}$$

and

$$g'(0) = \frac{f[\gamma(\theta(d))]}{q[\gamma(\theta(d))]} \gamma'(\theta(d)) \tag{42}$$

where $f(y) = -dq(y)/dy$.

If we make the reasonable assumption that the ordinate of the tolerance distribution, $f[\gamma(X)]$, is not zero for any positive dose $\gamma(X)$, then equation (42) would again imply that a zero slope $g'(0) = 0$ occurs only if $\gamma'(X)$ has a root at argument $X = \theta(d)$ precisely, which is most unlikely. Thus, a positive background will again imply a positive slope $g'(0) > 0$. However, equation (42) also shows that, for a zero background $\theta(d) = 0$, we shall have $g'(0) = 0$ if either $\gamma'(0) = 0$ or

$f(0) = 0$ or both. Now for many cancer incidence distributions $f(0) = 0$ (e.g. both logit and probit for log X have this property and so have many curves such as the multihit curve for $k > 1$). The exponential (single hit) curve is, however, an exception to this rule since for it $g(0) = 0$ and $g'(0) > 0$. Again, therefore, the question of a zero background must be left open although we would expect a more frequent occurrence of $g'(0) = 0$ situations.

The literature on the functional relations $\gamma(X)$ and $\theta(d)$ is, as yet, limited. However, a number of special cases have been considered. The question as to whether $\gamma'(0) = 0$ or $\gamma'(0) > 0$ is, of course, related to (though not identical with) the question as to whether there is a tolerance for the carcinogen X (that is, $\gamma'(X) = 0$ for a residual range of small X values) or whether there is an instantaneous response to doses X however small.

It has been argued (see e.g. Cornfield and Mantel (1977)) that the assumption of the additivity of background dose, d, and applied dose, x, is not acceptable in certain situations. The nonlinear dose metameter transformation $\theta(d)$ may well dispell some of such objections. Nevertheless, the problem remains. However, the following is a more general (non-additive) assumption which will lead to the same results: The effective body dose is assumed to be a <u>general</u> function $g(x, d)$ of the two doses x and d satisfying the following conditions: (a) $g(0, 0) = 0$, (b) $g(x, d) > 0$ for $x > 0$ and/or $d > 0$, (c) $\frac{\partial g}{\partial x}(x, d) \geq 0$ and (d) $\frac{\partial g}{\partial x}(x, d) > 0$ for $d > 0$. Conditions (a) and (b) simply state that, if both background and applied dose are zero, then the effective body dose is zero, but it is positive if at least one of them is positive. Condition (c) implies that the effective body dose can not decrease if the applied dose, x, increases. Condition (d) is perhaps more questionable. It states that for a positive background, d, the differential response of the effective body dose to the applied dose, x, must be at least linear.

As an example of the relation $\gamma(X)$ between the total dose X and the effective body dose we may quote Gehring, Watanabe and Park (1978), *Toxicology and Applied Pharmacology 44*, 581-591. These authors have conducted a special radioactive tracer experiment in which Vinyl Chloride (V.C.) was inhaled by rats to establish a relation between

$$X \text{ (their S)} = \mu\text{g of V.C./liter of air}$$
$$\gamma(X) \text{ (their v)} = \mu\text{g of V.C. metabolized/6 hours.}$$

The relation fitted to associated data of S, v is based on the Michaelis-Menten kinetics and is of the form

$$v = V_m S/(K_m + S) \tag{43}$$

where V_m and K_m are constants. Clearly

$$\left.\frac{dv}{dS}\right|_{S=0} = V_m/K_m > 0$$

This interesting paper raises another question: the function $\gamma(X)$ (and hence $g(x)$) is clearly not convex.

On the other hand the relation between the total (administered) dose X and the effective body dose is (at least approximately) known. In such situations it is advisable to fully utilize this knowledge by introducing the dose metameter $u = \gamma(X) = \gamma(x + \theta(d))$. A trial value for the unknown parameter $\theta(d)$ and the knowledge of the applied doses x_i permit the computation of the $u_i = \gamma(x_i + \theta(d))$ where $\gamma(X)$ is the known approximate relation between X and u. Assume now that the true effective body dose is given by $u = \gamma(X) + \delta u(X)$. It is convenient to write $\delta u(X) = \delta(u)$ as a function of $u = \gamma(X)$ provided that $\gamma' \geq 0$. Further, we introduce the function

$$\ell(u) = \begin{cases} u & \text{For the compartment analytic model} \\ -\log q(u) & \text{For the multiple attack model} \end{cases} \tag{44}$$

where $q(u)$ is replacing $q(x)$ in (31). The function $g(x)$ in model (1) can now be replaced by

$$g(u) = \ell(u + \delta(u)). \tag{45}$$

It follows that

$$\frac{dg}{du} = \ell'(u + \delta(u))\ (1 + \delta'(u)) \geq 0 \tag{46}$$

and

$$\begin{aligned} \frac{d^2g}{du^2} &= \ell''(u + \delta(u))\ (1 + \delta'(u))^2 \\ &\quad + \ell'(u + \delta(u))\ \delta''(u) \geq -\varepsilon \end{aligned} \tag{47}$$

where in (46) it is assumed that $|\delta'(u)| << 1$ and in (47) that $\ell''(u) \geq 0$ and $\delta''(u)$ is small.

The use of the M.L. algorithm by Hartley and Sielken (1977) is therefore applicable with suitable modifications.

STATISTICS AND RELATED TOPICS
M. Csörgö, D.A. Dawson, J.N.K. Rao, A.K.Md.E. Saleh (eds.)

DOSE RESPONSE MODELS FOR QUANTAL RESPONSE TOXICITY DATA

Daniel Krewski
Health Protection Branch
Health and Welfare Canada
Ottawa, Ontario
Canada K1A 1B8

John Van Ryzin[1]
School of Public Health
Columbia University
New York, New York
U.S.A. 10032

A number of existing dose response models for quantal response toxicity data are reviewed, including the probit, logit, extreme value, gamma multi-hit and multi-stage models. The incorporation of background response assuming either independence or additivity is discussed and the low dose behaviour of the corresponding dose response curve is examined. Statistical procedures which may be employed in fitting these models to experimental data are presented, along with procedures for assessing goodness-of-fit. The application of these techniques in low dose extrapolation is illustrated using twenty sets of quantal response toxicity data taken from the literature.

1. INTRODUCTION

As a result of the increasing awareness of the potential health hazards of environmental chemicals, considerable effort is currently being devoted to the identification and regulation of those compounds which are carcinogenic as well as those which have other toxic effects. Information on the toxic effects of such chemicals is necessarily derived primarily from bioassay studies conducted using animal models. Since the number of animals used must be compatible with the practical limitations imposed by space, time and cost, the dose levels employed must be sufficiently large so as to induce an observable rate of response. In order to estimate risks in the human population, it is thus necessary to first extrapolate from the results obtained at these high dose levels to lower levels consistent with the levels anticipated in the environment and then to make the conversion from animal to man. Once this has been done, regulatory agencies are still faced with the difficult task of establishing tolerances which will adequately protect the public and at the same time give proper consideration to social and economic factors such as consumer expectations, benefits conferred and costs to industry.

Here, we will address only the problem of low dose extrapolation and refer the reader to Krewski & Brown (1981) for references on the broader issues involved in risk assessment. In addition, we will confine ourselves to quantal response toxicity data, since many of the existing statistical procedures for low dose extrapolation are applicable only in this case. Although outside the scope of this paper, we note that time to response data may also be utilized in these procedures (Chand & Hoel, 1974; Daffer, Crump & Masterman, 1980; Hartley, Tolley & Sielken, 1981). While we are concerned here only with the analysis of experimental results, we note that some recent results on optimal experimental designs for low dose extrapolation have been given by Hoel & Jennrich (1979), Krewski, Kovar & Arnold (1980) and Krewski & Kovar (1981).

Traditional toxicological procedures define a safe level of exposure as some arbitrary fraction of that dose level at which no effects are observed in any of the animals tested. This procedure has been criticized on the grounds that the observed no effect level will depend on the sample size, with response rates of

0/10, 0/100 and 0/1000 obviously having different interpretations (Cornfield, Carlborg & Van Ryzin, 1978). Implicit in this approach is the assumption of the existence of a threshold dose below which no adverse effects will occur. While metabolic activation and deactivation mechanisms may result in the existence of thresholds in certain cases, such thresholds are likely to vary among individual animals. The determination of a population threshold thus presents the difficult statistical problem of determining the minimum of the individual thresholds, a minimum which may well be effectively zero (Brown, 1976). In the case of carcinogens, moreover, even the existence of individual thresholds is subject to debate (Truhaut, 1979). Less generally recognized is the fact that the application of a standard safety factor does not take into account the slope of the dose response curve for the particular response of interest. Clearly, a moderate safety factor may provide an adequate margin of safety if the dose response relationship is relatively steep but may be far from satisfactory if the dose response curve is actually relatively shallow. Conversely, the universal application of a very large safety factor will result in tolerances which will often be excessively low.

Statistical procedures for low dose extrapolation involve a mathematical model under which the probability $P(d)$ of an induced response at dose d is assumed to have a particular functional form, say $P(d) = f(d)$. Here, we will consider only nonthreshold models for which $f(d)$ is strictly increasing in d for all $d \geq 0$. While absolute safety can only be guaranteed in this case when $d = 0$, a virtually safe level of exposure may be defined by $d^* = f^{-1}(p)$, where p is some suitably low level of risk. The actual determination of an acceptable level of risk is a societal decision which will depend on the severity of the hazard involved as well as a variety of socioeconomic considerations which we will not discuss here.

In section 2, existing mathematical models and their biological bases are reviewed, with particular attention devoted to the probit, logit, extreme value, gamma multi-hit and multi-stage models. The accomodation of background response assuming either independence, additivity or a combination thereof is described and the impact of this assumption on the shape of the dose response curve at low doses is explored. Alternative approaches to defining a virtually safe level of exposure in the presence of background response are also compared.

Maximum likelihood procedures which may be employed in fitting these models are given in section 3 along with procedures for assessing goodness-of-fit. Methods for obtaining upper confidence limits on the risk at a given dose and lower confidence limits on the dose corresponding to a specified risk are presented, along with a conservative linear extrapolation procedure. (The underlying large sample theory for these likelihood based estimation procedures are discussed in detail in the Appendix to this paper.)

In section 4, the application of these techniques is illustrated using twenty sets of quantal response toxicity data taken from the literature. A brief discussion of the significance of these results and summary of the current state of the art of low dose extrapolation is presented in section 5.

2. DOSE RESPONSE MODELS

Mathematical Models

Statistical or tolerance distribution models are based on the notion that each animal in the population has its own tolerance to the test compound. Any dose d not exceeding the tolerance t for an individual will have no effect on that individual while any dose exceeding t will result in a positive response. Letting $G(t)$ $(0 \leq t < \infty)$ denote the cumulative distribution of tolerances in the population, the probability that an individual selected at random will respond at dose d is simply $P(d) = Pr(\text{tolerance} \leq d) = G(d)$.

A general class of such statistical models is defined by $G(t) = F(\alpha + \beta \log t)$ for $t > 0$ and $G(0) = F(-\infty) = 0$, where F denotes any standardized cumulative distribution function and α and $\beta > 0$ are unknown parameters (Chand & Hoel, 1974). Perhaps the best known model in this class is the <u>probit model</u> (Finney, 1971), where F is the standard normal distribution function given by

$$F(x) = (2\pi)^{-\frac{1}{2}} \int_{-\infty}^{x} \exp(-u^2/2)\,du \ . \tag{2.1}$$

In this case, the distribution of tolerances is lognormal. For the <u>logistic model</u> (Ashton, 1972)

$$F(x) = [1 + \exp(-x)]^{-1} \tag{2.2}$$

while for the <u>extreme value model</u> (Gumbel, 1958)

$$F(x) = 1 - \exp\{-\exp(x)\} \ . \tag{2.3}$$

Since $G(t) = 1 - \exp(-at^b)$ under the extreme value model, where $a = \exp(\alpha)$ and $b = \beta$, this model is sometimes called the Weibull model.

Stochastic or mechanistic models are based on the notion that for each animal, a positive response is the result of the random occurence of one or more biological events. The <u>one-hit model</u> is based on the concept that a response will be induced after the target site has been hit by a single biologically effective unit of dose within a specified time interval. (More generally, the concept of a hit may be thought of as the occurence of one or more of a variety of possible fundamental biological events.) If the number of hits during this period follows a generalized homogeneous Poisson process, then the probability that an individual will respond at dose d is

$$\begin{aligned} P(d) &= \Pr(\text{at least 1 hit}) \\ &= 1 - \exp(-\lambda d) \ , \end{aligned} \tag{2.4}$$

where λd $(\lambda > 0)$ is the expected number of hits during this time period. If $k \geq 1$ hits are required, the probability that an individual will respond at dose d is given by

$$\begin{aligned} P(d) &= 1 - \sum_{j=0}^{k-1} \exp(-\lambda d) \frac{(\lambda d)^j}{j!} \\ &= \int_0^d [\Gamma(k)]^{-1} \lambda^k t^{k-t} \exp(-\lambda t)\,dt \end{aligned} \tag{2.5}$$

(Rai & Van Ryzin, 1979, 1981), where $\Gamma(k)$ denotes the gamma function. Using the latter expression for $P(d)$ in (2.5), this multi-hit stochastic model may also be interpreted as a tolerance distribution model where the tolerance distribution is gamma with scale parameter $1/\lambda$ and shape parameter $k > 0$. This extension of the one-hit model is thus called the <u>gamma multi-hit model</u>.

Another stochastic model is based on the assumption that the induction of irreversible self-replicating toxic effects such as carcinogenesis is the result of the occurence of a number of different random biological events, with the age-specific rate of occurence of each event linearly related to dose. This and other assumptions (see Crump, Hoel, Langley & Peto (1976) or Crump (1979) for details) leads to a multi-stage model of the form

$$P(d) = 1 - \exp\left(-\sum_{i=1}^{k} \beta_i d^i\right) \quad (\beta_i \geq 0) , \tag{2.6}$$

where k denotes the number of events or stages.

Incorporation of Background Response

All of the models discussed previously assume that the background response rate $P(0)$ is zero. In many experiments, however, the response of interest also occurs spontaneously in control animals. This background may be assumed to be independent of the induced responses or additive in a mechanistic manner (Hoel, 1980).

If the spontaneous and induced responses are assumed to be independent, then the probability of observing a response of either type at dose d is given by

$$P^*(d) = \gamma + (1-\gamma)P(d) , \tag{2.7}$$

where $0 < \gamma < 1$ denotes the spontaneous response rate. Under the additivity assumption, the background response rate may be considered as arising from an effective gackground dose $\delta > 0$, with

$$P^*(d) = P(d + \delta) . \tag{2.8}$$

For the gamma multi-hit model, for example, it may be postulated that the spontaneous rate is determined by the expected number of background hits $\lambda\delta$ (Cornfield & Mantel, 1977). A combination of both independent and additive background may be represented by the model

$$P^*(d) = \gamma + (1-\gamma)P(d+\delta) . \tag{2.9}$$

We note that assuming independent background, the multi-stage model may be expressed in the form

$$P^*(d) = 1 - \exp\left\{-\sum_{i=0}^{k} \beta_i d^i\right\} \quad (\beta_i \geq 0) , \tag{2.10}$$

where $\beta_o = -\log(1-\gamma)$. Under additive background, the model may be written as

$$P^*(d) = 1 - \exp\left\{-\sum_{i=0}^{k} \alpha_i d^i\right\} \quad (\alpha_i \geq 0) , \tag{2.11}$$

where $\alpha_o = \Sigma_{i=1}^{k} \beta_i \delta^i$ and $\alpha_r = \Sigma_{s=r}^{k} \beta_s \binom{s}{r} \delta^{s-r}$ $(r = 1,\ldots,k)$, although the α_i are subject to additional constraints. When $k = 2$, for example, positive values of α_0, α_1 and α_2 correspond to positive values of δ, β_1 and β_2 only when $\alpha_1^2 - 4\alpha_0\alpha_2 \geq 0$. Such nonlinear constraints will have to be taken into account when fitting the multi-stage with additive background.

Although the manner in which background response is accomodated is crucual, the extent to which independence or additivity is indicated by either biological theory or experimental data is somewhat unclear at this time (Food Safety Council, 1980a). While we make no attempt to settle this issue here, we will discuss in some detail the consequences of assuming either independent or additive background in low dose extrapolation.

Shape of the Dose Response Curve at Low Doses

Under the assumption of independent background, the excess risk over background $\Pi(d) = P^*(d) - P^*(0)$ for the probit model approaches zero very rapidly, more rapidly than any power of dose; that is,

$$\lim_{d \downarrow 0} \Pi(d)/d^m = 0 , \tag{2.12}$$

no matter how large the value of m. The excess risk for the logit and extreme value models approach zero more slowly since

$$\lim_{d \downarrow 0} \Pi(d)/d^\beta = (1-\gamma)e^\alpha > 0 \tag{2.13}$$

in both cases. It follows immediately from (2.13) that the excess risk for these two models will be linear at low doses when $\beta = 1$, convex with $\beta > 1$ and concave when $\beta < 1$. The excess risk for the gamma multi-hit model also approaches zero more slowly than that for the probit model since

$$\lim_{d \downarrow 0} \Pi(d)/d^k = (1-\gamma)\lambda^k/\Gamma(k+1) > 0 . \tag{2.14}$$

Since $\Pi(d)$ is thus linear at low doses when $k = 1$, the one-hit model is sometimes referred to as the linear model. For small d, the excess risk for the gamma multi-hit model will be convex when $k > 1$ and concave when $k < 1$.

In the case of the logit, extreme value and gamma multi-hit models, it follows from (2.13) and (2.14) that a log-log plot of $\Pi(d)$ versus d will be nearly linear at low doses, as illustrated in Figure 1. Here, the models have

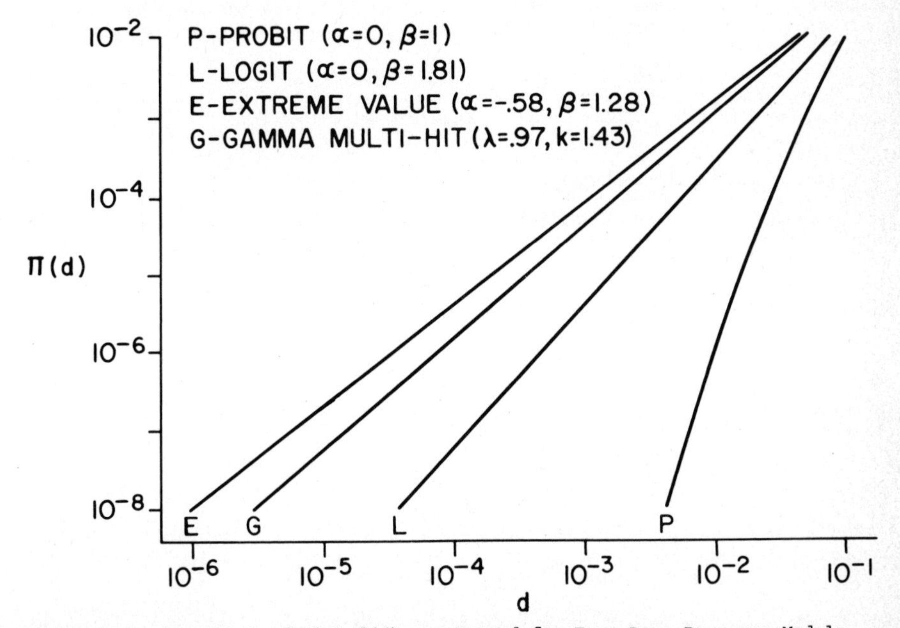

Figure 1. Log-Log Plot of Risk $\Pi(d)$ vs. Dose d for Four Dose Response Models.

been parametrized so that the mean and variance of the distributions of the logarithms of the individual tolerances all have mean zero and variance one (Johnson & Kotz, 1970ab) with a background rate of zero. As indicated in (2.12), the excess risk for the probit model decreases more rapidly than that for the other three tolerance distribution models.

For the multi-stage model,

$$\lim_{d \downarrow 0} \Pi(d)/d^{\ell} = \beta^{\ell} \exp(-\beta_0) > 0 , \qquad (2.15)$$

where ℓ is the smallest integer (≥ 1) for which $\beta_{\ell} > 0$. The excess risk for this model is thus linear at low doses when $\ell = 1$ and convex when $\ell > 1$ with the relationship between $\log \Pi(d)$ and $\log d$ again nearly linear for small d.

While the multi-stage model does not admit the possibility of low dose concavity, it does provide for dose response curves which are both linear at low doses and convex at moderate doses. (This would be the case when both β_1 and higher order coefficients are positive.) Low dose linearity in the logit, extreme value and gamma multi-hit models, on the other hand, is compatible only with dose response curves which are concave at moderate and high doses.

Assuming $\Pi''(d)$ exists on an interval $[0,D]$ encompassing the low dose region of interest, we may in general write

$$\Pi(d) = d\Pi'(0) + O(d^2) \qquad (2.16)$$

so that $\Pi(d)$ will be nearly linear in d whenever $\Pi'(0) > 0$ (Crump, Hoel, Langley & Peto, 1976; Peto, 1978). Under the assumption of additive background, this will be the case whenever $P'(\delta) > 0$. This latter condition will be satisfied whenever $P(d)$ is strictly increasing in d as is the case for all those models considered here. The same argument may be used to establish the low dose linearity of $\Pi(d)$ when the background includes both independent and additive components.

We note that low dose linearity is characterized by the slope of $\Pi(d)$ being positive at $d = 0$, whereas convexity and concavity in the low dose region correspond to zero and infinite slopes respectively.

Virtually Safe Levels of Exposure

In the presence of a positive background level of response $(c = P^*(0) > 0)$, several different definitions of a virtually safe dose (VSD) have been proposed. Crump, Guess & Deal (1977) define the VSD d_o^* by the equation

$$\pi = \Pi(d_o^*) , \qquad (2.17)$$

where $0 < \pi < 1\text{-}c$ is the allowable increment in the spontaneous response rate. Hartley & Sielken (1977a) define virtual safety on the basis of the allowable proportionate reduction T in the spontaneous tumor free rate $1\text{-}c$. The VSD d_1^* in this case is given by

$$T = (1 - P^*(d_1^*))/(1\text{-}c) \quad \text{or} \quad (1\text{-}T)(1\text{-}c) = \Pi(d_1^*) . \qquad (2.18)$$

When the spontaneous and induced responses are assumed to be independent, the dose d_2^* which corresponds to a risk p of an induced response is defined by (Mantel, Bohidar, Brown, Ciminera & Tukey, 1975)

$$p = P(d_2^*) \quad \text{or} \quad p(1\text{-}c) = \Pi(d_2^*) . \qquad (2.19)$$

When $c = 0$, setting $\pi = (1-T) = p$ yields $d_0^* = d_1^* = d_2^*$.

When $c > 0$, setting $\pi = (1-T) = p$ yields $d_0^* > d_1^* = d_2^*$. Under the logit and extreme value models with independent background, it follows from (2.13) that

$$d_0^*/d_1^* \doteq (1-d)^{-1/\beta} \tag{2.20}$$

for small π. Similarly, $d_0^*/d_1^* \doteq (1-c)^{-1/k}$ and $d_0^*/d_1^* \doteq (1-c)^{-1/\ell}$ for the gamma multi-hit and multi-stage models respectively. The difference between d_0^* and d_1^* will thus increase as the background rate c increases or as the parameters β, k or ℓ (≥ 1) decrease. As indicated in Table 1, however, d_0^* will exceed d_1^* by less than a factor of two whenever $c \leq 0.25$ and β or $k \geq 0.5$. Because the the extreme flatness of the dose response curve for the probit model in the low dose region, the difference between d_0^* and d_1^* will be less than that for the logit and extreme value models.

Table 1. Values of $(1-c)^{-1/\beta}$ for Selected β and c

β	c			
	0.01	0.10	0.25	0.50
0.25	1.04	1.52	3.16	16.00
0.5	1.02	1.23	1.78	4.00
1.0	1.01	1.11	1.33	2.00
10.0	< 1.01	1.01	1.03	1.07

When the background includes an additive component, $d_0^*/d_1^* \doteq (1-c)^{-1}$ for any dose response model for which (2.16) holds.

Because of the generally small differences between d_0^* , d_1^* and d_2^* , we will focus our attention in what follows on $d^* \equiv d_0^*$.

3. ESTIMATION PROCEDURES

Point Estimates and Confidence Limits

Suppose that a total of n animals are used in an experiment involving dose levels $0 = d_0 < d_1 < \ldots < d_m$ and that x_i of the n_i animals at dose d_i respond ($i = 0,1,\ldots,m$). (Here, we will require that $n_i > 0$ for $i = 1,\ldots,m$ but will allow the possibility that $n_o = 0$. The effective number of dose levels will be denoted by $m^* = m+1$ for $n_o > 0$ and $m^* = m$ for $n_o = 0$.) We now consider the estimation of the excess risk $\Pi(d)$ and the VSD d^* on the basis of these data.

Assuming that each animal responds independently of all other animals in the experiment[2], the likelihood of the observed outcome under any dose response model $P^*(d) = P^*(d;\, \underset{\sim}{\theta})$ depending on the (t-dimensional) vector of parameters $\underset{\sim}{\theta} \in \Omega$ is given by

$$L(\underset{\sim}{\theta}) = \Pi_{i=0}^{m} \binom{n_i}{x_i} (P_i^*)^{x_i} (Q_i^*)^{n_i - x_i} , \tag{3.1}$$

where $P_i^* = P^*(d_i)$ and $Q_i^* = 1 - P_i^*$. Since maximization of $L(\underset{\sim}{\theta})$ or $\ell(\underset{\sim}{\theta}) = \log L(\underset{\sim}{\theta})$ using direct analytical procedures is generally not possible, the maximum likelihood estimator $\hat{\underset{\sim}{\theta}}$ of $\underset{\sim}{\theta}$ must be obtained using iterative numerical procedures. The large sample properties of $\hat{\underset{\sim}{\theta}}$ given in Theorems 1 and 2 in the Appendix provide a basis for constructing point estimates and confidence limits for $\Pi(d)$ and d^*. The regularity conditions required in order to establish these general results will be satisfied for the probit, logit, Weibull and gamma multi-hit models, provided the model is identifiable[3] and that the number of dose levels employed is at least as great as the number of unknown parameters in the model.

Letting $\hat{P}^*(d) = P^*(d;\hat{\underset{\sim}{\theta}})$ denote the fitted dose response curve, we have that $\hat{\Pi}(d) = \hat{P}^*(d) - \hat{P}^*(0)$ is a strongly consistent estimator of $\Pi(d)$ and that $\hat{d}^* = \hat{\Pi}^{-1}(\pi)$ is a strongly consistent estimator of d^*. In order to construct an upper confidence limit for $\Pi(d)$, we note that $\sqrt{n}(\hat{\underset{\sim}{\theta}} - \underset{\sim}{\theta})$ is approximately normally distributed with mean $\underset{\sim}{0}$ and dispersion matrix $\underset{\sim}{\Sigma} = ((\sigma_{rs})) = ((\sigma^{rs}))^{-1}$ defined by

$$\sigma^{rs} = \Sigma_{i=0}^{m} c_i \frac{\partial P_i^*}{\partial \theta_r} \frac{\partial P_i^*}{\partial \theta_s} / (P_i^* Q_i^*) , \tag{3.2}$$

where $c_i = \lim_{n\to\infty} n_i/n$ with $c_i > 0$ for $i = 1,\ldots,m$ and $c_o \geq 0$.[4] It now follows that $\sqrt{n}(\hat{\Pi}(d) - \Pi(d))$ is approximately normally distributed with mean zero and variance

$$\sigma_1^2 = \Sigma_{r=1}^{t} \Sigma_{s=1}^{t} \frac{\partial \Pi}{\partial \theta_r} \frac{\partial \Pi}{\partial \theta_s} \sigma_{rs} \tag{3.3}$$

so that an approximate $100(1-\rho)\%$ upper confidence limit for $\Pi(d)$ at any fixed dose $d > 0$ is given by

$$U_1(d) = \hat{\Pi}(d) + z_{1-\rho} \frac{\hat{\sigma}_1}{\sqrt{n}} . \tag{3.4}$$

Here, $\Phi(z_\rho) = \rho$ and $\hat{\sigma}_1^2$ is a strongly consistent estimator of σ_1^2 obtained by evaluating (3.2) and (3.3) at $\underset{\sim}{\theta} = \hat{\underset{\sim}{\theta}}$ and $c_i = n_i/n$.

An upper confidence limit on $\Pi(d)$ may also be obtained using the asymptotic distribution of the likelihood ratio statistic given in Theorem 4 in the Appendix. This approach leads to the upper limit defined by

$$U_2(d) = \sup_{\underset{\sim}{\theta} \in \Omega} \{\Pi(d;\underset{\sim}{\theta}) : 2(\ell(\hat{\underset{\sim}{\theta}}) - \ell(\underset{\sim}{\theta})) \leq \chi^2_{t,1-2\rho}\} \tag{3.5}$$

(Cox & Hinkley, 1974), where $\chi^2_{t,\rho}$ denotes the $100\rho^{th}$ percentage point of the chi-squared distribution with t degrees of freedom. We note that the use of $\chi^2_{t,1-2\rho}$ rather than $\chi^2_{t,1-\rho}$ in (3.5) is due to the fact that a one sided rather than a two-sided confidence limit on $\Pi(d)$ is required (Fleiss, 1973, pp.20-21).

If $\Pi(d)$ is strictly monotone in θ_r for $r = 1,\ldots,t$, it is also possible to set an upper confidence limit on $\Pi(d)$ by first setting confidence limits on each θ_r. For the probit, logit and extreme value models with independent background, $\Pi(d) = (1-\gamma)F(\alpha + \beta \log d)$ is decreasing in γ and increasing in α and β, provided $d > 1$. In this case, a $100(1-\rho/3)\%$ lower confidence limit $\gamma_L = \hat{\theta}_1 - z_{1-\rho/3}(\hat{\sigma}_{11}/n)^{\frac{1}{2}}$ on $\gamma = \theta_1$ and $100(1-\rho/3)\%$ upper confidence limits α_U and β_U on α and β would first be calculated. (For $d < 1$, a lower limit on β is required.) Then

$$U_3(d) = (1-\gamma_L)P(d;\ \alpha_U,\ \beta_U) \tag{3.6}$$

is a $100(1-\rho)\%$ upper confidence limit on $\Pi(d)$ since

$$\begin{aligned}\Pr\{U_3(d) \geq \Pi(d)\} &\geq \Pr\{\gamma_L \leq \gamma,\ \alpha_U \geq \alpha,\ \beta_U \geq \beta\} \\ &\geq 1 - [\Pr\{\gamma_L > \gamma\} + \Pr\{\alpha_U < \alpha\} + \Pr\{\beta_U < \beta\}] \\ &= 1 - \rho\ . \end{aligned} \tag{3.7}$$

For the gamma multi-hit model,an analogous procedure based on the fact that $\Pi(d)$ is decreasing in k for $k \geq 1$ and increasing in λ for all k may be employed (Van Ryzin & Rai, 1980). While this approach to setting confidence limits is computationally slightly simpler than the previous two, it is conservative in that the actual confidence level is <u>at least</u> the nominal $100(1-\rho)\%$. A second disadvantage of this approach is that it is not readily applicable in the case of additive background since $\Pi(d)$ is not in general strictly monotone in δ.

As in (3.4), an approximate lower confidence limit on d^* is given by

$$L_1(\pi) = \hat{d}^* - z_{1-\rho}\frac{\hat{\sigma}_2}{\sqrt{n}}\ , \tag{3.8}$$

where $\hat{\sigma}_2^2 = (\partial\Pi/\partial d|_{d=\hat{d}^*})^{-2}\ \hat{\sigma}_1^2$ with $\hat{\sigma}_1^2$ evaluated at $d = \hat{d}^*$. Alternative lower limits for d^* may be obtained by first constructing a lower limit for $\log d^*$ or an upper limit for $1/d^*$ in a similar fashion. These latter confidence limits, denoted by L_1' and L_1'' respectively, have the advantage of not assuming negative values as is possible in the case of L_1.

The likelihood based lower confidence limit on d^* is given by

$$L_2(\pi) = \inf_{\underset{\sim}{\theta} \in \Omega} \{d^* = \Pi^{-1}(\pi;\underset{\sim}{\theta}) : 2(\ell(\hat{\underset{\sim}{\theta}}) - \ell(\underset{\sim}{\theta})) \leq \chi^2_{t,1-2\rho}\}. \tag{3.9}$$

The confidence limit procedure illustrated in (3.6) is not applicable in the case of the probit, logit and extreme value models since $d^* = \exp\{(F^{-1}(\pi/(1-\gamma))-\alpha/\beta\}$ is not in general monotone in β. For the gamma multi-hit model, however, conservative lower confidence limits $L_3(\pi)$ may be based on the approximation

$$d \doteq \frac{1}{\lambda}\left[\frac{\pi}{(1-\gamma)}\ \Gamma(k+1)\right]^{1/k} \tag{3.10}$$

for small π, which is decreasing in λ and increasing in both γ and k since $\pi < 1-\gamma$. (This approximation has also been used by Rai & Van Ryzin (1981) to obtain lower confidence limits on d^* based on the approximate asymptotic distribution of $\hat{d}^*$ derived via (3.10). In light of Corollary 1, however, the use of this approximation for this purpose is no longer necessary.)

Another approach which has been employed is to take $L(\pi) = U^{-1}(\pi)$ as a lower confidence limit on d^*, where $U(d)$ denotes some upper confidence limit on $\Pi(d)$ (Rai & Van Ryzin, 1979; Van Ryzin & Rai, 1980). This procedure is based on the observation that $\Pr\{U(d^*) \geq \pi\} = \Pr\{d^* \geq U^{-1}(\pi)\}$ for $U(d)$ strictly increasing in d (Hartley & Sielken, 1977b). Here we will take $L_3'(\pi) = U_3^{-1}(\pi)$ as a conservative lower confidence limit on d^* for the probit, logit and extreme value models with independent background, noting that the required monotonicity of $U_3(d)$ follows from the continuity of $U_3(d)$ at $d = 1$.

Estimation procedures for the multi-stage model are complicated by the constraints on the parameter space. However, efficient algorithms for obtaining the restricted maximum likelihood estimators in the case of independent background have been developed by Guess & Crump (1976) and Hartley & Sielken (1977a). The former algorithm is applicable in the case of the more general model

$$P^*(d) = 1 - \exp\{-Q(d)\} \tag{3.11}$$

where $Q(d) = \Sigma_{i=0}^{\infty} \beta_i d^i$ $(\beta_i \geq 0)$ and all but finitely many of the β_i are zero. This avoids specifying the degree k of the polynomial $Q(d)$. Conditions for the existence and uniqueness of this infinite parameter maximum likelihood problem and for the strong consistency of the corresponding maximum likelihood estimators have been given by Guess & Crump (1976, 1978).

Because of the constraints $\beta_i \geq 0$, the asymptotic distribution of the vector of maximum likelihood estimators will generally not be normal (Guess & Crump, 1978). However, when k ($\leq m^* - 1$) is known and $\beta_i > 0$ for $i=0,1,\ldots,k$, the usual asymptotic distribution given in Theorem 2 will apply. If some $\beta_i = 0$ for $i < k$, the limited simulation studies conducted by Crump, Guess & Deal (1977) suggest that confidence limits based on Corollary 1 will be nearly correct for $k = 1$ or 2 and somewhat conservative for larger k.

Hartley & Sielken (1977a) employ an alternative approach to obtaining confidence limits which circumvents the difficulties imposed by the nonstandard asymptotic distribution of the maximum likelihood estimators. Their procedure for obtaining a lower confidence limit $L_4(\pi)$ on d^* involves splitting the data at random into G groups of equal size and calculating an estimate $\hat{d}_g^*$ of d^* for each group $g = 1,\ldots,G$ in order to obtain an estimate of the variance of $\hat{d}^*$. A disadvantage of this random group method of obtaining confidence limits is that it depends on the particular random split obtained.

Goodness-of-Fit

We now consider statistical goodness-of-fit tests which may be applied in order to determine whether a given model provides a reasonable fit to the experimental data in the observable range. Statistical procedures for discriminating between models will also be discussed briefly.

The usual chi-square statistic

$$\chi^2 = \Sigma(x_i - n_i\hat{P}_i^*)^2/(n_i\hat{P}_i^*\hat{Q}_i^*) \tag{3.12}$$

may be used to assess the goodness-of-fit of the probit, logit, extreme value and gamma multi-hit models. (As shown in Theorem 3 in the Appendix, the asymptotic distribution of this statistic is chi-squared with m^*-t degrees of freedom, provided that the assumed model is correct and $m^* > t$.)

In assessing goodness-of-fit, we will refer to the observed level of significance defined by $p = \Pr\{\chi^2 \geq \chi^2_{obs}\}$, where χ^2_{obs} denotes the observed value of the chi-square goodness-of-fit test statistic. This p-value measures the degree to which the observed data are consistent with the assumed model, with small values of p suggesting lack of fit.

Because of the nonnegativity constraints on the parameters, the chi-square goodness-of-fit test discussed above will generally not be applicable in the case of the multi-stage model. However, a Monte Carlo goodness-of-fit test based on bootstrapping the likelihood function (Efron, 1979) has been proposed by Crump, Guess & Deal (1977). After fitting the multi-stage model to the observed data, this model is used to generate 99 additional data sets by computer simulation. The rank S of the likelihood of the original data among the likelihoods for all 100 data sets arranged in increasing order then provides a measure of goodness-of-fit, with small values of S suggesting lack of fit. Here, we will take $(S-1)/100$ as an approximate p-value for the Monte Carlo goodness-of-fit test.

As shown in Theorem 4 in the Appendix, the usual likelihood ratio statistic may be used to test whether or not either the extreme value model or the gamma multi-hit model provides a significant improvement in fit over the one-hit model. (Recall that the one-hit model is a special case of both the extreme value and gamma multi-hit models.) Tests for comparing the relative fits of two models not in the same class may also be constructed (Chambers & Cox, 1967; Crump, 1981). However, because even a moderately large experiment designed specifically for this purpose will have relatively low power for discriminating between two plausible models, such tests will not be discussed explicitly here.

Linear Extrapolation

Linear extrapolation is based on the assumption that the true dose response curve is convex at low doses. In this case, a straight line $\Pi_x(d) = d\Pi(d_1)/d_1$ joining the origin and any point $(d_1, \Pi(d_1))$ on the curve $\Pi(d)$ in the low dose region will lie above the curve and thus provide an upper limit on the excess risk at any dose $d \leq d_1$. The use of such a conservative model has been proposed as a means of providing for the uncertainties involved in low dose extrapolation.

The application of linear extrapolation in cases where only a control and single test group have been used has been discussed by Gaylor & Shapiro (1979). For studies involving more than one positive dose level, Gaylor & Kodell (1980) suggest fitting a particular model to the experimental data and then extrapolating linearly from that point on the curve corresponding to the lowest positive dose level employed. Provided that this dose corresponds to a measurable level of risk, the results obtained with this approach should be largely independent of the model selected. In order to minimize any such model dependency, we propose extrapolating from some point on the fitted curve where the excess risk π_o is still within the observable range, such as $\pi_o = 10^{-2}$ (Van Ryzin, 1980).

Letting $\hat{d}_o^* = \hat{\Pi}^{-1}(\pi_o)$, an estimate of the upper limit $\Pi_x(d)$ on the excess risk at any dose $d \leq \hat{d}_o^*$ is given by $\hat{\Pi}_x(d) = d\pi_o/\hat{d}_o^*$. An estimate of the corresponding lower limit $d_x^* = \Pi_x^{-1}(\pi)$ on d^* is given by $\hat{d}_x^* = (\pi/\pi_o)\hat{d}_o^*$. Upper and lower confidence limits on $\Pi_x(d)$ and d_x^* respectively may be obtained through the use of a lower confidence limit on d_o^* in place of the point estimate $\hat{d}_o^*$.

4. EMPIRICAL RESULTS

Application of the dose response models and statistical procedures discussed in sections 2 and 3 will be illustrated using actual experimental data on the toxic responses induced by the twenty substances listed in Table 2. Data on the first fourteen compounds has been assembled by the Food Safety Council (1980a). The last six data sets are derived from a recently completed large-scale study conducted by the National Center for Toxicological Research designed to accurately define the dose response curve for a known chemical carcinogen, 2-actelyamino-fluorine, at relatively low doses. Data on the induction on bladder and liver

Table 2. Toxic Responses Induced by Twenty Substances

Substance	Response	Period of Exposure[a] (Dose Units)
1. NTA (Nitrilotriacetic Acid)	Kidney Tumors	2 years (%)
2. Aflatoxin B_1	Liver Tumors	2 years (ppb)
3. ETU (Ethylenethiourea)	Fetal Anomalies	Days 6-15 of gestation (mg/kg body weight/day)
4. TCDD (2,3,7,8-tetrachlorodibenzo-p-dioxin)	Fetal Intestinal Anomalies	Days 6-15 of gestation (mg/kg body weight/day)
5. DMN (Dimethylnitrosamine)	Liver Tumors	ppm
6. Vinyl Chloride	Liver Angiosarcoma	12 months (ppm in atmosphere)
7. HCB (Hexachlorobenzene)	Fetal 14th Rib Anomalies	Days 6-16 of gestation (mg/kg body weight/day)
8. Botulinum Toxin - Type A	Death	Single Oral Dose (ng)
9. Bischloromethyl Ether	Respiratory Tumors	Inhalation of 100 ppb for 6 hours (number of exposures)
10. Sodium Saccharin	Bladder Tumors	2 years (%)
11. ETU (Ethylenethiourea)	Thyroid Carcinoma	2 years (ppm)
12. Dieldrin	Liver Tumors	128 weeks (ppm)
13. DDT	Liver Tumors	130 weeks (ppm)
14. Span Oil	Cardiac Lesions	16 weeks (%)
15. 2-AAF	Bladder Tumors	18 months (ppm)
16. 2-AAF	Bladder Tumors	24 months (ppm)
17. 2-AAF	Bladder Tumors	33 months (ppm)
18. 2-AAF	Liver Tumors	18 months (ppm)
19. 2-AAF	Liver Tumors	24 months (ppm)
20. 2-AAF	Liver Tumors	33 months (ppm)

[a] Unless indicated otherwise, oral exposure with dose units specifying concentration in the diet.

tumors in this important study has been summarized by Littlefield, Farmer, Gaylor & Sheldon (1979). The animal model used in all cases is either the rat or the mouse. With the exception of span oil and 2-AAF at 33 months, the number of animals at each dose level used is generally well in excess of twenty.

Independent Background

The probit, logit, extreme value, gamma multi-hit and multi-stage models with independent background were fitted to each of these twenty data sets. The gamma multi-hit model was fitted using the computer program MULTI80 developed by Rai & Van Ryzin (1980). This program utilizes a modified Fletcher-Reeves conjugate gradient method of maximizing the likelihood function. A similar approach was used in the case of the probit, logit and extreme value models, using the transformation $\gamma = F(\gamma^*)$ in order to provide for the constraint $0 < \gamma < 1$. (Whenever $\hat{\gamma}$ was found to be less than 10^{-6}, γ was assumed to be zero.) The multi-stage model was fitted using both the GLOBAL program developed by Crump, Guess & Deal (1977) and a program developed by Hartley & Sielken (1978) called MRS.T. (Statistical Methodology for Toxicological Research). (Restrictions on the number of doses in MRS.T. were relaxed so that the program could be used with botulinum toxin, bischloromethyl ether and 2-AAF.) The former program uses a modified gradient projection technique while the latter program is based on convex programming techniques. The GLOBAL program also provides for fitting the multi-stage model where the degree k of the polynomial $Q(d)$ is not fixed in advance. With the restriction $k \leq m^*-1$, the estimates of the β_i obtained using the GLOBAL program and MRS.T. were virtually identical.

As reported earlier (Food Safety Council, 1980a), both the extreme value and gamma multi-hit models provide satisfactory fits to the first fourteen data sets, with the p-values for the chi-square goodness-of-fit test ranging from $p = .19$ to $p > .99$. The present analysis indicated that the probit and logit models also fit these fourteen data sets, with $p > .14$ in all cases. The multi-stage model was found earlier to provide a good fit to these same results except in the case of vinyl chloride, where the Monte Carlo goodness-of-fit p-value was $p = .03$. (This lack of fit may be attributed to the fact that the multi-stage model is necessarily convex or linear at low doses whereas the observed dose response curve for vinyl chloride is concave.) All five models were found to fit the six 2-AAF data sets reasonably well, with $p > .20$ in all cases except no.16 (bladder tumors at 33 months), where $p = .06$ for the multi-stage model.

The one-hit model is not considered here other than as a special case of the multi-stage, extreme value and gamma multi-hit models since it is necessarily linear at low doses and concave throughout the entire dose range and will thus provide a poor fit to those data sets exhibiting strong upward curvature. This is reflected in a significant lack of fit ($p < .05$ in the chi-square goodness-of-fit test) for nine data sets and a significant improvement in fit with the extreme value and gamma multi-hit models ($p < .05$ in the likelihood ratio test) for these and five other data sets. Of the remaining six cases, (compounds no.4, 7,13,14,18 and 20), visual examination of the raw data suggests that the one-hit model may be appropriate only for substance no.7, hexachlorobenzene, where the dose response curve is linear throughout the entire range of doses used in the experiment.

Information on the parameters determining the low dose behaviour of each of the five models fitted is given in Table 3, along with the estimated background response rate. As shown in section 2, the multistage model will be linear at low doses whenever $\hat{\beta}_1 > 0$ and sublinear or convex otherwise. Low dose linearity is reflected by $\hat{\beta} \doteq 1$ in the extreme value and logit models and by $\hat{k} \doteq 1$ in the gamma multi-hit model as is the case for hexachlorobenzene. With the exception of vinyl chloride, the fitted extreme value, logit and gamma multi-hit models are

Table 3. Maximum Likelihood Estimates of the Model Parameters Assuming Independent Background

Substance	Multi-Stage		Extreme Value		Logit		Gamma Multi-Hit		Probit	
	$\hat{\gamma}$	$\{i:\hat{\beta}_i>0\}$	$\hat{\gamma}$	$\hat{\beta}$	$\hat{\gamma}$	$\hat{\beta}$	$\hat{\gamma}$	$\hat{k}$	$\hat{\gamma}$	$\hat{\beta}$
1. NTA	0	1 5	.003	9.3	.003	9.8	.003	28.4	.003	4.9
2. Aflatoxin	.042	0 1 2	.045	2.0	.049	3.0	.047	3.2	.050	1.6
3. ETU	0	3	0	3.4	0	4.4	0	7.2	0	2.5
4. TCDD	0	2	0	2.0	0	2.2	0	2.5	0	1.1
5. DMN	0	2	0	2.0	0	2.4	0	2.8	0	1.4
6. Vinyl Chloride	.035	0 1	0	0.4	0	0.5	0	0.4	0	0.3
7. HCB	0	1 2	0	1.0	0	1.1	0	1.0	0	0.6
8. Botulinum Toxin	0	6	0	6.1	0	8.9	0	27.6	0	5.2
9. Bischloromethyl Ether	0	1 2 5	.009	1.7	.016	2.0	.025	1.8	.023	1.2
10. Sodium Saccharin	0	4	0	4.7	0	5.0	0	9.2	0	2.6
11. ETU	.021	0 3 4	.021	3.3	.022	4.8	.022	8.2	.023	2.9
12. Dieldrin	.107	0 1 2	.107	1.7	.107	2.2	.107	2.3	.108	1.3
13. DDT	.045	0 1 2	.049	1.4	.052	1.8	.050	1.7	.056	1.1
14. Span Oil	.086	0 1 2	.087	1.4	.086	1.6	.088	1.6	.087	1.0
15. 2-AAF	.021	0 1 3 7	.003	7.3	.003	8.1	.003	20.8	.003	4.5
16. 2-AAF	.001	0 6 7	.001	6.1	.002	7.6	.002	20.5	.002	4.4
17. 2-AAF	0	1 6	.004	5.7	.006	7.8	.006	21.6	.006	4.5
18. 2-AAF	.002	0 1 2	.002	1.3	.002	1.3	.002	1.3	.003	0.5
19. 2-AAF	.021	0 1 2	.021	1.5	.022	1.7	.022	1.7	.023	0.9
20. 2-AAF	.356	0 1 3 4	.365	1.7	.371	2.1	.363	2.1	.368	1.2

sublinear at low doses in the remaining cases with $\hat{\beta}$ and $\hat{k}$ being greater than unity. For vinyl chloride, $\hat{\beta}$ and $\hat{k}$ are less than unity with the dose response curves for these three models approaching zero at a faster than linear or supralinear rate. The probit model is sublinear at low doses regardless of the value of $\hat{\beta}$.

The estimates of the background rate of response γ for the probit, logit, extreme value and gamma multi-hit models are in close agreement. The values of $\hat{\gamma}$ for the multi-stage model are generally reasonably close to those for these four models, although zero background is estimated in the case of NTA, bischloromethyl ether and 2-AAF (no.17) with the multistage model while a small background rate of response (less than 1%) is estimated with each of the remaining models. The positive background indicated for the multi-stage model in the case of vinyl chloride reflects the lack of fit noted earlier.

The actual data is displayed in graphical form in Figure 2, along with the fitted extreme value model. In addition, the estimated excess risk over background is depicted graphically in Figure 2. Also shown are the results of linear extrapolation from the dose estimated to induce an excess risk of $\pi_o = 10^{-2}$ using the multi-stage model. (Because of the concavity of the dose response curve for vinyl chloride, linear extrapolation was not applied in that case.[5]) The estimated virtually safe levels of exposure at an excess risk of $\pi = 10^{-5}$ based on these six extrapolation procedures are given in Table 4.

These results clearly demonstrate that there are substantial differences among the six procedures, with the differences increasing as π decreases. With the exception of vinyl chloride, linear extrapolation leads to the most conservative results in all cases, followed by the multi-stage and extreme value models respectively. (While linear extrapolation represents the most conservative procedure, it should be noted that the results for the multi-stage model are comparable whenever $\hat{\beta}_1 > 0$.) The probit model leads to the least conservative results with the logit and gamma multi-hit models falling between the extreme value and probit models. The results for the logit and gamma multi-hit models are close, however, with the VSD's down to $\pi = 10^{-8}$ always within a factor of three. Because of the linearity of the observed dose response curve for hexachlorobenzene throughout the entire dose range, all models except the probit provide results close to those based on linear extrapolation in this case. This example vividly reflects the inability of the probit model to accomodate low dose linearity, with the VSD at $\pi = 10^{-5}$ for this model being about 50 fold greater than that based on linear extrapolation.

We note that linear extrapolation from $\pi_o = 10^{-2}$ based on the probit, logit, extreme value or gamma multi-hit models will lead to slightly higher VSD's than those based on the multi-stage model, with the probit model providing results a factor of 3-5 higher in several cases. Although linear extrapolation from $\pi_o = 10^{-1}$ leads to VSD's which are largely independent of the model employed, these VSD's can be an order of magnitude lower than those based on linear extrapolation from $\pi_o = 10^{-2}$ with the multistage model.

Lower confidence limits on the VSD's for the multi-stage and extreme value models are given in Table 5. Whenever the linear term β_1 in the multi-stage model is estimated to be zero, the direct lower confidence limits (L_1) are roughly comparable to the point estimates based linear extrapolation. (This is because the confidence limit procedure admits the possibility that this linear term, which is dominant at low doses, may in fact be positive.) These confidence limits are generally somewhat comparable to those based on the random group method (L_4), although the latter procedure can lead to quite different results depending on the particular random split obtained.

For the extreme value model, the lowest confidence limits are provided by the conservative confidence limit procedure (L_3'), with these limits being notably lower than the point estimates based on linear extrapolation. For the direct confidence limit procedures, we have $L_1 < L_1' < L_1''$, although L_1 frequently assumes negative values. The likelihood based confidence limit procedure (L_2) can be somewhat conservative, although it is often comparable to the direct procedure (L_1'') in several cases. This same pattern was found to hold in the case of the logit and probit models.

Additive Background

We also fit the probit, logit and extreme value models with additive background in the thirteen cases for which a positive background was indicated with these same models assuming independent background. (The transformation $\delta = \exp(\delta^*)$ was used in order to provide for the constraint $\delta > 0$.) While convergence was considerably slower than in the independent background case due to the apparent flatness of the likelihood surface near its maximum, reasonable fits were eventually obtained ($p > .15$ in the chi-square goodness-of-fit test) escept with three of the 2-AAF data sets (no.'s 15,16 and 20) and ethylene thiourea (no.11).

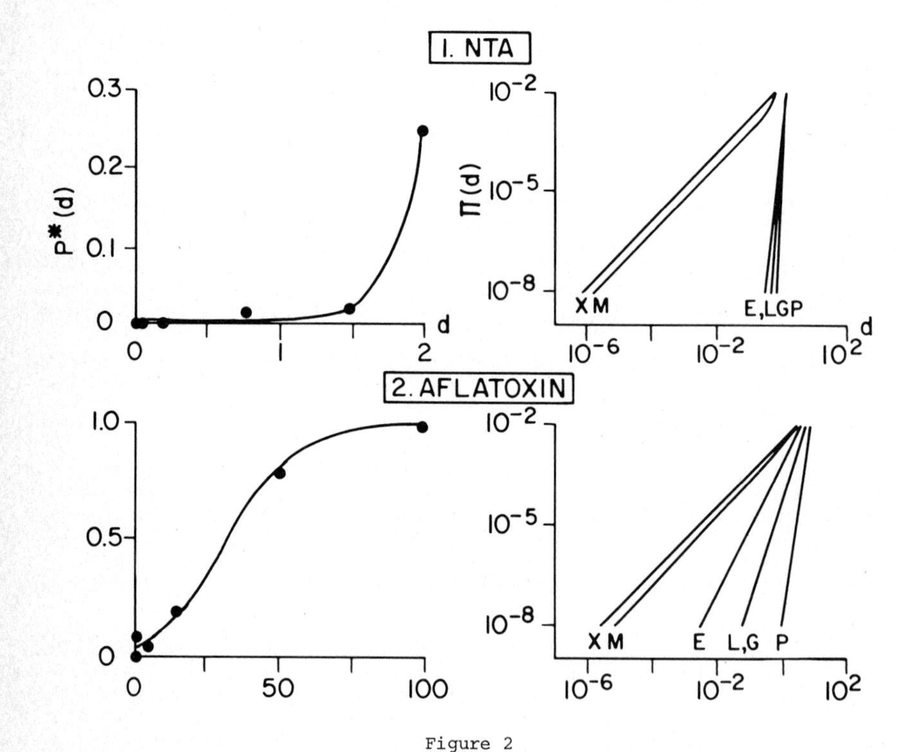

Figure 2

Fitted Extreme Value Dose Response Curves

and Estimates of Excess Risk

Based on Six Extrapolation Procedures

X - Linear Extrapolation
M - Multi-Stage
E - Extreme Value
L - Logit
G - Gamma Multi-Hit
P - Probit

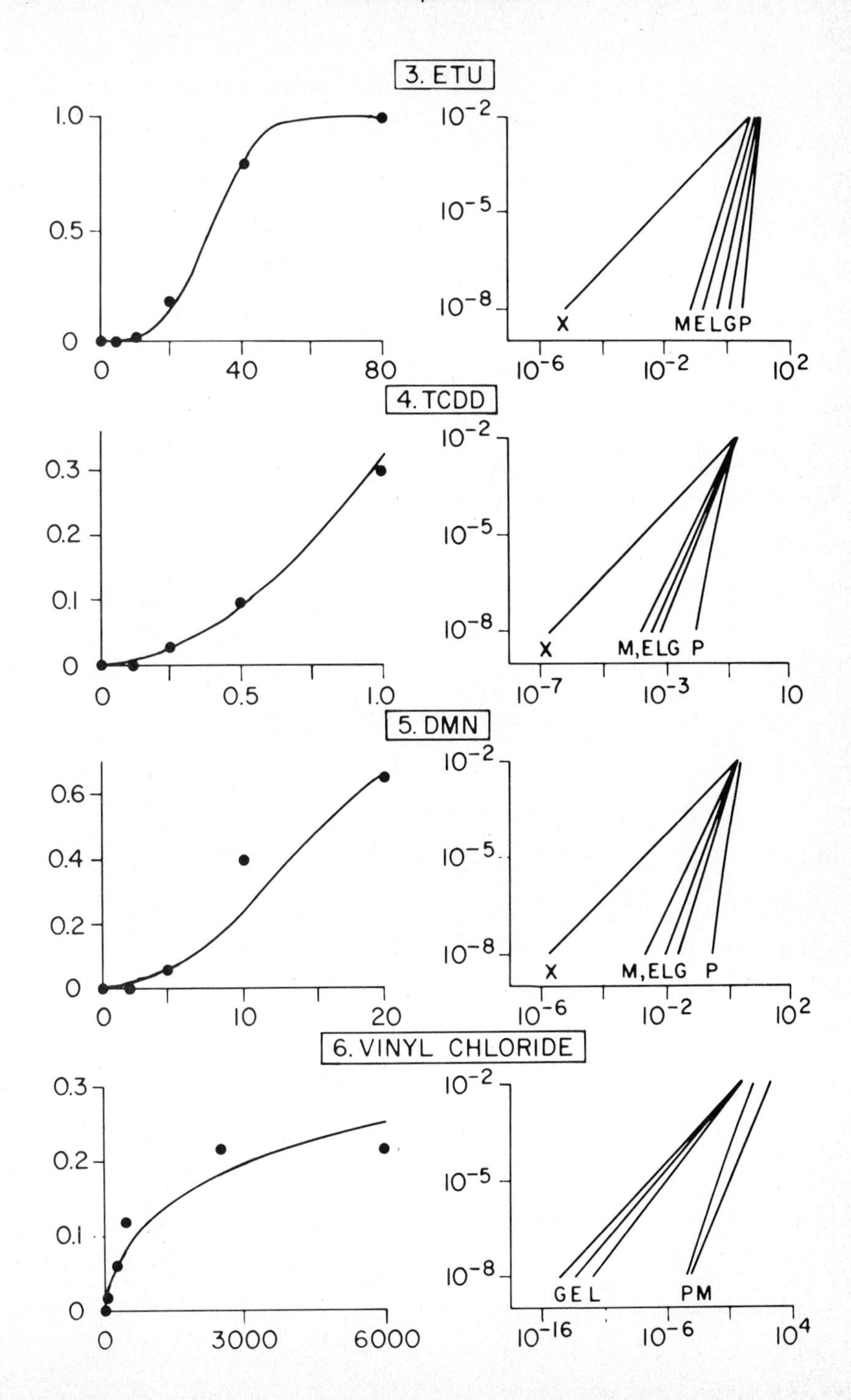

3. ETU
1.0
0.5
0
0
40
80
10^-2
10^-5
10^-8
X
MELGP
10^-6
10^-2
10^2
4. TCDD
0.3
0.2
0.1
0
0
0.5
1.0
10^-2
10^-5
10^-8
X
M,ELG P
10^-7
10^-3
10
5. DMN
0.6
0.4
0.2
0
0
10
20
10^-2
10^-5
10^-8
X
M,ELG P
10^-6
10^-2
10^2
6. VINYL CHLORIDE
0.3
0.2
0.1
0
0
3000
6000
10^-2
10^-5
10^-8
GEL
PM
10^-16
10^-6
10^4

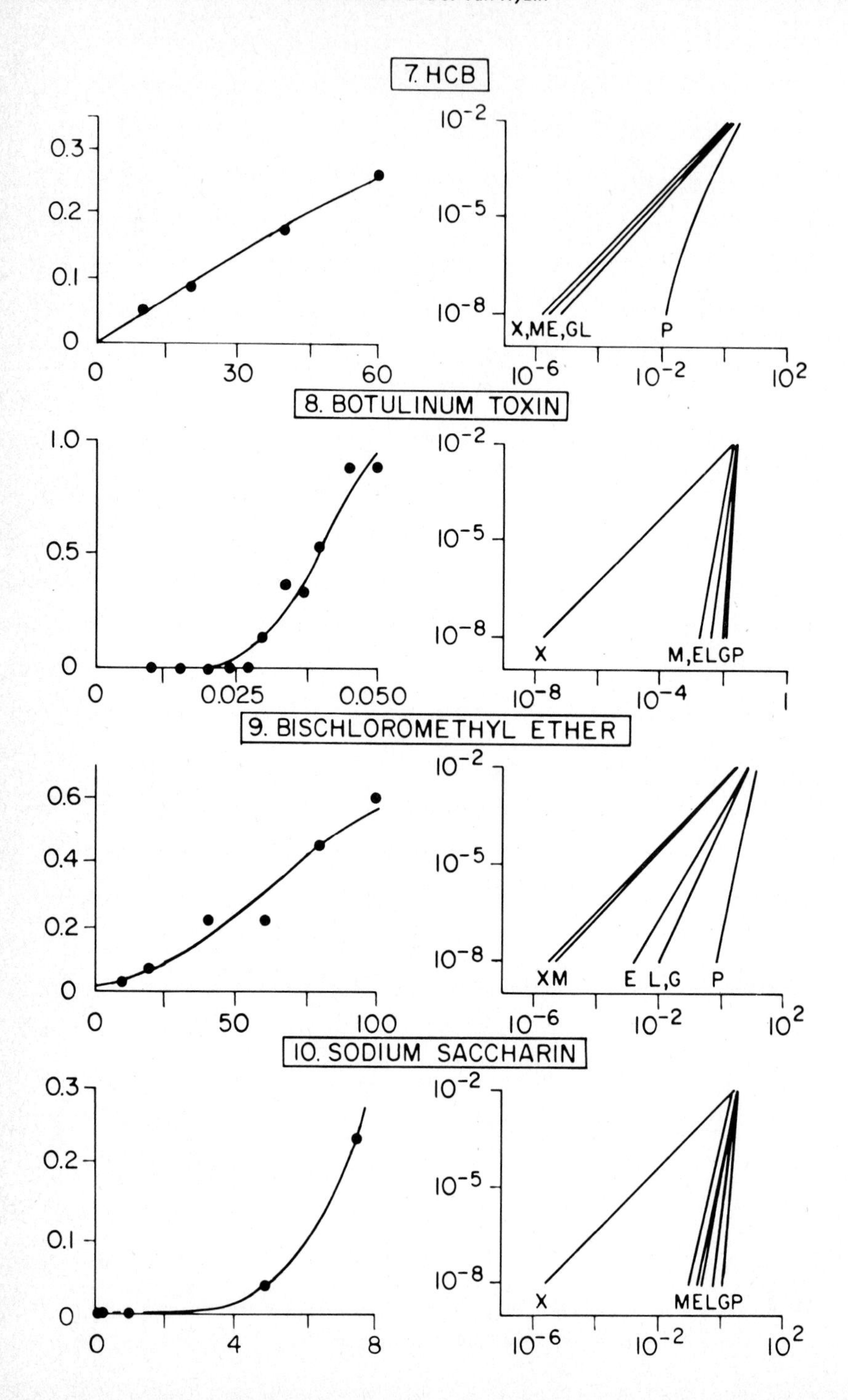

7. HCB
0.3
0.2
0.1
0
0
30
60
10^{-2}
10^{-5}
10^{-8}
X,ME,GL
P
10^{-6}
10^{-2}
10^{2}
8. BOTULINUM TOXIN
1.0
0.5
0
0
0.025
0.050
10^{-2}
10^{-5}
10^{-8}
X
M,ELGP
10^{-8}
10^{-4}
1
9. BISCHLOROMETHYL ETHER
0.6
0.4
0.2
0
0
50
100
10^{-2}
10^{-5}
10^{-8}
XM
E L,G
P
10^{-6}
10^{-2}
10^{2}
10. SODIUM SACCHARIN
0.3
0.2
0.1
0
0
4
8
10^{-2}
10^{-5}
10^{-8}
X
MELGP
10^{-6}
10^{-2}
10^{2}

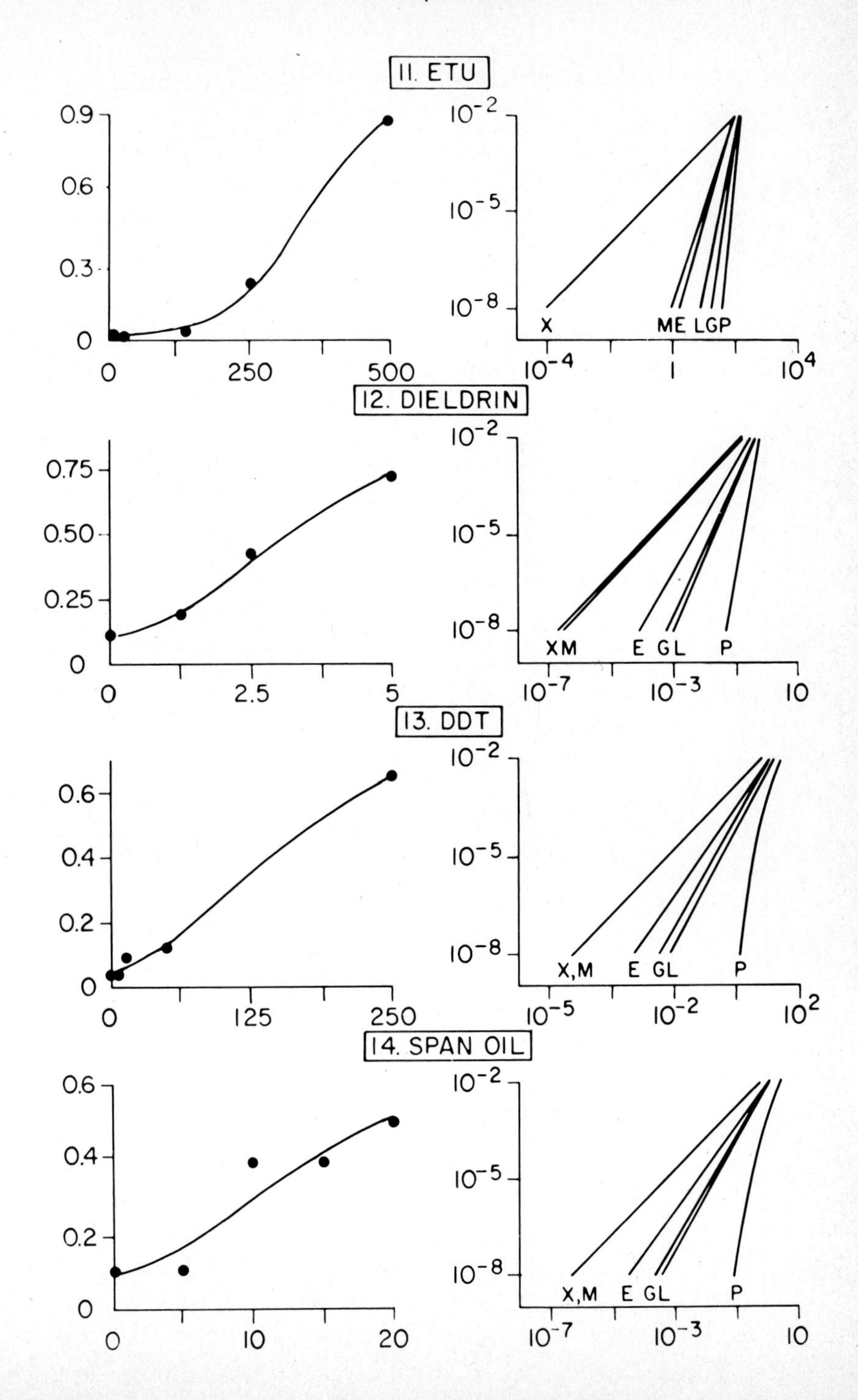
11. ETU
0.9
0.6
0.3
0
0
250
500
10^{-2}
10^{-5}
10^{-8}
X
ME LGP
10^{-4}
1
10^{4}
12. DIELDRIN
0.75
0.50
0.25
0
0
2.5
5
XM
E GL
P
10^{-7}
10^{-3}
10
13. DDT
0.6
0.4
0.2
0
125
250
X,M
E GL
P
10^{-5}
10^{-2}
10^{2}
14. SPAN OIL
0.6
0.4
0.2
0
10
20
X,M
E GL
P
10^{-7}
10^{-3}
10

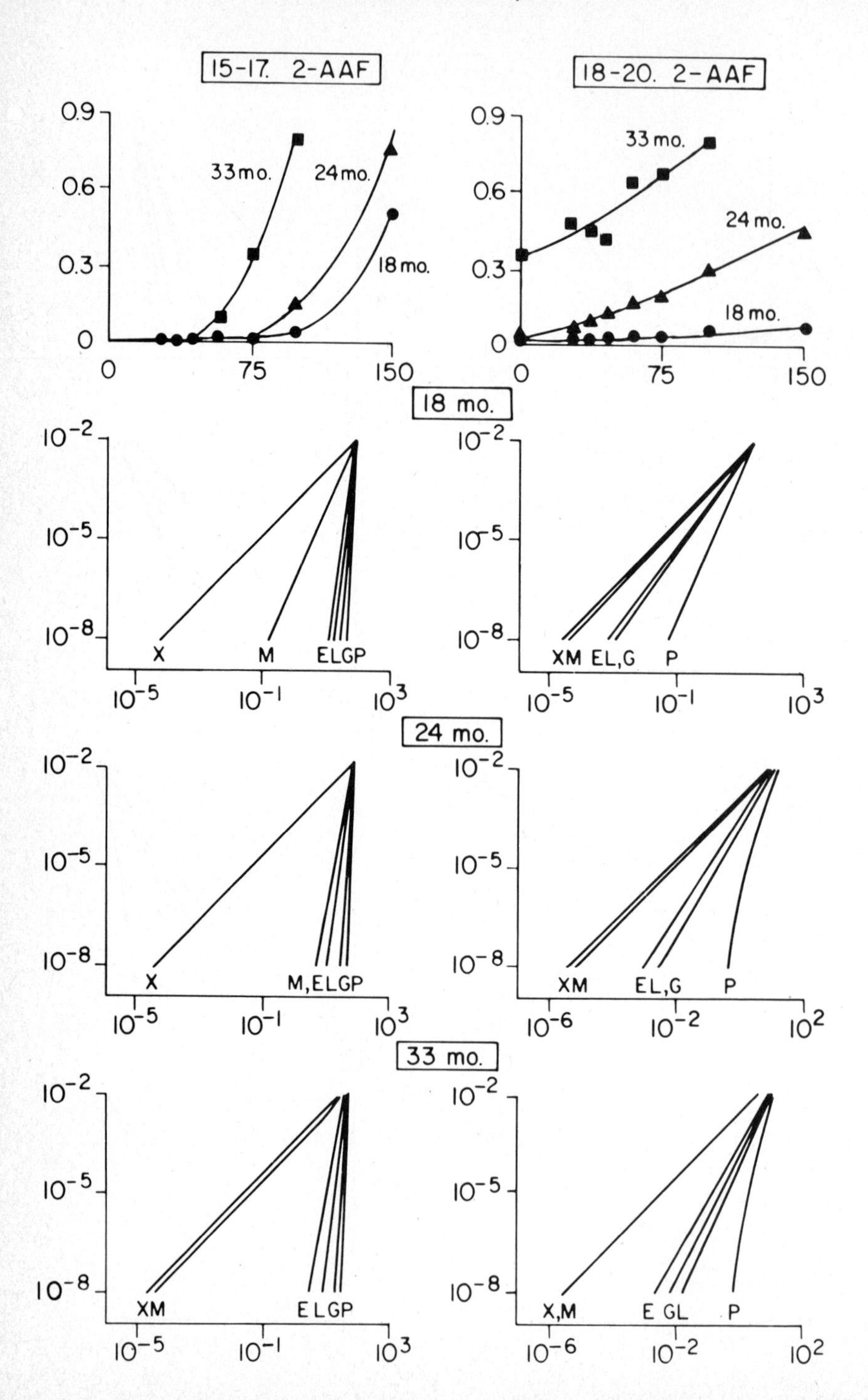
15-17. 2-AAF
18-20. 2-AAF
0.9
0.6
0.3
0
33 mo.
24 mo.
18 mo.
0
75
150
18 mo.
24 mo.
33 mo.
10^-2
10^-5
10^-8
X
M
ELGP
XM
EL,G
P
M,ELGP
X,M
E GL
10^-5
10^-1
10^3
10^-6
10^-2
10^2

Table 4. Estimated Virtually Safe Levels of Exposure at an Excess Risk of $\pi = 10^{-5}$ Based on Six Extrapolation Procedures[a]

	Substance	Linear Extrapolation ($\pi_o = 10^{-2}$)	Multistage Model	Extreme Value Model	Logit Model	Gamma Multi-Hit Model	Probit Model
1.	NTA	9.5-4	1.9-3	.67	.70	.90	.96
2.	Aflatoxin	3.2-3	7.9-3	.13	.63	.57	2.5
3.	ETU	7.5-3	.74	1.2	2.1	3.3	5.2
4.	TCDD	1.6-4	5.1-3	5.2-3	6.8-3	9.7-3	3.1-2
5.	DMN	1.9-3	6.0-2	6.1-2	.13	.18	.70
6.	Vinyl Chloride	b	.20	3.9-7	1.8-6	8.4-8	4.8-3
7.	HCB	2.2-3	2.2-3	2.5-3	4.4-3	2.5-3	.12
8.	Botulinum Toxin	2.0-5	6.2-3	6.3-3	1.1-2	1.5-2	1.7-2
9.	Bischloromethyl Ether	3.7-3	4.0-3	.12	.29	.30	2.4
10.	Sodium Saccharin	3.3-3	.59	.86	.95	1.5	1.9
11.	ETU	9.4-2	9.7	12.	30.	46.	74.
12.	Dieldrin	1.9-4	2.2-4	4.6-3	2.0-2	1.7-2	.14
13.	DDT	6.2-3	6.4-3	8.3-2	.34	.19	4.0
14.	Span Oil	5.6-4	5.7-4	5.9-2	1.8-2	1.5-2	.28
15.	2-AAF	7.9-2	5.6	32.	36.	49.	58.
16.	2-AAF	6.6-2	21.	21.	28.	42.	48.
17.	2-AAF	3.6-2	5.8-2	12.	19.	28.	32.
18.	2-AAF	3.6-2	4.3-2	.17	.18	.18	.95
19.	2-AAF	6.9-3	7.5-3	9.1-2	.17	.17	1.2
20.	2-AAF	2.9-3	2.9-3	.13	.39	.25	2.5

a The notation 9.5-4 means 9.5×10^{-4}.

b Not shown because of the concavity of the observed dose response curve.

Table 5. Estimated Virtually Safe Levels of Exposure ($\pi = 10^{-5}$) and 95% Lower Confidence Limits for the Multistage and Extreme Value Models[a]

	Substance	Linear Extra-polation	Multistage Model			Extreme Value Model					
			L_1	L_4	$\hat{d}^*$	L'_3	L_2[b]	L_1	L'_1	L''_1	$\hat{d}^*$
1.	NTA	9.5-4	6.9-5	1.1-4	1.9-3	5.5-2	.11	.28	.37	.42	.67
2.	Aflatoxin	3.2-3	6.2-4	4.1-4	7.9-3	5.7-5	4.0-2	0	9.5-3	3.5-2	.13
3.	ETU	7.5-3	9.3-3	.46	.74	.52	.53	.53	.67	.75	1.2
4.	TCDD	1.6-4	1.2-4	4.5-4	5.1-3	1.4-8	1.7-3	0	3.4-4	1.4-3	5.2-3
5.	DMN	1.9-3	1.4-3	1.8-3	6.0-2	2.4-3	2.7-3	0	1.1-2	2.2-2	6.1-2
6.	Vinyl Chloride		.14	7.8-2	.20	1.9-17	1.5-17	0	2.6-11	3.6-8	3.9-7
7.	HCB	2.2-3	1.3-3	1.6-4	2.2-3	1.8-3	1.7-3	0	4.4-5	4.9-4	2.5-3
8.	Botulinum Toxin	2.0-5	4.3-9	4.5-3	6.2-3	6.8-4	4.5-3	4.2-3	4.5-3	4.7-3	6.3-3
9.	Bischloromethyl Ether	3.8-3	6.1-5	8.4-5	4.0-3	6.2-17	1.2-3	0	1.6-3	2.3-3	.12
10.	Sodium Saccharin	3.3-3	2.1-2	.14	.59	c	2.3-4	0	.13	.29	.86
11.	ETU	9.5-2	1.6-2	1.1	9.7	1.9	4.1	28.	5.6	6.9	12.
12.	Dieldrin	1.9-4	7.1-5	1.2-5	2.2-4	1.7-5	8.0-5	0	4.4-4	1.4-3	4.6-3
13.	DDT	6.2-3	3.1-3	3.8-4	6.4-3	4.0-4	9.0-4	0	7.2-3	2.4-2	8.3-4
14.	Span Oil	5.7-4	1.3-4	2.3-5	5.7-4	c	0	0	2.3-7	5.2-4	5.9-3
15.	2-AAF	7.9-2	1.2-3	9.9-5	5.6	4.9	25.	21.	23.	24.	32.
16.	2-AAF	6.6-2	4.6-2	d	21.	7.8	17.	17.	17.	18.	21.
17.	2-AAF	3.6-2	3.6-5	6.5-4	5.8-2	.72	3.6	1.2	5.0	6.5	12.
18.	2-AAF	3.6-2	2.3-2	0	4.3-2	1.7-4	1.1-3	0	1.4-2	4.8-2	.17
19.	2-AAF	6.9-5	7.5-3	2.0-3	7.5-5	1.8-2	2.4-2	0	3.3-2	4.5-2	9.1-2
20.	2-AAF	2.9-3	1.5-4	7.2-6	2.9-5	c	7.1-16	0	2.5-4	1.7-2	.13

a The notation 9.5-4 means 9.5×10^{-4}.
b Approximate limits based on a rough grid search.
c $\beta_L < 0$.
d MRS.T. did not converge.

For the nine cases in which good fits were realized, it was found that the background rate was similar to that estimated assuming independence, except for NTA and 2-AAF (no.17) where the small background found assuming independence was notably larger than that found assuming additivity. Whenever the background was greater than about 1%, the VSD's obtained with all three models were close and all within a factor of two of those obtained previously using linear extrapolation. This same result was found to hold in the case of 2-AAF (no.18), where the background rate is estimated to be less than 1% but the dose response curve appears somewhat linear throughout the entire dose range. In the remaining two cases where the background was found to be less than 1%, the low dose linearity imposed by the additivity assumption did not become apparent as quickly, as illustrated in Table 6 below with NTA.

Table 6. Estimated Virtually Safe Levels of Exposure of NTA Assuming Additive Background[a]

Excess Risk	Linear Extrapolation	Extreme Value Model	Logit Model	Probit Model
10^{-2}	.95	1.1	1.1	1.1
10^{-4}	9.5 - 3	.14	.15	.25
10^{-6}	9.5 - 5	1.8 - 3	2.0 - 3	6.5 - 3
10^{-8}	9.5 - 7	1.9 - 5	2.0 - 5	6.7 - 5

[a] 9.5 - 3 means 9.5×10^{-3} .

Independent and Additive Background

Finally, we fit the probit, logit and extreme value models to these same thirteen data sets with provision for both independent and additive background components. This result is either a completely independent or completely additive background model being selected, except in the case of NTA and 2-AAF (no.15). In these two cases, the VSD's were found to be between those obtained assuming either independent or additive background.

5. DISCUSSION

The empirical results presented in section 4 assuming independent background indicate that of the dose response models considered, the multi-stage and probit models generally provide the most conservative and least conservative results respectively upon extrapolation to low doses. Intermediate results were obtained using the extreme value, logit and gamma multi-hit models, with extrapolations based on the latter two models usually in fairly close agreement.

The exception to this general pattern noted in the case of vinyl chloride was attributed to the concave nature of the observed dose response curve for that compound. Although this example represents perhaps the first serious suggestion of concavity or supralinearity in the low dose region, Ghering, Watanabe and Park (1978) have shown that the metabolism of vinyl chloride does not increase in proportion to the level of exposure. Using the effective dose at the target tissue (the liver) rather than the administered dose as the dose metameter, Van Ryzin & Rai (1980) subsequently found the resulting dose response curve to be very nearly linear at low doses using the gamma multi-hit model. This example points to the need for supplementary metabolic and pharmacokinetic studies in order to determine the appropriate dose metameter for extrapolation.

When conservative results are required, linear extrapolation from that point on the dose response curve estimated to induce an excess risk of 1% would appear to be a reasonable procedure, given the assumption that the true dose response curve is convex or sublinear at low doses. When the background rate is greater than about 1%, the empirical results obtained using the probit, logit, and extreme value models suggest that the assumption of additive background (and hence low dose linearity), may be expected to provide results close to those based on linear extrapolation, regardless of the model employed.

The examples presented in this paper clearly indicate the divergent results that are obtained with different models upon extrapolation to low doses. Since these models all provide generally good fits to the data in the observable range, however, statistical tests for discriminating between models will be of limited sensitivity. Moreover, until the mechanisms of carcinogenesis and other toxicological phenomena are better understood, biological considerations cannot fully resolve this issue.

The inability to accurately extrapolate to low doses continues to present serious problems in regulatory applications. One approach that has been recently proposed (Food Safety Council, 1980a) is to consider estimates based on several models, including linear extrapolation, as in the log-log plots in Figure 2. (Because of the extreme steepness of the dose response curve for the probit model in the low dose region, however, this model might be excluded from consideration.) Confidence limits might also be considered, although the experimental error reflected by these limits may be less than the variation among models. The final decision should then take into account all available information on the biological characteristics of the test compound, nature and severity of the toxic effects induced, interspecies differences, and any relevant socioeconomic factors (Munro & Krewski, 1981; Food Safety Council, 1980b).

6. APPENDIX: MAXIMUM LIKELIHOOD ESTIMATION

Identifiability

A dose response model $P^*(d) = P^*(d;\underset{\sim}{\theta})$ depending on the vector of parameters $\underset{\sim}{\theta}$ is said to be identifiable if $P^*(d;\underset{\sim}{\theta}) = P^*(d;\underset{\sim}{\theta}')$ for all $d \geq 0$ implies $\underset{\sim}{\theta} = \underset{\sim}{\theta}'$. If the model is not identifiable, then the maximum likelihood estimator of $\underset{\sim}{\theta}$ discussed below may not be unique.

In the case of zero background, it is easy to show that all of the models discussed previously are identifiable. For the probit, logit and extreme value models, for example, suppose that $P(d;\alpha,\beta) = P(d;\alpha',\beta')$ for all $d \geq 0$. The monotonicity of $P(d)$ then implies that $\alpha + \beta \log d = \alpha' + \beta' \log d$ for all $d > 0$, from which it follows that $\alpha = \alpha'$ and $\beta = \beta'$. In the case of completely independent background, the identifiability of all models considered in this paper follows from the following result.

Lemma 1. If $P(d;\underset{\sim}{\theta})$ is identifiable, then $P^*(d;\gamma,\underset{\sim}{\theta}) = \gamma + (1-\gamma)P(d;\underset{\sim}{\theta})$ is identifiable provided $P(d;\underset{\sim}{\theta})$ is continuous at $d=0$ for all $\underset{\sim}{\theta}$.

Proof. Suppose $P^*(d;\gamma,\underset{\sim}{\theta}) = P^*(d;\gamma',\underset{\sim}{\theta}')$ for all $d \geq 0$. Then $\gamma = \lim_{d\downarrow 0} P^*(d;\gamma,\underset{\sim}{\theta}) = \lim_{d\downarrow 0} P^*(d;\gamma',\underset{\sim}{\theta}') = \gamma'$ from which it follows that $P(d;\underset{\sim}{\theta}) = P(d;\underset{\sim}{\theta}')$ and hence that $\underset{\sim}{\theta} = \underset{\sim}{\theta}'$.

When the background is assumed to be completely additive, the identifiability of the probit, logit and extreme value models follows from the monotonicity of $P^*(d)$. Identifiability of the gamma multi-hit model follows from certain properties of the incomplete gamma function while the argument for the multi-stage model follows from repeated differentiation of $-\log(1-P^*(d))$.

When the background contains both independent and additive components, only the probit model is identifiable in general. As will be illustrated below, the one-hit model is not identifiable, with the extreme value, gamma multi-hit and multi-stage models being identifiable only if the model parameters do not lie in that subspace of the parameter space which includes the one-hit model. Similarly, the logit model is identifiable only when $\beta \neq 1$. For brevity, we omit the details of the proofs of these statements since they are fairly straightforward but tedious.

To show that the one-hit model is not identifiable in the case of mixed (independent and additive) background, we note that

$$P^*(d;\gamma,\delta,\lambda) = \gamma + (1-\gamma)\{1 - \exp(-\lambda(d+\delta))\}$$

$$= P^*(d;\gamma',\delta',\lambda) \qquad (A.1)$$

whenever $\exp\{-\lambda(\delta - \delta') = (1-\gamma)/(1-\gamma')$. The nonidentifiability of the one-hit model in this general case now follows from the observation that there are infinitely many choices of $\gamma \neq \gamma'$ and $\delta \neq \delta'$ satisfying this last relationship.

Asymptotic Properties

Let $P_i^* = P^*(d_i;\underset{\sim}{\theta})$ denote the probability of a positive response at dose d_i under any dose response model $P^*(d;\underset{\sim}{\theta})$ with parameter $\underset{\sim}{\theta} = (\theta_1,\ldots,\theta_t)' \in \Omega$, where the parameter space Ω is an open subset of t dimensional Euclidean space. (In all cases of interest here, Ω will be the Cartesian product of intervals of the real line. For the probit, logit, extreme value and gamma multi-hit models, moreover, $t = 2,3$ or 4 according to whether no background, either additive or independent background, or mixed background is present.)

In order to establish the desired asymptotic properties of the maximum likelihood estimator $\hat{\underset{\sim}{\theta}}$, we will require the following regularity conditions.

C1. $\lim_{n\to\infty} \frac{n_i}{n} = c_i$ $(0 \leq c_0 < 1$ and $0 < c_i < 1$ for $i = 1,\ldots,m)$.

C2. $P^*(d;\underset{\sim}{\theta})$ is identifiable.

C3. The information matrix $\Sigma^{-1} = ((\sigma^{rs}))$ defined by

$$\sigma^{rs} = \Sigma_{i=0}^{m} c_i \frac{\partial P_i^*}{\partial\theta_r} \frac{\partial P_i^*}{\partial\theta_s} /(P_i^* Q_i^*)$$

$(r,s = 1,\ldots,t)$ is positive definite.

C4. For all $d \geq 0$, second partial derivatives $\partial^2 P^*/(\partial\theta_r \partial\theta_s)$

$(r,s = 1,\ldots,t)$ are continuous for all $\underset{\sim}{\theta} \in \Omega$.

C5. $P^*(d;\underset{\sim}{\theta})$ is continuously differentiable in d and $\underset{\sim}{\theta}$ for all $d > 0$ and $\underset{\sim}{\theta} \in \Omega$.

Although the use of a control group is strongly recommended in practice, C1 reflects the fact that this is not required in theory. (When $P^*(0) = 0$, however, we must assume $c_o = 0$ and take $i = 1,\ldots,m$ in the summation in C3 in order to obtain the asymptotic results presented below.) As noted previously, C2 will be satisfied for the probit, logit, extreme value and gamma multi-hit models provided $\beta \neq 1$ for the logit and extreme value models and $k \neq 1$ for the gamma multi-hit model in the case of mixed background. In addition, it may be verified that C4 and C5 are both satisfied for these four models.

Letting $m^* = m+1$ for $c_o > 0$ and $m^* = m$ for $c_o = 0$ denote the effective number of dose levels employed, we now show that C3 holds for the probit, logit, extreme value and gamma multi-hit models provided $m^* \geq t$. We will also show that this condition holds in the case of the multi-stage model in (2.10) provided $\beta_i > 0$ for $i = 0,1,\ldots,k$ and $m^* \geq t = k+1$. These results follow quite naturally from the following considerations concerning Tchebycheff sets of functions (Rice, 1964, pp.54-55, p.91).

<u>Definition</u>. A set of functions $\{\phi_r(d);\ r = 1,\ldots,t\}$ is said to form a <u>Tchebycheff set</u> on $[0,D]$ if $\phi(d;\underset{\sim}{a}) = \Sigma a_r\phi_r(d)$ has at most $t-1$ zeros in $[0,D]$ for all $\underset{\sim}{a} = (a_1,\ldots,a_t)' \neq \underset{\sim}{0}$.

<u>Lemma 2</u>. The set of functions $\{\phi_r(d);\ r = 1,\ldots,t\}$ forms a Tchebycheff set on $[0,D]$ if and only if the matrix $((\phi_r(d_i)))$ $(r,i = 1,\ldots,t)$ is of full rank for every set of t distinct points $\{d_i\}$ in $[0,D]$.

<u>Lemma 3</u>. The information matrix Σ^{-1} in C3 is positive definite provided $\{\partial P^*(d)/\partial\theta_r;\ r = 1,\ldots,t\}$ forms a Tchebycheff set on $[0,d_m]$ and $m^* \geq t$.

<u>Proof</u>. Provided $m^* \geq t$, it follows from Lemma 2 that the $t \times m^*$ matrix $\underset{\sim}{A} = ((a_{ri}))$ defined by $a_{ri} = \partial P_i^*/\partial\theta_r$ is of rank $R(\underset{\sim}{A}) = t$. Letting $\underset{\sim}{B}$ denote the $m^* \times m^*$ diagonal matrix whose i^{th} diagonal element is $(c_i/P_i^*Q_i^*)^{1/2}$, we have that $R(\underset{\sim}{A}\,\underset{\sim}{B}) = R(\underset{\sim}{A}) = t$ since $\underset{\sim}{B}$ is nonsingular (Rao, 1973, p.30). It now follows that $\underset{\sim}{\Sigma}^{-1} = \underset{\sim}{C}\,\underset{\sim}{C}'$ is positive definite since $\underset{\sim}{C} = \underset{\sim}{A}\,\underset{\sim}{B}$ is of full row rank (Searle, 1971, p.37).

For the multi-stage model in (2.10) with $\beta_i > 0$ $(i = 0,1,\ldots,k)$ we have

$$\partial P^*(d)/\partial\beta_i = [1 - P^*(d)]d^i \tag{A.2}$$

where $\{1,d,\ldots,d^k\}$ forms a well known Tchebycheff set on any interval $[0,D]$. Using the fact that any Tchebycheff set multiplied by a positive function is still a Tchebycheff set (Rice, 1964, p.55), it now follows from Lemma 3 that C3 holds in this case provided $m^* \geq k+1$.

The derivatives of $P^*(d)$ with respect to the parameters α,β,γ and δ in the probit, logit and extreme value models are, omitting a multiplicative factor $1-P^*(d)$ which appears as in (A.2), given by $h(d+\delta)$, $h(d+\delta)\log(d+\delta)$, $(1-\gamma)^{-1}$ and $\beta h(d+\delta)/(d+\delta)$ respectively, where

$$\begin{aligned} h(d) &= f(\alpha + \beta\log d)/[1 - F(\alpha + \beta \log d)] \\ &= -\frac{d}{\beta}\,\frac{\partial \log\,(1 - F(\alpha + \beta \log d))}{\partial d} > 0 \end{aligned} \tag{A.3}$$

for $d > 0$ and $f(x) = F'(x)$. For purely additive background $(\gamma = 0)$, C3 will thus be satisfied for $m^* \geq 3$ provided we can show that $\{1, \log(d+\delta), (d+\delta)^{-1}\}$ is a Tchebycheff set on $[0,D]$ or, letting $x = \log(d+\delta)$, if $\{1,x,e^{-x}\}$ is a Tchebycheff set on $[\log\delta, \log(D+\delta)]$. In this case, the required result follows from the following lemma.

Lemma 4. If $g(x)$ is strictly convex on $[a,b]$, then $\{1,x,g(x)\}$ forms a Tchebycheff set on $[a,b]$.

Proof. We first note that $\phi(x;\underset{\sim}{a}) = a_1 + a_2x + a_3g(x)$ has at most one zero in $[a,b]$ if $a_3 = 0$ and thus satisfies the Tchebycheff condition given in the preceeding definition. For $a_3 \neq 0$, $\phi(x;\underset{\sim}{a}) = 0$ if and only if $g(x) = a_1' + a_2'x$, where $a_i' = -a_i/a_3 (i = 1,2)$. Since $a_1' + a_2'x$ will intersect the convex function $g(x)$ at at most two points, it follows that $\phi(x;\underset{\sim}{a})$ has at most two zeros in any interval $[a,b]$ for all $\underset{\sim}{a} \neq \underset{\sim}{0}$.

For purely independent background $(\delta = 0)$, C3 will be satisfied for $m^* \geq 3$ provided we can show that $\{h(d), h(d)\log d, (1-\gamma)^{-1}\}$ is a Tchebycheff set on $[0,D]$. In order to avoid problems with $h(0) = \lim_{d\downarrow 0} h(d) = 0$, we will first show that $\{1, \log d, 1/h(d)\}$ is a Tchebycheff set on $[\varepsilon,D]$ for all $\varepsilon > 0$ or, equivalently, that $\{1, x, g(x)\}$ is a Tchebycheff set on $[\log\varepsilon, \log D]$ where $x = \log d$ and $g(x) = 1/h(e^x)$. Since $g(x)$ can be shown to be convex for the probit, logit and Weibull models, the result in this case now follows from Lemma 4. The extension to the interval $[0,D]$ follows by checking the Tchebycheff condition given in Lemma 2 after noting that $\lim_{d\downarrow 0} h(d)\log d = 0$ for these three models.

Similar arguments apply in the case of the gamma multi-hit model with either additive or additive background. The case of mixed background follows from the fact that $\{1, \log(d+\delta), 1/h(d+\delta), (d+\delta)^{-1}\}$ can be shown to be a Tchebycheff set on $[0,D]$ provided $\beta \neq 1$ in the logit and extreme value models and $k \neq 1$ in the gamma multi-hit model.

Under the above conditions, $\hat{\underset{\sim}{\theta}}$ may be shown to have the asymptotic properties indicated in Theorems 1 and 2 below. The proofs of these two theorems as well as Theorem 3 below follow along the same lines as those given by Rai & Van Ryzin (1981) for the gamma multi-hit model with independent background and are omitted. Under C5, Corollary 1 follows from Theorem 2 with the use of the implicit function theorem.

Theorem 1. Under C1-C4, the maximum likelihood equations $\partial\Pi/\partial\underset{\sim}{\theta} = \underset{\sim}{0}$ have a unique root $\hat{\underset{\sim}{\theta}}$ with probability one which is strongly consistent for $\hat{\underset{\sim}{\theta}}$.

Since $P^*(d;\underset{\sim}{\theta})$ is a continuous function of $\underset{\sim}{\theta}$, it follows from Theorem 1 that $\hat{\Pi}(d) = \hat{P}^*(d) - \hat{P}^*(0)$ is a strongly consistent estimator of $\Pi(d)$, where $\hat{P}^*(d) = P^*(d;\hat{\underset{\sim}{\theta}})$. Similarly, $\hat{d}^* = \hat{\Pi}^{-1}(\pi)$ is a strongly consistent estimator of d^* whenever C5 holds.

Theorem 2. Under C1 - C4, $\sqrt{n}(\hat{\underset{\sim}{\theta}} - \underset{\sim}{\theta})$ is normally distributed with mean $\underset{\sim}{0}$ and covariance matrix $\underset{\sim}{\Sigma}$.

Corollary 1. Under C1 - C5, $\sqrt{n}(\hat{\Pi}(d) - \Pi(d))$, $\sqrt{n}(\hat{d}^* - d^*)$, $\sqrt{n}(\log\hat{d}^* - \log d^*)$ and $\sqrt{n}(1/\hat{d}^* - 1/d^*)$ are all normally distributed with zero means and variances

$$\sigma_1^2 = \Sigma_{r=1}^t \Sigma_{s=1}^t \frac{\partial\Pi}{\partial\theta_r}\frac{\partial\Pi}{\partial\theta_s}\sigma_{rs},$$

$\sigma_2^2 = (\partial\Pi/\partial d|_{d=d^*})^{-2}\sigma_1^2$ with σ_1^2 evaluated at $d = d^*$, $\sigma_3^2 = (d^*)^{-2}\sigma_2^2$ and

$\sigma_4^2 = (d^*)^{-4} \sigma_2^2$ respectively.

The use of the usual chi-square goodness-of-fit test in the case of any dose response model $P^*(d;\underset{\sim}{\theta})$ satisfying C1-C4 may be justified on the basis of the following theorem.

Theorem 3. Under C1-C4, the statistic $\chi^2 = \Sigma(x_i - n_i\hat{P}_i^*)^2/(n_i\hat{P}_i^*\hat{Q}_i^*)$ has a chi-squared distribution with m^*-t degrees of freedom, provided that the assumed model $P^*(d;\underset{\sim}{\theta})$ is correct and $m^* > t$.

In discriminating between two dose response models, we consider only the situation where one model is a special case of a more general model, with the parameter space Ω_o for the reduced model defined by restricting the range of t_o of the θ_j to be a single point θ_j^o. This is the case with the one-hit model, which is a special case of both the extreme value and gamma multi-hit models. Letting $\hat{\underset{\sim}{\theta}}_o$ denote the maximum likelihood estimator of θ within the restricted parameter space Ω_o, the following theorem shows that the usual likelihood ratio statistic may be used to test the null hypothesis $H_o : \underset{\sim}{\theta} \in \Omega_o$ versus the alternative $H_A : \underset{\sim}{\theta} \in \Omega - \Omega_o$. (The proof of this result follows from Theorem 2 using standard arguments (Rao, 1973, p. 419).)

Theorem 4. Under C1-C4, $2(\ell(\hat{\underset{\sim}{\theta}}) - \ell(\hat{\underset{\sim}{\theta}}_o))$ has a chi-squared distribution with t_o degrees of freedom.

We note that due to the restriction that $\underset{\sim}{\theta}$ cannot lie on the boundary of Ω (C3), this result cannot be used to test the hypothesis $H_o : \gamma = 0$ or $H_o : \delta = 0$.

ACKNOWLEDGEMENTS

The authors would like to thank Mr. John Kovar for his efficient programming and for several useful suggestions concerning computational procedures. We would also like to thank Dr. R.L. Sielken for several constructive suggestions concerning the presentation of this material.

FOOTNOTES

[1] Research supported by the National Institute of Environmental Health Sciences under grant 7R01-ES-02557.

[2] This assumption may be called into question in experiments where littermates are used, particularly in studies involving prenatal exposure (Grice, Munro, Krewski & Blumenthal, 1981). However, since estimation procedures for dependent observations remain to be developed, we will nonetheless make this assumption in the applications presented in section 4.

[3] The one-hit model is not identifiable when the background includes both independent and additive components as is the logit model when $\beta = 1$.

[4] In the absence of a control group $(c_o = 0)$, the first term would be omitted in the summation in (3.2).

[5] The multi-stage models fitted to the NTA and 2-AAF (no.17) data are concave at very low doses, as are those fitted to the HCB and 2-AAF (no.20) data even at moderate doses. This was however ignored for purposes of linear extrapolation since the downward curvature is slight.

7. REFERENCES

Ashton, W.D. (1972). The Logit Transformation. Griffin, London.

Brown, C. (1976). Mathematical aspects of dose-response studies in carcinogenesis-the existence of thresholds. Oncology 33, 62-65.

Chambers, E.A. & Cox, D.R. (1967). Discrimination between alternative binary response models. Biometrika 54, 573-578.

Chand, N. & Hoel, D.G. (1974). A comparison of models for determining safe levels of environmental agents. In Reliability and Biometry. F. Proschan and R.J. Serfling (eds.), SIAM, Philadelphia, 681-700.

Cornfield, J., Carlborg, F. & Van Ryzin, J. (1978). Setting tolerances on the basis of mathematical treatment of dose-response data extrapolated to low doses. In Proceedings of the First International Congress on Toxicology. G.L. Plaa & W.A.M. Duncan (eds.), Academic Press, New York, 143-164.

Cornfield, J. & Mantel, N. (1977). Discussion of Hartley and Sielken (1977a). Biometrics 33, 21-24.

Cox, D.R. & Hinkley, D.V. (1974). Theoretical Statistics. Chapman and Hall, London.

Crump, K.S. (1979). Dose response problems in carcinogenesis. Biometrics 35, 157-167.

Crump, K.S. (1981). Designs for discriminating between binary dose response models with applications to animal carcinogenicity experiments. Submitted for publication.

Crump, K.S., Guess, H.A. & Deal, K.L. (1977). Confidence intervals and test of hypotheses concerning dose response relations inferred from animal carcinogenicity data. Biometrics 33, 437-451.

Crump, K.S., Hoel, D.G., Langley, C.H. & Peto, R. (1976). Fundamental carcinogenic processes and their implications for low dose risk assessment. Cancer Research 36, 2973-2979.

Daffer, P.Z., Crump, K.S. & Masterman, M.D. (1980). Asymptotic theory for analyzing dose response survival data with application to the low dose extrapolation problem. Mathematical Biosciences (to appear).

Efron, B. (1979). Bootstrap methods: another look at the jackknife. Annals of Statistics 7, 1-26.

Fleiss, J.L. (1973). Statistical Methods for Rates and Proportions. Wiley, New York.

Finney, D.J. (1971). Probit Analysis (3rd edition). Cambridge University Press, London.

Food Safety Council (1980a). Quantitative risk assessment. Food & Cosmetics Toxicology 18, 711-734.

Food Safety Council (1980b). Principles and Processes for Making Food Safety Decisions. Food Safety Council, Washington, D.C.

Gaylor, D.W. & Kodell, R.L. (1980). Linear interpolation algorithm for low dose risk assessment of toxic substances. Journal of Toxicology and Environmental Health (to appear).

Gaylor, D.W. & Shapiro, R.E. (1979). Extrapolation and risk estimation for carcinogenesis. In Advances in Modern Toxicology (Vol. 1), New Concepts in Safety Evaluation (Part 2). M.A. Mehlman, R.E. Shapiro & H. Blumenthal (eds.), Wiley, New York, 65-87.

Gehring, P.J., Watanabe, P.G. and Park, C.N. (1978). Resolution of dose response toxicity data for chemicals requiring metabolic activation: Example - vinyl chloride. Toxicology and Applied Pharmacology 44, 581-591.

Grice, H.G., Munro, I.C., Krewski, D. & Blumenthal, H. (1981). In utero exposure in chronic toxicity/carcinogenicity studies. Food and Cosmetics Toxicology (in press).

Guess, H.A. & Crump, K.S. (1976). Low dose extrapolation of data from animal carcinogenicity experiments - analysis of a new statistical technique. Mathematical Biosciences 32, 15-36.

Guess, H.A. & Crump, K.S. (1978). Maximum likelihood estimation of dose-response functions subject to absolutely monotonic constraints. Annals of Statistics 6, 101-111.

Gumbel, E.J. (1958). Statistics of Extremes. Columbia University Press, New York.

Hartley, H.O. & Sielken, R.L. (1977a). Estimation of "safe doses" in carcinogenic experiments. Biometrics 33, 1-30.

Hartley, H.O. & Sielken, R.L. (1977b). Estimation of "safe doses" in carcinogenic experiments. Journal of Environmental Pathology and Toxicology 1, 241-278.

Hartley, H.O. & Sielken, R.L. (1978). Development of statistical methodology for risk estimation. Institute of Statistics, Texas A & M University. (Report prepared for National Center for Toxicological Research.)

Hartley, H.O., Tolley, H.D. & Sielken, R.L. (1980). The product form of the hazard rate model in carcinogenic testing. In Current Topics in Probability and Statistics. M. Csörgő, D. Dawson, J.N.K. Rao & E. Saleh (eds.), North Holland, New York (to appear).

Hoel, D.G. (1980). Incorporation of background in dose-response models. Federation Proceedings 39, 73-75.

Hoel, P.G. & Jennrich, R.I. (1979). Optimal designs for dose response experiments in cancer research. Biometrika 66, 307-316.

Johnson, N.L. & Kotz, S. (1970a). Continuous Univariate Distributions - 1. Wiley, New York.

Johnson, N.L. & Kotz, S. (1970b). Continuous Univariate Distributions - 2. Wiley, New York.

Krewski, D. & Brown, C. (1981). Carcinogenic risk assessment: a guide to the literature. Biometrics (in press).

Krewski, D. & Kovar, J. (1981). Low dose extrapolation under single parameter dose response studies. Communications in Statistics, Series B (in press).

Krewski, D., Kovar, J. & Arnold, D.L. (1980). Optimal experimental designs for dose response studies. (Presented at Joint Statistical Meetings, Houston, Texas, Aug. 11-14).

Littlefield, N.A., Farmer, J.H., Gaylor, D.W. & Sheldon, W.G. (1979). Effects of dose and time in a long-term, low-dose carcinogenic study. In Innovations in Cancer Risk Assessment. J.A. Staffa & M.A. Mehlman (eds.). Pathotox Publishers, Park Forest South, Illinois, 17-34.

Mantel, N., Bohidar, N., Brown, C., Ciminera, J. & Tukey, J. (1975). An improved Mantel-Bryan procedure for "safety" testing of carcinogens. Cancer Research 34, 865-872.

Munro, I.C. & Krewski, D. (1981). The role of risk assessment in regulatory decision making. In Health Risk Analysis. P.J. Walsh & C.R. Richmond (eds.). Franklin Institute Press, Philadelphia (in press).

Peto, R. (1978). Carcinogenic effects of chronic exposure to very low levels of toxic substances. Environmental Health Perspectives 22, 155-159.

Rai, K. & Van Ryzin, J. (1979). Risk assessment of toxic environmental substances based on a generalized multi-hit model. In Energy and Health. N. Breslow & A. Whittemore (eds.). SIAM, Philadelphia, 99-117.

Rai, K. & Van Ryzin, J. (1980). MULTI80: A computer program for risk assessment of toxic substances. Technical Report No. N-1512-NIEHS, Rand Corporation, Santa Monica, California.

Rai, K. & Van Ryzin, J. (1981). A generalized multi-hit dose response model for low-dose extrapolation. Biometrics 37 (in press).

Rao, C.R. (1973). Linear Statistical Inference and its Applications. Wiley, New York.

Rice, J.R. (1964). The Approximation of Functions. Addison-Wesley, Boston.

Searle, S.R. (1971). Linear Models. Wiley, New York.

Truhaut, R. (1979). An overview of the problem of thresholds for chemical carcinogens. In Carcinogenic Risks/Strategies for Intervention. W. David & C. Rosenfeld (eds.), IARC, Lyon.

Van Ryzin, J. & Rai, K. (1980). The use of quantal response data to make predictions. In The Scientific Basis of Toxicity Assessment. H. Witschi (ed.). Elsevier/North Holland, New York, 273-290.

Van Ryzin, J. (1980). Quantitative risk assessment. Journal of Occupational Medicine 22, 321-326.

PART IV
NONPARAMETRIC METHODS AND ROBUST INFERENCE

STATISTICS AND RELATED TOPICS
M. Csörgö, D.A. Dawson, J.N.K. Rao, A.K.Md.E. Saleh (eds.)

THE CONDITIONAL APPROACH TO ROBUSTNESS

George A. Barnard

University of Waterloo
Waterloo, Ontario
Canada

1. CONDITIONAL APPROACH

There seems to have grown up in the statistical world a notion that it makes sense for a statistician to demand of a set of data that it should provide an answer, in a form specified in advance by him, to any question he sees fit to put to the data. Thus a colleague of mine wrote a paper on the problem of 'estimating' the value of x_o, the independent variable, corresponding to a specified value y_o of the dependent variable, based on data providing an estimated regression:

$$y = ax + b + \text{error}.$$

My colleague's method had the common-sense property, that when the estimate of the regression coefficient failed to differ significantly from zero, his method of estimation failed to give a definite answer. His paper was rejected by one referee with the comment that "informative or not one must use the data to come up with a decent estimate" -- no matter whether or not the data allow a reasonable estimate to be made. We all, of course, prefer the situation where we can say that the conclusion which our clients wish to draw is in fact justified by the data; but it is an important part of our duty also to say so clearly, when such a conclusion is not justified.

The 'decision theoretic', or, in Neyman's phrase, the 'inductive behaviour' approach to statistics encourages the a priori laying down of the kind of inference to be drawn -- since in principle the space of possible decisions must be given in advance. A theory of statistical inference, by contrast, must show, in some way, that the conclusions follow from the data, and their form cannot always be laid down in advance. As an example of the absurdity to which the former approach can lead, we may cite the demand for 'unbiasedness' in a significance test to be applied to the 2x2 table:

	A	not-A	Total
Population I	a	b	m
Population II	c	d	n
Total	r	s	N

when it is required to test the hypothesis H_o that $p_1 = p_2 = p$ against, say, the alternative $p_1 > p_2$, at a level $\alpha = 0.05$. Here p_i is the probability of A in population i, for i=1,2. The requirement of unbiasedness says that we must reject H_o, when true, 5% of the time whatever the value of the nuisance parameter p. Let us suppose, for definiteness, that m = 100 and n = 10, and consider what happens when $p = 10^{-6}$. On 99.99% of occasions when H_o is true, our data will take the form:

	A	not-A
I	0	100
II	0	10

and if we are to meet the condition of unbiasedness we must, 5% of the time, with such data, reject H_o in favour of $p_1 > p_2$. But such an 'inference' would be absurd.

Let us apply the point I am making to the problem of 'robust estimation' of, for definiteness, a location parameter λ, say. We suppose we have observations x_i, $i = 1,2,\ldots,n$ with a joint density

$$f(x_1-\lambda, x_2-\lambda,\ldots,x_n-\lambda) \tag{1}$$

and we are asking for a 'point estimate' $T(x_1,\ldots,x_n)$ of λ, or, better, a family of interval estimates, which shall be 'efficient', in some sense, even though we do not know the form of f. If we take my point, we should approach such a request bearing in mind the possibility that the correct answer is to deny the possibility of answering in the form requested. It may be that we do not know enough about the form of f to answer the question as put.

If we put

$$p_i = x_i - \lambda \tag{2}$$

then (1) tells us that the joint density of the p_i is

$$f(p_1,\ldots,p_n) \tag{3}$$

which does not contain λ, no matter what the form of f. Further, the facts embodied in (2) and (3) are logically equivalent to those embodied in (1). We preserve this logical equivalence if we make the 1-1 transformations on the p_i, setting

$$p_i = \bar{p} + c_i, \text{ subject to } \sum_i c_i = 0 \tag{4}$$

in which we regard c_n as equal to $-(c_1+c_2+\ldots+c_{n-1})$. The Jacobian is $\partial(p_1,p_2,\ldots,p_n)/\partial(\bar{p},c_1,\ldots,c_{n-1})=n$, so that the joint density of $(\bar{p},c_1,\ldots,c_{n-1})$ is

$$nf(\bar{p} + c_1, \bar{p} + c_2,\ldots,\bar{p} + c_n) \tag{5}$$

and this in turn is logically equivalent to saying that the marginal density of $(c_1,c_2,\ldots,c_{n-1})$ is

$$\phi(\underline{c}) = \int_{-\infty}^{\infty} n\, f(u + c_1, u + c_2,\ldots,u + c_n)du \tag{6}$$

while the conditional density of $\bar{p}$, given $c_i=k_i$ is

$$\xi(p|\underline{k}) = f(\bar{p} + k_1, \bar{p} + k_2,\ldots,\bar{p} + k_n)/\phi(\underline{k}). \tag{7}$$

In terms of the observations, and the parameter λ,

$$\bar{p} = \bar{x} - \lambda,\ c_i = x_i - \bar{x}. \tag{8}$$

Thus to say that $\bar{x} = m$ and $c_i=k_i$, $i = 1,2,\ldots$, and that $\underline{c}$ has density $\phi(\underline{c})$ while, given $\underline{c}$, $\bar{p} = \bar{x} - \lambda$ has density $\xi(\bar{p}|\underline{c})$ is logically equivalent to saying that $x_i = m+k_i$, $i = 1,2,\ldots,n$, and the x_i have the density (1). This is true, whatever f may be.

Now it may happen, for the sample we have to hand, that the uncertainty about f leaves $\xi(\bar{p}|\underline{k})$ relatively unaffected. To illustrate this we consider a simple unrealistic example, with n=2 and (a) $x_1 = 0.75$, $x_2 = 1.25$ and (b) $x_1 = -1$, $x_2 = 3$. In both cases we have $\bar{x}=1$, but for (a) we have $c_1 = -c_2 = 0.5$, while for (b) we have $c_1 = -c_2 = 2.0$. Suppose we know about f that the x_i are independent, and either (i) x_i has a Cauchy distribution, density $1/\pi(1+(x_i-\lambda)^2)$, or (ii) x_i is normally distributed with mean λ and variance 6. In case (i) c has the standard Cauchy distribution centred on zero, with density $\phi_1(c_1) = 1/\pi(1+c_1)^2$, and we then have

$$\text{(a)}\quad \xi_1(\bar{p}|c_1 = 0.5) = (1 + 0.5^2)/\pi(1+(\bar{p}+0.5)^2)(1+(\bar{p}-0.5)^2),$$

$$\text{(b)}\quad \xi_1(\bar{p}|c_1 = 4.0) = (1+2^2)/\pi(1+(\bar{p}+2)^2)(1+(\bar{p}-2)^2).$$

In case (ii), c_1 is normal with zero mean and variance 3, and in both (a) and (b) $\xi_2(p|c_1) = (1/\sqrt{3}\sqrt{2\pi})\exp-\tfrac{1}{2}p^2/3$. We have tabulated the densities of $\bar{p}$ in Table 1. They all differ; but for $c_1 = 0.5$, ξ_1 is much closer to ξ_2 than for $c_1 = 2$. When $c_1 = 0.5$ the most plausible single value for λ corresponds to $\bar{p}=0$, i.e. is

$\lambda=1$, as in the normal case; but when $c_1 = 2.0$, there are two 'most plausible' values, $\bar{p} = \pm 1.9$, giving two 'point estimates' for λ, 2.9 and -0.9. If we take 1 ± 1.3 as a confidence interval for λ, for case (ii) the confidence coefficient is about 56%, while for (i) (a) it is about 43%, and for (i) (b) it is less than 20%.

Table I

$\bar{p}$	$\xi_1(\bar{p}\mid c_1)$ $c_1 = 0.5$	$\xi_1(\bar{p}\mid c_1)$ $c_1 = 2.0$	$\xi_2(\bar{p}\mid c_1)$ Any c_1
0.0	0.2546	0.0637	0.2303
0.3	0.2333	0.0650	0.2269
0.6	0.1783	0.0693	0.2169
0.9	0.1159	0.0765	0.2012
1.2	0.0686	0.0863	0.1812
1.5	0.0398	0.0961	0.1583
1.8	0.0235	0.0991	0.1342
2.1	0.0144	0.0885	0.1104
2.4	0.0092	0.0674	0.0882
2.7	0.0061	0.0463	0.0683
3.0	0.0041	0.0306	0.0514

Note: In all three cases $\xi(-\bar{p}) = \xi(\bar{p})$.

Thus we can say that when $c_1 = 0.5$, the sample is reasonably robust relative to our ignorance of the distributional form; but when $c_1 = 4.0$ the sample is not at all robust, and the kind of inference we can draw depends crucially on which of the two possible distributions is the true one. The inference which follows from the data, in this latter case, must take the form: 'If f is normal then ..., but if f is Cauchy, then ...' -- with the implication that if the client wishes for a more specific statement, he must discover something more about f, or, perhaps, take a larger sample. This latter helps in two ways -- a larger sample will provide some information helping to discriminate between the two possible f's, and it is also likely to be more robust in itself.

A more realistic situation arises when not only the location λ but also the scale σ is unknown. In this case we can represent what we know about n observations x_i by putting

$$p_i = (x_i - \lambda)/\sigma \tag{9}$$

and representing the density by

$$f(p_1, p_2, \ldots, p_n)$$

as before. We now transform to $(z, t, c_1, \ldots, c_{n-2})$ where

$$p_i = z(t+c_i) \text{ subject to } \sum_i c_i = 0, \quad \sum_i c_i^2 = n(n-1) \tag{10}$$

the Jacobian of which transformation is shown in the Appendix to be

$$z^{n-1}J(\underline{c}) = z^{n-1}n(n-1)(n-2)/|c_n - c_{n-1}|$$

so that the joint density of $(z, t, c_1, \ldots, c_{n-2})$ is

$$z^{n-1}J(\underline{c})f(z(t+c_1), \ldots, z(t+c_n)) \tag{11}$$

from which it follows that the joint density of the c_i is

$$\phi(\underline{c}) = \int_{-\infty}^{\infty}\int_0^{\infty} u^{n-1}J(c)f(u(v+c_1), \ldots, u(v+c_n))\,du\,dv \tag{12}$$

while the conditional density of (t,z), given $c_i = k_i$, is

$$\xi(t,z|\underline{k}) = z^{n-1}f(z(t+k_1), \ldots, z(t+k_n))/\phi(\underline{k}) \tag{13}$$

In terms of the observations,

$$t = (\bar{x} - \lambda)\sqrt{n}/s_x, \quad z = s_x/\sigma\sqrt{n}, \quad c_i = (x_i - \bar{x})\sqrt{n}/s_x \tag{14}$$

so that the values k_i of the c_i will be known when the observations are known and the density (13) becomes available for estimates and confidence statements about λ and σ. If, further, we want confidence statements about λ alone, based only on the sample, we can use the marginal density

$$\xi(t|\underline{k}) = K\int_0^\infty z^{n-1} f(z(t+k_1),\ldots,z(t+k_n))dz \qquad (15)$$

where K is determined by the condition that $\xi(t|\underline{k})$ should integrate to 1.

Applying our general mode of argument, we obtain $1-\alpha$ confidence limits for λ by finding a set $\mathcal{C}$ such that

$$\int_{\mathcal{C}} \xi(t|\underline{k})dt = 1-\alpha \quad \text{and then the set} \quad C = \{\lambda : (\bar{x}_o-\lambda)\sqrt{n}/s_{xo} \in \mathcal{C}\} \qquad (16)$$

where x_o and s_{xo} are the observed values of $\bar{x}$ and s_x, is a $1-\alpha$ confidence set for λ. If we want this set to be as short as possible we should take $\mathcal{C}$ to be of the form

$$\mathcal{C} = \{t: \xi(t|\underline{k}) \geq d(\alpha)\}$$

where $d(\alpha)$ is such that (16) is satisfied. If, on the other hand, we want C to be of the form $\{\lambda: \lambda \leq U\}$ we should take $\mathcal{C}$ to be of the form $\{t: t \geq \ell(\alpha)\}$ where now $\ell(\alpha)$ is such that (16) is satisfied.

The sensitivity of these inferences about λ to changes in the form of f will depend on the sensitivity of (15) to such changes; and it can be seen that this will depend on the values observed for the k_i. Thus, as with the unrealistic example given above, we may well make quite large changes in f which have, for particular k_i, very little influence on ξ; while, conversely, we may have, for other values of k_i, the same changes in f producing large changes in ξ. In the first case we have a robust sample; in the second we have a non-robust sample, and in this case we should point to the non-robustness as part of our inference so as to warn our clients that they should, if possible, obtain more information about distribution form, or take a larger sample. Whether or not our sample is robust can be examined by evaluating (15), for the observed k_1, over a range of possible f's regarded as plausible. This evaluation can be performed, for arbitrary f, quite simply on a hand-held calculator. And when we know the observations to be independent, f takes the form

$$f(p_1,p_2,\ldots,p_n) = K^* \exp -\sum_i \ell(p_i)$$

where $\ell(p_i)$ is (up to an additive constant absorbed in K^*) the logarithm of the density function for a single observation, and then, as Dr. H.E. Daniels has pointed out, (15) is well suited to evaluation by Laplace's method:
Putting $u = \ln z$, $du = dz/z$, and setting $t+k_i = b_i$, $(b_i) = \underline{b}$,

$$I(t) = \int_0^\infty z^{n-1} \exp -\sum_i \ell(z(t+k_i))dz = \int_{-\infty}^\infty \exp -h(u,\underline{b})du$$

where

$$h(u,\underline{b}) = nu - \sum_i \ell(e^u b_i).$$

The first approximation by Laplace's method is obtained by expanding $h(u,\underline{b})$ around its maximum value $h(\hat{u},\underline{b})$,

$$h(u,\underline{b}) \approx h(\hat{u},\underline{b}) + h''(\hat{u},\underline{b})u^2/2$$

from which

$$I(t) \approx e^{-h(\hat{u},\underline{b})}\sqrt{(2\pi h''(\hat{u},\underline{b})}.$$

The accuracy of this approximation depends on how near $h(u,\underline{b})$ is to being quadratic in u -- how near its graph is to being parabolic. When the underlying distribution has the Cauchy form I have found that $h(u,\underline{b})$ is better represented by a hyperbolic graph in which case the error integral involved in the approximation to I(t) is replaced by a Bessel function. There is clearly work here for numerical analysts; but Simpson's rule for I(t) is always available, and its use is not excessively laborious.

Dr. D.A. Sprott (1978) has examined the Darwin data quoted by Fisher (1960, page 37) and has shown that the inference to be drawn depends quite critically on what we suppose to be known concerning the shape of the distribution. He points out that this non-robustness of the sample arises mainly from the two 'outliers' at -67 and -48, when all the other observations are positive and range between 6 and 75. If these two observations had not been quite so low -- if they had been around -10, for example, the data would have been much more robust with respect to distributional form. In fact, a reading of Darwin's meticulous account of his observations shows that at least one of the plants involved was probably being eaten by an insect; and another clear possibility is, that one pair of plants was reversed, so that the difference is given the wrong sign. Whatever may be the truth, it is clear from Sprott's analysis, confirming and extending an earlier study by Box and Tiao (1973), that information about the distributional shape forms an essential part of the data in this case.

The attitudes of statisticians towards assumptions of normality form an interesting history. Without going farther back than Fisher, he was greatly annoyed in 1928 by a review of his 'Statistical Methods' by E.S. Pearson, in which Pearson had criticised Fisher for Fisher's failure to emphasise the extent to which his methods relied on the normality assumption. Characteristically, while reacting somewhat belligerently towards Pearson, Fisher took Pearson's point to the extent of suggesting to A.L. O'Toole that he should look into some non-normal distributions. This resulted in a couple of papers in the earliest volumes of 'Annals of Mathematical Statistics'; but (as Fisher told me) it became clear that the computational facilities then available to statisticians would not suffice to deal adequately with the problem raised. Furthermore, Fisher was primarily concerned with biological applications, where there was a great deal of evidence, from the work of Quetelet, Galton, Weldon and Karl Pearson, to support normality assumptions for univariate data. It is interesting that Fisher himself noted the absence of corresponding data on multivariate normality in all the editions of 'Statistical Methods' (Section 30), from the first to the last. E.S. Pearson was primarily concerned with industrial applications of statistics, where there was less evidence for univariate normality -- and he and Walter Shewhart went to a lot of trouble to collect data on distributional forms. Fisher introduced 'non-parametric' tests in 1935, without advocating their use in cases where there is information about distributional form; indeed, he later pointed to the dangers of a wrong assumption of ignorance. In the 1950's non-parametric tests enjoyed a widespread vogue, particularly after the work of Hoeffding and Hodges and Lehmann showed that little in the way of power was lost by their use; but their difficulty in complex situations, their lack of easy adaptability to estimation, and their strong dependence on independence of observations held back their general use. Box introduced the term 'robust' in 1952, but it needed the powerful advocacy of Tukey to draw the attention of many statisticians to the difficulties of robust estimation. He and his coworkers, and especially Huber and Hampel, have done a great deal towards developing estimation methods which are relatively insensitive to departures from normality assumptions. But their approach has been a 'marginal' rather than a 'conditional' one in that they have looked for robustness over a series of repeated samples, without recognising that some samples may be more robust than others. A conditional approach to the problems was first adopted by Box and Tiao in the work referred to above. Such an approach is, of course, natural from the Bayesian point of view adopted by them. One way of summing up the present paper is to say that one can adopt a conditional approach to these problems without being Bayesian. The inferential methods we have suggested go back to Fisher's (1934) paper, but the fact that they can, with modern computers, be readily applied in practice, was first pointed out in print by D.A.S. Fraser (1976).

What is needed now, on the empirical data, is a revival of the work of Quetelet, Galton, Weldon, the two Pearsons, and others, on the types of distributional form likely to be met with in various contexts. And on the theoretical side there are related outstanding problems. The empirical data on distributional form is more likely to take the form of large numbers of sets of residuals left after fitting location and scale parameters than of very large raw samples. But we do not know what can, and what cannot be learned about distributional form from such sets.

2. APPENDIX

If $p_i = z(t+c_i)$, $i = 1,2,\ldots,m$, subject to $\sum_i c_i = 0$ and $\sum_i c_i^2 = m(m-1)$, and $z \geq 0$, and the new variables are taken to be $t, z, c_1, \ldots, c_{m-2}$, we have

$$\partial(p_1,p_2,p_3,\ldots,p_m)/\partial(t,z,c_1,\ldots,c_{m-2}) = J_1$$

$$= \begin{vmatrix} z & (t+c_1) & z & 0 & 0 & \cdots & 0 \\ z & (t+c_2) & 0 & z & 0 & \cdots & 0 \\ z & (t+c_3) & 0 & 0 & z & \cdots & 0 \\ \cdots & \cdots & \cdots & \cdots & \cdots & \cdots & \cdots \\ z & (t+c_{m-2}) & 0 & 0 & 0 & & z \\ z & (t+c_{m-1}) & za_1 & za_2 & za_3 & \cdots & za_{m-2} \\ z & (t+c_m) & zb_1 & zb_2 & zb_3 & \cdots & zb_{m-2} \end{vmatrix}$$

Putting $a_i = \partial c_{m-1}/\partial c_i$, $b_i = \partial c_m/\partial c_i$, so that, from the conditions on the c_i,

$$1+a_i+b_i = 0 \qquad (1)$$
$$c_i+a_ic_{m-1}+b_ic_m = 0$$

$= z^{m-1}\Delta(c)$ where, by multiplying the first column by t and subtracting from the second,

$$\Delta(c) = \begin{vmatrix} 1 & c_1 & 1 & 0 & 0 & \cdots & 0 \\ 1 & c_2 & 0 & 1 & 0 & \cdots & 0 \\ 1 & c_3 & 0 & 0 & 1 & \cdots & 0 \\ \cdots & \cdots & \cdots & \cdots & \cdots & \cdots & \cdots \\ 1 & c_{m-2} & 0 & 0 & 0 & \cdots & 1 \\ 1 & c_{m-1} & a_1 & a_2 & a_3 & \cdots & a_{m-2} \\ 1 & c_m & b_1 & b_2 & b_3 & \cdots & b_{m-2} \end{vmatrix}$$

Multiplying the ith row by c_i and adding to the mth multiplied by c_m gives

$$c_m\Delta(c) = \begin{vmatrix} 1 & c_1 & 1 & 0 & 0 & \cdots & 0 \\ 1 & c_2 & 0 & 1 & 0 & \cdots & 0 \\ 1 & c_3 & 0 & 0 & 1 & \cdots & 0 \\ \cdots & \cdots & \cdots & \cdots & \cdots & \cdots & \cdots \\ 1 & c_{m-2} & 0 & 0 & 0 & \cdots & 1 \\ 1 & c_{m-1} & a_1 & a_2 & a_3 & \cdots & a_{m-2} \\ 0 & m(m-1) & 0 & 0 & 0 & \cdots & 0 \end{vmatrix} \quad \text{using (1)}$$

$$= (-1)^{m-1} m(m-1) \sum_{i=1}^{m-2} a_i .$$

Similarly, after interchanging the last two rows,

$$c_{m-1}\Delta(c) = (-1)^m m(m-1) \sum_{i=1}^{m-2} b_i .$$

Subtracting, taking absolute values, and using (1) again gives

$$|c_m - c_{m-1}|\Delta(c) = m(m-1)(m-2).$$

The Jacobian for the q_j is treated similarly, and the result follows.

3. REFERENCES

[1] Box, G.E.P. and Tiao, G.C., Bayesian Influence in Statistical Analysis (Addison-Wesley, 1973).

[2] Fraser, D.A.S., Necessary analysis and adaptive inference. J. Amer. Stat. Assoc. 71 (1976) 99-113.

[3] Fisher, R.A., Two new properties of mathematical likelihood. Proc. Roy. Soc. A, 144 (1934) 285-307.

[4] Fisher, R.A., The Design of Experiments (Oliver and Boyd, Edinburgh, 1960).

[5] Sprott, D.A., Robustness and non-parametric procedures are not the only or the safe alternatives to normality. Canad. J. Psychol. 32 (1978) 180-185.

Research supported by NSERC Grant A-3086.

STATISTICS AND RELATED TOPICS
M. Csörgö, D.A. Dawson, J.N.K. Rao, A.K.Md.E. Saleh (eds.)

THE STRONG CONVERGENCE OF EMPIRICAL NEAREST NEIGHBOR ESTIMATES OF INTEGRALS

Luc Devroye*
McGill University

SUMMARY

Let $X_1,\ldots,X_n$ be independent random variables with common probability measure μ on the Borel sets of R^1, and let $A_{n1},\ldots,A_{nn}$ be the nearest neighbor partition of the real line obtained from $X_1,\ldots,X_n$. When $f\epsilon L^1(\mu)$, then it is known that $\sum_{i=1}^{n} f(X_1)\,\mu(A_{ni}) \to \int f(x)\,\mu(dx)$ in probability as $n \to \infty$ (Stone, 1977). We show that "in probability" can be replaced by "almost surely" whenever $f\epsilon L^\infty(\mu)$. No conditions are placed upon μ.

1. INTRODUCTION

We consider the problem of the approximation of $\int f(x)\,\mu(dx)$ by $I_n = \int f(x)\,\mu_n(dx)$ where μ is an arbitrary probability measure on the Borel sets of R^1, f is a Borel measurable function and μ_n is an empirical probability measure. We assume throughout that $X_1,\ldots,X_n$ are independent identically distributed random variables with probability measure μ. When μ_n is the classical empirical measure, then

$$\int f(x)\,\mu_n(dx) = \frac{1}{n}\sum_{i=1}^{n} f(X_i)$$

and $I_n \to I=\int f(x)\,\mu(dx)$ a.s. as $n \to \infty$ for all $f\epsilon L^1(\mu)$. We are interested here in the same type of result for the empirical nearest neighbor measure μ_n. In section 3 we will highlight the impact of this result on the study of the <u>nearest neighbor method in discrimination</u>. Yakowitz (1977) has suggested the use of the empirical nearest neighbor estimate I_n in <u>Monte Carlo integration</u>, and he has given evidence showing that I_n converges faster to I when μ satisfies some regularity conditions.

The <u>empirical nearest neighbor estimate</u> I_n is defined by

---

* This research was sponsored by U.S. Air Force Grant AFOSR 76-3062 while the author was at the University of Texas during the summer of 1979.

$$\sum_{i=1}^{n} f(X_1)\ \mu(A_{ni})$$

where $A_{ni} = \{x|x \in R^1,\ X_i$ is the nearest neighbor of x among $X_1,\ldots,X_n\}$.
We say that X_i is closer to x than X_j when

$$\text{either}\quad |x-X_i| < |x-X_j|$$

$$\text{or}\quad |x-X_i| = |x-X_j|\ ,\ X_i < X_j$$

$$\text{or}\quad X_i = X_j\ ,\ i < j\ .$$

Thus, $X_i=X_j$, $i < j$, implies that A_{nj} is empty.

Stone (1977) has shown that when $f \in L^p(\mu)$, $p \geq 1$,

$$E(|I_n-I|^p) \to 0 \quad \text{as} \quad n \to \infty\ ,$$

for all probability measures μ . In particular, when $f \in L^1(\mu)$, it is true $I_n \to I$ in probability as $n \to \infty$. The almost sure convergence of I_n to I cannot be established by the methods employed by Stone. The main result of this paper is the following Theorem:

Theorem 1. Let $|f| \leq c < \infty$ be a Borel measurable function and let μ be a probability measure on the Borel sets of R^1 . Then the empirical nearest neighbor estimate I_n satisfies

$$|I_n-I| \leq \sum_{i=1}^{n} \int_{A_{ni}} |f(X_i)-f(x)|\ \mu(dx) \to 0 \text{ a.s.} \quad \text{as} \quad n \to \infty\ . \tag{1}$$

2. PROOFS

Lemma 1. Theorem 1 is true whenever μ is nonatomic.

Proof of Lemma 1.

Replace all X_i's by $F(X_i)$'s where F is the distribution function corresponding to μ . Let $X_{(1)} \leq X_{(2)} \leq \ldots \leq X_{(n)}$ be the order statistics of $F(X_1),\ldots,F(X_n)$ and let $B_{ni}=(X_{(i-1)},X_{(i)}]$ where $X_{(0)}=0$, $X_{(n+1)}=1$. Clearly

$$\sum_{i=1}^{n} \int_{A_{ni}} |f(X_i)-f(x)|\mu(dx)$$

$$\leq \sum_{i=1}^{n} \int_{B_{ni} \cup B_{n\ i+1}} |g(X_{(i)})-g(x)|\ dx \tag{2}$$

where $g(u)=f(F^{-1}(u))$ and $F^{-1}(u)=\inf\{y|F(y)=u\}$, and $0 \leq u \leq 1$.

Consider a sample of size $2n$, and define

$$V_{2n} = \sum_{i=1}^{2n} \int_{B_{2n\,i} \cup B_{2n\,i+1}} |g(X_{(i)}-g(x)|\, dx$$

which can be split up into two sums $V'_{2n}+V''_{2n} = \sum_{i \text{ even}} + \sum_{i \text{ odd}}$. It suffices to show that $V'_{2n} \to 0$ a.s. as $n \to \infty$. Lemma 1 then follows by symmetry. Let us define $C_i = B_{2n\,2i} \cup B_{2n\,2i+1}$, $1 \le i \le n$, and $D_n=(X_{(1)},X_{(3)},\ldots,X_{(2n-1)})$. We will show that

$$V'_{2n} - E(V'_{2n}|D_n) \to 0 \text{ a.s. as } n \to \infty , \tag{3}$$

and

$$E(V'_{2n}|D_n) \to 0 \text{ a.s. as } n \to \infty . \tag{4}$$

Since $C_1,\ldots,C_n$ are determined by D_n, we have for all $r \ge 2$ and some constant $a_r < \infty$, by a result of Dharmádhikari and Jogdeo (1969), after defining

$$A_i = \int_{C_i} |g(X_{(2i)})-g(x)|\, dx , \ 1 \le i \le n ,$$

$$E(|V'_{2n}-E(V'_{2n}|D_n)|^r) = E(|\sum_{i=1}^{n} (Z_i-E(Z_i|D_n))|^r)$$

$$\le a_r\, n^{\frac{r}{2}-1} \sum_{i=1}^{n} E(|Z_i-E(Z_i|D_n)|^r)$$

$$\le 2^r a_r n^{\frac{r}{2}-1} \sum_{i=1}^{n} E(|Z_i|^r)$$

$$\le (4c)^r a_r n^{\frac{r}{2}-1} \sum_{i=1}^{n} E(\mu^r(C_i))$$

$$= (4c)^r a_r n^{\frac{r}{2}-1} E(\mu^r(C_1))$$

$$= O(n^{-\frac{r}{2}})$$

which is summable in n for $r > 2$; (3) then follows by the Borel-Cantelli lemma.

Let us next define

$$\rho^+(x,b) = b^{-1} \int_{y > x, ||y-x|| \le b} |f(y)-f(x)|dy$$

for $x \in [0,1]$ and $b > 0$. Define $\rho^-(x,b)$ similarly on the set $y<x, ||y-x|| \le b$, and let

$$\rho(x,a) = \sup_{0<b\le a} (\rho^+(x,b), \rho^-(x,b)) , \quad a > 0 .$$

By the Lebesque density theorem (Stein, 1970), we know that $\rho(x,a) \to 0$ as $a \downarrow 0$ for almost all x . Also, $\rho(x,a)$ is nonincreasing as $a \downarrow 0$. Now,

$$E(V'_{2n} | D_n) \le \sum_{i=1}^{n} \int_{C_i} \rho(x,U_n)\, dx = \int_0^1 \rho(x,U_n)\, dx \tag{5}$$

where $U_n = \sup_{0\le i\le n} |X_{(2i+1)} - X_{(2i-1)}|$. Because $U_n \to 0$ a.s. as $n \to \infty$ (Slud, 1978) and $0 \le \rho \le 2c$, we have by a generalization of the dominated convergence theorem (Glick, 1974), $\int \rho(x,U_n)\, dx \to 0$ a.s. as $n \to \infty$.

When μ has atoms, we consider the decomposition of μ into its atomic part (μ_1) and its nonatomic part (μ_2) : $\mu=\mu_1+\mu_2$. Let A be the set of of atoms of μ . If A is empty, theorem 1 follows from lemma 1. If $\mu(A)=1$, theorem 1 is almost trivial. For the other cases, the following lemma will be useful.

<u>Lemma 2.</u> $W_n = \sum_{i:X_i \in A} \mu_2(A_{ni}) \to 0$ a.s. as $n \to \infty$,

<u>Proof of Lemma 2.</u>

Consider the two subsequences of random variables from $X_1, X_2, \ldots,$ defined by membership in A . Let $Y_1, \ldots, Y_{M_n}$ be the collection of X_i's , $1 \le i \le n$, belonging to A^c , and let $Z_1, \ldots, Z_{N_n}$ be the corresponding collection for A . Y_i's and Z_i's are added in order of their appearance in the X_i sequence. Clearly, $M_n + N_n = n$ for all n . If $F(x)=\mu_2((-\infty,x])$, then

$$W_n \le \sum_{i=0}^{M_n} (F(Y_{(i+1)}) - F(Y_{(i)}))\, T_{ni} = W_n^*$$

where $Y_{(1)} \le \ldots \le Y_{(M_n)}$ are the order statistics corresponding to $Y_1, \ldots, Y_{M_n}$; $Y_{(0)} = -\infty$; $Y_{(M_n+1)} = +\infty$; T_{ni} is the indicator function of the event that at least one Z_j , $1 \le j \le N_n$, belongs to $(Y_{(i)}, Y_{(i+1)}]$. Assume that $\mu_2(R) = q > 0$.

If $E_0, \ldots, E_{M_n}$ are independent identically distributed exponential random variables with sum S_{M_n} , then W_n^* is distributed as

$$\sum_{i=1}^{M_n} E_i T_{ni} / S_{M_n} \le S_{M_n}^{-1} \left[\sum_{i=0}^{M_n} E_i^2\, Q_n \right]^{1/2}$$

where Q_n is the number of different values among $Z_1,\ldots,Z_{N_n}$ and T_{ni} is the indicator function of the event that one or more of the $F(Z_j)$'s belongs to $(S_{M_n}^{-1}(E_0+\ldots+E_{i-1}), S_{M_n}^{-1}(E_0+\ldots+E_i)]$, with $E_{-1}=0$. To conclude that $W_n^* \to 0$ a.s. as $n \to \infty$, it suffices to show that

(i) $Q_n/n \to 0$ a.s. as $n \to \infty$,

(ii) $M_n/n \to q$ a.s. as $n \to \infty$,

and

(iii) $\sum_{n=1}^{\infty} P(|(n+1)^{-1} \sum_{i=0}^{n} E_i^2 - 2| > \varepsilon) < \infty$, all $\varepsilon > 0$.

Here $E_0, E_1, \ldots, E_n$ are i.i.d. exponential random variables with sum S_n. Statement (ii) is true by the strong law of large numbers, and statement (iii) can be proved without difficulty by using exponential inequalities for sums of independent random variables (e.g., see Baum, Katz and Read, 1962). Lemma 2 will follow if we can show (i).

It is clear that

$$E(Q_n) \le \sum_{i=1}^{\infty} (1-(1-a_i)^n)$$

for some sequence of a_i's with $\sum a_i \le 1$, $a_i \ge 0$. Thus, for $c_2 > 0$,

$$\begin{aligned} E(Q_n) &\le \sum_{a_i<c_2/n} na_i + \sum_{a_i \ge c_2/n} 1 \\ &\le n \sum_{a_i<c_2/n} a_i + 1 + \frac{n}{c_2} \\ &= n\, o(1) + 1 + \frac{n}{c_2} \\ &\le \frac{2n}{c_2}, \quad n \text{ large enough.} \end{aligned}$$

Since $c_2 > 0$ was arbitrary, we conclude that $E(Q_n)/n \to 0$ as $n \to \infty$. If $Q_{(k,\ell]}$ is the number of different values among $Z_{N_k+1},\ldots,Z_{N_\ell}$, then obviously

$$0 \le Q_{(k,\ell]} \le Q_{(k,s]} + Q_{(s,\ell]}, \text{ all } k < s < \ell,$$

and the distribution of $Q_{(k,\ell]}$ only depends upon $\ell-k$. By the subadditive ergodic theorem (Kingman, 1968, 1973), we may conclude that $Q_{(0,n]}/n \to c_1 = \lim_{n\to\infty} E(Q_{(0,n]})/n$ a.s., $n \to \infty$. Since $Q_{(0,n]}=Q_n$, we have $c_1=0$ and $Q_n/n \to 0$ a.s. as $n \to \infty$.

Proof of Theorem 1.

Let M_n , N_n , $Y_1,Y_2,\ldots$ and $Z_1,Z_2,\ldots$ be as in the proof of Lemma 2, and let B_{ni} , $1 \le i \le M_n$, be the nearest neighbor partition of R^1 corresponding to $Y_1,\ldots,Y_{M_n}$. Let C_n be $\{x|\ x\in A\ ,\ X_i \ne x\ ,\ \text{all}\ i \le n\}$. Then,

$$\sum_{i=1}^{n} \int_{A_{ni}} |f(X_i)-f(x)|\ \mu(dx)$$
$$\le \sum_{i=1}^{M_n} \int_{B_{ni}} |f(Y_i)-f(x)|\ \mu_2(dx) + \sum_{i:X_i\in A} 2c\mu_2(A_{ni})+2c\mu_1(C_n)\ . \tag{6}$$

Since $M_n \to \infty$ a.s. as $n \to \infty$, the first term on the right hand side of (6) tends to 0 a.s. as $n \to \infty$ (Lemma 1). The second term tends to 0 a.s. as $n \to \infty$ by Lemma 2. Finally $\mu_1(C_n)$ is monotone $\downarrow$, and

$$E(\mu_1(C_n)) = \sum_{x\in A} \mu_1(\{x\})\ \ (1-\mu_1(\{x\}))^{\,n} \to \infty \quad \text{as} \quad n \to \infty\ ,$$

so that $\mu_1(C_n) \to 0$ a.s. as $n \to \infty$. This concludes the proof of Theorem 1.

3. THE NEAREST NEIGHBOR RULE

We will now consider the implications of Theorem 1 in nonparametric discrimination. Let (X,Y) , $(X_1,Y_1),\ldots,(X_n,Y_n),\ldots$ be independent identically distributed $R^d\times\{0,1\}$-valued random vectors, and let Y be estimated by $\hat{Y}_n=Y_i$ when $X \in A_{ni}$. Thus, $\hat{Y}_n$ depends upon X , and $(X_1,Y_1),\ldots,(X_n,Y_n)$. It is called the *nearest neighbor estimate of* Y (Fix and Hodges, 1951; Cover and Hart, 1967). We define

$$L_n = P(\hat{Y}_n \ne Y | X_1,Y_1,\ldots,X_n,Y_n)$$

and

$$L^* = \inf_{g:R^d\to\{0,1\}} P(g(X)\ne Y)\ .$$

L^* is called the Bayes probability of error, and L_n is the probability of error for the nearest neighbor estimate and the given data. Clearly, $L_n \ge L^*$, all n . Under some restrictions on the distribution of (X,Y) , Cover and Hart have shown that

$$\lim_{n\to\infty} E(L_n) = 2E(\eta(X)(1-\eta(X))) \le 2L^*(1-L^*) \tag{7}$$

where

$$\eta(x) = P(Y=1|X=x)\ ,\quad x \in R^d\ .$$

Stone (1977) and Devroye (1980) showed that (7) remains valid for *all* distributions of (X,Y) . In general, L_n does not converge to a constant in probability. For example, when $X=0$ a.s., and $\eta(0) = \frac{1}{3}$, then

$$L_n = \frac{1}{3} I_{[Y_1=0]} + \frac{2}{3} I_{[Y_1=1]} ,$$

so that convergence to a constant is excluded. Devroye (1980) has shown recently that

$$L_n \to 2E(\eta(X)(1-\eta(X))) \quad \text{in probability as} \quad n \to \infty$$

whenever the probability measure μ of X is nonatomic. This result can be strengthened now for $d=1$:

Theorem 2. For $d=1$ and all nonatomic probability measures μ , we have, a.s.,

$$\lim_{n\to\infty} L_n = 2E(\eta(X)(1-\eta(X))) \le 2L^*(1-L^*) .$$

Proof of Theorem 2.

Clearly,

$$L_n = \sum_i I_{[Y_i=1]} \int_{A_{ni}} (1-\eta(x))\mu(dx)$$

$$+ \sum_i I_{[Y_i=0]} \int_{A_{ni}} \eta(x)\ \mu(dx)$$

$$= L_{n1}+L_{n0} .$$

We will show that $L_{n1} \to E(\eta(X)(1-\eta(X)))$ a.s. as $n \to \infty$. By symmetry, we can then conclude that $L_n \to 2E(\eta(X)(1-\eta(X)))$ a.s. as $n \to \infty$. The inequality in Theorem 2 is a simple consequence of Jensen's inequality when one notices that $L^*=E(\min(\eta(X),1-\eta(X)))$. Let

$$C_{ni} = \int_{A_{ni}} (1-\eta(x))\ \mu(dx) ,$$

$$Z_i = I_{[Y_i=1]}-\eta(X_i) .$$

Then,

$$|L_{n1}-\int\eta(x)(1-\eta(x))\mu(dx)|$$

$$\le \sum_i C_{ni}Z_i + \sum_i \int_{A_{ni}} |\eta(X_i)-\eta(x)|\mu(dx) . \tag{8}$$

The last term of (8) tends to 0 a.s. for all probability measures μ (Theorem 1). Check that $\sum C_{ni} \le 1$, $C_{ni} \ge 0$, $E(C_{ni}|X_1,\ldots,X_n)=C_{ni}$ a.s., $|Z_i| \le 1$ and $E(Z_i|X_1,\ldots,X_n)=0$ a.s. Thus, by an inequality of Dharmadhikari and Jogdeo (1969), for all $r > 1$, there exists a constant $a=a(r) > 0$ such that

$$E(|\sum_i C_{ni}Z_i|^{2r}) = E(E|\sum_i C_{ni}Z_i|^{2r}|X_1,\ldots,X_n))$$

$$\leq E(an^{r-1}\sum_i E(|C_{ni}Z_i|^{2r}|X_1,\ldots,X_n))$$

$$= a\,n^r\,E(|C_{n1}Z_1|^{2r})$$

$$\leq a\,n^r\,E(\mu^{2r}(A_{n1}))$$

$\leq a\,n^r\,E(W^{2r})$ (where W is the second largest of a sample of n independent identically distributed uniform (0,1) random variables)

$$\leq a\,n^r \int_0^1 x^{2r}\,n(n-1)(1-x)^{n-2}x\,dx$$

$$= a\,n^r\,n(n-1)\,\frac{\Gamma(n-1)\Gamma(2r+2)}{\Gamma(n+2r+1)}$$

$$= O(n^{-r})\ .$$

Thus, by the Borel-Cantelli lemma, (8) tends to 0 a.s. as $n \to \infty$.

4. REFINEMENTS

In Lemma 1, we established the strong convergence to I of the empirical nearest neighbor estimate I_n when f is bounded and μ is nonatomic. The condition that f is bounded can be dropped without much trouble.

Lemma 3. (1) is valid whenever $f \in L^p(\mu)$ for some $p > 1$, and μ is nonatomic.

Proof of Lemma 3.

By Theorem 1, (1) is valid for the function $f(x)I_{[|f(x)|\leq M]}$ and any constant M. Let us fix a constant $M > 0$, and define $g(x) = f(x)I_{[|f(x)|>M]}$. We have,

$$|I_n - I| \leq \sum_{i=1}^n \int_{A_{ni}} |g(X_i)-g(x)|\ \mu(dx) + o(1) \text{ a.s. as } n \to \infty\ .$$

Also,

$$\sum_{i=1}^n \int_{A_{ni}} |g(X_i)-g(x)|\mu(dx)$$

$$\leq \int |g(x)|\mu(dx) + \sum_{i=1}^n |g(X_i)|\mu(A_{ni})\ . \tag{9}$$

For all $g \in L^1(\mu)$, the first term on the right-hand-side of (9) is small by the choice of M. By Hölder's inequality, the last term of (9) is not greater than

$$\left[\sum_{i=1}^{n} |g(X_i)|^p\right]^{\frac{1}{p}} \left[\sum_{i=1}^{n} \mu^q(A_{ni})\right]^{\frac{1}{q}} \tag{10}$$

where $p,q > 1$, $\frac{1}{p} + \frac{1}{q} = 1$. Now, $\sum_{i=1}^{n} |g(X_i)|^p \sim n\, E(|g(X_1)|^p)$ a.s. as $n \to \infty$. Let S_{ni}, $0 \le i \le n$, be the spacings of n i.i.d. uniform $(0,1)$ random variables. Clearly, for all n and all $u > 0$,

$$P\left(\sum_{i=1}^{n} \mu^q(A_{ni}) > u\right) \le P\left(2^{q-1} \sum_{i=0}^{n} S_{ni}^q > u\right). \tag{11}$$

Now, we will show that $\frac{1}{n} \sum_{i=0}^{n} (nS_{ni})^q$ converges completely to $\Gamma(q+1)$ as $n \to \infty$. This in turn implies that for all $\varepsilon > 0$, $\sum_{i=1}^{n} \mu^q(A_{ni}) \le 2^{q-1}\Gamma(q+1)\ (1+\varepsilon)/n^{q-1}$ except possibly for finitely many n, a.s. Thus, almost surely, (10) is smaller than

$$E(|g(X_1)|^p)^{\frac{1}{p}}\ 2\ \Gamma(q+1)^{\frac{1}{q}},$$

except possibly for finitely many n. This too can be made arbitrarily small by choosing M large enough. This would complete the proof of Lemma 3.

It is known that $S_{n0},\dots,S_{nn}$ are distributed as $E_0/S, E_1/S,\dots,E_n/S$ where $E_0,E_1,\dots,E_n$ are i.i.d. exponential random variables and $S=E_0+\dots+E_n$. Thus, Thus, $\frac{1}{n} \sum_{i=0}^{n} (nS_{ni})^q \to \Gamma(q+1)$ completely when

$$\text{(i)}\quad \frac{1}{n} \sum_{i=0}^{n} E_i^q \to \Gamma(q+1) \quad \text{completely},$$

$$\text{(ii)}\quad \frac{1}{n} \sum_{i=0}^{n} E_i \to 1 \quad \text{completely}.$$

The latter two results follow from the fact that $E(E_0^q) = \Gamma(q+1)$, all $q \ge 0$, and that $E(E_0^{3q}) < \infty$ (apply for example, Theorem 28, p. 286 of Petrov (1975) which states that under the said conditions $P\left(\left|\frac{1}{n} \sum_{i=0}^{n} (E_i^q - \Gamma(q+1))\right| > \varepsilon\right) = o(n^{-2})$ for all $\varepsilon > 0$).

5. REFERENCES

[1] Baum, L.E., Katz, M. and Read, R.R., "Exponential convergence rates in the law of large numbers", Transactions of the American Mathematical Society 120, 108-123, 1962.

[2] Cover, T.M. and Hart, P.E., "Nearest neighbor pattern classification", IEEE Transactions on Information Theory 13, 21-27, 1967.

[3] Devroye, L., "On the inequality of Cover and Hart in nearest neighbor discrimination", to appear in IEEE Transactions on Pattern Analysis and Machine Intelligence, 1980.

[4] Dharmadhikari, S.W. and Jogdeo, K., "Bounds on moments of certain random variables", Annals of Mathematical Statistics 40, 1506-1508, 1969.

[5] Fix, E. and Hodges, J.L., "Discriminatory analysis. Nonparametric discrimination, consistency properties", Project 21-49-004, Report No. 4, School of Aviation Medicine, Randolph Field, Texas, 1951.

[6] Glick, N., "Consistency conditions for probability estimators and integrals of density estimators", Utilitas Mathematica 6, 61-74, 1974.

[7] Kingman, J.F.C., "The ergodic theory of subadditive stochastic processes", Journal of the Royal Statistical Society Series B 30, 499-510, 1968.

[8] Kingman, J.F.C., "Subadditive ergodic theory", Annals of Probability 1, pp. 883-909, 1973.

[9] Petrov, V.V., Sums of Independent Random Variables, Springer-Verlag, Berlin, 1975.

[10] Slud, E., "Entropy and maximal spacings for random partitions", Zeitschrift fur Wahrscheinlichkeitstheorie und verwandte Gebiete 41, 341-352, 1978.

[11] Stein, E.M., Singular Integrals and Differentiability Properties of Functions, Princeton University Press, Princeton, New Jersey, 1970.

[12] Stone, C.J., "Consistent nonparametric regression", Annals of Statistics 5, 595-645, 1977.

[13] Yakowitz, S., Computational Probability and Simulation, Addison-Wesley, Mass., 1977.

STATISTICS AND RELATED TOPICS
M. Csörgö, D.A. Dawson, J.N.K. Rao, A.K.Md.E. Saleh (eds.)

ON ADAPTIVE ROBUST INFERENCES

Robert V. Hogg

The University of Iowa
Iowa City, Iowa

This is an expository paper in which the general nature of adaptive inference is first reviewed. A brief summary of major methods of adapting spans unrestrictive to restrictive ones. The latter procedures include those based on preliminary selector statistics. Recent studies are cited that indicate substantial gains in power and efficiency can be made with adaptive nonparametric and/or robust inferences over traditional non-adaptive ones.

1. INTRODUCTION

In adaptive inference, we use the sample to help us select the appropriate type of statistical procedure needed for the situation under consideration. Since there is an element of model building in the selection, the resulting procedures are usually robust. That is, if the selection of the model (and hence the procedure) is reasonably accurate, the resulting statistical inference would be quite efficient, not necessarily the best for each underlying model, but reasonably good for all.

For illustration, say we wish to estimate the center of a symmetric univariate distribution of the continuous type, but we do not know the form of the underlying density f. If we knew f, of course we could find one good estimator by maximizing the likelihood function $L = \prod_{i=1}^{n} f(x_i-\theta)$. Suppose we can find this maximum by differentiation; that is, we can determine $\hat{\theta}$ by solving

$$\sum_{i=1}^{n} - \frac{f'(x_i-\hat{\theta})}{f(x_i-\hat{\theta})} = 0.$$

Since we do not know f, we can replace it by an appropriate estimate, say $\hat{f}$, and solve

$$\sum_{i=1}^{n} - \frac{\hat{f}'(x_i-\tilde{\theta})}{\hat{f}(x_i-\tilde{\theta})} = 0$$

to obtain an adaptive estimator $\tilde{\theta}$. This is essentially a procedure of Stone [23].

In nonparametrics methods, various persons have also tried to obtain some reasonable estimates of f'/f to determine pseudo linear rank statistics. Among these efforts are those of Hájek[6], van Eeden[25], Beran[2], and Jones[13]. The latter two have possibly been more successful because they have assumed something about the form of $-f'/f$. Nevertheless, it is extremely difficult to estimate the density f to say nothing of its derivative f'. Thus, while the asymptotic results in this area are outstanding, there is some question about the practical meaning when $n = 20$ or 30.

Accordingly, there have been some more restrictive approaches to adaptation. To estimate a middle point θ, Takeuchi[24] uses a pseudo linear function of k of the n order statistics in which the "best" coefficients have been determined using all n sample items. On the other hand, Jaeckel[11] restricts his adaptive estimator to be a trimmed mean $\bar{X}_p$, where p is selected to minimize the estimate of the variance of $\bar{X}_p$. The procedure of Johns[12] essentially finds the weighted average of two or more trimmed means, in which the weights have been estimated from the sample.

2. ESTIMATION OF THE CENTER

While obviously there has been a great deal done on adaptive techniques, most of our comments will be made about a very restrictive method. Suppose that we have a family (not necessarily finite) of possible underlying distributions. From this family of distributions, consider a few representative ones, say $F_1, F_2, \cdots, F_k$. Now, for each of these k distributions, say we can find a good statistic with which to make the inference under consideration. Suppose these corresponding statistics are $T_1, T_2, \cdots, T_k$. We then observe the sample and, with a selector statistic, say Q, we determine which one of $F_1, F_2, \cdots, F_k$ seems to be closest to the underlying distribution from which the sample arose. If Q picks F_i, then we would use the inference based upon T_i. Or possibly Q might indicate some distribution between F_i and F_j, in which case we could use a combination of T_i and T_j for the inference. More generally, Q could dictate a statistic that is a combination (linear for this example) of $T_1, T_2, \cdots, T_k$, say

$$T = \sum_{i=1}^{k} W_i T_i \quad , \quad \sum_{i=1}^{k} W_i = 1,$$

where the weights $W_1, W_2, \cdots, W_k$ are functions of the statistic Q. If, using Q, it looks more like the sample arises from F_i, then the weight W_i would be large.

As a simple illustration of this, say the family of distributions has the family of densities $f(x) = c \exp(-|x|^{\tau})$, $-\infty < x < \infty$, $\tau > 0$. We wish to estimate θ in $f(x-\theta)$, $-\infty < \theta < \infty$. Select the $k = 3$ representative distributions as $\tau = 1$ (double exponential), $\tau = 2$ (normal), and $\tau = \infty$ (uniform). The respective maximum likelihood estimators are the median (T_1), the mean (T_2) and the midrange (T_3). If for the selector statistic Q we use the sample kurtosis

K, a possible T is

$$T = \begin{cases} T_1 = \text{median} , & 4 \le K \\ T_2 = \text{mean} , & 2 < K < 4 \\ T_3 = \text{midrange} , & K \le 2. \end{cases}$$

Hogg[7] proved that the asymptotic properties of T are excellent provided the sample arises from one of the representative distributions: the double exponential, the normal, or the uniform distribution. But what is the situation if τ is in between the representative values $\tau = 1$, $\tau = 2$, and $\tau = \infty$? Or, as a matter of fact, what happens if the underlying distribution is outside of the family?

Clearly, in this type of adaptation, the selector statistic plays a major role and must be chosen carefully. For the family of this previous example, Hogg[8] showed that the best scale invariant test of $H_0: \tau = \tau_1$ against $H_1: \tau = \tau_2$ is given by

$$\frac{(\frac{1}{n} \Sigma |x_i|^{\tau_2})^{1/\tau_2}}{(\frac{1}{n} \Sigma |x_i|^{\tau_1})^{1/\tau_1}} \le c.$$

Incidentally, if $\tau_1 = 2$ (normal) and $\tau_2 = 4$ (lighter tails than normal), we obtain a statistic that is something like the fourth root of the sample kurtosis K.

In testing $\tau = 1$ (double exponential) against $\tau = 2$ (normal), we have a ratio that suggests, when the middle is unknown, the inequality

$$\frac{s}{\frac{1}{n} \Sigma |x_i - \text{median}|} \le c$$

as a test. The sample standard deviation s can be approximated very well by a constant times the difference $\overline{U}_\alpha - \overline{L}_\alpha$ when $\overline{U}_\alpha$ $(\overline{L}_\alpha)$ is the average of the upper (lower) $n\alpha$ order statistics, using fractional items in case $n\alpha$ is not an integer. Usually α is about 0.2, 0.1, or 0.05. Thus the ratio of s to the mean deviation from the sample median suggests the ratio

$$Q_1(\alpha, \beta) = \frac{\overline{U}_\alpha - \overline{L}_\alpha}{\overline{U}_\beta - \overline{L}_\beta}$$

as a selector statistic, where usually $\beta = 0.5$ and $\alpha = 0.05, 0.1,$ or 0.2.

Moreover, the best scale and location invariant test for $\tau = \infty$ (uniform) against the asymmetric density e^{-x}, $0 < x < \infty$, is given by

$$\frac{\text{range}}{\frac{1}{n} \Sigma [x_i - \min(x_i)]} \le c,$$

which suggests

$$\frac{\bar{U}_\alpha - \bar{L}_\alpha}{\bar{M}_\gamma - \bar{L}_\alpha} = Q_2(\alpha,\gamma) + 1,$$

where $\bar{M}_\gamma$ is the average of the middle $n\gamma$ order statistics and

$$Q_2(\alpha,\gamma) = \frac{\bar{U}_\alpha - \bar{M}_\gamma}{\bar{M}_\gamma - \bar{L}_\alpha}.$$

Usual values are $\gamma = 0.5$ and $\alpha = 0.05, 0.1, 0.2$. Hogg, Fisher, and Randles[10] have used $Q_1(\alpha,\beta)$ and $Q_2(\alpha,\gamma)$ as selector statistics for tail length and skewness rather than the usual measures of those characteristics

De Wet and van Wyk[4,5] consider a modification of Hogg's proposal for the estimation of the center of a symmetric distribution using $Q_1 = Q_1(\alpha = .2, \beta = .5)$. Their trimmed mean $\bar{X}_p$ is selected by taking the trimming proportion p to be the following function of Q_1:

$$p = \begin{cases} .05 & , \quad Q_1 \le 1.75 \\ .05 + \left(\frac{.40-.05}{1.95-1.75}\right)(Q_1 - 1.75), & 1.75 < Q \le 1.95 \\ .40 & , \quad 1.95 < Q_1. \end{cases}$$

Under some conditions for a given distribution, they show that the random p is reasonably close to the correct nonrandom trimming proportion π, and $\bar{X}_p$ and $\bar{X}_\pi$ have the same asymptotic distribution. An abridgement of one of their tables using a Monte Carlo study $(n = 20)$ to compare this adaptive trimmed mean to two of the best robust estimators, H15 and 25A, of the Princeton study of Andrews at al.[1] is:

$$100(\sigma^2_{\text{Princeton estimator}}/\sigma^2_{\text{Adaptive estimator}})$$

Princeton Est. \ Dist.	N	5% 3N	10% 3N	T(3)	D-Ex	C
H15	96	96	97	104	116	204
25A	97	96	97	101	112	132

In fairly standard notation N, 5% 3N, and 10% 3N stand for the normal and two contaminated normal distributions while T(3), D-Ex, and C are respectively the student's t (3 d.f.), double exponential, and Cauchy distributions. The adaptive does better in the latter three distributions, with great improvement in the Cauchy case.

Prescott[15,16] has some results similar to those of De Wet and van Wyk, although he includes the possibility of using the average of the trimmings (an outmean) in his first adaptive estimator. Then he considers a compromise adaptive trimmed mean $\bar{X}_p$ in case the family of densities consists of the union of two families: $c \exp[-|x|^\tau]$, $1 \le \tau \le 2$, and student's t family. Here p is determined by, with $Q_1 = Q_1(.1,.5)$,

$$p = \begin{cases} 0 & , \quad Q_1 < 2.2 \\ Q_1 - 2.2, & 2.2 \le Q_1 < 2.6 \\ 0.4 & , \quad 2.6 \le Q_1 . \end{cases}$$

Note, with different α's in the Q_1 function, the similarity of this and the De Wet-van Wyk statistic.

It is difficult to think about this problem and not consider a Bayesian solution. Smith[21] and Spiegelhalter[22] have provided that. We obtain, with a square error loss function,

$$E(\theta|\underset{\sim}{x}) = \sum_{\tau} E(\theta|\underset{\sim}{x},\tau)h(\tau|\underset{\sim}{x}),$$

where $\underset{\sim}{x}$ is the data, τ is the parameter of the family of densities under consideration, $E(\theta|\underset{\sim}{x},\tau)$ is the individual Bayes estimator for that fixed τ, and $h(\tau|\underset{\sim}{x})$ is the posterior probability of τ. That is, the solution is a weighted average of the individual Bayes estimators. Using our exponential power family, $c \exp[-|x|^{\tau}]$, and the representative values $\tau = 1$, $\tau = 2$, and $\tau = \infty$, we obtain, with $\tilde{\theta}_i$ being the respective individual Bayes estimates,

$$\tilde{\theta} = \tilde{\theta}_1 \, h(\tau = 1|\underset{\sim}{x}) + \tilde{\theta}_2 \, h(\tau = 2|\underset{\sim}{x}) + \tilde{\theta}_\infty \, h(\tau = \infty|\underset{\sim}{x}).$$

If we are not concerned about light tailed distributions (that is, $\tau = \infty$ has prior, and hence post, probability of zero), we obtain

$$\tilde{\tilde{\theta}} = \tilde{\theta}_1 \, h(\tau = 1|\underset{\sim}{x}) + \tilde{\theta}_2 \, h(\tau = 2|\underset{\sim}{x}).$$

An abridgement of the Monte Carlo $(n = 20)$ comparing $\tilde{\theta}$ and $\tilde{\tilde{\theta}}$ (with equal priors) to another good robust estimator 21A of the Princeton study is:

$$100(\sigma^2_{\text{Reference estimator}}/\sigma^2_{\text{Estimator}})$$

Est. \ Dist.	N	10% 3N	D-Ex	C
$\tilde{\theta}$	90	86	85	60
$\tilde{\tilde{\theta}}$	95	92	95	61
21A	93	96	84	70.

Of course, with the light-tailed considerations in its development, $\tilde{\theta}$ is not expected to do well here (but would be excellent if light-tailed distributions were used in the Monte Carlo study). Of course, $\tilde{\tilde{\theta}}$ compares very favorably to 21A.

The results of De Wet, van Wyk, Prescott, Smith, and Spiegelhalter seem very encouraging to additional work in adaptation.

3. DISTRIBUTION-FREE METHODS

To illustrate some of the adaptive techniques in distribution-free methods, we consider the two sample problem in which we are interested in the test of equality of locations. That is, if $F(x)$ is a distribution function of the continuous type and if $G(x) = F(x-\theta)$ is that associated with an independent distribution, we wish to test the hypothesis $H_0: \theta = 0$. Moreover we will make the test with a linear rank statistic. Thus if R_i is the rank of Y_i in the combined sample consisting of the two independent samples $X_1, X_2, \cdots, X_m$ and $Y_1, Y_2, \cdots, Y_n$ from the respective distributions, then the statistic is of the form

$$\sum_{i=1}^{n} a(R_i),$$

where the score function $a(\cdot)$ enjoys the property $a(1) \le a(2) \le \cdots \le a(m+n)$. Of course, an asymptotically good score function is given by

$$a(i) = \varphi(\frac{i}{m+n}), \quad \text{where} \quad \varphi(u) = \frac{-f'[F^{-1}(u)]}{f[F^{-1}(u)]},$$

where $f = F'$. So again it seems desirable to estimate f'/f from the sample values, but this is an extremely difficult task with relatively small sample sizes, like 20 or 30.

To make the estimation of $\varphi(u)$ easier, $\varphi(u)$ can be restricted somewhat. For example, Jones[12] did this by requiring $F^{-1}(u)$ to be in the lambda family of Tukey, namely,

$$F^{-1}(u) = \frac{u^{\lambda} - (1-u)^{\lambda}}{\lambda}, \quad 0 < u < 1.$$

In some of our work, we have found that restricting $\varphi(u)$ to be a "stick" function, something like that in Figure 1, is very satisfactory. These functions include the φ's associated with the

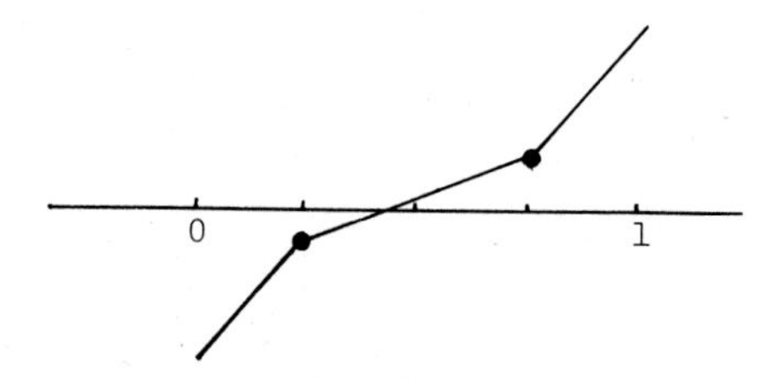

Figure 1

Wilcoxon and median tests. Moreover, due to the robust nature of the Wilcoxon statistic (it works well over a wide range of distributions), stick functions should be extremely robust.

If an appropriate stick function [and thus $a(\cdot)$] is selected by observing the order statistics of the combined sample (as if the null hypothesis was true), then this selection and the resulting test statistic are independent under H_0. The reason for this independence is that these order statistics are complete sufficient statistics for the "parameter" $F(x) = G(x)$, and the selected linear rank statistic has a distribution that is free of $F = G$. Hence the order statistics (and a selection based upon them) and the resulting distribution-free test are independent under H_0. Thus if a test has nominal significance level α, the overall test (selection and testing) will have significance level α.

Hogg, Fisher, and Randles[10] used one of the distribution-free tests associated with the stick functions depicted in Figure 2. The test statistics are denoted by W_L (for light tails), W (the Wilcoxon for heavier tails), and W_S (for distributions skewed to the right).

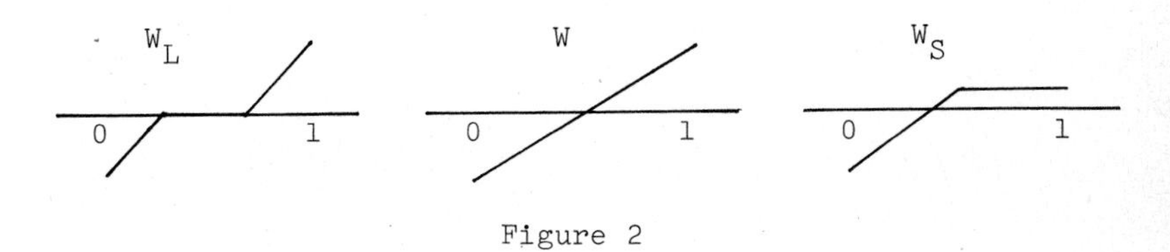

Figure 2

Our selection scheme is given by:

use W_L if $Q_1(.05,.5) \le 2$, $Q_2(.05,.5) \le 2$;
use W_S if $Q_1(.05,.5) \le 3$, $Q_2(.05,.5) > 2$;
use W otherwise.

In a Monte Carlo study with $m = n = 15$ and significance level $\alpha = .05$, we compared the powers of the following three tests: Student's T, the Wilcoxon W, and this adaptive test A, when $\theta = .6\sigma$, for various underlying distributions having given skewness α_3 and kurtosis α_4. An abridgement of the results is given in the following table.

Percentage Power at $\theta = .6\sigma$

Test \ α_3 / α_4	0 1.8	0 3	0 6	0 11.6	.7 2.1	1.5 5.8	1.7 8.0	2.0 21.0
T	48	49	51	54	48	52	52	53
W	44	46	55	60	48	64	60	61
A	53	46	54	60	52	73	63	61

While Student's T is not distribution-free, it is almost such and held $\alpha = .05$ quite well over this range of distributions. Of course, the fine quality of the Wilcoxon shows up in this study. However, the adaptive test A is clearly the winner except for distributions around the normal characteristics of $\alpha_3 = 0$, $\alpha_4 = 3$. So again the value of adaptive procedures is dramatized.

4. THE REGRESSION SITUATION

Of course, robust M- and R-estimators can be rather easily extended to the regression case. To adapt here, we could use the residuals from a preliminary estimate of the middle to compute Q_1 and Q_2 functions; these, in turn, could select an appropriate ψ function for the M-estimator or an appropriate φ function for the R-estimator.

But rather than consider these in any detail, let us briefly consider one rather easy scheme of extending the L-estimators to the regression case. Incidentally, Bickel[3] has provided a more complicated procedure.

Consider the usual model, $\underset{\sim}{Y} = \underset{\sim}{X}\underset{\sim}{\beta} + \underset{\sim}{Z}$, where the elements of $\underset{\sim}{Z}$ are independent and have identical symmetric distributions. If $\underset{\sim}{x}_i$ is a row of $\underset{\sim}{X}$, Koenker and Bassett[14] found an estimate of the pth $(0 < p < 1)$ regression quantile through

$$\min_{\underset{\sim}{\beta}}\left[\sum_{Y_i \geq \underset{\sim}{x}_i\underset{\sim}{\beta}} (p)(|Y_i - \underset{\sim}{x}_i\underset{\sim}{\beta}|) + \sum_{Y_i < \underset{\sim}{x}_i\underset{\sim}{\beta}} (1-p)(|Y_i - \underset{\sim}{x}_i\underset{\sim}{\beta}|)\right].$$

This, in effect, modifies least absolute values by replacing the absolute value function by a "check" function with a slope of $-(1-p)$ when the argument is less than zero and a slope of p when the argument is greater than zero. That is, if this check function is denoted by $\rho(\cdot)$, the Koenker-Bassett procedure is to find

$$\min_{\underset{\sim}{\beta}} \sum_{i=1}^{n} \rho(Y_i - \underset{\sim}{x}_i\underset{\sim}{\beta}).$$

These regression quantiles have asymptotic behaviors similar to the sample quantiles. This suggests that we could use L-estimators like the median, Tukey's trimean, or Gastwirth's estimator. But we could also construct regression L-estimators like the trimmed mean.

Koenker and Bassett suggest that, with $0 < p < 1/2$, we should remove items outside the pth and (1-p)th regression quantiles and then compute the least squares regression surface with the remaining interior points. Ruppert and Carroll[19] proved their conjecture that the resulting estimator, say $\hat{\beta}_{KB}$, is asymptotically normal with mean $\underset{\sim}{\beta}$ and covariance matrix

$$\sigma_p^2\left[\lim_{n\to\infty}\left(\frac{1}{n}\underset{\sim}{X}'\underset{\sim}{X}\right)\right]^{-1},$$

when σ_p^2 is the asymptotic variance of the trimmed mean $\bar{X}_p$. This, of course, permits the construction of confidence ellipsoids for the unknown $\underset{\sim}{\beta}$.

Ruppert and Carroll[19] also made another proposal that is similar to the one of Koenker and Bassett, but possibly closer to the approach of many applied statisticians. Find a preliminary estimate of $\underset{\sim}{X}\underset{\sim}{\beta}$, say $\underset{\sim}{X}\underset{\sim}{\beta}_0$, and remove the largest np and the smallest np residuals. Then determine the least squares solution using the remaining points. They found that their estimator, say $\hat{\underset{\sim}{\beta}}_{RC}$, depends a great deal on the preliminary estimate $\underset{\sim}{\beta}_0$; and, in particular, least squares and least absolute value "starts" were not good in general. If the start is the average of the pth and (1-p)th regression quantile, then $\hat{\beta}_{RC}$ has a distribution similar to that of $\hat{\beta}_{KR}$.

One way in which we could create adaptive estimators like those trimmed means of de Wet and van Wyk, would be as follows. Begin with a reasonable amount of trimming, say $p_0 = 0.1$, and use the average 0.1th and 0.9th regression quantiles as a preliminary estimate. Determine the residuals from this estimate and determine Q_1 and Q_2 with these residuals. These, in turn, would suggest a different amount of trimming, say p_1. Using this value, determine either $\hat{\beta}_{KB}$ or $\hat{\beta}_{RC}$. As a matter of fact, it would be possible to iterate any number of times before computing $\hat{\beta}_{KB}$ or $\hat{\beta}_{RC}$.

More generally, it would be possible to construct adaptive estimates corresponding to any adaptive L-estimate since we can find the estimate of any pth regression quantile. Moreover, the advantage of getting estimates of several regression quantiles (say $p = 0.1, 0.25, 0.5, 0.75,$ and 0.9) is the possibility of checking the model: are the β's equal for each p or are the variances the same at different x_i?

One final remark that might be useful in determining adaptive regression estimates. Seely and Hogg[20] assumed that

$$Y - X\beta \overset{d}{=} X\beta - Y,$$

where $\overset{d}{=}$ means "distributed the same as"; that is, this requirement involves symmetrical distributions. Consider an estimator T of β that enjoys the properties

$$T(Y + Xh) = T(Y) + h, \quad T(Y) = -T(-Y),$$

for all h. Moreover, say $W(Y)$ is a random weight such that

$$W(Y + Xh) = W(-Y) = W(-Y).$$

Then

$$[T(Y) - \beta, W(Y)] \overset{d}{=} [\beta - T(Y), W(Y)]$$

and thus

$$E[T(Y)|W(Y)] = \beta, \quad E[T(Y)] = \beta,$$

and

$$E\{[X' W(Y)X]^{-1}X' W(Y)Y\} = \beta.$$

Incidentally, M- and R-estimators enjoy these properties.

These results mean that we can select weights $W(Y)$ after considering functions of the preliminary residuals. Moreover, the estimator could be the weighted average of several such T's, say

$$T(Y) = \sum_{i=1}^{k} w_i(Y) T_i(Y),$$

and these should have the same advantages as those of adaptive location estimators.

5. CONCLUSIONS

We see that some of the work of de Wet, van Wyk, Prescott, Smith, and Spiegelhalter is quite encouraging for the use of adaptive estimators. Also adaptive nonparametric methods have certain advantages. Then consideration is given to one general method of extending L-estimation to the regression case, with suggestions about adaptation. Clearly there is much more to be done here and other questions naturally arise.

1. Do adaptive "Hubers, Hampels, etc." have a property similar to that of adaptive trimmed means? Yohai[26] has considered this problem.
2. Is there a selector, say Q_3, that classifies well among different "shapes"? This might permit us to select one family of distributions over another; for example, choose between an exponential power family and Student's t family.
3. Should more robust work be done on the scale problem in life testing, particularly with a censored sample?
4. Are there better measures of skewness, kurtosis, and shape than Q_1, Q_2, and a certain proposed Q_3? For example, consider the U statistic to measure asymmetry as did Randles, Fligner, Policello, and Wolfe[17].
5. How effective are stick functions? That is, how close can we come to the best linear rank test using a stick function?

Each serious reader of this paper can probably think of five more questions to ask. So clearly there is plenty to be done in this exciting and practical area of adaptive estimation. Join in the fun!

6. REFERENCES

[1] Andrews, D.F. et al., Robust estimates of location (Princeton University Press, Princeton, N.J., 1972).

[2] Beran, Rudolf, Asymptotically efficient adaptive rank estimates in location models, Ann. of Statist. 2 (1974) 63-74.

[3] Bickel, Peter J., On some analogues to linear combinations of order statistics in the linear model, Ann. of Statist. 1 (1973) 597-616.

[4] De Wet, T. and van Wyk, J.W.J., Some large sample properties of Hogg's adaptive trimmed means, South African Statistics Jrnl. 13 (1979) 53-69.

[5] De Wet, T. and van Wyk, J.W.J., Efficiency and robustness of Hogg's adaptive trimmed means, Comm. Statist. A8 (1979) 117-128.

[6] Hájek, J., Asymptotically most powerful rank-order tests, Ann. Math. Statist. 33 (1962) 1124-1147.

[7] Hogg, Robert V., Some observations on robust estimation, Jrnl. Amer. Statist. Assoc. 62 (1967) 1179-1186.

[8] Hogg, Robert V., More light on the kurtosis and related statistics, Jrnl. Amer. Statist. Assoc. 67 (1972) 422-424.

[9] Hogg, Robert V., Adaptive robust procedures: A partial review and some suggestions for future applications and theory, Jrnl. Amer. Statist. Assoc. 69 (1974) 909-927.

[10] Hogg, Robert V., Fisher, Doris M., and Randles, Ronald H., A two-sample adaptive distribution-free test, Jrnl. Amer. Statist. Assoc. 70 (1975) 656-661.

[11] Jaeckel, Louis, Robust estimates of location: Symmetric and asymmetric contamination, Ann. Math. Statist. 42 (1971) 1020-1034.

[12] Johns, M.V., Nonparametric estimation of location, Jrnl. Amer. Statist. Assoc. 69 (1974) 453-460.

[13] Jones, Douglas H., An efficient adaptive distribution-free test of location, Jrnl. Amer. Statist. Assoc. 74 (1979) 822-828.

[14] Koenker, Roger and Bassett, Gilbert, Regression quantiles, Econometrica 46 (1978) 33-50.

[15] Prescott, P., Selection of trimming proportions for robust adaptive trimmed means, Jrnl. Amer. Statist. Assoc. 73 (1978) 133-140.

[16] Prescott, P., The robustness of adaptive trimmed means, unpublished manuscript, presented at Oslo meeting, 1978.

[17] Randles, Ronald H., Fligner, Michael A., Policello, George E., and Wolfe, Douglas A., An asymptotically distribution-free test for symmetry versus asymmetry, to appear in March issue, Jrnl. Amer. Statist. Assoc. 75 (1980).

[18] Ruppert, David and Carroll, Raymond J., Robust regression by trimmed least-squares estimation, Manuscript #1186, Dept. of Statistics, Univ. of North Carolina (1978).

[19] Ruppert, David and Carroll, Raymond J., Trimming the least squares estimator in the linear model by using a preliminary estimator, Manuscript #1220, Dept. of Statistics, Univ. of North Carolina (1979).

[20] Seely, Justus and Hogg, Robert V., Unbiased estimation in linear models, Technical Report #57, Dept. of Statistics, University of Iowa (1977).

[21] Smith, A., Personal communication (1979).

[22] Spiegelhalter, D.J., Sampling properties of a fixed mixture model, unpublished manuscript, Univ. of Nottingham (1979).

[23] Stone, Charles J., Adaptive maximum likelihood estimators of a location parameter, Ann. of Statist. 3 (1975) 267-284.

[24] Takeuchi, Kei, A uniformly asymptotically efficient estimator of a location parameter, Jrnl. Amer. Statist. Assoc. 66 (1971) 292-301.

[25] Van Eeden, Constance, Efficiency-robust estimation of location, Ann. Math. Statist. 41 (1970) 172-181.

[26] Yohai, Victor J., Robust estimation in the linear model, Ann. of Statist. 3 (1974) 562-567.

STATISTICS AND RELATED TOPICS
M. Csörgö, D.A. Dawson, J.N.K. Rao, A.K.Md.E. Saleh (eds.)

AN APPLICATION OF RANK INVARIANT MULTIPLE REGRESSION AND VARIABLE SELECTION

D. L. McLeish

Department of Mathematics
University of Alberta
Edmonton, Alberta
Canada

A permutation technique using Kendall's Tau is developed for testing for the addition or deletion of variables in stepwise regression. This is related to the rank-invariant of Theil. These techniques are applied to variable selection for hail seeding data.

A significant seeding effect, apparent by parametric techniques, vanishes when these non-parametric methods are applied.

Seeding in order to suppress hail has been carried out in many jurisdictions including the Alberta Hail Project carried out by the Research Council of the Province of Alberta. One of the major problems in evaluating the success of a program like this is the non-normality in the usual response variables (which measure aspects of the potential for crop damage in a hail storm). This may result in the failure of parametric techniques such as those employed in stepwise regression. It is the principle purpose of this paper to investigate non-parametric analogues to some of these techniques and discuss their impact on the problem of evaluating hail seeding data.

Rank invariant procedures for multiple regression are discussed by several authors, including Jaeckel (1972), Jurečková (1971), Sen (1969), and Koul (1969). The methods we derive here are not obtained from a linear rank statistic as these are, but are based on a vector of Kendall's Tau; a measure of the association between the response variable and each of the predictor variables. When specialized to the case of simple linear regression, this method reduces to estimates like the weighted median regression estimates of Jaeckel (1972).

Unlike most of the other methods for multiple regression, our estimates and tests of regression affect are functions of the ranks of the predictor and response variable and not of the magnitude of either. For this reason, this methodology may be more appropriate than that of Jaeckel, for example, when there is the possibility of errors in the independent variables, or when we want a test that is also sensitive to a non-linear monotonic relationship between predictors and predictands. This is fairly important for the data at hand, since some of the predictors are based on the subjective judgement of observers, and others are values output from a cloud physical model that attempts to estimate hail size.

Consider a model of the form

$$\underset{\sim}{Y} = \beta_0 \underset{\sim}{1} + X\underset{\sim}{\beta} + \underset{\sim}{\varepsilon}$$

where $\underset{\sim}{Y}$, $\underset{\sim}{\varepsilon}$ are n-dimensional random column vectors, $\underset{\sim}{1}$ is an n-dimensional vector of 1's, X is an n by p "design matrix" of the regression, and $\underset{\sim}{\beta}$ a p-dimensional vector of regression coefficients. We discuss initially a test for regression affect for this model (i.e. a test of the hypothesis $\underset{\sim}{\beta} = 0$) and the estimator of $\underset{\sim}{\beta}$ obtained by inverting this test.

Let $\underset{\sim}{R}(B)$ be the vector of residuals,

$$\underset{\sim}{R}(B) = \underset{\sim}{Y} - XB$$

and

$$\underset{\sim}{C}(B) = \sum_i \sum_j \underset{\sim}{W}_{ij} g(R_j - R_i)$$

where $\underset{\sim}{W}_{ij}$ is a p-dimensional vector of weights and $g(\cdot)$ is an odd function.

We will usually assume

$$W_{ij}^{(k)} \geq 0 \text{ with equality if } X_{ik} \geq X_{jk} . \tag{1.1}$$

although an equivalent statistic would be obtained setting

$$W_{ij}^{(k)} \begin{array}{ll} > 0 & \text{if } X_{ik} < X_{jk} \\ = -W_{ji}^{(k)} & \text{if } X_{ij} > X_{jk} \\ = 0 & \text{if } X_{ik} = X_{jk} \end{array} .$$

We assume that the "errors", the components of $\underset{\sim}{\varepsilon}$, are absolutely continuous, independent, identically distributed (for some results, it is sufficient to assume the weaker condition that each of the 4! rankings of any four distinct components of $\underset{\sim}{\varepsilon}$ are equally likely).

Although we could use an arbitrary odd function in some of the following we will restrict ourselves to the sign function:

$$g(x) = \begin{cases} 1 & \text{if } x > 0 \\ 0 & \text{if } x = 0 \\ -1 & \text{if } x < 0 \end{cases} \tag{1.2}$$

We now consider the behaviour of $\underset{\sim}{C}(B)$ as $n \to \infty$. This means that we will have to add rows to $\underset{\sim}{1}$, X, $\underset{\sim}{Y}$, and $\underset{\sim}{\varepsilon}$, but otherwise the model (and the true regression coefficients $\underset{\sim}{\beta}$) remain unchanged.

Under reasonably weak regularity conditions governing the growth of X, we have:

Result 1: $\underset{\sim}{C}(B)$ is asymptotically distributed as a multivariate normal with mean vector $\underset{\sim}{0}$ and covariance matrix S given by:

$$3S_{k\ell} = \sum_i \sum_j W_{ij}^{(k)} \left[W_{ij}^{(\ell)} + 2W_{ji}^{(\ell)}\right] + 2\sum_i (M_{ik} - m_{ik})(M_{i\ell} - m_{i\ell})$$

where $M_{ik} = \sum_j W_{ij}^{(k)}$ and $m_{ik} = \sum_j W_{ji}^{(k)}$.

This asymptotic normality follows from the Cramer-Wold device and a one dimensional version of the same result such as that contained in the proof of Theorem 1 or Scholz (1978). The form of the covariance matrix is obtained in the appendix.

In order to test a hypothesis of the form H_0; $\underset{\sim}{\beta} = \underset{\sim}{\beta}_0$, we may evaluate $C(\underset{\sim}{\beta}_0)$.

Under the null hypothesis, this random vector should be close to $\underset{\sim}{0}$ (since its expectation is $\underset{\sim}{0}$) whereas if the hypothesis fails, we expect components some distance from 0. It is therefore natural to test this hypothesis using the chi-squared statistic:

$$\chi^2(\underset{\sim}{\beta}_0) = \underset{\sim}{C}'(\underset{\sim}{\beta}_0) S^{-1} \underset{\sim}{C}(\underset{\sim}{\beta}_0) . \tag{1.3}$$

Asymptotically (as $n \to \infty$), this is chi-squared with p degrees of freedom (where we must assume that S is asymptotic to a non-random multiple of some non-singular matrix) and the null hypothesis is rejected for large values of this statistic.

We now consider inverting this test to obtain confidence regions and point estimates for β. The $1 - \alpha$ (large sample) confidence region is the set of satisfying

$$\chi^2(B) \leq \chi^2_{p,1-\alpha} \ . \tag{1.4}$$

An estimator $\hat{B}$ minimizing the significance of the above test is one satisfying

$$\chi^2(\hat{B}) = \min_B \chi^2(B) \ . \tag{1.5}$$

In many cases, the minimum of (1.5) is 0 or very close to 0. For example in simple linear regression when the W_{ij} are either 0 or 1, the "median" estimator of Theil (1950) results in $C(\hat{B}) = 0$. In the appendix we show, under some natural regularity conditions:

<u>Result 2:</u> When $p = 2$, $\chi^2(\hat{B}) \leq \frac{2}{\lambda_2}$ where λ_2 is the smaller eigenvalue of the matrix S.

It should be noted in connection with this result that the matrix S normally has elements that grow very fast with sample size. For example, if there are no ties in the columns of X, the diagonal elements of S in the case

$$W_{ij}^{(k)} = \begin{cases} 1 & \text{if } X_{jk} > X_{ik} \\ 0 & \text{otherwise} \end{cases} \tag{1.6}$$

are asymptotic to a constant times n^3. Therefore, the regularity conditions imposed by the convergence to normality require that S is asymptotic to $n^3 \mathcal{S}$ where $\mathcal{S}$ is a nonsingular matrix. This, in turn, implies

$$\chi^2(\hat{B}) \leq \frac{2}{n^3 \lambda_2} \tag{1.7}$$

where λ_1 is the smaller eigenvalue of $\mathcal{S}$.

It seems to hold in general that the minimum value of (1.5) can be made very small (in fact it can often be made equal to 0). However, when p is greater than 2, a reasonable upper bound such as that in (1.7) seems rather more complicated to obtain. Such an upper bound can be useful in algorithms designed to minimize $\underset{\sim}{C}(\underset{\sim}{B})$. For example (1.7) implies

$$\|C(\hat{B})\|^2 \leq \frac{2\lambda_1}{n^3 \lambda_2}$$

where λ_1 is the larger eigenvalue of $\mathcal{S}$ and $\|\cdot\|$ denotes Euclidean norm.

1. EXAMPLES

<u>EXAMPLE 1:</u> <u>Simple Linear Regression.</u>

We consider $g(\cdot)$ defined by (1.2) and the case $p = 1$. Then the estimator of the regression slope is the value of B minimizing $C(B)$. Let ξ be a random variable taking all values of the form:

$$\frac{Y_i - Y_j}{x_i - x_i} \quad \text{with probability} \quad \frac{W_{ij}}{\sum_i \sum_j W_{ij}} \ .$$

Then we wish to find B minimizing

$$|P[\xi < B] - P[\xi > B]|.$$

Such a B is often the median of the distribution of ξ and by the following lemma (proved in the appendix) is always either the median or a point adjacent to it.

Lemma 1. Let ξ be a discrete random variable taking finitely many values. Then

$$\min_B |P[\xi < B] - P[\xi > B]|$$

occurs for B the median or a point immediately adjacent to the median.

Weighted median estimators for the slope in simple linear regression were proposed by Jaeckel (1972) and discussed further by Scholz (1978) and Sievers (1978). In these papers, some of the ramifications of the choice of weights W_{ij} on asymptotic efficiency are discussed, indicating that for non-equally spaced data, the choice:

$$W_{ij}^{(k)} = (X_{jk} - X_{ik})^+$$

provides efficient estimators.

Note also that for the unweighted median estimator of Theil (1950) W_{ij} is defined as in (1.6). In this case, $\chi^2(\hat{B})$ is 0 for $\hat{B}$ = Theil's estimator.

EXAMPLE 2.

Let $r(X_{jk})$ denote the rank (in ascending order) of X_{jk} among $X_{1k}, X_{2k}, \ldots, X_{nk}$. Assume for simplicity that there are no ties within columns of X. Define:

$$W_{ij}^{(k)} = [r(X_{jk}) - r(X_{ik})]^+ \tag{2.1}$$

where

$$[x]^+ = \begin{cases} x & \text{if } x > 0 \\ 0 & \text{if } x \le 0 \end{cases}.$$

Then $C_k(B) = \frac{1}{6}n(n^2 - 1)\rho_{R,X_{\cdot k}}$ where $\rho_{R,X_{\cdot k}}$ is Spearman's rank correlation between the residuals and the elements of the kth column of X.

Thus regression estimates based on this statistic minimize a quadratic form involving the vector of Spearman's correlation between the dependent variable and each of the independent variables. According to problem 2, page 125 of Hajek (1960), tests performed using Spearman's rho will be asymptotically equivalent to the tests involving Kendall's tau outlined in example 3.

EXAMPLE 3.

Define W_{ij} as in (1.6). Then $C_k(B)$ is the number of "concordant" pairs between the residuals and the kth column of X minus the number of discordant pairs (the terminology here is that of Kendall (1948)). This, then is a constant times τ_{R,X_k}. Estimation involves minimizing a quadratic function of the p correlations of this form. We now discuss a specific numerical example to illustrate these methods in the case $p = 2$.

Let $\underset{\sim}{Y}' = (3, 0, 3, 2)$, $X = \begin{pmatrix} 3 & 1 \\ 3 & 2 \\ 2 & 0 \\ 1 & 2 \end{pmatrix}$, W is defined by (1.6) and

$S = \begin{pmatrix} 7^2/_3 & 1^2/_3 \\ 1^2/_3 & 7^2/_3 \end{pmatrix}$. Figure 1 shows the lines (in the (B_1, B_2) plane) of the form ℓ_{ij}: $R_i(\underset{\sim}{B}) = R_j(\underset{\sim}{B})$. Each of the 18 regions is labelled with $\underset{\sim}{C}(\underset{\sim}{B})$ for $\underset{\sim}{B}$ in that region, and immediately below, $\chi^2(B)$. The determination of the coefficients $\underset{\sim}{B}$ such that $C_1(\underset{\sim}{B}) = 0$, $C_2(\underset{\sim}{B}) = 0$ in this case is fairly easy, there being a whole line segment for which this holds. This is the segment

$$\{B;\ B_1 + 2B_2 = -3 \text{ and } -1 < B_1 < -1/5\} .$$

A natural estimator of the coefficient β is the midpoint of this segment $(-^3/_5, -^6/_5)$. This can be compared with the least squares estimate $(-^8/_{15}, -^{17}/_{15})$. In this case, of the 4! ranks of the residuals that are *a priori* equiprobable, exactly 12 lead to values of $\chi^2(\beta)$ less than 1.54. Therefore a 50% confidence region will take the form: $\{\underset{\sim}{B};\ \chi^2(B) \leq 1.54\}$. In figure 1, the 6 (out of 18) regions for which this occurs are outlined. Note that use of the limiting distribution, a chi-squared with 2 degrees of freedom, would lead to a 50% confidence region of the form $\{B;\ \chi^2(B) \leq 1.39\}$. One property of $\chi^2(B)$ useful in labelling this figure is that it depends only on the ranks of the residuals, which are subject to a transposition whenever we cross one of the lines ℓ_{ij}. Under such a transposition, each of the components of $\underset{\sim}{C}(\underset{\sim}{B})$ may change by at most 2.

Moreover, suppose one travels from B^* to B by crossing each of the lines of the form ℓ_{ij} exactly once. Then the ranks of the residuals $R(B)$ are the "antiranks" of the residuals $R(B^*)$, and $C(B) = -C(B^*)$. The values of chi-squared, then are the same: $\chi^2(B^*) = \chi^2(B)$.

2. NUMERICAL PROCEDURES

In the proof of result 2, contained in the appendix, we show that the number of distinct regions in the B plane in the case $p = 2$ is

$1 + \frac{n(n-1)(3n^2-7n+14)}{24} \sim \frac{n^4}{8}$ with probability 1. An exhaustive search of all of these regions for the one with minimum $\chi^2(B)$ is possible only for very small n. For large n, we will require an alternate procedure which, with less computation, will at least approximate the minimum of (1.5).

Various algorithms for this minimization were considered. For example, one might select k such that $|C_k(B)| = \max_{1 \leq i \leq p} |C_i(B)|$ and then use medians to adjust B_k holding B_j, $j \neq k$ fixed, until $|C_k| \leq 1$. A procedure based on this observation was programmed and often converged, although only rather slowly, since we can only adjust one component of $\underset{\sim}{B}$ at a time.

An alternative that seems to converge more often and more rapidly is to observe distinct $\underset{\sim}{B}_1, \underset{\sim}{B}_2, \ldots, \underset{\sim}{B}_m$ and the concomitant values of $\underset{\sim}{C}(\underset{\sim}{B}_1), \underset{\sim}{C}(\underset{\sim}{B}_2), \ldots, \underset{\sim}{C}(\underset{\sim}{B}_m)$, and then to use linear regression of the form

$$E\underset{\sim}{C}(B) = A(\underset{\sim}{B} - \underset{\sim}{B}_0)$$

where A is a p by p non-singular matrix and $\underset{\sim}{B}_0$, a p-dimensional vector. $\underset{\sim}{B}_0$ is estimated using weighted regression (the weights we used for the point $(\underset{\sim}{B}_i, \underset{\sim}{C}(\underset{\sim}{B}_i))$ were inversely proportional to $\chi^2(B_i)$), and was taken to be the next estimator of the desired value of B. More precisely, the algorithm chosen was as follows:

Numerical Algorithm:

1. Pick $m \geq p$ arbitrary values for $\underset{\sim}{B}$, $\underset{\sim}{B}_1, \underset{\sim}{B}_2, \ldots, \underset{\sim}{B}_m$ say.
2. Compute the corresponding values $\underset{\sim}{C}_1 = C(\underset{\sim}{B}_1)$, $\underset{\sim}{C}_2 = C(\underset{\sim}{B}_2)$, ..., etc. and $\omega_1 = \frac{1}{\chi^2(\underset{\sim}{B}_1)}$, $\omega_2 = \frac{1}{\chi^2(\underset{\sim}{B}_2)}$, ...
3. Define the vector $\Omega' = \left(\frac{\omega_1}{\Sigma\omega_i}, \frac{\omega_2}{\Sigma\omega_i}, \ldots, \frac{\omega_m}{\Sigma\omega_i}\right)$ and matrices

$$\underset{m \times m}{\mathcal{D}} = \text{Diag}(\Omega)$$

$$\underset{p \times m}{\mathcal{B}'} = \left(\underset{\sim}{B}_1, \underset{\sim}{B}_2, \ldots, \underset{\sim}{B}_m\right)$$

$$\underset{p \times m}{\mathcal{C}'} = \left(\underset{\sim}{C}_1, \underset{\sim}{C}_2, \ldots, \underset{\sim}{C}_m\right)$$

4. Define $\underset{\sim}{B}_{m+1} = \mathcal{B}'\Omega = [\mathcal{B}'\mathcal{D}\mathcal{B} - \mathcal{B}'\Omega\Omega'\mathcal{B}][\mathcal{C}'\mathcal{D}\mathcal{B} - \mathcal{C}'\Omega\Omega'\mathcal{B}]^{-1}\mathcal{C}'\Omega$ and compute the corresponding $\underset{\sim}{C}_{m+1}$ and ω_{m+1}.
5. If $\chi^2(\underset{\sim}{B}_{m+1})$ is sufficiently small (i.e. ω_{m+1} sufficiently large) or $m \geq$ some predetermined number of iterations go to 6. Otherwise set $m = m + 1$ and return to 3.
6. Let the estimator be $\underset{\sim}{B}_j$ corresponding to the minimum value of $\chi^2(\underset{\sim}{B}_j)$, $j = 1, 2, \ldots, m$.

This regression method can be viewed from several different perspectives. One applies when there are no ties within columns of X. Suppose the rows of both Y and X have been subjected to the same (random) permutation so that the elements of $\underset{\sim}{R}(B)$ are now arranged in increasing order. Under this new ordering, let r_{ik} be the rank of X_{ik} among $X_{1k}, X_{2k}, \ldots, X_{ik}$. Then the statistic $\underset{\sim}{C}(B)$ written in terms of these ranks is:

$$C_k(B) = 2\sum_{j=2}^{n} r_{jk} - \frac{n(n+3)}{2} .$$

We can also regard the estimation of B from a geometric point of view. Consider Consider $n(n-1)$ points of the form:

$$(Y_i - Y_j, X_{i1} - X_{j1}, X_{i2} - X_{j2}, \ldots, X_{ip} - X_{jp})$$

in $p + 1$ dimensional space. Assign to each such point the corresponding p-dimensional vector of weights $\underset{\sim}{W}_{ij}$. Our problem is to find a plane through the origin (i.e. find B) of the form:

$$\{\xi;\ \xi'\begin{pmatrix}1\\-B\end{pmatrix} = 0\}$$

such that the sum of the weights for points above the plane in $p + 1$ dimensional space (above = larger value for the first component) is close to the sum of the weights of the points below (actually we minimize a quadratic form in the vector of differences).

3. TESTS OF PARTIAL CORRELATION

We now consider a regression model of the form:

$$\underset{\sim}{Y} = \beta_0 + X\underset{\sim}{\beta} + \underset{\sim}{Z}\gamma + \underset{\sim}{\varepsilon}$$

where $\underset{\sim}{Z}$ is a column vector, γ is a scalar. We wish to test a hypothesis of the form:

$$H_0 : \gamma = 0$$
$$H_1 : \gamma \neq 0 .$$

A permutation test such as the following might be employed for this test:

1. Regress Y on $\underset{\sim}{X}$ by the methods of section 2, obtaining an estimate $\hat{B}$ of β.
2. Set $\underset{\sim}{R}(\hat{B}) = \underset{\sim}{Y} - \underset{\sim}{X}\hat{B}$ and compute Kendall's tau for the correlation between $\underset{\sim}{R}(\hat{B})$ and $\underset{\sim}{Z}$.
3. Permute the rows of $\underset{\sim}{Z}$ randomly and repeat 2.
4. Compare the values of Tau for the original data with that for a number of replications of the permuted model (of step 3) to obtain an approximate significance probability for the test.

The large sample approximation to this test is just Theil's (1950) test for trend in the regression of $R(\hat{B})$ on Z. This statistic is standardized, and the normal approximation is used. In our case, the statistic appears in squared form and is therefore compared with a chi-squared variate on one degree of freedom. This is the process we use to test for the addition or the deletion of a variable. The stepwise algorithm used is as follows (all tests are at the 5% level):

1. Compute the standardized Theil statistic for testing for trend in Y against each of the (33) predictors. Compute the squares of each of these and include the variable corresponding to the largest square in the regression model if the square exceeds 3.84. If no value exceeds 3.84, stop.
2. Regress Y on the variables currently selected and obtain the residuals. Then repeat 1 with Y replaced by the vector of residuals, then go on to 3.
3. Suppose we have currently included $k \geq 2$ of the p variables in our regression model. Select a subset of $k - 1$ of these including the variable just added in step 2. Regress Y on this subset and compute the residuals. Then compute the squared standardized Theil statistic for association between these residuals and the remaining predictors.
4. Repeat 3 for all $k - 1$ such subsets. Compute the smallest of the $k - 1$ resulting chi-squared statistics and delete the variable that this model excludes if its value is less than 3.84. Return to step 2.

This algorithm is guaranteed to eventually terminate. However, like stepwise regression, it does not guarantee that the resulting set of variables is the "best" set of that size.

4. APPLICATION TO THE HAIL DATA

The data studied by these methods was provided by the cloud modeling group of the Alberta Research Council (cf [1]). The data obtains from a program of silver iodide cloud seeding design for hail suppression carried out in 1975 and 1976. Various predictors are listed in Table 1. Some of these (numbered 2 to 19) are measurements on meteorological variables or output from the LMA model which processes such variables to obtain a predictor of maximum hailsize. These predictors would be considered covariates in a study designed to evaluate the seeding program. The variables 1, and those numbered 20-33 are seeding predictors. They indicate whether seeding was carried out, the amount, and the rate, etc. of seeding.

There are two functions of a stepwise regression that might be considered useful here. The first is to determine which of the meteorological variables have the greatest effect on the hailfall (and the resulting crop damage). The other is to determine whether any of the seeding predictors are significant in the multiple regression, i.e. to evaluate the success of the seeding program.

The response variable we selected was AMHS - average maximum of the reported maximum hailsizes (in cm). This is the average over all reports for a storm of the two largest hailstone size categories. This response variable was such that there were very few missing observations (there are approximately 100 observations from each year) and its distribution seemed reasonably close to continuous. It also showed some indications of non-normality. The predictors are specified in Table 1. Table 2 shows the results of both the parametric and non-parametric variable selections. The most obvious difference in the results is the inclusion of the seeding predictors 28 and 32 using ordinary stepwise regression for the 1975 and 21 for the 1976 data, but the failure for any seeding predictors to appear using the non-parametric techniques. This significance may be due to the non-normality of the errors. To check this for the 1975 data, the non-parametric techniques above were used to regress Y on the predictors 11, 14, and 32 using the rank invariant techniques discussed above and the residuals studied. Indeed of the 96 residuals, 5 values were more than 2.5 standard deviations from the mean and one was over 5 standard deviations away. This seems to indicate that the apparent significance by parametric techniques is due to the presence of outliers. In order to facilitate comparison between the variable sets selected, Table 3 contains the (Tau) correlations between all pairs of the non-seeding predictors.

Table 1. Predictors

Symbol	Description	Units
1.	1 = seeded, 0 = not seeded	
2. PCB	Pressure at cloud base	mb
3. HCB	Height at cloud base	km MSL
4. TCB	Temperature at cloud base	C
5. HCTM	Height of cloud top, i.e. the height where the calculated updraft speed has decreased to 0	km MSL
6. HETM	Height of the radar echo top, i.e. the height where the calculated updraft speed has decreased to 10 ms^{-1}	km MSL
7. MAXU	Maximum updraft speed	ms^{-1}
8. TMAXU	Temperature at the level where the updraft speed is a maximum	C
9. MAXLWC	Maximum liquid water content	gm^{-3}
10. MAXPE	Maximum potential energy	jg^{-1}
11. MAXLWCPE	Maximum liquid water content multiplied by the maximum potential energy	j(g of water) (g of air^{-1}m^{-3}
12. THETAE	Equivalent potential temperature	K
13. CLDPT	Cloud depth	km
14. HFZL	Height of freezing level in cloud	km MSL
15. PMXSZ	Predicted maximum hailstone size	cm
16. WSCL	Horizontal wind speed in sub-cloud layer	ms^{-1}
17. DSCL	Directional wind shear in sub-cloud layer	deg m^{-1}
18. WSSCL	Horizontal wind speed shear in cloud layer	s^{-1}
19. MWCL	Horizontal wind in mid-cloud layer	ms^{-1}
20. CTF	Number of cloud-top flares	
21. TCTF	Time of cloud-top seeding	s
22. RCTF	Rate of cloud-top seeding	flares s^{-1}
23. GCTF	Total grams of AG1 dispensed by cloud-top seeding	g
24. RGCTF	Rate of Ag1 dispensed by cloud-top seeding	gs^{-1}
25. CBF	Number of cloud-base flares	
26. TCBF	Time of cloud-base seeding	s
27. RCBF	Rate of cloud-base seeding	flares s^{-1}
28. GCBF	Total grams of Ag1 dispensed by cloud-base seeding	g
29. RGCBF	Rate of Ag1 dispensed by cloud-base seeding	gs^{-1}
30. TGFL	Total amount of Ag1 dispensed by both cloud-top and cloud-base seeding	g
31. RGTFL	Rate of Ag1 dispensed for both cloud-top and cloud-base seeding, i.e. RGCTF + RGCBF	gs^{-1}
32. TIME	Total time of seeding run	s
33. R2FL	Rate of Ag1 dispensed for total seeding run, i.e. TGFL/TIME	gs^{-1}

Table 2

Selected Predictors (Significance of partial correlation)

YEAR	NON-PARAMETRIC	BY STEP-WISE REGRESSION
1975	14. HFZL	11. MAXLWCPE
	15. PMXSZ	14. HFZL
		28. GCBF (.01)
		32. R2FL (.03)
1976	11. MAXLWCPE	11. MAXLWCPE
	15. PMXSZ	15. PMXSZ
	18. WSSCL	16. WSCL
		19. MWCL
		21. TCTF (.001)

5. APPENDIX

Proof of Result 1:

$$C(\underset{\sim}{B}) = \sum_i \sum_j \underset{\sim}{W}_{ij} g(\varepsilon_j - \varepsilon_i)$$

Clearly for each $i \neq j$, since $g(\cdot)$ is odd and $\varepsilon_j - \varepsilon_i$ is symmetrically distributed, $Eg(\varepsilon_j - \varepsilon_i) = 0$. Furthermore, it is easily verified for distinct i, j, k, ℓ that:

$$E[g(\varepsilon_j - \varepsilon_i)g(\varepsilon_\ell - \varepsilon_k)] = 0$$

$$E[g(\varepsilon_j - \varepsilon_i)g(\varepsilon_k - \varepsilon_i)] = \frac{1}{3}$$

$$E[g(\varepsilon_j - \varepsilon_i)g(\varepsilon_k - \varepsilon_j)] = \frac{-1}{3}$$

$$E[g(\varepsilon_k - \varepsilon_i)g(\varepsilon_k - \varepsilon_j)] = \frac{1}{3} \quad .$$

Therefore
$$S_{k\ell} = \sum_{i,j}\sum W_{ij}^{(k)} W_{ij}^{(\ell)} - \frac{1}{3}\sum_{i,j,m}\sum\sum W_{ij}^{(k)} W_{jm}^{(\ell)} - \frac{1}{3}\sum_i\sum_j\sum_m W_{ij}^{(k)} W_{mi}^{(\ell)}$$
$$+ \frac{1}{3}\sum_i\sum_j\sum_m W_{ij}^{(k)} W_{im}^{(\ell)} + \frac{1}{3}\sum_i\sum_j\sum_m W_{ij}^{(k)} W_{mj}^{(\ell)} .$$

Where summations are over all distinct choices of i, j, m. This reduces to the form in result 1.

Proof of Result 2:

Consider lines of the form

$$\ell_{ij} = \{\underset{\sim}{\beta} \; ; \; R_i(\underset{\sim}{\beta}) = R_j(\underset{\sim}{\beta})\}$$

for distinct k, j. When we pass from one side of such a line to the other in the (β_1, β_2) plane, the order of R_i and R_j are reversed: i.e. a transposition is applied to the ranks of the residuals. There are $\binom{n}{2}$ such lines. However, note that ℓ_{ij}, ℓ_{jk} and ℓ_{ik} all intersect at a single point. Therefore, the $N = \frac{n(n-1)}{2}$ lines will cut the (β_1, β_2) plane into $1 + \frac{N(N+1)}{2} - \binom{n}{3}$ regions where $\binom{n}{3}$ is the number of points at the intersection of 3 lines. This reduces to $1 + \frac{n(n-1)(3n^2 - 7n + 14)}{24}$ regions.

Now observe that the value of χ^2 depends only on the order of the residuals, not on their magnitude. Therefore, when we cross a line of the form ℓ_{ij}, in the $\underset{\sim}{\beta}$

Table 3
"Tau" between pairs of predictors 2-19

Year	Variable																		
1975	2	1																	
	3	-0.91	1																
	4	0.38	-0.39	1															
	5	-0.04	0.03	0.49	1														
	6	-0.03	0.02	0.48	0.97	1													
	7	-0.04	0.03	0.36	0.71	0.73	1												
	8	-0.02	0.04	-0.32	-0.68	-0.69	-0.66	11											
	9	0.41	-0.43	0.97	0.46	0.46	0.34	-0.31	1										
	10	-0.04	0.03	0.36	0.71	0.73	1	-0.66	0.34	1									
	11	0.04	-0.05	0.47	0.75	0.77	0.89	-0.65	0.45	0.89	1								
	12	-0.02	0.02	0.59	0.65	0.64	0.46	-0.41	0.57	0.46	0.52	1							
	13	0.11	-0.12	0.6	0.84	0.84	0.66	-0.65	0.58	0.67	0.73	0.63	1						
	14	-0.02	0.02	0.6	0.68	0.67	0.46	-0.41	0.58	0.47	0.53	0.9	0.64	1					
	15	-0.06	0.04	0.41	0.71	0.73	0.87	-0.65	0.39	0.87	0.86	0.54	0.67	0.54	1				
	16	-0.16	0.16	-0.13	-0.06	-0.06	-0.07	-0.04	-0.14	-0.07	-0.09	-0.02	-0.1	-0.05	-0.07	1			
	17	0.05	-0.05	0.08	0	-0.01	0.06	0.04	0.08	0.06	0.07	0.03	0	0.03	0.08	-0.24	1		
	18	0.08	-0.07	0.02	-0.04	-0.04	0.01	0.1	0.02	0.01	0.01	-0.02	-0.06	-0.01	0.05	-0.18	0.07	1	
	19	0.01	-0.01	0.15	0.17	0.17	0.19	-0.14	0.14	0.19	0.19	0.18	0.17	0.17	0.24	0.03	0.06	0.48	1
1976	2	1																	
	3	-0.9	1																
	4	0.56	-0.51	1															
	5	0.12	-0.07	0.47	1														
	6	0.13	-0.08	0.47	0.96	1													
	7	0.05	-0.01	0.2	0.56	0.58	1												
	8	0	0.03	-0.04	-0.4	-0.42	-0.47	1											
	9	0.59	-0.54	0.97	0.44	0.44	0.19	-0.04	1										
	10	0.05	-0.01	0.2	0.56	0.58	1	-0.47	0.19	1									
	11	0.19	-0.15	0.39	0.66	0.68	0.81	-0.42	0.38	0.81	1								
	12	0.2	-0.13	0.62	0.64	0.62	0.28	-0.11	0.59	0.28	0.42	1							
	13	0.31	-0.26	0.62	0.82	0.83	0.51	-0.36	0.6	0.51	0.68	0.62	1						
	14	0.15	-0.08	0.6	0.65	0.63	0.26	-0.13	0.56	0.26	0.39	0.85	0.62	1					
	15	0.07	-0.03	0.27	0.63	0.65	0.88	-0.47	0.26	0.88	0.85	0.36	0.6	0.33	1				
	16	-0.05	0.07	-0.12	-0.12	-0.12	0.02	0.05	-0.11	0.02	-0.02	-0.15	-0.12	-0.13	0	1			
	17	-0.01	0.06	0.01	-0.01	-0.01	-0.05	0.07	0.01	-0.05	-0.05	0.03	-0.03	0.04	-0.05	-0.1	1		
	18	0.03	0.01	0.07	-0.05	-0.04	-0.06	0.24	0.07	-0.06	-0.02	0.06	-0.02	0.02	-0.02	-0.07	0.11	1	
	19	-0.08	0.1	0.01	0.11	0.12	0.22	0.07	0	0.22	0.2	0.07	0.09	0.02	0.25.	0.07	-0.12	0.29	1

plane, this has the effect of interchanging the order of R_i and R_j, i.e. replacing $g(R_j - R_i)$ by its negative. The other terms in the expansions for $\underset{\sim}{C}$ remain unchanged. Therefore the maximum change in any one of the components of $\underset{\sim}{C}$ is 2. It is not difficult to show that the value of C_k at a point $\underset{\sim}{\beta}$ on one (or more) of the lines is its average on all adjacent regions.

We now show there exists a point β_0 such that $|C_1(\beta_0)| \leq 1$ and $|C_2(\beta_0)| \leq 1$. Let ℓ be the horizontal line $\{(\beta_1, \beta_2); \beta_2 = \beta_2^*\}$ where β_2^* is some constant. Observe that for $\underset{\sim}{\beta} \varepsilon \ell$, $R_j(\beta) - R_i(\beta)$ is a constant minue $(X_{j1} - X_{i1})\beta_1$ which is a non-increasing function of β_1 if $X_{j1} > X_{i1}$. Therefore, on ℓ, $C_1(\underset{\sim}{\beta})$ is a non-increasing function of β_1. Furthermore, $g(R_j - R_i)$ approaches $+1$ as $\beta_k \to -\infty$ and -1 as $\beta_k \to +\infty$ for each such i, j. Therefore, if $\sum_i \sum_j W_{ij}^{(1)} > 1$, there are points on ℓ for which $C_1(\underset{\sim}{\beta}) > 1$ and other points for which $C_1(\beta) < -1$. We now show there is at least one point with $|C_1(\underset{\sim}{\beta})| \leq 1$. If this fails, there is a point β^* on ℓ such that $C_1(\beta) \geq 2$ for β on one side and $C_1(\beta) \leq -2$ on the other. Since the maximum jump in $C(\beta)$ as we cross one of the lines ℓ_{ij} is 2, this jump of at least 4 units implies β^* lies at the intersection of either two or three lines of the form ℓ_{ij}. If there are two such lines, we must have the following configuration:

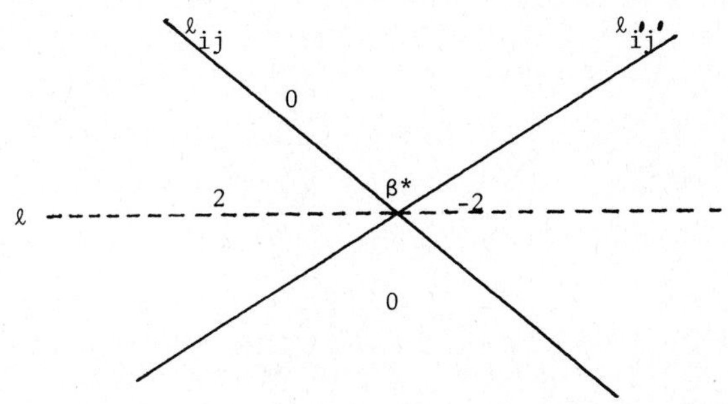

The labels on the regions are the values of $C_1(\beta)$ there, and since $C_1(\beta^*)$ is the average of the 4 adjacent regions, it is 0. If β^* is at the intersection of 3 lines, the configuration is one of a few possible such as the following:

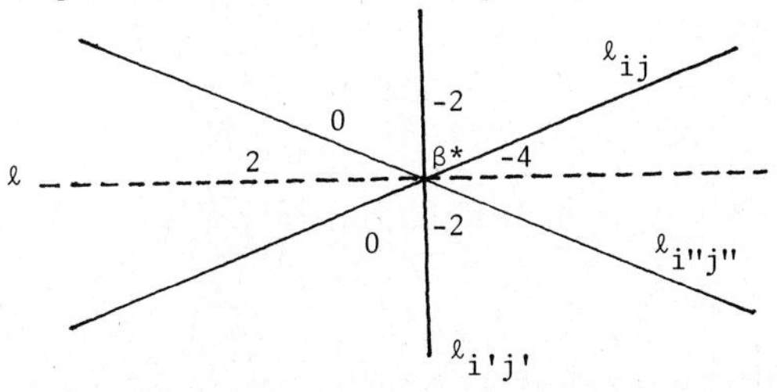

For all such configurations, $|C_1(\beta^*)| \leq 1$.

We have thus shown that for any value of β_2^*, there exists β_1^* such that $|C_1(\beta_1^*, \beta_2^*)| \leq 1$, i.e. $\{\underset{\sim}{\beta}; |C_1(\underset{\sim}{\beta})| \leq 1\}$ contains a curve. Similarly $\{\underset{\sim}{\beta}; |C_2(\underset{\sim}{\beta})| \leq 1\}$ contains a curve. Therefore if all four types of behaviour

$$C_1(\underset{\sim}{\beta}) > 1 \text{ and } C_2(\underset{\sim}{\beta}) < -1$$
$$C_1(\underset{\sim}{\beta}) > 1 \text{ and } C_2(\underset{\sim}{\beta}) > 1$$
$$C_1(\underset{\sim}{\beta}) < -1 \text{ and } C_2(\underset{\sim}{\beta}) < -1$$
$$C_1(\underset{\sim}{\beta}) < -1 \text{ and } C_2(\underset{\sim}{\beta}) > 1$$

are possible, then these curves must intersection (at a point for which both $|C_1(\beta^*)| \leq 1$ and $|C_2(\beta^*)| \leq 1$).

Therefore the minimum possible value of $\chi^2(B)$ is less than

$$\max_{\substack{|C_1| \leq 1 \\ |C_2| \leq 1}} C'S^{-1}C \leq \max_{\underset{\sim}{C}} C'S^{-1}C$$

where the second maximum is subject to constraint $C_1^2 + C_2^2 \leq 2$:

$$= \frac{2}{\lambda_2} \text{ where } \lambda_2 \text{ is the smaller eigenvalue of } S.$$

Proof of Lemma 1. Let $G(x) = \frac{1}{2}(P[\xi < x] - P[\xi > x]) = P[\xi < x] - \frac{1}{2} + \frac{1}{2}[\xi = x]$. If m denotes the median and if $P[\xi = m] > 0$, then for a mass point $y > m$.

$$G(y) = P[\xi < y] - \frac{1}{2} + \frac{1}{2}P[\xi = y] \geq P[\xi \leq m] - \frac{1}{2} + \frac{1}{2}P[\xi = y] \geq 0.$$

Similarly for $y < m$, $G(y) \leq 0$. If x and y are mass points and $x > y$, $G(x) - G(y) = \frac{1}{2}(P[\xi=x] - P[\xi=y]) + P[y \leq \xi < x] \geq \frac{1}{2}P[\xi=x] + P[y < \xi < x] \geq 0.$

Therefore the function $G(x)$ is non-decreasing. This implies $\min_y |G(y)|$ occurs either at m or at the point on either side of m.

Acknowledgements: The author gratefully acknowledges the Alberta Weather Modification Board and M. English for the provision and elucidation of the data, M. D. McLeish for bringing the problem to his attention, help with the reduction of the data, and some useful suggestions, and finally K. Lam and D. Bates for several helpful discussions.

6. REFERENCES

[1] Barlow, F. D., Kochtubajda, B. and English, M. Predictor and Response Variables for the Hailstorms of 1975 and 1976. Natural Resources Notes 80-1, Alberta Research Council.

[2] English, M., Kochtubajda, B., and Barlow, F. (1980). The Cloud Modelling Programs. Tech. Report 79-1, Vol. II.

[3] Hajek, J. (1969). A Course in Non-Parametric Statistics. Holden-Day.

[4] Jaeckel, L.A. (1972). Estimating Regression Coefficients by Minimizing the Dispersion of the Residuals. Annals of Math. Statist. 43, 1449-1458.

[5] Kendall, M. (1948). Rank Correlation Methods. Griffin.

[6] Koul, (1969). Asymptotic Behaviour of Wilcoxen Type Confidence Regions in Multiple Linear Regression. Ann. Math. Statist. 40, 1950-1979.

[7] Sen, P. K. (1968). Estimates of the Regression Coefficient based on Kendall's Tau. Ann. Math. Statist. 63, 1379-1389.

[8] Sen. P. K. (1969). A class of rank order tests for a general linear hypothesis. Ann. Math. Statist. 40, 1325-1343.

[9] Scholz, F. (1978). Weighted Median Regression Estimates. Ann. Statist. 6, 603-609.

[10] Sievers, G. L. (1978). Weighted Rank Statistics for Simple Linear Regression. J.A.S.A. 73, 628-631.

[11] Theil, H. (1950). A Rank Invariant Method of Linear and Polynomial Regression Analysis, I, II, III. Proc. Kon. Nederl. Akad. Wetensch.A. 53, 368-392, 521-525, 1397-1412.

STATISTICS AND RELATED TOPICS
M. Csörgö, D.A. Dawson, J.N.K. Rao, A.K.Md.E. Saleh (eds.)

A NONPARAMETRIC TEST FOR EQUALITY AGAINST ORDERED ALTERNATIVES IN THE CASE OF SKEWED DATA, WITH A BIOMEDICAL APPLICATION

A.R. Padmanabhan (1)
Monash University
Clayton, Victoria
Australia

M.L. Puri (2)
Indiana University
Bloomington, U.S.A.

A.K.Md. Ehsanes Saleh (3)
Carleton University
Ottawa, Canada

For testing the equality of k-samples against ordered alternatives, a new nonparametric test is proposed based on the ideas of Gastwirth (1965), Hogg et al (1975) and Puri (1965). The test is shown to be superior to the Jonckheere (1945) test in the case of skewed distributions. An application to lung cancer data illustrates the theory.

1. INTRODUCTION

Suppose k independent samples are drawn from the k continuous distributions $F(x-\theta_1),\ldots,F(x-\theta_k)$. We consider the problem of testing the hypothesis $H_o : \theta_1 = \ldots = \theta_k$ against the ordered alternatives $H_A : \theta_1 \le \theta_2 \le \ldots \le \theta_k$ (or $\theta_1 \ge \theta_2 \ge \ldots \ge \theta_k$). Probably, the best known nonparametric test is the Jonckheere test (1945) which is essentially a Wilcoxon type test. However, there are many situations where F is known to be skewed though the form may not be known. Such cases are the distribution of (i) life-times of cancer patients; (ii) radii of aerosals; (iii) age at death of infants dying from respiratory distress syndrome; (iv) Biomass contained in unit volume of water; (v) toxicity of some drugs, food additives, etc. Thus, for skewed distributions, Jonckheere test may not be suitable and the object of this paper is to propose a new test superior to Jonckheere test based on the ideas of Gastwirth (1965), Hogg et al (1975) and Puri (1965) which incorporates the information on skewness of the data in the test.

In section 2, we propose the test and in section 3, we present the ARE of the test relative to the Jonckheere test. We complete the paper with section 4 where we illustrate the theory by an example using the survival data of lung cancer patients.

2. THE PROPOSED TEST

Let $X_i = (X_{i1},\ldots,X_{in_i})$, $i=1,2,\ldots,k$ be k independent samples from continuous cdf's $F(x-\theta_i)$, $i=1,\ldots,k$ respectively. Consider the ith and the jth samples i.e. $X_i = (X_{i1},\ldots,X_{in_i})$ and $X_j = (X_{j1},\ldots,X_{jn_j})$. Denote by $N_{ij} = [\frac{n_i+n_j+1}{2}]$ and define the scores

$$a_{n_i+n_j}(R) = (R - N_{ij} - 1) \quad \text{if} \quad R \le N_{ij} \tag{2.1}$$

$$= 0 , \qquad \text{otherwise.}$$

following Hogg et al (1975). Then, consider the rank tests

$$S_{ij}^{(i)} = \frac{1}{n_i} [a_{n_i+n_j}(R_{i1}) + \ldots + a_{n_i+n_j}(R_{in_i})] \tag{2.2}$$

and

$$S_{ij}^{(j)} = \frac{1}{n_j} [a_{n_i+n_j}(R_{j1}) + \ldots + a_{n_i+n_j}(R_{jn_j})] \tag{2.3}$$

where $R_{i\ell}$ and R_{jm} are the ranks of $X_{i\ell}$ and X_{jm} in the combined order samples (X_i, X_j). Now, following Puri (1965), we define the statistic V given below for testing H_o against H_A.

$$V = \sum_{i<j} \sum n_i n_j (S_{ij}^{(i)} - S_{ij}^{(j)}) . \tag{2.4}$$

We note that $N = \sum_{i=1}^{k} n_i$ and $\rho_i = n_i/N$ and define the general scores

$$a_N(R) = \frac{1}{N}(R - [\frac{N+1}{2}] - 1) \quad \text{if} \quad R \leq [\frac{N+1}{2}] .$$
$$= 0 , \qquad \text{otherwise.} \tag{2.5}$$

Also,

$$\bar{a}_N = \frac{1}{N}(a_N(1) + \ldots + a_N(N)) . \tag{2.6}$$

Then, by Theorem 5.2 of Puri (1965), $N^{-3/2}V$ (under H_o) is asymptotically normal with zero mean and variance

$$\sigma^2(N^{-3/2}V) = \frac{1}{3}[1 - (\sum_{i=1}^{k} \rho_i^3)]A^2 \tag{2.7}$$

where

$$A^2 = \int_0^1 J^2(u)du - (\int_0^1 J(u)du)^2 . \tag{2.8}$$

Further, the J-functions satisfy the regularity conditions of lemma 7.2 of Puri (1965). In the definition of the test, the J-function is given by

$$J_2(u) = u - \frac{1}{2} \quad \text{if} \quad 0 < u \leq \frac{1}{2} ,$$
$$= 0 , \qquad \text{otherwise.} \tag{2.9}$$

The J-function used in the Jonckheere tests is given by

$$J_1(u) = u , \quad 0 < u < 1 . \tag{2.10}$$

Therefore, if there are no ties in the samples, the above results validate the test using the normal distribution. In practice, ties do occur, therefore, we modify the statistics $S_{ij}^{(i)}$ and $S_{ij}^{(j)}$ by the average score method. For such procedures see Conover (1971). Then, V is modified to $\tilde{V}$ using the average scores $\tilde{a}_{n_i+n_j}(R)$ instead of $a_{n_i+n_j}(R)$ and defining

$$\tilde{A}^2 = \frac{1}{N-1}\left[\sum_{i=1}^{N} \tilde{a}_N(i) - \bar{\tilde{a}}_N\right]^2 \tag{2.11}$$

where $\tilde{a}_N(i)$ is a modification of $a_N(i)$. Then, by Theorem 4 of Vorlickova (1970) and arguments similar to Puri (1965) $N^{-3/2}\,\tilde{V}$ (under H_o) is asymptotically normal with zero mean and variance

$$\sigma^2(N^{-3/2}\,\tilde{V}) = \frac{1}{3}\left[1 - \left(\sum_{i=1}^{k} \rho_i^3\right)\right]\tilde{A}^2 . \tag{2.12}$$

This result allows us to perform the test when ties are present.

3. ASYMPTOTIC RELATIVE EFFICIENCY (ARE) OF V-TEST

In order to obtain the ARE of V-test relative to Jonckheere test (henceforth to be called W-test), we need the asymptotic normality of V-test under H_A. This is provided by Theorem 5.2 of Puri (1965). Under suitable regularity conditions (which are satisfied for the V-test), as given in Theorem 5.2 of Puri (1965), $N^{-3/2}\,V$ is asymptotically normal with mean

$$\sum_{i<j}\sum \rho_i\rho_j(\theta_j - \theta_i)\int[dJ(F(x))/dx]dF(x) \tag{3.1}$$

and variance in expression (4.3) of Puri (1965). The variance expression reduces to our expression (2.7) under H_o which is required for the computation of the ARE. Thus,

$$\text{ARE}(V:W) = \frac{A_1^2\,B_2^2}{A_2^2\,B_1^2} , \tag{3.2}$$

where

$$A_i^2 = \int_0^1 J_i^2(u)\,dx - \left(\int_0^1 J_i(u)\,du\right)^2 \tag{3.3}$$

$$B_i = \int [dJ_i(F(x))/dx]dF(x) , \tag{3.4}$$

for $i = 1,2$.

Now for the W-test $J_1(u) = u$, $0 < u < 1$ and the V-test $J_2(u) = u - \frac{1}{2}$ for $0 < u \leq \frac{1}{2}$ and 0, elsewhere. Therefore, we have $A_1^2 = \frac{1}{12}$ and $A_2^2 = \frac{5}{192}$.

We now evaluate the ARE(V : W) for the two well-known distribution viz, exponential and lognormal used in Biomedical analysis.

Case 1. Exponential. $F(x) = 1 - e^{-x}$ and $f(x) = e^{-x}$ for $x \geq 0$.

Then,
$$B_1 = \int_0^\infty f^2(x)\,dx = \int_0^\infty e^{-2x}dx = \frac{1}{2}$$

$$B_2 = \int_0^{\ln 2} f^2(x)\,dx = \int_0^{\ln 2} e^{-2x}dx = \frac{3}{8} .$$

Thus, for $F(x) = 1 - e^{-x}$,

$$\text{ARE}(V : W) = 1.8 . \tag{3.5}$$

Therefore, the V-test is superior to W-test.

Case 2. <u>lognormal distribution</u>.

$$f(x) = \frac{1}{x\sqrt{2\pi}\, t} \exp\left\{-\frac{1}{2}(\log x - m)^2/t^2\right\}, \quad x > 0,\ t > 0.$$

In this case, by routine computation

$$B_1 = \frac{\exp(t^2/4)}{2\sqrt{\pi}\, t \exp(m)}$$

$$B_2 = \frac{\exp(t^2/4)\Phi(t/\sqrt{2})}{2\sqrt{\pi}\, t \exp(m)}$$

where $\Phi(\cdot)$ is cdf of standard normal distribution. Consequently, for the lognormal distribution

$$\text{ARE}(V : W) = \frac{16}{5}\Phi^2(t/\sqrt{2}) . \tag{3.6}$$

It is easy to verify that the $\text{ARE}(V : W)$ satisfies the inequality

$$.80 < \text{ARE}(V : W) < 3.2 \tag{3.7}$$

and V-test is superior to W-test as soon as $t > .2017$.

The same ARE results hold for the two-sample problem. It may be noted that we have considered the case of right skewed distribution. In case of left skewed distribution the scores are reversed.

4. APPLICATION TO LUNG CANCER DATA

The following data represents the survival days of cancer patients. The data have been provided by Professor Van Belle of the Department of Biostatistics, University of Washington, Seattle.

Days of Survival

Group 1 Standard Squamous	72	411	228	126	118	10	82	110	314	100	42	8	144	25	11
Group 2 Standard Small	30	384	4	54	13	23	97	153	59	117	16	151	22	56	21
	18	133	20	31	52	287	18	51	122	27	54	7	63	392	10
Group 3 Standard Adena	8	92	35	117	132	12	162	3	95	X	X	X	X	X	X
Group 4 Standard Large	177	162	216	553	278	12	260	200	156	182	143	105	103	250	100

Since, ties occur in the data, we use average score method to compute $\tilde{V} = -99.4615$ with $N = 69$ so that $N^{-3/2}\tilde{V} = -.1735$ and variances equal to 0.0084. A two-sided test has been carried out as the direction was unknown.

Hence the critical region is given by $(-\infty, -.1801) \cup (.1801, \infty)$. The hypothesis H_0 is accepted, i.e., that the mean survival days of the four populations are equal.

Acknowledgement. We are grateful to Professor G. Van Belle of the Department of Biostatistics, University of Washington, Seattle, for the data and to Professor Ewens of Monash University for some useful discussion of the material.

5. REFERENCES

[1] Conover, N.J. (1971). Practical Nonparametric Statistics. (John Wiley & Sons. New York)

[2] Gastwirth, J.L. (1965). Percentile modifications of two-sample rank tests. JASA, 60, 1127-1141.

[3] Hogg, R.V., Fisher, D.M., and Randles, R.H. (1975). A two-sample adaptive distribution-free test. JASA, 70, 656-661.

[4] Jonckheere, A.R. (1945). A distribution-free k sample test against ordered alternatives. Biometrika 41, 135-145.

[5] Puri, M.L. (1964). Asymptotic efficiency of c-sample tests. Ann. Math. Statist. 35, 102-121.

[6] Puri, M.L. (1965). Some distribution-free k-sample rank tests for homogeneity against ordered alternatives. Communications on Pure and Applied Mathematics. XVIII, 51-63.

[7] Vorlickova, D. (1970). Asymptotic properties of rank tests under discrete distributions. Z. Wahrscheinlichkeitstheorie und Verw. Geb. 14, 275-289.

---

(1) Research supported by NSF Grant no. MCS76-00951 at Indiana University.

(2) Research supported by AIR FORCE OFFICE of Scientific Research, AFSC, USAF under grant no. AFOSR 76-1927. Reproduction in whole or part permitted for any purpose of United States Government.

(3) Research supported by NSERC grant no. A3088.

STATISTICS AND RELATED TOPICS
M. Csörgö, D.A. Dawson, J.N.K. Rao, A.K.Md.E. Saleh (eds.)

RANK ANALYSIS OF COVARIANCE UNDER PROGRESSIVE CENSORING, II

Pranab Kumar Sen†

Department of Biostatistics
University of North Carolina
Chapel Hill, North Carolina 27514
U.S.A.

For some general analysis of covariance models, a class of progressively censored nonparametric tests based on suitable rank order statistics is considered. The proposed procedure is a generalization of an one (for a simpler model) in Sen (1979b). Along with some invariance principles for progressively censored (multivariate) linear rank statistics, asymptotic properties of the proposed tests are studied.

1. INTRODUCTION

In *clinical trials* or *life testing problems*, a *progressively censoring scheme* (PCS) incorporates a *continuous monitoring* of experimentation from the beginning with the objective of an *early termination* (contingent on the accumulating statistical evidence) without increasing the margin of the *type I error*. Chatterjee and Sen (1973) have formulated a general class of PCS *nonparametric tests* for a simple regression model (which includes the classical two-sample problem as a special case); their theory rests on some *invariance principles* for some PCS linear rank statistics. Sen (1979a) has developed some invariance principles for some related quantile processes arising in PCS; some generalizations of these are due to Sen (1981b). Sen (1979b) has incorporated *concomitant variates* in the simple regression model and studied some PCS *analysis of covariance* (ANOCOVA) tests; references to other relevant works are cited in these papers. PCS nonparametric tests for *multiple regression* (containing the several sample problem as a special case) have been considered by Majumdar and Sen (1978). The object of the present investigation is to incorporate (stochastic) concomitant variates in some general multi-parameter models and to formulate appropriate PCS rank tests for such general ANOCOVA models.

Section 2 deals with the basic ANOCOVA model and the preliminary notions. The proposed PCS tests are then developed in Section 3. Section 4 is devoted to the asymptotic distribution theory of the allied test-statistics, both under the null and some local alternative hypotheses. Since our proposed procedure is a natural extension of Sen (1979b), in the sequel, often, to minimize the technical

†Work supported by the National Heart, Lung and Blood Institute, Contract No. NIH-NHLBI-71-2243 from the National Institutes of Health.

manipulations, we shall omit some details by suitable cross references to the earlier paper. Some general remarks are made in the concluding section.

2. PRELIMINARY NOTIONS

Let $\underset{\sim}{X}_i^* = (X_{0i}, X_{1i}, \ldots, X_{pi})' = (X_{0i}, \underset{\sim}{X}_i')'$, $i = 1,\ldots,n$ be independent random vectors (r.v.) with continuous (p + 1)-variate distribution functions (d.f.) F_i^*, $i = 1,\ldots,n$, where the X_{0i} are the *primary variates* with (marginal) d.f. F_{0i}, defined on $E(=(-\infty,\infty))$ and the $\underset{\sim}{X}_i$ are the *concomitant variates* with marginal (joint) d.f. F_i, defined on E^p, for some $p \geq 1$. Let $F_i^o(y|\underset{\sim}{x})$ be the conditional d.f. of X_{0i}, given $\underset{\sim}{X}_i = \underset{\sim}{x}$, $i = 1,\ldots,n$, and, as is usually the case in an ANOCOVA model, we assume that $F_1 \equiv \cdots \equiv F_n \equiv F$ (unknown). Our basic problem is to test for the null hypothesis

$$H_0:\ F_1^o \equiv \cdots \equiv F_n^o \equiv F^o \text{ (unknown)}, \tag{2.1}$$

against an alternative that they are not all the same. Generalizing the model in Sen (1979b), we let (for every $y \in E$, $\underset{\sim}{x} \in E^p$):

$$F_i^o(y|\underset{\sim}{x}) = F^o(y;\ \underset{\sim}{\Delta}(\underset{\sim}{c}_i - \bar{\underset{\sim}{c}}_n)|\underset{\sim}{x}),\quad i = 1,\ldots,n, \tag{2.2}$$

where $\underset{\sim}{\Delta}$ is an $m \times q$ matrix $(m \geq 1,\ q \geq 1)$ of unknown parameters, the $\underset{\sim}{c}_i$ are specified q-vectors and $\bar{\underset{\sim}{c}}_n = n^{-1}\sum_{i=1}^n \underset{\sim}{c}_i$. The classical ANOCOVA model relating to the one way layout is a special case of (2.2) where $m = 1$, $q \geq 1$ and $F^o(y;\ \underset{\sim}{\Delta}(\underset{\sim}{c}_i - \bar{\underset{\sim}{c}}_n)|\underset{\sim}{x}) \equiv F^o(y - \underset{\sim}{\Delta}(\underset{\sim}{c}_i - \bar{\underset{\sim}{c}}_n)|\underset{\sim}{x})$, $\forall\ i \geq 1$. Further, in this special case, we have $k(= q + 1)$ samples of sizes $n_1,\ldots,n_k$, respectively $(n = \sum_{i=1}^k n_i)$, $\underset{\sim}{c}_i = \cdots = \underset{\sim}{c}_{n_1} = \underset{\sim}{0}$, $\underset{\sim}{c}_{n_1+1} = \cdots = \underset{\sim}{c}_{n_1+n_2} = (1,0,\ldots,0)'\ldots,$ $\underset{\sim}{c}_{n_1+\cdots+n_q+1} = \cdots = \underset{\sim}{c}_{n_1+\cdots+n_k} = (0,\ldots,0,1)'$ and $\underset{\sim}{\Delta}$ stands for the vector of treatment-effects. More general models may be conceived by allowing the location and scales to vary under alternatives. Now, under (2.2), we may recast (2.1) as

$$H_0:\ \underset{\sim}{\Delta} = \underset{\sim}{0} \text{ against } H_1:\ \underset{\sim}{\Delta} \neq \underset{\sim}{0}. \tag{2.3}$$

In a life testing situation, though the concomitant variates $\underset{\sim}{X}_1,\ldots,\underset{\sim}{X}_n$ may be observable at the beginning of the experimentation, the primary variates are not so. If $Z_{n1}^o < \cdots < Z_{nn}^o$ be the ordered rv's corresponding to $X_{01},\ldots,X_{0n}$ (ties neglected, with probability 1, by virtue of the assumed continuity of F_0) and if we define the *anti-ranks* $S_1,\ldots,S_n$ by

$$Z_{nj}^o = X_{0S_j},\ \text{for } j = 1,\ldots,n, \tag{2.4}$$

then, at the k-th failure Z_{nk}^o, the observable rv's are

$$\underset{\sim}{Q}_i = (S_i, Z_{ni}^o, \underset{\sim}{X}_{S_i}),\ i = 1,\ldots,k,\ \text{for } k = 1,\ldots,n. \tag{2.5}$$

[Though the complementary sets of concomitant variates are known, their anti-ranks are not specified, in advance.]

Nonparametric ANOCOVA tests (based on the entire set $\{\underset{\sim}{X}_1^*,\ldots,\underset{\sim}{X}_n^*\}$) have been considered by Quade (1967), Puri and Sen (1969a) and Sen and Puri (1970), among others. Under PCS, a special case of a simple regression model, has been studied by Sen (1979b). In this paper, the general model in (2.2) (for $m \geq 1$, $q \geq 1$) will be considered. We introduce the following notations. Let R_{ji} be the rank of X_{ji} among $X_{j1},\ldots,X_{jn}$, for $i = 1,\ldots,n$; $j = 0,1,\ldots,p$. These yield the *rank-collection matrix* $\underset{\sim}{R}_n = ((R_{ji}))$ (of order $(p + 1) \times n$), and permuting the columns of $\underset{\sim}{R}_n$, so that the top row is in the natural order, we obtain the *reduced rank-collection matrix* $\underset{\sim}{R}^*_n = ((R^*_{ji}))$, where by (2.4),

$$R_{0S_i} = R^*_{0i} = i \text{ and } R_{jS_i} = R^*_{ji}, \text{ for } 1 \leq j \leq p,\ i = 1,\ldots,n, \tag{2.6}$$

For each $j(= 0,\ldots,p)$, let $\{\underset{\sim}{a}_{n,j}(i) = (a^{(1)}_{n,j}(i),\ldots,a^{(m)}_{n,j}(i))',\ i = 1,\ldots,n\}$ be a set of *scores* (vectors), which are chosen with the model in (2.2) in mind. For example, if we restrict ourselves to the multiple linear regression model, we have then $m = 1$ and we may choose $a_{n,j}(i) = i/(n + 1)$ [Wilcoxon scores] or $\Phi^{-1}(\frac{i}{n + 1})$ [normal scores], $1 \leq i \leq n$, where Φ is the standard normal d.f. If we have the joint location/scale model, then $m = 2$, and besides the above scores (suitable for location alternatives), we need to choose [for $a^{(2)}_{n,j}(i)$] some scores suitable for scale alternatives. For example, for $a^{(2)}_{n,j}(i)$, either $[i/(n + 1) - \frac{1}{2}]^2$ or $[\Phi^{-1}(i/(n + 1))]^2$ may be used in practice.

As in Puri and Sen (1969b), we consider the (multi-variate) linear rank statistics

$$\underset{\sim}{T}_n = \sum_{i=1}^{n}(\underset{\sim}{c}_i - \bar{\underset{\sim}{c}}_n)[\underset{\sim}{a}'_{n,0}(R_{0i}),\ldots,\underset{\sim}{a}'_{n,p}(R_{pi})] \tag{2.7}$$

where $\bar{\underset{\sim}{c}}_n = n^{-1}\sum_{i=1}^{n}\underset{\sim}{c}_i$. By (2.6), we may rewrite $\underset{\sim}{T}_n$ as

$$\underset{\sim}{T}_n = \sum_{i=1}^{n}(\underset{\sim}{c}_{S_i} - \bar{\underset{\sim}{c}}_n)[\underset{\sim}{a}'_{n,0}(i),\underset{\sim}{a}'_{n,1}(R^*_{1i}),\ldots,\underset{\sim}{a}'_{n,p}(R^*_{pi})]. \tag{2.8}$$

Keeping in mind the observable r.v's in (2.5), as in Sen (1979b), we let P_n be the conditional (permutational) probability measure generated by the $n!$ (conditionally) equally likely column permutations of $\underset{\sim}{R}_n$ (given $\underset{\sim}{R}^*_n$) and let $\underset{\sim}{S}_{n,k} = (S_1,\ldots,S_k)$, for $k = 1,\ldots,n$. Then, we let

$$\begin{aligned}\underset{\sim}{T}_{n,k} &= E_{P_n}(\underset{\sim}{T}_n|\underset{\sim}{S}_{n,k})\\ &= \sum_{i=1}^{k}(\underset{\sim}{c}_{S_i} - \bar{\underset{\sim}{c}}_n)[\underset{\sim}{a}'_{n,0}(i) - \underset{\sim}{a}^*_{n,0}(k),\ldots,\underset{\sim}{a}'_{n,p}(R^*_{pi}) - \underset{\sim}{a}^{*\prime}_{n,p}(k)]\end{aligned} \tag{2.9}$$

where

$$\underset{\sim}{a}^{*\prime}_{n,j}(k) = \begin{cases}(n - k)^{-1}\{n\bar{\underset{\sim}{a}}'_{n,j} - \sum_{i=1}^{k}\underset{\sim}{a}'_{n,j}(R^*_{ji})\}, & 1 \leq k \leq n - 1\\ 0, & k = n,\end{cases} \tag{2.10}$$

and

$$\bar{\underset{\sim}{a}}'_{n,j} = n^{-1}\sum_{i=1}^{n}\underset{\sim}{a}'_{n,j}(i), \text{ for } j = 0,1,\ldots,p. \tag{2.11}$$

Conventionally, we let $\underset{\sim}{T}_{n,0} = 0$, $\forall\, n \geq 1$. The proposed PCS tests rest on the partial sequence $\{\underset{\sim}{T}_{n,k};\ k \leq n\}$ and are formulated in the next section.

3. THE PROPOSED PCS TESTS

Let us define

$$\underset{\sim}{C}_n = \sum_{i=1}^{n}(\underset{\sim}{c}_i - \underset{\sim}{c}_n)(\underset{\sim}{c}_i - \underset{\sim}{c}_n)' \tag{3.1}$$

and assume that $\underset{\sim}{C}_n$ is positive definite (p.d). Further, the statistics in (2.9) are location invariant, and hence, without any loss of generality, we may set

$$\bar{\underset{\sim}{a}}_{n,j} = \underset{\sim}{0}, \text{ for } j = 0,1,\ldots,p. \tag{3.2}$$

For every k: $1 \le k \le n$, and $j,\ell(= 0,1,\ldots,p)$, we define

$$\underset{\sim}{V}^{(k)}_{n,j\ell} = n^{-1}\left\{\sum_{i=1}^{k} \underset{\sim}{a}_{n,j}(R^*_{ji})\underset{\sim}{a}'_{n,\ell}(R^*_{\ell i}) + (n-k)\underset{\sim}{a}^*_{n,j}(k)\underset{\sim}{a}^{*\prime}_{n,\ell}(k)\right\}, \tag{3.3}$$

and consider then the $m(p+1) \times m(p+1)$ matrices

$$\underset{\sim}{V}_{n,k} = \begin{pmatrix} \underset{\sim}{V}^{(k)}_{n,00} & \cdots & \underset{\sim}{V}^{(k)}_{n,0p} \\ \vdots & & \\ \underset{\sim}{V}^{(k)}_{n,p0} & \cdots & \underset{\sim}{V}^{(k)}_{n,pp} \end{pmatrix} = \begin{pmatrix} \underset{\sim}{V}^{(k)}_{n,00} & \underset{\sim}{V}^{(k)}_{n,0*} \\ m \times m & m \times mp \\ \underset{\sim}{V}^{(k)}_{n,*0} & \underset{\sim}{V}^{(k)}_{n,**} \\ mp \times m & mp \times mp \end{pmatrix}, \text{ say}, \tag{3.4}$$

for $k = 1,\ldots,n$. Also, we rewrite $\underset{\sim}{T}_{n,k}$ in (2.9) as

$$\underset{\sim}{T}_{n,k} = (\underset{\sim}{T}^{(0)}_{n,k}, \underset{\sim}{T}^{(1)}_{n,k}, \ldots, \underset{\sim}{T}^{(p)}_{n,k}) = (\underset{\sim}{T}^{(0)}_{n,k}, \underset{\sim}{T}^{(*)}_{n,k}), \text{ say}. \tag{3.5}$$

Then, following the line of attack of Sen (1979b), we have by (2.9),

$$E_{P_n}\underset{\sim}{T}_{n,k} = E\{E_{P_n}(\underset{\sim}{T}_n|\underset{\sim}{S}_{n,k})\} = E_{P_n}(\underset{\sim}{T}_n) = \underset{\sim}{0}, \tag{3.6}$$

$$V_{P_n}(\underset{\sim}{T}_{n,k}) = \underset{\sim}{C}_n \otimes \underset{\sim}{V}_{n,k}, \quad \forall\, 1 \le k \le n. \tag{3.7}$$

To eliminate the effects of the concomitant variates, we take into account the asymptotic multinormality of $\underset{\sim}{T}_{n,k}$ (insuring the asymptotic linearity of regression of $\underset{\sim}{T}^{(0)}_{n,k}$ on $\underset{\sim}{T}^{(*)}_{n,k}$) and work with the residuals:

$$\begin{aligned} \underset{\sim}{T}^{o}_{n,k} &= \underset{\sim}{T}^{(0)}_{n,k} - (\text{fitted value of } \underset{\sim}{T}^{(0)}_{n,k} \text{ on } \underset{\sim}{T}^{(*)}_{n,k}) \\ &= \underset{\sim}{T}^{(0)}_{n,k} - \underset{\sim}{V}^{(k)}_{n,0*}(\underset{\sim}{V}^{(k)}_{n,**})^{-}\underset{\sim}{T}^{(*)}_{n,k}, \quad (1 \le k \le n), \end{aligned} \tag{3.8}$$

where $\underset{\sim}{A}^{-}$ stands for the generalized inverse of $\underset{\sim}{A}$. Note that

$$E_{P_n}\underset{\sim}{T}^{o}_{n,k} = \underset{\sim}{0} \text{ and } V_{P_n}(\underset{\sim}{T}^{o}_{n,k}) = \underset{\sim}{C}_n \otimes [\underset{\sim}{V}^{(k)}_{n,00} - \underset{\sim}{V}^{(k)}_{n,0*}(\underset{\sim}{V}^{(k)}_{n,**})^{-}\underset{\sim}{V}^{(k)}_{n,*0}]. \tag{3.9}$$

We let

$$\underset{\sim}{V}^{(k)}_{n,00\cdot*} = \underset{\sim}{V}^{(k)}_{n,00} - \underset{\sim}{V}^{(k)}_{n,0*}(\underset{\sim}{V}^{(k)}_{n,**})^{-}\underset{\sim}{V}^{(k)}_{n,*0}, \quad i \le k \le n; \tag{3.10}$$

$$\underset{\sim}{G}_{nk} = \underset{\sim}{T}^{o}_{n,k}\underset{\sim}{T}^{o\prime}_{n,k} \text{ and } \underset{\sim}{H}_{nk} = \underset{\sim}{C}_n \otimes \underset{\sim}{V}^{(k)}_{n,00\cdot*}, \quad 1 \le k \le n; \tag{3.11}$$

$$\xi_{nk} = \text{Trace}[\underset{\sim}{G}_{nk}\underset{\sim}{H}^{-}_{nk}]; \quad 1 \le k \le n. \tag{3.12}$$

Then, ξ_{nk} is the covariate-adjusted rank test statistic (for testing H_0 in (2.1)) based on $\underset{\sim}{Q}_1,\ldots,\underset{\sim}{Q}_k$, for $k = 1,\ldots,n$. Suppose now that $r(= r_n)$ is some prefixed positive integer, such that $r/n \to \tau$, $0 < \tau \le 1$. Then, the proposed PCS can be formulated as follows:

Continue experimentation as long as $k \ge n_0$ and $\xi_{nk} \le \ell^{(\alpha)}_n$; if, for the first time, for some $k = N(\le r)$, ξ_{nN} is $> \ell^{(\alpha)}_n$ then experimentation is curtailed at the N-th failure Z^o_{nN} along with the rejection of H_0. If no such $k(\le r)$ exists, then experimentation is stopped at the preplanned r-th failure Z^o_{nr} along with the acceptance of H_0. Here $\alpha(0 < \alpha < 1)$ is the desired *level of significance* of the PCS test and the critical value $\ell^{(\alpha)}_n$ (and the initial number

n_0) are to be chosen in such a way that

$$P\{\xi_{nk} > \ell_n^{(\alpha)} \text{ for some } k: n_0 \le k \le r | H_0\} \le \alpha. \tag{3.13}$$

Note that N is the *stopping number* and Z_{nN}^o is the *stopping time*. Our task is to develop theory leading to the choice of $\ell_n^{(\alpha)}$ (and n_0) satisfying (3.13) and to study the (asymptotic) properties of the proposed tests. This is done in the next section.

4. ASYMPTOTIC PROPERTIES OF THE PROPOSED TESTS

Note that if in (2.7) - (2.9), we replace the $\underset{\sim}{c}_i$ by $\underset{\sim}{d}_i = \underset{\sim}{D}\underset{\sim}{c}_i$ where $\underset{\sim}{D}$ is any p.d. matrix and proceed as in (3.11) - (3.12) (with $\underset{\sim}{C}_n$ replaced by $\underset{\sim}{D}\underset{\sim}{C}_n\underset{\sim}{D}'$), then the ξ_{nk} remain invariant under any choice of $\underset{\sim}{D}$. Hence, without any loss of generality, in (2.7) - (2.9), and elsewhere, we may replace $(\underset{\sim}{c}_i - \overline{\underset{\sim}{c}}_n)$ by $\underset{\sim}{c}_{ni}$, $1 \le i \le n$, where

$$\sum_{i=1}^n \underset{\sim}{c}_{ni} = \underset{\sim}{0}, \quad \sum_{i=1}^n \underset{\sim}{c}_{ni}\underset{\sim}{c}_{ni}' = \underset{\sim}{C}_n = \underset{\sim}{I}_q \tag{4.1}$$

and we assume that

$$\max_{1\le i\le n} (\underset{\sim}{c}_{ni}'\underset{\sim}{c}_{ni}) \to 0, \text{ as } n \to \infty. \tag{4.2}$$

Secondly, defining $r = r_n$, as in after (3.12), we assume that

$$\lim_{n\to\infty} n^{-1} r_n = \tau: 0 < \tau \le 1. \tag{4.3}$$

Concerning the scores $\{\underset{\sim}{a}_{n,j}(i)\}$, we assume that for each $j(= 0,1,\ldots,p)$,

$$\underset{\sim}{a}_{n,j}(i) = E\underset{\sim}{\phi}_j(U_{ni}) \text{ or } \underset{\sim}{\phi}_j(i/(n+1)), \ 1 \le i \le n, \tag{4.4}$$

where $U_{n1} < \cdots < U_{nn}$ are the ordered r.v.'s of a sample of size n from the rectangular (0,1) d.f. and $\underset{\sim}{\phi}_j = (\phi_{1j},\ldots,\phi_{mj})'$ is expressible as

$$\phi_{sj}(u) = \phi_{sj,1}(u) - \phi_{sj,2}(u), \ 0 < u < 1, \ 1 \le s \le m, \ 0 \le j \le p, \tag{4.5}$$

where the $\phi_{sj,k}$ are all nondecreasing, absolutely continuous and square integrable inside (0,1).

Let now $F_{[j]}$, $F_{[j\ell]}$ and $F_{[0j\ell]}$ be respectively the marginal d.f. of X_{ji}, the joint d.f. of $(X_{ji}, X_{\ell i})$ and the trivariate (if $p \ge 2$) d.f. of $(X_{0i}, X_{ji}, X_{\ell i})$, for $0 \le j \le p$, $0 \le j \ne \ell \le p$ and $1 \le j \ne \ell \le p$, respectively, all under H_0 in (2.1). Further, we assume that $F_{[0]}(x) = t$ has a unique solution ζ_t^o, $\forall\, t \in (0,1)$, and let $\zeta_0^o = -\infty$ and $\zeta_1^o = +\infty$. Let then

$$\overline{\underset{\sim}{\Phi}}_{0t} = (1-t)^{-1}\int_t^1 \underset{\sim}{\phi}_0(u)du, \ 0 \le t < 1, \ \overline{\underset{\sim}{\Phi}}_{00} = \underset{\sim}{0} = \overline{\underset{\sim}{\phi}}_0, \ \overline{\underset{\sim}{\Phi}}_{01} = \underset{\sim}{0}; \tag{4.6}$$

$$\overline{\underset{\sim}{\Phi}}_{jt} = \frac{1}{1-t}\int_{\zeta_t^o}^\infty \int_{-\infty}^\infty \underset{\sim}{\phi}_j(F_{[j]}(y))dF_{[0j]}(x,y), \ 1 \le j \le p, \ t \in [0,1], \tag{4.7}$$

so that $\overline{\underset{\sim}{\Phi}}_{j0} = \int_0^1 \underset{\sim}{\phi}_j(u)du = \overline{\underset{\sim}{\phi}}_j = \underset{\sim}{0}$ (by assumption), and let $\overline{\underset{\sim}{\Phi}}_{j1} = 0$, $1 \le j \le p$.

Also, let

$$\underset{\sim}{\nu}_{00}(t) = \int_0^t [\underset{\sim}{\phi}_0(u)][\underset{\sim}{\phi}_0(u)]'du + (1-t)\overline{\underset{\sim}{\Phi}}_{0t}\overline{\underset{\sim}{\Phi}}_{0t}', \ t \in [0,1], \tag{4.8}$$

$$\underset{\sim}{\nu}_{0j}(t) = [\underset{\sim}{\nu}_{j0}(t)]' = \int_{-\infty}^{\zeta_t^0}\int_{-\infty}^{\infty} [\phi_{\sim 0}(F_{[0]}(x))][\underset{\sim}{\phi}_j(F_{[j]}(y))]' dF_{[0j]}(x,y)$$

$$+ (1-t)\bar{\underset{\sim}{\phi}}_{0t}\bar{\underset{\sim}{\phi}}_{0t}', \ 1 \le j \le p, \ t \in [0,1], \tag{4.9}$$

$$\underset{\sim}{\nu}_{j\ell}(t) = \int_{-\infty}^{\zeta_t^0}\int_{-\infty}^{\infty}\int_{-\infty}^{\infty} [\underset{\sim}{\phi}_j(F_{[j]}(y))][\underset{\sim}{\phi}_\ell(F_{[\ell]}(z))]' dF_{[0j\ell]}(x,y,z)$$

$$+ (1-t)\bar{\underset{\sim}{\phi}}_{jt}\bar{\underset{\sim}{\phi}}_{\ell t}', \ 1 \le j, \ell \le p, \ t \in [0,1], \tag{4.10}$$

$$\underset{\sim}{\nu}(t) = \begin{pmatrix} \underset{\sim}{\nu}_{00}(t) & \cdots & \underset{\sim}{\nu}_{0p}(t) \\ \vdots & & \vdots \\ \underset{\sim}{\nu}_{p0}(t) & \cdots & \underset{\sim}{\nu}_{pp}(t) \end{pmatrix} = \begin{pmatrix} \underset{\sim}{\nu}_{00}(t) & \underset{\sim}{\nu}_{0*}(t) \\ & \\ \underset{\sim}{\nu}_{*0}(t) & \underset{\sim}{\nu}_{**}(t) \end{pmatrix}, \ 0 \le t \le 1, \tag{4.11}$$

$$\underset{\sim}{\nu}_{00}^*(t) = \underset{\sim}{\nu}_{00}(t) - \underset{\sim}{\nu}_{0*}(t)[\underset{\sim}{\nu}_{**}(t)]^- \underset{\sim}{\nu}_{*0}(t), \ 0 \le t \le 1. \tag{4.12}$$

By a direct generalization of Theorem 4.1 of Sen (1979b), we have then under (2.1), (4.1)-(4.5) and for continuous F^*,

$$\max_{k \le n}\{||\underset{\sim}{V}_{n,k} - \underset{\sim}{\nu}(k/n)||\} \xrightarrow{p} 0, \text{ as } n \to \infty, \tag{4.13}$$

where $||\underset{\sim}{A}||$ stands for the maximum of the elements of $\underset{\sim}{A}$ [and the proof of (4.13) is omitted as it runs parallel to the proof of Theorem 4.1 of Sen (1979b)]. As such, by (3.10), (4.12) and (4.13), we obtain by some standard steps that under (2.1), (4.1)-(4.5) and for continuous F^*,

$$\max_{k \le n}\{||\underset{\sim}{V}_{n,00\cdot *}^{(k)} - \underset{\sim}{\nu}_{00}^*(k/n)||\} \xrightarrow{p} 0, \text{ as } n \to \infty. \tag{4.14}$$

We assume that

$$\underset{\sim}{\nu}_{00}^*(t) \text{ is p.d., for every } t \in (0,1], \tag{4.15}$$

and denote the reciprocal matrix by

$$\underset{\sim}{\nu}_{00}^{*-1}(t) = ((\nu_*^{rs}(t)))_{r,s=1,\dots,m}, \text{ for } 0 < t \le 1. \tag{4.16}$$

Note that by (2.9)-(2.11), for every $n(\ge 1)$, under P_n, $\{\underset{\sim}{T}_{n,k}, 0 \le k \le n\}$ is a martingale sequence, so that by (3.7), for every $0 \le k \le n-1$,

$$\underset{\sim}{V}_{n,k+1} - \underset{\sim}{V}_{n,k} \text{ is positive semi-definite (p.s.d.)} \tag{4.17}$$

Also, using (3.8), (3.10) and the martingale property of $\{\underset{\sim}{T}_{n,k}, 0 \le k \le n\}$, we get that

$$E[\underset{\sim}{T}_{n,k+1}^0 - \underset{\sim}{T}_{n,k}^0][\underset{\sim}{T}_{n,k+1}^0 - \underset{\sim}{T}_{n,k}^0]' = \underset{\sim}{V}_{n,00\cdot *}^{(k+1)} - \underset{\sim}{V}_{n,00\cdot *}^{(k)} \text{ is p.s.d.} \tag{4.18}$$

for every $0 \le k \le n-1$. Now, by virtue of (4.4)-(4.5) and the assumed continuity of F^*, it follows that

$$\sup\{||\underset{\sim}{\nu}(t) - \underset{\sim}{\nu}(s)||: 0 \le s \le t \le s+\delta \le 1\} \to 0 \text{ as } \delta \downarrow 0, \tag{4.19}$$

$$\sup\{||\underset{\sim}{\nu}_{00}^*(t) - \underset{\sim}{\nu}_{00}^*(s)||: 0 \le s \le t \le s+\delta \le 1\} \to 0 \text{ as } \delta \downarrow 0, \tag{4.20}$$

so that by (3.10), (4.13), 4.15), (4.19) and (4.20)-(4.21), as $\delta \downarrow 0$,

$$\max\{||\underset{\sim}{V}_{n,00\cdot *}^{(q)} - \underset{\sim}{V}_{n,00\cdot *}^{(k)}||: |q-k| \le \delta n\} \xrightarrow{p} 0, \ n \to \infty. \tag{4.21}$$

Theorem 4.1. *Under* (2.1), (4.2), (4.4) *and* (4.5), *for every (fixed)* $s(\ge 1)$,

$(0 \le)t_1 < \cdots < t_s(\le 1)$ *and* $\{k_1,\ldots,k_s\}$, *satisfying* $\lim_{n\to\infty} n^{-1}k_j = t_j$, $1 \le j \le s$,

$$\underset{\sim}{C}_n^{-1/2}(\underset{\sim}{T}_{n,k_1},\ldots,\underset{\sim}{T}_{n,k_s}) \xrightarrow{\mathcal{D}} N(\underset{\sim}{0},\underset{\sim}{I}_q \otimes \underset{\sim}{\Gamma}), \tag{4.22}$$

where

$$\underset{\sim}{\Gamma} = \begin{pmatrix} \underset{\sim}{\Gamma}_{11} & \cdots & \underset{\sim}{\Gamma}_{1s} \\ \vdots & & \vdots \\ \underset{\sim}{\Gamma}_{s1} & & \underset{\sim}{\Gamma}_{ss} \end{pmatrix}; \; \underset{\sim}{\Gamma}_{j\ell} = \underset{\sim}{\nu}(t_j \wedge t_\ell),\; j,\ell = 1,\ldots,s. \tag{4.23}$$

Outline of the proof. Note that each of the $\underset{\sim}{T}_{n,k_j}$ is a $qm(p+1)$-vector, so to prove (4.22), one may use the Cramér-Wold theorem, whereby one needs to consider an arbitrary linear combination of these $sqm(p+1)$ random variables and to prove the asymptotic normality of the same. Once we consider such a linear combination, the case becomes very similar to the one treated in the proof of Theorem 4.3 of Sen (1979b) (dealing with $q = m = 1$), and the same martingale-approach works out nicely. Therefore, the details are omitted.

Next, by (3.8), (3.10), (4.12), (4.14) and Theorem 4.1, we arrive at the following.

Theorem 4.2. *Under the hypothesis of Theorem* 4.1, *for every (fixed)* $s(\ge 1)$, $(0 \le)t_1 < \cdots < t_s(\le 1)$ *and* $\{k_1,\ldots,k_s\}$, *satisfying* $\lim_{n\to\infty} n^{-1}k_j = t_j$, $1 \le j \le s$,

$$\underset{\sim}{C}_n^{-1/2}(\underset{\sim}{T}^o_{n,k_1},\ldots,\underset{\sim}{T}^o_{n,k_s}) \xrightarrow{\mathcal{D}} N(\underset{\sim}{0},\underset{\sim}{I}_q \otimes \underset{\sim}{\Gamma}^*), \tag{4.24}$$

where

$$\underset{\sim}{\Gamma}^* = \begin{pmatrix} \underset{\sim}{\Gamma}^*_{11} & \cdots & \underset{\sim}{\Gamma}^*_{1s} \\ \vdots & & \vdots \\ \underset{\sim}{\Gamma}^*_{s1} & & \underset{\sim}{\Gamma}^*_{ss} \end{pmatrix}; \; \underset{\sim}{\Gamma}^*_{j\ell} = \underset{\sim}{\nu}^*_{00}(t_j \wedge t_\ell),\; j,\ell = 1,\ldots,s. \tag{4.25}$$

By virtue of (4.15) and (4.18), we conclude that for every $\varepsilon(0 < \varepsilon \le 1)$,

$$P\{V^{(k)}_{n,00\cdot *} \text{ is p.d.}, \forall\, k \ge n\varepsilon\} \to 1, \text{ as } n \to \infty. \tag{4.26}$$

Now, for every $\varepsilon(0 < \varepsilon < 1)$ and (r,n) satisfying (4.3), we consider a mq-variate stochastic process $_{\varepsilon}\underset{\sim}{W}_{n,r} = \{\underset{\sim}{W}_{n,r}(t),\ \varepsilon \le t \le 1\}$ by letting

$$\underset{\sim}{W}_{n,r}(t) = [\underset{\sim}{C}_n \otimes \underset{\sim}{V}^{(k)}_{n,00\cdot *}]^{-1/2}\underset{\sim}{T}^o_{n,k} \quad \text{for } \frac{k}{r} \le t < \frac{k+1}{r},\ t \in [\varepsilon,1], \tag{4.27}$$

so that it belongs to the space $D^{qm}[\varepsilon,1]$. Also, let $_{\varepsilon}\underset{\sim}{W} = \{\underset{\sim}{W}(t),\ \varepsilon \le t \le 1\}$ be a qm-variate Gaussian function with no drift and covariance function

$$E\{[\underset{\sim}{W}(s)][\underset{\sim}{W}(t)]'\} = \underset{\sim}{I}_q \otimes ([(\underset{\sim}{\nu}^*_{00}(\tau(s \wedge t))]^{1/2}[\underset{\sim}{\nu}_{00}(\tau(s \vee t))]^{-1/2}), \tag{4.28}$$

for $s,t \in [\varepsilon,1]$. Then, we have the following.

Theorem 4.3. *Under the hypothesis of Theorem* 4.1, *as* $n \to \infty$,

$$_{\varepsilon}\underset{\sim}{W}_{n,r} \xrightarrow{\mathcal{D}} {_{\varepsilon}\underset{\sim}{W}}, \textit{ in the (extended) } J_1\textit{-topology on } D^{mq}[\varepsilon,1], \tag{4.29}$$

for every $\varepsilon \in (0,1)$.

Outline of the proof. The convergence of the finite dimensional distributions of $_{\varepsilon}\underset{\sim}{W}_{n,r}$ to those of $_{\varepsilon}\underset{\sim}{W}$ follows readily from Theorem 4.2, (4.14) and (4.26). Also, note that for $s,t \in [\varepsilon,1]$, $k = [ns]$ and $k' = [nt]$,

$$||\underset{\sim}{W}_{n,r}(t) - \underset{\sim}{W}_{n,r}(s)|| \le ||[\underset{\sim}{C}_n \otimes \underset{\sim}{V}^{(k)}_{n,00\cdot*}]^{-1/2}(\underset{\sim}{T}^{o}_{n,k} - \underset{\sim}{T}^{o}_{n,k'})||$$

$$+ ||([\underset{\sim}{C}_n \otimes \underset{\sim}{V}^{(k)}_{n,00\cdot*}]^{-1/2} - [\underset{\sim}{C}_n \times \underset{\sim}{V}^{(k')}_{n,00\cdot*}]^{-1/2})\underset{\sim}{T}_{n,k'}|| \tag{4.30}$$

where $||\cdot||$ stands for the Euclidean norm. For each coordinate of $\underset{\sim}{T}^{o}_{n,k}$ (or $\underset{\sim}{T}^{o}_{n,k} - \underset{\sim}{T}^{o}_{n,k'}$), we may proceed as in the proof of Theorem 4.5 of Sen (1979b), and hence, using (4.21), (4.26) and the above, the *tightness* of $\{{}_{\varepsilon}\underset{\sim}{W}_{n,r}\}$ follows. For intended brevity, the details are omitted.

Let us now define

$$W^*_{\varepsilon} = \sup\{[\underset{\sim}{W}(t)]'[\underset{\sim}{W}(t)]: \varepsilon \le t \le 1\} \tag{4.31}$$

and, in (3.13), we let $n_0 = [\varepsilon r]$, where $0 < \varepsilon < 1$. Then, by virtue of Theorem 4.3, under the null hypothesis (2.1),

$$\max_{n_0 \le k \le r} \xi_{nk} \xrightarrow[\mathcal{D}]{} W^*_{\varepsilon}, \text{ as } n \to \infty, \tag{4.32}$$

and hence, in (3.13), for large n, $\ell^{(\alpha)}_n$ may be replaced by

$$\omega^*_{\alpha\varepsilon}, \text{ where } P\{W^*_{\varepsilon} \ge \omega^*_{\alpha\varepsilon}\} = \alpha. \tag{4.33}$$

Thus, the task reduces to that of finding $\omega^*_{\alpha\varepsilon}$ and, toward this, some developments will be reported in the next section.

Let us next proceed to study the non-null distribution theory. We confine ourselves to some local alternatives for which the asymptotic distributions are non-degenerate. For every n, we conceive of a sequence $\{\underset{\sim}{X}^*_{n1},\ldots,\underset{\sim}{X}^*_{nn}\}$ of $(p + 1)$-vectors, where $\underset{\sim}{X}^*_{ni}$ has the d.f. F^*_{ni}, $1 \le i \le n$ and keeping in mind (2.1)-(2.2), we consider the model where F^*_{ni} has an absoluately continuous p.d.f. f^*_{ni} and

$$f^*_{ni}(x^*) = f^*(\underset{\sim}{x}^*; \Delta\underset{\sim}{d}_{ni}), \; 1 \le i \le n, \;\; \underset{\sim}{x}^* = (x_0,\underset{\sim}{x}')' \in E^{p+1} \tag{4.34}$$

and as in Section 2, we have $\int f^*_{ni}(\underset{\sim}{x}^*)dx_0 = f_{ni}(\underset{\sim}{x}) = f(\underset{\sim}{x})$, $\forall\, \underset{\sim}{x} \in E^p$ and $1 \le i \le n$. For the triangular array $\{\underset{\sim}{d}_{ni}\}$, we let $\underset{\sim}{D}_n = \sum_{i=1}^{n}\underset{\sim}{d}_{ni}\underset{\sim}{d}'_{ni}$ and assume that

$$\sum_{i=1}^{n}\underset{\sim}{d}_{ni} = \underset{\sim}{0}, \; \sup_n \operatorname{Tr}(\underset{\sim}{D}_n) < \infty, \tag{4.35}$$

$$\underset{\sim}{D}_n \text{ is p.d. for every } n(\ge n_0), \tag{4.36}$$

$$\lim_{n\to\infty}\{\max_{1\le k\le n} \underset{\sim}{d}'_{nk}\underset{\sim}{D}^{-1}_n\underset{\sim}{d}_{nk}\} = 0. \tag{4.37}$$

Further, we define the $\underset{\sim}{c}_{ni}$ as in (4.1)-(4.2) and assume that

$$\lim_{n\to\infty}(\sum_{i=1}^{n}\underset{\sim}{d}_{ni}\underset{\sim}{c}'_{ni}) = \underset{\sim}{P} \text{ exists.} \tag{4.38}$$

Let J be an m-dimensional rectangle containing $\underset{\sim}{0}$ as an inner point and for every $\underset{\sim}{\theta} \in J$, we assume that $f^*(\underset{\sim}{x}^*;\underset{\sim}{\theta})$ satisfies the following:

(i) $f^*(\underset{\sim}{x}^*;\underset{\sim}{\theta})$ is absolutely continuous in $\underset{\sim}{\theta} = (\underset{\sim}{\theta}_1,\ldots,\underset{\sim}{\theta}_n)'$, for almost every $\underset{\sim}{x}^*$, (4.39)

(ii) For every $j(= 1,\ldots,m)$ the limits

$$\dot{f}_j^*(\underset{\sim}{x}^*;\underset{\sim}{0}) = \lim_{\theta_j\to 0} \theta_j^{-1}[f^*(\underset{\sim}{x}^*;(0,\ldots,\theta_j,0\cdots 0)) - f^*(\underset{\sim}{x}^*;\underset{\sim}{0})], \tag{4.40}$$

exist for almost every $\underset{\sim}{x}^*$, and

(iii)
$$\lim_{\underset{\sim}{\theta}\to\underset{\sim}{0}} \int_{-\infty}^{\infty}\cdots\int_{-\infty}^{\infty}[\dot{\underset{\sim}{f}}^*(\underset{\sim}{x}^*;\underset{\sim}{\theta})][\dot{\underset{\sim}{f}}^*(\underset{\sim}{x}^*;\underset{\sim}{\theta})]'\{f^*(\underset{\sim}{x}^*;\underset{\sim}{\theta})\}^{-1}d\underset{\sim}{x}^*$$
$$= \int_{-\infty}^{\infty}\cdots\int_{-\infty}^{\infty}[\dot{\underset{\sim}{f}}^*(\underset{\sim}{x}^*;\underset{\sim}{0})][\dot{\underset{\sim}{f}}^*(\underset{\sim}{x}^*;\underset{\sim}{0})]'\{f^*(\underset{\sim}{x}^*;\underset{\sim}{0})\}^{-1}d\underset{\sim}{x}^* \tag{4.41}$$

is p.d. and finite, where $\dot{\underset{\sim}{f}}^* = (\dot{f}_1^*,\ldots,\dot{f}_m^*)'$. We denote the sequence of alternatives in (4.34) by $\{K_n\}$ and assume that (4.35)-(4.41) hold. Note that H_0 in (2.1) relates to $\underset{\sim}{\Delta} = \underset{\sim}{0}$. The *contiguity* of the sequence of probability measures under $\{K_n\}$ with respect to one under H_0 is insured by (4.35)-(4.41) [c.f., Hájek and Šidák (1967, pp. 238-239) and Patel (1973) in this context] and will not be proved in detail.

With the same notations as in (4.4) through (4.12), we now let

$$\underset{\sim}{\mu}_j(t) = \int_{-\infty}^{\zeta_t^0}\int_{E^p}\cdots\int\underset{\sim}{\phi}_j(F_{[j]}(x_j))[\dot{\underset{\sim}{f}}^*(\underset{\sim}{x}^*;0)]'d\underset{\sim}{x}^* + f_{[0]}(\zeta_t^0)\overline{\Phi}_{jt} \tag{4.42}$$

for $j = 0,1,\ldots,p$ and $0 < t \le \tau$ and let

$$\underset{\sim}{\mu}_\tau^*(t) = [\underset{\sim}{I}_q \times \underset{\sim}{\nu}_{00}^*(\tau t)]^{-1/2}\underset{\sim}{P}'\underset{\sim}{\Delta}'\{\underset{\sim}{\mu}_0(\tau t) - \underset{\sim}{\nu}_{0*}(\tau t)\underset{\sim}{\nu}_{**}^-(\tau t)\underset{\sim}{\mu}(\tau t)\} \tag{4.43}$$

for $0 < t \le 1$, where $\underset{\sim}{\mu}(\underset{\sim}{\alpha})$ is the $pm \times m$ matrix with components $\underset{\sim}{\mu}_j(\alpha)$, $1 \le j \le p$. Then, we have the following.

<u>Theorem 4.4.</u> *Under* (4.1)-(4.5) *and* $\{K_n\}$ *in* (4.34) *through* (4.41),

$$_{\varepsilon}\underset{\sim}{W}_{n,r} \xrightarrow[\mathcal{D}]{} {}_{\varepsilon}\underset{\sim}{W} + {}_{\varepsilon}\underset{\sim}{\mu}^*, \text{ in the } J_1\text{-topology on } D^{qm}[\varepsilon,1], \tag{4.44}$$

for every $\varepsilon > 0$, *where* ${}_{\varepsilon}\underset{\sim}{\mu}^* = \{\underset{\sim}{\mu}_\tau^*(t),\ \varepsilon \le t \le 1\}$ *is defined by* (4.43).

<u>Outline of the proof.</u> Note that the tightness of $\{{}_{\varepsilon}W_{n,r}\}$ (under H_0), established in Theorem 4.3 and the contiguity of the sequence of probability measures under $\{K_n\}$ to that under H_0 insure the tightness of $\{{}_{\varepsilon}W_{n,r}\}$ under $\{K_n\}$ as well [c.f., Theorem 2 of Sen (1976)]. Hence, to prove (4.44), it suffices to establish the convergence of finite dimensional distributions of $\{{}_{\varepsilon}W_{n,r}\}$ to those of ${}_{\varepsilon}\underset{\sim}{W} + {}_{\varepsilon}\underset{\sim}{\mu}^*$. For this, we may readily extend the proof of Theorem 5.1 of Sen (1979b) [to the case of $q \ge 1$, $m \ge 1$] by considering an arbitrary linear compound of the $\underset{\sim}{W}_n(t_j)$, and hence, for intended brevity, the details are omitted.

By virtue of (3.12), (3.13), (4.33) and Theorem 4.4, we conclude that under $\{K_n\}$ in (4.34)-(4.41), the asymptotic power of the proposed PCS test is given by

$$\lim_{n\to\infty} P\{\max_{[n\varepsilon]\le k\le r} |\xi_{nk}| > \ell_n^{(\alpha)} | K_n\}$$
$$= P\{[\underset{\sim}{W}(t) + \underset{\sim}{\mu}_\tau^*(t)]'[\underset{\sim}{W}(t) + \underset{\sim}{\mu}_\tau^*(t)] > \omega_{\alpha\varepsilon}^* \text{ for some } t \in [\varepsilon,1]\}. \tag{4.45}$$

Also, we have for $k/n \to u$: $\tau\varepsilon < u \le \tau$,

$$\lim_{n\to\infty} P\{N > k|K_n\} = P\{[\underset{\sim}{W}(t) + \underset{\sim\tau}{\mu}^*(t)]'[\underset{\sim}{W}(t) + \underset{\sim\tau}{\mu}^*(t)] \le \omega^*_{\alpha\varepsilon},\ \forall\ \varepsilon \le t \le u|\tau\}, \quad (4.46)$$

while $P\{N > n\tau|K_n\} = 0$, by definition in (3.13) and (4.3). Further, noting that $Z^o_{n[n\alpha]} \to \zeta^o_\alpha$, in the s-th mean $(s > 0)$, $\forall\ 0 \le \alpha < 1$ and that $n[\zeta^o_{k/n} - \zeta^o_{(k-1)/n}] - [f_{[0]}(\zeta^o_{k/n})]^{-1} \to 0$, $\forall\ k \le r$, we obtain from (4.45) and some routine steps that

$$\lim_{n\to\infty} E(Z^o_{nN}|K_n) = \tau\int_0^1 \frac{1}{f_{[0]}(\zeta^o_{\tau u})} P\{[W(t) + \underset{\sim\tau}{\mu}^*(t)]'[\underset{\sim}{W}(t) + \underset{\sim\tau}{\mu}^*(t)] \le \omega^*_{\alpha\varepsilon},\ \forall\ \varepsilon \le t \le u\}du. \quad (4.47)$$

Therefore the *asymptotic power* as well as the *expected stopping time* of the proposed PCS test depends on the boundary crossing probability of ${}_{\varepsilon}\underset{\sim}{W} + {}_{\varepsilon}\underset{\sim}{\mu}^*$. We shall discuss more on this in the next section.

5. SOME GENERAL REMARKS

As has been noted after (2.2), the classical ANOCOVA model is a special case of (2.2) where $m = 1$ and $q \ge 1$. In the case, by virtue of (4.18), (4.28) and (4.31), we can also write

$$W^*_\varepsilon = \sup\{t^{-1}[\underset{\sim}{Y}(t)]'[\underset{\sim}{Y}(t)]: \varepsilon \le t \le 1\} \quad (5.1)$$

where $\underset{\sim}{Y} = \{\underset{\sim}{Y}(t),\ 0 \le t \le 1\}$ is a vector of q independent copies of a standard Wiener process, so that $\{[\underset{\sim}{Y}(t)]'[\underset{\sim}{Y}(t)],\ 0 \le t \le 1\}$ is a q-parameter Bessel process. Some tabulation of the critical values of W^*_ε is due to Majumdar (1977) and a more analytical approach to this is due to DeLong (1981). Thus, for this special case, we have no problem in applying the proposed PCS tests. A very similar case holds when we have an ANOCOVA model relating to the scale parameter when we have in (2.2)

$$F^o(y;\underset{\sim}{\Delta}(\underset{\sim i}{c} - \underset{\sim n}{\bar{c}})|\underset{\sim}{x}) = F^o(y(1 + \underset{\sim}{\Delta}(\underset{\sim i}{c} - \underset{\sim n}{\bar{c}}))|\underset{\sim}{x}) \quad (5.2)$$

and the $\underset{\sim i}{c}$ and $\underset{\sim}{\Delta}$ satisfy the condition that $1 + \underset{\sim}{\Delta}(\underset{\sim i}{c} - \underset{\sim}{\bar{c}}) > 0$, $\forall\ i$. Though this scale model has been considered for the analysis of variance models [c.f., Hájek and Šidák (1967)], it has not been considered explicitly for the ANOCOVA model, even in the case of no censoring.

In the general case of $m \ge 2$ (e.g., joint location-scale model), the computation of the (simulation or analytical) critical values of $W^*_{\alpha\varepsilon}$ seems to be quite involved (because it depends on the covariance structure in (4.28)). Simplification is of course possible when we have for some scalar function $\{\gamma_t(s),\ s \le t\}$

$$\underset{\sim}{\nu}^*_{00}(s) = \gamma_t(s)\underset{\sim}{\nu}^*_{00}(t) \quad \text{for every } s \le t, \quad (5.3)$$

where $\gamma_t(s)$ is $\nearrow$ in s. In this case, again we have the solution given by (5.1) with q being replaced by qm.

Study of the asymptotic power properties of the PCS tests [viz. (4.45)-(4.47)] demands the knowledge of the boundary crossing probabilities for drifted Bessel processes. The prospect for this rests heavily on the simulation techniques (as the analytical tools are not yet properly available). For the analysis of variance model, some studies are due to Majumdar (1977) and DeLong (1980, 1981), among others. There is some pressing needs for such studies for the ANOCOVA problem.

Further, it is usually difficult to employ the concept of Pitman-efficiency in PCS tests. Hence, some alternative, meaningful measures of asymptotic efficiency should be explored and, if needed, numerical studies should be made.

Finally, throughout the paper, we have considered the case of PCS rank statistics for the ANOCOVA problem. Cox (1972, 1975) has considered some quasi-nonparametric regression models for covariate-adjustments in survival analysis. Though Cox has not considered the case of repeated significance tests arising in PCS models, his theory can be extended to cover such repeated significance testing. In fact, for a class of simple hypotheses relating to the Cox regression model, such repeated significance tests are considered in Sen (1981a). The theory of such tests rests on some invariance principles similar to the ones considered in this paper. It is intended to extend the theory to the case of composite hypotheses (which requires a somewhat different approach as the simple unweighted random sampling theory from a finite population will no longer be applicable for this complicated situation) and the same will be taken up in a subsequent communication. Unlike the Cox model, the models in the current paper do not require the *proportional hazard* assumption and therefore remain valid for a wider class of situations. If, however, the proportional hazard model can be justified, then, one would expect that the Cox regression model based theory would perform better.

6. REFERENCES

[1] Chatterjee, S. K. and Sen, P. K., Nonparametric testing under progressive censoring, *Calcutta Statist. Assoc. Bull. 22* (1973) 13-50.

[2] Cox, D. R., Regression models and life tables, *Jour. Royal Statistic. Soc. Ser. B 34* (1972) 187-220.

[3] Cox, D. R., Partial likelihoods, *Biometrika 62* (1975) 269-276.

[4] DeLong, D. M., Some asymptotic properties of a progressively censored nonparametric test for multiple regression, *Jour. Multivar. Anal. 10* (1980).

[5] DeLong, D. M., Crossing probabilities for a square root boundary, *Comm. Statist., Ser. A 10* (1981).

[6] Hájek, J. and Šidák, Z., *Theory of Rank Tests* (Academic Press, New York, 1967).

[7] Majumdar, H., Rank order tests for multiple regression for grouped data under progressive censoring, *Calcutta Statist. Assoc. Bull. 26* (1977) 1-16.

[8] Majumdar, H. and Sen, P. K., Nonparametric tests for multiple regression under progressive censoring, *Jour. Multivar. Anal. 8* (1978) 73-95.

[9] Patel, K. M., Hájek-Šidák approach to the asymptotic distribution of multivariate rank order statistics, *Jour. Multivar. Anal. 3* (1973) 57-70.

[10] Puri, M. L. and Sen, P. K., Analysis of covariance based on general rank scores, *Ann. Math. Statist.* *40* (1969) 610-618.

[11] Puri, M. L. and Sen, P. K., A class of rank order tests for a general linear hypothesis, *Ann. Math. Statist.* *40* (1969) 1325-1343.

[12] Quade, D., Rank analysis of covariance, *Jour. Amer. Statist. Assoc.* *62* (1967) 1187-1200.

[13] Sen, P. K., A two-dimensional functional permutational central limit theorem for linear rank statistics, *Ann. Probability* *4* (1976) 13-26.

[14] Sen, P. K., Weak convergence of some quantile processes arising in progressively censored tests, *Ann. Statist.* *7* (1979) 414-431.

[15] Sen, P. K., Rank analysis of covariance under progressive censoring, *Sankhyà, Ser. A* *41* (1979) 147-169.

[16] Sen, P. K., The Cox regression model, invariance principles for some induced quantile processes and some repeated significance tests, *Ann. Statist.* *9* (1981).

[17] Sen, P. K., Asymptotic theory of some time-sequential tests based on progressively censored quantile processes, in: Kallianpur, G. et al. (eds.), *Statistics and Probability: Essays in Honor of C. R. Rao* (North-Holland, Amsterdam, 1981).

[18] Sen, P. K. and Puri, M. L., Asymptotic theory of likelihood ratio and rank order tests in some multivariate linear models, *Ann. Math. Statist.* *41* (1970) 87-100.

STATISTICS AND RELATED TOPICS
M. Csörgö, D.A. Dawson, J.N.K. Rao, A.K.Md.E. Saleh (eds.)

ROBUST TWO-SAMPLE TEST AND ROBUST REGRESSION AND ANALYSIS-OF-VARIANCE VIA MML ESTIMATORS

M.L. TIKU

Department of Mathematical Sciences
McMaster University, Hamilton, Canada

We investigate the joint efficiencies of Tiku's (1967) MML (modified maximum likelihood) estimators of the location and scale parameters of symmetric distributions and show that these estimators are jointly at least as efficient as the celebrated robust estimators (bisquare, wave and Hampel; Gross, 1976). We define a statistic (based on the MML estimators) for testing that two distributions are identical and show that this statistic is robust and generally more powerful than the prominent non-parametric statistics (Wilcoxon, normal-score, Smirnov-Kolmogorov-Massey). We also investigate the efficiencies of Tiku's (1978) MML estimators of the regression coefficients in linear and multiple regression; these estimators are shown to be more efficient than Atkinson and Cox (1977) robust estimators. We develop analysis-of-variance procedures (based on the MML estimators); these procedures are robust and powerful.

1. INTRODUCTION

Let $x_1, x_2, \ldots, x_n$ be a random sample from the normal $N(\mu,\sigma)$ distribution, and let

$$X_{r+1}, X_{r+2}, \ldots, X_{n-r} \tag{1.1}$$

be the Type II symmetrically censored sample obtained by arranging the above n random observations in ascending order of magnitude and censoring the r smallest and the r largest observations; the value of r will be specified later. Tiku's (1967, 1970) MML estimators (defined formally by Tiku and Stewart, 1977) of μ and σ are given by

$$\mu_c = \{\sum_{i=r+1}^{n-r} X_i + r\beta(X_{r+1} + X_{n-r})\}/m \tag{1.2}$$

and

$$\sigma_c = \{B + \surd(B^2 + 4AC)\}/2\surd\{A(A - 1)\}, \tag{1.3}$$

where

$$m = n - 2r + 2r\beta,\ A = n - 2r,\ B = r\alpha(X_{n-r} - X_{r+1})$$

and

$$C = \sum_{i=r+1}^{n-r} X_i^2 + r\beta(X_{r+1}^2 + X_{n-r}^2) - m\mu_c^2;$$

α and β are simple constants and are given by Tiku (1967, 1970). For $n \geq 10$, α and β are obtained from the following simpler equations ($q = r/n$)

$$\beta = -f(t)\{t - f(t)/q\}/q \quad \text{and} \quad \alpha = \{f(t)/q\} - \beta t, \tag{1.4}$$

where $f(t) = \{1/\surd(2\pi)\} \exp(-t^2/2)$, $-\infty < t < \infty$, and t is determined by $F(t) = \int_{-\infty}^{t} f(x)dx = 1 - q$. Note that $0 < \alpha < 1$ and $0 < \beta < 1$.

For normal samples, Tiku (1978) investigated the distributions and the efficiencies of μ_c and σ_c. Suffice it to say here that μ_c and σ_c are an ideal pair of estimators, for censored normal samples; see also Smith, Zeis and Syler (1973). Note that for large $A = n - 2r$, the term AC in (1.3) dominates and $B^2/AC = 0$ in which case $\sigma_c = \surd\{C/(A-1)\}$; $(\mu_c - \mu)\surd m/\sigma$ and $(A - 1)\sigma_c^2/\sigma^2$ are independently distributed as normal $N(0,1)$ and chi-square with $A - 1$ df, for large A; see Tiku (1978, Lemmas 1 and 2).

For $r = 0$ (no censoring), μ_c and σ_c reduce to the sample mean and standard deviation $\bar{x}$ and s, respectively.

The question is how efficient are the estimators μ_c and σ_c for estimating the location parameter μ and scale parameter σ of symmetric non-normal distributions. Note that μ_c is unbiased and uncorrelated with σ_c, for symmetric distributions.

2. JOINT EFFICIENCY OF μ_c and σ_c

Tiku (1980) investigated (through Monte Carlo simulations) in detail the joint deficiency (efficiency = 1/deficiency) $\mathrm{Def}(\mu_c, \sigma_c) = \mathrm{MSE}(\mu_c) + \mathrm{MSE}(\sigma_c)$, of the pair (μ_c, σ_c) for a large number of symmetric distributions and compared the values of $\mathrm{Def}(\mu_c, \sigma_c)$ with the corresponding values for the celebrated robust estimators (Gross, 1976), namely, the trimmed mean and the matching sample estimate of σ: (μ_T, σ_T), the Hampel estimators H22: (μ_H, σ_H), the wave estimators w24: (μ_w, σ_w), and the bisquare estimators BS82: (μ_B, σ_B). He concluded that with the right choice of r (the number of observations censored on either side of the ordered sample), namely, (i) $r = 0$ (no censoring) if the sample comes from a short-tailed (kurtosis $\beta_2^* = \mu_4/\mu_2^2 < 3$) distribution, (ii) $r = [0.5 + 0.1n]$ if the sample either comes from a long-tailed ($\beta_2^* > 3$) distribution with finite variance or has only a few (not more than 20%) outliers, and (iii) $r = [0.5 + 0.3n]$ if the sample either comes from a distribution with infinite variance (Cauchy, for example) or has more than 20% outliers, the pair (μ_c, σ_c) is on the whole considerably more efficient than the celebrated robust estimators mentioned above; $[g]$ denotes the integer part of g. Note that (μ_c, σ_c) and (μ_T, σ_T) are both applicable in the censoring as well as in the robustness framework but, for a given r, the pair (μ_c, σ_c) is always jointly more efficienct than the pair (μ_T, σ_T); see Tiku (1980). However, if one has no knowledge whether the situation is that of (i), (ii), or (iii) above, one may use the adaptive estimators μ_c^* and σ_c^*; the adaptive estimators μ_c^* and σ_c^* are defined to be exactly the same as μ_c and σ_c, respectively, but with the constant fraction $q = r/n$ replaced by the random fraction $q^* = r^*/n$, where $r^* = 0$ if $m^*/n = 0$, $r^* = [0.5 + 0.1n]$ if $m^*/n \leq 0.1$, and $r^* = [0.5 + 0.3n]$ if $m^*/n > 0.1$; m^* is the number of values of $|z_i| = |x_i - \mathrm{median}(x_i)|/1.48\,\mathrm{median}|x_i - \mathrm{median}(x_i)|$, $i = 1, 2, \ldots, n$, which exceed 3.0. The adaptive estimators μ_c^* and σ_c^* are on the whole at least as efficient (jointly) as the celebrated robust estimators H22, w24 and BS82; see Tiku (1980). For example, for the distributions (1) Uniform, (2) Normal, (3) $f(p;x)$ with $p = 5$, (4) $f(p;x)$ with $p = 2.5$, (5) $(n - 5)N(0,1)$ & $5N(0,3)$, (6) Student's t_2, (7) Cauchy, and (8) Normal/Uniform, we have the following values; $n = 20$, $f(p;x) = C\{1 + (x - \mu)^2/k\sigma^2\}^{-p}$, $-\infty < x < \infty$ $(k = 2p - 3,\ p \geq 2)$, $\sigma = 1$. Note that all the above estimators of μ are uncorrelated with the corresponding estimators of σ, for symmetric distributions.

In calculating the following means and mean square errors, the random deviates were divided by the standard deviation (whenever it exists) of the population. For example, the random deviates generated from the population $(n-5)N(0,1)$ & $5N(0,3)$ were divided by $\sqrt{(1+40/n)}$.

	Distributions							
	(1)	(2)	(3)	(4)	(5)	(6)	(7)	(8)
Mean σ_c	1.01	.99	.95	.84	1.15	1.15	1.41	2.09
σ_c^*	.99	.96	.91	.81	1.29	1.31	1.60	2.36
σ_B	1.03	.98	.93	.82	1.32	1.36	1.79	2.51
MSE μ_c	.050	.052	.048	.040	.092	.095	.167	.329
μ_c^*	.053	.051	.049	.040	.109	.121	.222	.444
μ_B	.060	.052	.048	.037	.097	.105	.183	.359
MSE σ_c	.012	.034	.040	.062	.158	.190	.516	1.81
σ_c^*	.013	.035	.048	.078	.193	.278	.880	2.72
σ_B	.016	.031	.041	.066	.190	.274	1.07	3.02
Def. (μ_c, σ_c)	.062	.086	.088	.102	.250	.285	.682	2.14/3.695
(μ_c^*, σ_c^*)	.066	.086	.097	.118	.302	.399	1.10	3.16/5.328
BSB2	.076	.083	.089	.103	.287	.379	1.25	3.38/5.647

For the estimators μ_c and σ_c, $r = 0$ for distribution (1), $r = 2$ for (2), (3) and (4), and $r = 6$ for (5), (6), (7) and (8).

Note that the values 3.695, 5.328 and 5.647 are the sums of the deficiencies of the three pairs (μ_c, σ_c), (μ_c^*, σ_c^*) and BS82 (generally the most efficient of the celebrated robust estimators w24, H22, w18, BS74, BS90, etc.); see Gross (1976). However, there is a crucial difficulty with all the above adaptive estimators (μ_c^*, σ_c^*), w24, BS82, etc., and that is that the distributions of the Studentized statistics based on them are not tractable for finite n; see also Gross (1976).

On the other hand, the distribution of the Studentized statistic $t_c = \sqrt{m}(\mu_c - \mu)/\sigma_c$ based on the pair (μ_c, σ_c) is closely approximated by Student's t distribution (see the next Section) having A-1 df. In the rest of the paper, therefore, we consider only the pair (μ_c, σ_c). Note that the pair (μ_c, σ_c) has the distinct advantage that it is applicable in the censoring as well as in the robustness framework.

3. ROBUST TWO-SAMPLE TEST

Let $x_{11}, x_{12},\ldots,x_{1n_1}$ and $x_{21},x_{22},\ldots,x_{2n_2}$ be two independent random samples from two symmetric distributions $(1/\sigma_1)g_1((x-\mu_1)/\sigma_1)$ and $(1/\sigma_2)g_2((x-\mu_2)/\sigma_2)$, respectively. One wants to test the null hypothesis that g_1 and g_2 are identical in all respects. Of considerable practical interest, however, is the location-shift alternative

$$H_1: \quad d = \mu_1 - \mu_2 > 0;$$

$\sigma_1 = \sigma_2$ and g_1 and g_2 have the same functional form. Under H_0, $d = 0$. Consider the statistic

$$T_c = (\mu_{1c} - \mu_{2c})/\sigma_c\surd\{(1/m_1) + (1/m_2)\}, \tag{3.1}$$

where $\sigma_c^2 = \{(A_1 - 1)\sigma_{1c}^2 + (A_2 - 1)\sigma_{2c}^2\}/(A_1 + A_2 - 2)$; $A_i = n_i - 2r_i$ and $m_i = n_i - 2r_i + 2r_i\beta_i$, $i = 1,2$. The value of r_i is chosen to be $r_i = [0.5 + 0.1n_i]$, $i = 1,2$, unless stated otherwise. Note that the constants α_i and β_i are the same as α and β, respectively, with n replaced by n_i and $q = r/n$ replaced by $q_i = r_i/n_i$, $i = 1,2$. The estimators (μ_{ic}, σ_{ic}), $i = 1,2$, are the estimators (μ_c, σ_c) calculated from the censored samples

$$X_{1,r_1+1}, X_{1,r_1+2},\ldots,X_{1,n_1-r_1} \tag{3.2}$$

and

$$X_{2,r_2+1}, X_{2,r_2+2},\ldots,X_{2,n_2-r_2}, \tag{3.3}$$

respectively. Large values of T_c lead to the rejection of H_0.

For normal samples the null distribution of T_c is Student's t having $v = \sum_{i=1}^{2}(n_i - 2r_i - 1)$ df, for large A_i, $i = 1,2$; see Tiku (1978, Lemmas 1 and 2). For any non-normal distribution which satisfies the very general regularity conditions listed by Shorack (1974, p. 662), the asmptotic $(A_i \to \infty, i = 1,2)$ null distribution of T_c is normal $N(0,1)$; see also Tiku (1980), Huber (1970) and Shorack (1974). For small samples $(n_i \geq 6, i = 1,2)$, the probabilities $P(T_c \geq h)$ are closely approximated (under H_0) by the corresponding probabilities of the Student's t distribution having $v = \sum_{i=1}^{2}(n_i - 2r_i - 1)$ df, for symmetric distributions (and, surprisingly, also for skew distributions; Tiku, 1980); see the values for $d = 0$ in Table I.

Note that for normal samples $E(\sigma_c) = \sigma$ for large A_1 and A_2 (Tiku, 1978) and the power of T_c is, say for equal n_1 and n_2, given by ($n_1 = n_2 = n$ and $r_1/n = r_2/n = 0.1$)

$$P\{Z \geqslant Z_\delta - 0.983\Delta\}, \tag{3.4}$$

where Z is a standard normal variate; Z_δ is the $100(1-\delta)\%$ point (δ is Type I error) and $\Delta = d\sqrt{n}/\sigma\sqrt{2}$. The power of the two-sample Student's t statistic for large normal samples is given by ($n_1 = n_2 = n$)

$$P\{Z \geqslant Z_\delta - \Delta\}. \tag{3.5}$$

The values of the power of the statistics T_c and t (calculated from the Eqs. 3.4 and 3.5) are given by

	Δ= 0	.5	1	1.5	2	2.5	3	3.5	4
t	.05	.13	.26	.44	.64	.80	.91	.97	.99
T_c	.05	.12	.25	.43	.63	.79	.90	.96	.99
t	.01	.03	.09	.20	.37	.57	.75	.88	.95
T_c	.01	.03	.09	.20	.36	.55	.73	.87	.95

It is clear that for normal samples, the loss of power resulting from the use of T_c in place of the UMP Student's t statistic is not too serious.

Nonparametric statistics are often used to test the null hypothesis H_0, particularly if one suspects the validity of the normality assumption, and these are known to be generally more powerful than t for long-tailed distributions (Gastwirth, 1965). The most prominent nonparametric statistics are the Wilcoxon statistic

$$W = \sum_{i=1}^{n} i\, Z_i \qquad (n = n_1 + n_2) \tag{3.6}$$

the Terry-Hoeffding normal score statistic

$$C_1 = \sum_{i=1}^{n} Z_i\, E(V_i), \tag{3.7}$$

and the Smirnov-Kolmogorov-Massey statistic

$$D^+ = \max_{1,2,\ldots,n} \{ \sum_{i=1}^{j} [n_2(1 - Z_i) - n_1 Z_i\}; \tag{3.8}$$

see Milton (1970, pp. 23, 25 and 27). We simulated (from 4000 Monte Carlo runs) the values of the power of D^+, C_1, W and T_c, for testing H_0 against H_1, for sample sizes $(n_1, n_2) = (8,6)$, $(8,8)$, $(10,8)$ and $(10,10)$. The values for D^+ were obtained through radomization between the two appropriate commulative probabilities (Kim and Jenrich, 1973). The statistic T_c was found to be generally more powerful than these nonparametric statistics (see Table I), except for some extremely skewed distributions (χ_1^2, for example) in which case the statistic D^+ was found to be considerably more powerful. For the χ_1^2 distribution, the values of the power are given by ($n_1 = n_2 = 8$ and significance level is 1%)

d=	0	.5	1	1.5	2	2.5	3
D^+	.010	.21	.54	.76	.90	.95	.98
C_1	.010	.16	.35	.54	.66	.76	.82
W	.010	.17	.40	.62	.75	.86	.91
T_c	.007	.10	.34	.59	.76	.87	.93

Note that for symmetric distributions with infinite variance, i.e. 'disaster' distribution, the values of the power of T_c with $r_i = [0.5 + 0.3n_i]$ are considerably larger than the corresponding values of T_c with $r_i = [0.5 + 0.1n_i]$, $i = 1,2$. For example for the Cauchy, the values of the power of T_c with $n_1 = n_2 = 8$ and $r_1 = r_2 = 2$ are 0.006, 0.10, 0.36 and 0.64, and these are considerably larger than the corresponding values of T_c with $r_1 = r_2 = 1$ (Table I).

It is interesting to note that with Tiku's (1975, 1977) outlier model preserving symmetry, that is when the k smallest and the k largest observations of a random sample from the normal $N(0, \sigma)$ distribution are multiplied by a positive constant λ (or a positive constant λ is subracted from and added to the smallest k and the largest k observations, respectively, of a random sample from the normal $N(\mu, \sigma)$), the significance levels of the statistics D^+, C_1 and W (but not T_c) get substantially distorted from their preassigned values; see Table I. Note also that the significance levels of T_c, like those of D^+, C_1 and W, are not sensitive to small differences in the standard deviations, say $0.8 < \sigma_1/\sigma_2 < 1.25$, of the two distributions; see Table I.

Note that for large n_1 and n_2 (say $n_1 = n_2 = n$), the power of the Wilcoxon statistic W for normal samples is given by (Lehmann, 1975, p. 73)

$$P\{Z \geq Z_\delta - 0.977\Delta\} \tag{3.9}$$

and, therefore, the statistic W is less powerful than T_c; the power function of T_c for large normal samples is given by equation (3.4).

The statistic T_c can in a straightforward fashion be generalized to test that k symmetric distributions are identical. The generalized statistic is given by

$$T_c^* = \sum_{i=1}^{k} m_i(\mu_{c_i} - \bar{\mu}_c)^2/(k-1)\sigma_c^2, \tag{3.10}$$

where $\bar{\mu}_c = \sum_{i=1}^k m_i\mu_{c_i}/\sum_{i=1}^k m_i$ and $\sigma_c^2 = \sum_{i=1}^k (A_i - 1)\sigma_{c_i}^2/\sum_{i=1}^k (A_i - 1)$. We would expect T^*_c to be generally more powerful than, say, the Kruskal-Wallis statistic (Lehmann, 1975, p. 204).

4. ROBUST REGRESSION BASED ON GROUPED DATA

Consider the linear model

$$y_{ij} = \mu + \Theta x_i + e_{ij};\ j = 1, 2, \ldots, n_i,\ i = 1, 2, \ldots, k. \tag{4.1}$$

We assume that e_{ij}'s are iid, and $E(e_{ij}) = 0$ and $V(e_{ij}) = \sigma^2$. Let $(a_i = r_i + 1,\ b_i = n_i - r_i)$

$$Y_{i,a_i},\ Y_{i,a_i+1}, \ldots, Y_{i,b_i}, \tag{4.2}$$

$i = 1, 2, \ldots, k$, be the k Type II symmetrically censored samples corresponding to the k x-values; $r_i = [0.5 + 0.1n_i]$. If e_{ij}'s are normal, then the MML estimators based on the samples (4.2) are given by (Tiku, 1978)

$$\theta^* = \sum_{i=1}^{k} m_i(x_i - \bar{x})K_i / \sum_{i=1}^{k} m_i(x_i - \bar{x})^2,\ \mu^* = \bar{K} - \theta^* \bar{x}, \tag{4.3}$$

and

$$\sigma_c^* = \{B + \sqrt{(B^2 + 4AC)}\}/2\sqrt{\{A(A-2)\}}. \tag{4.4}$$

Here

$$m_i = n_i - 2r_i + 2r_i\beta_i,\ M = \sum_{i=1}^{k} m_i,\ \bar{K} = \sum_{i=1}^{k} m_i K_i/M, \bar{x} = \sum_{i=1}^{k} m_i x_i/M,$$

$$A = \sum_{i=1}^{k} (n_i - 2r_i),\ B = \sum_{i=1}^{k} r_i\alpha_i(Y_{i,b_i} - Y_{i,a_i}),\ \text{and}$$

$$C = \sum_{i=1}^{k} C_i - M\,\bar{K}^2 - \theta^*Q,\ Q = \sum_{i=1}^{k} m_i(x_i - \bar{x})K_i; \tag{4.5}$$

$$K_i = \{\sum_{j=a_i}^{b_i} Y_{i,j} + r_i\,\beta_i(Y_{i,a_i} + Y_{i,b_i})\}/m_i$$

and

$$C_i = \sum_{j=a_i}^{b_i} Y_{i,j}^2 + r_i\beta_i(Y_{i,a_i}^2 + Y_{i,b_i}^2). \tag{4.6}$$

The constants α_i and β_i, $i = 1, 2,\ldots,k$, are the same as α and β (Section 1) with n replaced by n_i and $q = r/n$ replaced by $q_i = r_i/n_i$; see Tiku (1978).

The efficiencies of μ^* and σ^* for normal and non-normal samples are similar to those of the estimators μ_c and σ_c (reported in Section 2) and we do not, therefore, reproduce them here for conciseness.

5. EFFICIENCY OF Θ^*

The variance of Θ^* is given by

$$V(\Theta^*) = \sum_{i=1}^{k} a_i^2\,V(K_i), \tag{5.1}$$

since K_i's are independent; $a_i = m_i(x_i - \bar{x})/\sum_{i=1}^{k} m_i(x_i - \bar{x})^2$. Note that Θ^* is unbiased, for symmetric distributions. The variance of K_i is of the form $V(K_i) = (\underset{\sim}{\ell}'\,\underset{\sim}{V}\,\underset{\sim}{\ell})\sigma^2$, where $\underset{\sim}{V}$ is the variance-covariance matrix of the standardized variates $Z_j = (Y_{i,j} - \mu - \theta x_i)/\sigma$, $j = a_i, a_i + 1,\ldots,b_i$ $(i = 1, 2,\ldots,k)$. The variances and covariances of these standardized variates are available for a number of symmetric distributions; see Tietjen et al. (1977), Barnett (1966), Govindarajulu (1966), David et al. (1977) and Tiku and Kumra (1979). Using these variances and covariances, we calculated the exact values of the variance $V(\Theta^*)$ for the linear model considered by Atkinson and Cox (1977), that is in the model (4.1), $k = 10$, $n_i = n = 5$ $(i = 1, 2,\ldots,k)$, and x_i are equally spaced on $(-9,9)$. The exact values of the ratio $V(\Theta^*)/0.000606\ \sigma^2$ ($0.000606\ \sigma^2$ is the variance $\sigma^2/\sum_{i=1}^{10} x_i^2$ of the ordinary least square estimator of Θ^*, for populations with finite variance) for the distributions (1) Normal, (2) Student's t_9 (which is effectively the Logistic; Tiku and Jones, 1971), (3) Student's t_7, (4) Laplace, (5) Student's t_4, (6) Student's t_3, (7)4N(0,1) & 1 N(0,4) for $x = 9$ and N(0,1) otherwise, (8) 4 N(0,1) & 1 N(0,4), (9) 4 N(0,1) & 1 N(0,10), and (10) Cauchy, are given below; $r_i = r(i = 1, 2,\ldots,k)$. Note that for $r_i = r = 1$ and $n_i = n = 5$, $i = 1, 2,\ldots,k$, α_i's are all equal and equal to $\alpha = 0.7540$ and β_i's are equal and equal to $\beta = 0.8108$.

Values of the ratio $V(\Theta^*)/0.000606\ \sigma^2$

$r = [0.5 + 0.1n] = 1$

(1)	(2)	(3)	(4)	(5)	(6)	(7)	(8)	(9)
1.129	1.016	0.975	0.829	0.805	0.619	0.844*	0.428	0.909**

$r = [0.5 + 0.3n] = 2$

(8)	(9)	(10)
0.502	0.103	6.107

* Here $r = 1$ for $x = 9$ and 0 otherwise

** Based on 4000 Monte Carolo runs, since the covariances of order statistic are not available (David et al., 1977).

Comparing these values with the few values given by Atkinson and Cox (1977, Table 1), Θ^* is clearly on the whole more efficient than Atkinson-Cox robust estimator. Besides, Atkinson-Cox estimator is rather restrictive in the sense that the number of observations n_i in the groups have to be equal (and there are computational difficulties with Atkinson-Cox estimator). The only desirable restriction on the MML estimator Θ^* is that $n_i \geq 4$ (preferably, $n_i \geq 5$), $i = 1,2,\ldots,k$; for $n_i < 4$, Θ^* would be based only on the medians and might, therefore, lose efficiency. Note that for the above Atkinson-Cox model, the variance of the BLUE (best linear unbiased estimator based on complete samples) of Θ for the Laplace distribution (Govindarajulu, 1966) is $0.00048\ \sigma^2$ and, therefore, the efficiency of Θ^* as compared with the BLUE (based on complete samples) is 95.5% which is remarkably high.

6. ROBUST REGRESSON

To test the null hypothesis H_0: $\Theta = 0$, and to test the linearity of the model (4.1), the component sums of squares are as follows:

$$\text{Linear fit:}\quad S_{\ell r} = \Theta^* Q,\ Q = \textstyle\sum_{i=1}^{k} m_i(x_i - \bar{x})K_i,\ \text{df} = 1$$

$$\text{Lack of fit:}\quad S_{dev} = \textstyle\sum_{i=1}^{k} m_i(K_i - \bar{K})^2 - \Theta^* Q,\ \text{df} = k - 2,$$

$$\text{Error:}\quad S_e = (A - 2)\,\sigma^{*2} - \textstyle\sum_{i=1}^{k} m_i(K_i - \bar{K})^2 + \Theta^* Q,\ \text{df} = A - k. \qquad (5.2)$$

For normal samples, the three sums of squares are asymptotically $(n_i - 2r_i \to \infty,$ $i = 1, 2, \ldots, k)$ independent and distributed as chi-square variates; see Tiku (1978). The F-statistics to test $\Theta = 0$ and to test lack of fit are, respectively,

$$F_1 = S_{\ell r}/\left(\frac{S_e}{A-k}\right) \quad\text{and}\quad F_2 = \left(\frac{S_{dev}}{k-2}\right)/\left(\frac{S_e}{A-k}\right). \qquad (5.3)$$

To verify the robustness of the null $(\Theta = 0)$ distribution of F_1, and the distribution of F_2, we simulated (from 4000 Monte Carlo runs using Box and Muller (1958) equations for generating random normal deviates) the probabilities of their falling to the left of the upper $100\,(1 - \delta)$ percent points of the approximate F-distributions having $(1, A - k)$ and $(k - 2, A - k)$ df, respectively, and found them remarkably robust. For example, we have the following probabilities for the above Atkinson-Cox model; $r_i = r = [0.5 + 0.1n]$:

Simulated values of the probabilities

δ	Normal	Laplace	t_3	Cauchy	Exponential
			F_1		
.90	.895	.895	.895	.896	.887
.95	.947	.945	.943	.949	.939
.99	.990	.990	.989	.993	.986
.995	.995	.995	.995	.996	.992
			F_2		
.90	.899	.891	.898	.913	.890
.95	.951	.950	.951	.962	.937
.99	.992	.991	.991	.992	.985
.995	.997	.995	.997	.996	.992

We also simulated the power of F_1 for testing $\Theta = 0$ against $\Theta \neq 0$ and found the statistic F_1 remarkably more powerful than the classical statistic based on the ordinary least square estimators, although slightly less powerful for the normal distribution; see Table II.

For censored normal samples, Tiku (1978, pp. 1227, 1228) gives the MML estimators of the regression coefficients in multiple linear regression with grouped data and defines the F-statistics for testing assumed values of the regression coefficients and testing linearity of the model. We determined the efficiencies of these MML estimators and investigated the robustness and power properties of the corresponding F-statistics, for all the above distributions. Like Θ^* and the statistics F_1 and F_2, these MML estimators and the F-statistics were found to be remarkably efficient, robust and powerful. We, however, omit details for conciseness.

It is clear that the above procedures only apply to grouped data. However, if the values of x are not initially grouped, a grouping (whenever feasible) can be imposed artificially and the above procedures become applicable. It is known that for least square analysis imposition of even course grouping has little effect, unless there are a few very extreme points; see Haitovsky (1973) and Atkinson and Cox (1977, pp. 15 and 19).

7. ROBUST ANALYSIS-OF-VARIANCE

In two-way classification for analysis-of-variance, assume the model

$$x_{ij\ell} = \mu + g_i + \tau_j + \delta_{ij} + e_{ij\ell}, \tag{7.1}$$

$\ell = 1, 2, \ldots, n_{ij}$, $j = 1, 2, \ldots, c$, $i = 1, 2, \ldots, k$; $e_{ij\ell}$'s are assumed to be iid with mean zero and variance σ^2. Let $(a_{ij} = r_{ij} + 1,\ b_{ij} = n_{ij} - r_{ij})$

$$X_{ij,a_{ij}},\ X_{ij,a_{ij}+1}, \ldots, X_{ij,b_{ij}} \tag{7.2}$$

be the Type II symmetrically censored sample for the (i,j)th cell; $j = 1, 2, \ldots, c$, $i = 1, 2, \ldots, k$. Writing

$$m_{ij} = n_{ij} - 2r_{ij} + 2r_{ij}\beta_{ij},\quad m_{i.} = \sum_{j=1}^{c} m_{ij},\quad m_{.j} = \sum_{i=1}^{k} m_{ij},\quad m_{..} = \sum_i \sum_j m_{ij},$$

and assuming the restrictions

$$\sum_i m_{i.} g_i = \sum_j m_{.j} \tau_j = \sum_i m_{ij} \delta_{ij} = \sum_j m_{ij} \delta_{ij} = 0, \tag{7.3}$$

the MML estimators of μ, g_i, τ_j, δ_{ij} and σ can be obtained from the kc censored samples (7.2), exactly on the same lines as in Tiku (1976, 1968, 1973, 1978), and are given by

$$\mu^* = \bar{K}_{..},\ g_i^* = \bar{K}_{i.} - \bar{K}_{..}, \tau_j^* = \bar{K}_{.j} - \bar{K}_{..},\ \delta_{ij}^* = K_{ij} - \bar{K}_{i.} - \bar{K}_{.j} + \bar{K}_{..}, \tag{7.4}$$

and

$$\sigma^* = \{B + \sqrt{(B^2 + 4AC)}\}/2\sqrt{\{A(A - kc)\}}. \tag{7.5}$$

Here

$$K_{ij} = \{\sum_{\ell=a_{ij}}^{b_{ij}} X_{ij,\ell} + r_{ij}\beta_{ij}(X_{ij,b_{ij}} + X_{ij,a_{ij}})\}/m_{ij};$$

$$\bar{K}_{i.} = \sum_j m_{ij}K_{ij}/m_{i.},\ \bar{K}_{.j} = \sum_i m_{ij}K_{ij}/m_{.j},\ \bar{K}_{..} = \sum_i \sum_j m_{ij}K_{ij}/m_{..},$$

$$A = \sum_i \sum_j (n_{ij} - 2r_{ij}),\ B = \sum_i \sum_j r_{ij}\alpha_{ij}(X_{ij,b_{ij}} - X_{ij,a_{ij}})$$

and

$$C = \sum_i \sum_j \{ \sum_{\ell=a_{ij}}^{b_{ij}} X_{ij,\ell}^2 + r_{ij}\,\beta_{ij}(X_{ij,b_{ij}}^2 + X_{ij,a_{ij}}^2) - m_{ij}K_{ij}^2\}.$$

The coefficients α_{ij} and β_{ij} are simple constants (α and β in Section 1); see also Tiku (1973).

For testing the null hypotheses $H_0(g_i = 0,\ \tau_j = 0$ and $\delta_{ij} = 0$, for all $i = 1, 2,\dots,k$ and $j = 1, 2,\dots,c)$, the component sums of squares are as follows:

Rows: $S_{Row} = \sum_i m_{i.}(\bar{K}_{i.} - \bar{K}_{..})^2$, df $= k - 1$

Columns: $S_{Col} = \sum_j m_{.j}(\bar{K}_{.j} - \bar{K}_{..})^2$, df $= c - 1$

Interactions: $S_I = \sum_i \sum_j m_{ij}(K_{ij} - \bar{K}_{i.} - \bar{K}_{.j} + \bar{K}_{..})^2$, df $= (k-1)(c-1)$

Error: $S_E = (A - kc)\sigma^{*2}$, df $= A - kc$. (7.6)

The F-statistics for testing H_0 are given by

$$F_1 = (A - kc)S_{Row}/(k - 1)S_E,\ F_2 = (A - kc)S_{Col}/(c - 1)S_E$$

and

$$F_3 = (A - kc)S_I/(k - 1)(c - 1)S_E. \tag{7.7}$$

For normal samples, the null distributions of the statistics F_1, F_2 and F_3 are closely approximated by F-distributions having $(k-1, A-kc)$, $(c-1, A-kc)$ and $((k-1)(c-1), A-kc)$ df, respectively; see also Tiku (1978, Table I). Like the statistics (5.3), the statistics F_1, F_2 and F_3 are robust and powerful.

For testing a linear contrast, say $\sum_{i=1}^{k} \ell_i g_i = 0$ $(\sum_i \ell_i = 0)$, the t_c-statistic is defined as

$$t_c = \frac{\sum_{i=1}^{k} \ell_i g_i^*}{\sigma^* \sqrt{\{\sum_{i=1}^{k} \ell_i^2 (\frac{1}{m_{i.}} - \frac{1}{m_{..}}) - \frac{1}{m_{..}} \sum_{i=1}^{k} \sum_{\substack{j=1 \\ j \neq \ell}}^{k} \ell_i \ell_j\}}}, \tag{7.8}$$

where the estimator σ^* is given by the equation (7.5); see also Tiku and Stewart (1977). For normal samples the null distribution of t_c, like that of T_c, is Student's t with A-kc df, for large $A_i = \sum_{j=1}^{c} A_{ij}$, $i = 1, 2, \ldots, k$; see also Tiku (1978, Lemma 1 and 2).

For non-normal samples, we simulated the probabilities $P(t_c \leqslant h)$ and found them very close to the corresponding probabilities of the Student's t distribution having A-kc df; h being the percentage points of the Student's t distribution. The statistic t_c has power properties similar to the statistic T_c (Eq. 3.1). We omit details for conciseness.

ACKNOWLEDGEMENT

Thanks are due to NSERC and McMaster University Research Board for research grants and to Barbara Holdcroft for typing the manuscript.

TABLE 1

Values of the Power, for 1% Significance Level; $d = \mu_1 - \mu_2$, $n_1 = 8$ and $n_2 = 8$, $r_1 = 1$ and $r_2 = 1$. Distributions $g_1 = g_2$ are (1) Normal, (2) (n-1) N(0,1) and 1 N(0,10), (3) k=1 and $\lambda = 2$ (Tiku's outlier model), (4) Student's t_2, (5) Cauchy, and (6) Exponential.

	d				d				d			
	0.0	1.0	2.0	3.0	0.0	1.0	2.0	3.0	0.0	1.0	2.0	3.0
	x and y are independent ($\sigma_1 = \sigma_2$)											
	(1)				(2)				(3)			
D^+	.010	.24	.81	.98	.010	.17	.60	.92	.007	.17	.67	.97
C_1	.010	.28	.88	1.00	.010	.15	.43	.59	.004	.08	.38	.78
W	.010	.29	.86	1.00	.010	.18	.55	.81	.006	.10	.46	.85
T_c	.011	.28	.85	1.00	.010	.19	.69	.97	.012	.27	.85	1.00
	(4)				(5)				(6)			
D^+	.010	.16	.55	.83	.010	.10	.34	.58	.010	.50	.94	1.00
C_1	.010	.14	.45	.67	.010	.08	.23	.36	.010	.44	.83	.95
W	.010	.15	.53	.78	.010	.10	.28	.48	.010	.48	.89	.98
T_c	.008	.17	.55	.84	.006	.08	.29	.54	.007	.42	.89	.99
	x and y are independent ($\sigma_1 \neq \sigma_2$)											
	$\sigma_1/\sigma_2 = 1.25$				$\sigma_1/\sigma_2 = 1.25$				$\sigma_1/\sigma_2 = 0.8$			
	(1)				(4)				(4)			
D^+	.012	.21	.70	.98	.009	.14	.45	.76	.009	.20	.60	.89
C_1	.009	.22	.79	.99	.011	.12	.38	.61	.012	.17	.50	.72
W	.012	.22	.79	.99	.012	.14	.43	.71	.013	.19	.53	.81
T_c	.013	.22	.76	.99	.011	.14	.48	.77	.013	.20	.64	.90

TABLE II

Simulated values (based on 2000 runs) of the power for testing $\theta = 0$ against $\theta \neq 0$ in case of the above Atkinson-Cox model; C=classical (i.e. $r_i = 0$ in Eq. 5.3) and R=robust ($r_i = [0.5 + 0.1n_i]$ in Eq. 5.3).

	$\theta = 0$		$\theta = 0.02$		$\theta = 0.05$		$\theta = 0.10$	
	C	R	C	R	C	R	C	R
				Normal N(0,1)				
10	.104	.105	.21	.21	.63	.61	.99	.98
5	.049	.053	.14	.14	.52	.48	.98	.96
1	.009	.010	.04	.04	.28	.23	.90	.84
				Student's t_3				
10	.094	.096	.22	.28	.70	.83	.97	1.00
5	.044	.051	.14	.18	.60	.74	.96	1.00
1	.006	.009	.04	.06	.35	.48	.92	.98
				4 N(0,1) & 1 N(0,4)				
10	.097	.103	.14	.18	.28	.47	.66	.92
5	.042	.050	.07	.10	.19	.36	.53	.85
1	.006	.010	.01	.03	.07	.15	.30	.65
				Exponential				
10	.091	.113	.21	.23	.65	.71	.99	.99
5	.042	.061	.14	.15	.53	.58	.97	.97
1	.006	.014	.04	.05	.29	.32	.90	.92

8. BIBLIOGRAPHY

[1] Atkinson, A.C. and Cox, D.R. (1977). Robust regression via discriminant analysis. Biometrika 64, 15-19.

[2] Barnett, V.D. (1966). Order statistics estimators of the location of the Cauchy distribution. J. Amer. Statist. Assoc. 61, 1205-18.

[3] Box, G.E.P. and Muller, M.E. (1958). A note on the generation of random normal deviates. Ann. Math. Statist. 29, 610-11.

[4] David H.A., Kennedy, W.J. and Knight R.D. (1977). Means, vriances and covariances of the normal order statistics in the presence of an outlier. Selected Tables In Mathematical Statistics, V. Rhode Island: American Mathematical Society.

[5] Gastwirth, J.L. (1965). Percentile modification of two sample rank tests. J. Amer. Statist. Assoc. 60, 1127-41.

[6] Govindarajulu, Z. (1966). Best linear estimates under symmetric censoring of the parameters of a double exponential population. J. Amer. Statist. Assoc. 61, 248-58.

[7] Gross, A.M. (1976). Confidence interval robustness with long-tailed symmetric distributions. J. Amer. Statist. Assoc. 71, 409-16.

[8] Haitovsky, Y. (1973). Regression estimation from grouped observations. London: Criffin.

[9] Huber, P.J. (1970). Studentizing robust estimates, in Nonparametric Techniques in Statistical Inferences, ed. M.L. Puri. Cambridge University Press, 453-63.

[10] Kim, P.J. and Jenrich, R.I. (1973). Tables of the exact sampling distribution of the two-sample Kolmogorov-Smirnov criterion D_{mn} $(m \leqslant n)$. Selected Tables in Mathematical Statistics - I. Rhode Island: American Mathematical Society.

[11] Lehmann, E.L. (1975). Nonparametrics: Statistical Methods Based on Ranks. New York: McGraw Hill.

[12] Milton, R.C. (1970). Rank Order Probabilities: New York: John Wiley and Sons, Inc.

[13] Shorack, G.R. (1974). Random means. Ann. Statist. 2, 661-75.

[14] Smith, W.B., Zeis C.D. and Syler, G.W. (1973). Three parameter lognormal estimation from censored data. J. Indian Statist. Assoc. 11, 15-31.

[15] Tietjen, G.L., Kahaner, D.K. and Beckman, R.J. (1977). Variances and covariances of the normal order statistics for sample sizes 2 to 50. Selected Tables In Mathematical Statistics V. Rhode Island: American Mathematical Society.

[16] Tiku, M.L. (1967). Estimating the mean and standard deviation from censored normal samples. Biometrika 54, 155-65.

[17] Tiku, M.L. (1968). Estimating the parameters of lognormal distribution from censored samples. J. Amer. Statist. Assoc. 63. 134-40.

[18] Tiku, M.L. (1970). Monte Carlo study of some simple estimators in censored normal samples. Biometrika 57, 207-10.

[19] Tiku, M.L. (1973). Testing group effects from Type II censored normal samples in experimental design. Biometrics 29, 25-33.

[20] Tiku, M.L. (1975). A new statistic for testing suspected outliers. Commun. Statist. 4(8), 737-52.

[21] Tiku, M.L. (1977). Rejoinder: "Comment on 'A new Statistic for testing suspected outliers'". Commun. Statist. A6(14), 1417-22.

[22] Tiku, M.L. (1978). Linear regression model with censored observations. Commun. Statist. A7(13), 1219-32.

[23] Tiku, M.L. (1980). Robustness of MML estimators based on censored samples and robust test statistic. J. Statistical Planning and Inferences 4, 123-43.

[24] Tiku, M.L. and Jones, P.W. (1971). Best linear unbiased estimators for a distribution similar to the logistic. Proc of the "Statistics 71 Canada", eds. Carter, C.S., Dwividi, T.D., Fellegi, I.P., Fraser, D.A.S., McGregor, J.R. and Sprott, D.A., 412-19.

[25] Tiku, M.L. and Stewart D. (1977). Estimating and testing group effects from Type I censored normal samples in experimental design. Commun. Statist. A6(15), 1485-1501.

[26] Tiku, M.L. and Kumra, S. (1979). Expected values and variances and covariances of order statistics for a family of symmetric distributions (Student's t). Likely to appear in Selected Tables In Mathematical Statistics.

PART V
STOCHASTIC PROCESSES

STATISTICS AND RELATED TOPICS
M. Csörgö, D.A. Dawson, J.N.K. Rao, A.K.Md.E. Saleh (eds.)

GALERKIN APPROXIMATION OF NONLINEAR MARKOV PROCESSES

D. A. Dawson

Carleton University, Ottawa, Canada

1. INTRODUCTION

The main objectives of this article are to describe a family of nonlinear Markov processes and to formulate the problem of finding numerical approximations to these processes. In recent years many applications of nonlinear Markov processes have arisen in statistical physics, chemical kinetics, stochastic control, nonlinear filtering and population biology. The importance of these processes is due to the fact that the combined effects of stochastic fluctuations and nonlinearity can give rise to new qualitative behavior having no deterministic or linear Markov analogues. This will be illustrated by the study of a simple model problem known as the stochastic logistic model. The method to be discussed is named the Gauss-Galerkin method since it combines elements of the methods of Gaussian quadrature and Galerkin approximation. This method provides both numerical approximations and theoretical bounds which can be used to study the bifurcation of steady states for nonlinear Markov processes. In this article we review the necessary background on nonlinear Markov processes, Gaussian quadrature and Galerkin methods, illustrate these methods in the case of a simple model problem and briefly indicate some of the theoretical foundations of the method.

2. BACKGROUND INFORMATION ON NONLINEAR MARKOV SYSTEMS

Let $D[0,\infty)$ denote the space of real-valued functions which are right continuous and have left limits, furnished with the Skorohod topology. Let $\underline{\underline{F}}$ denote the σ-algebra of Borel subsets of $D[0,\infty)$ and let $\{X(t):t \geq 0\}$ denote the canonical process on $D[0,\infty)$. Let $\underline{\underline{M}}$ denote the space of probability measures on R^1, furnished with the topology of weak convergence. Finally, let $\theta_t:D[0,\infty) \to D[0,\infty)$ denote the shift $\theta_t X(s) \equiv X(t+s)$, $s,t \geq 0$.

A <u>nonlinear Markov process</u> is prescribed by a family of probability measures $\{P_\mu : \mu \in \underline{\underline{M}}\}$ on $(D[0,\infty),\underline{\underline{F}})$ which satisfy:

(2.1) for each $B \in \underline{\underline{F}}$, $P_{\cdot}(B)$ is a Borel measurable function on $\underline{\underline{M}}$,

(2.2) for each $\mu \in \underline{\underline{M}}$, $P_\mu(X(0) \in B) = \mu(B)$,

(2.3) for $B \in \underline{\underline{F}}$, $0 \leq s \leq t$,

$$P_\mu(\theta_t X \in B|X(s) : 0 \leq s \leq t) = P_\mu(\theta_t X \in B|X(t))$$
$$= P_{\nu_t}(B|\theta_t X(0))$$

where $\nu_t \equiv \mathcal{L}(X(t)) \in \underline{\underline{M}}$ denotes the probability law of $X(t)$.

In the terminology of McKean (1966), $X(t)$ is Markov with nonconstant transition mechanism which depends upon t only via the distribution of $X(t)$.

If we define $T_t(\nu_0) \equiv \nu_t$, then $\{T_t : t \geq 0\}$ is a nonlinear semigroup in the sense that for $s,t \geq 0$,

$$(2.4) \qquad \nu_{t+s} = T_{t+s}(\nu_0) = T_t(\nu_s) = T_t(T_s(\nu_0)).$$

The generator, if it exists, is given by

$$(2.5) \qquad L^*\mu \equiv \lim_{t\downarrow 0} [T_t(\mu) - \mu]/t .$$

In this case the operator L^* is nonadditive. However ν_t is still a solution of the evolution equation

$$d\nu_t/dt = L^*\nu_t .$$

An important class of nonlinear Markov processes is given by the nonlinear Itô stochastic differential equation

$$(2.6) \qquad dX(t) = e(\mathcal{L}(X(t)),X(t))dt + \sigma(\mathcal{L}(X(t)),X(t))dw(t)$$

where $\{w(t) : t \geq 0\}$ denotes a standard Wiener process and

$$\sigma : C(R^2) \to C^2(R^1) , \quad e : C(R^2) \to C^1(R^1) .$$

If Equation (2.6) has a solution having a transition probability density function, $p(t,x)$, then it is a solution of the nonlinear Fokker-Planck equation

$$(2.7) \qquad \partial p(t,x)/\partial t = \tfrac{1}{2}\, \partial^2/\partial x^2[\sigma^2(p(t,x),x)p(t,x)] - \partial/\partial x[e(p(t,x),x)p(t,x)]$$

$$= L^*p(t,x) ,$$

$$P(0,x) = \delta(x - x_0) .$$

K. Itô and S. Watanabe (1978) showed that under reasonable hypotheses on $\sigma(\cdot)$ and $e(\cdot)$, the sequence of successive approximations

$$(2.8) \qquad X_0(t) \equiv x_0$$

$$X_{n+1}(t) = x_0 + \int_0^t \sigma(\mathcal{L}(X_n(s)),X_n(s))dw(s) + \int_0^t e(\mathcal{L}(X_n(s)),X_n(s))ds$$

converges to a continuous process which satisfies Equation (2.6). If $X(t)$ is such a solution, Itô's stochastic chain rule implies that for any function $\phi \in C^2(R^1)$,

$$(2.9) \qquad \phi(X(t)) - \phi(X(0)) - \int_0^t [\tfrac{1}{2}\sigma^2(\mathcal{L}(X(s)),X(s))\phi''(X(s)) + e(\mathcal{L}(X(s)),X(s))\phi'(X(s))]ds$$

is a P_μ-martingale where $\mu = \mathcal{L}(X(0))$. Taking expectations, this yields

$$(2.10) \qquad \int \phi(x)p(t,x)dx - \int\phi(x)\mu(dx)$$

$$= \int_0^t \{\int[\tfrac{1}{2}\sigma^2(p(s,x),x)\phi''(x) + e(p(s,x),x)\phi'(x)]p(s,x),x)dx\}ds$$

$$= \int \{\phi(x) \int_0^t L^*p(s,x)ds\}dx .$$

Remark 2.1. McKean (1966) gave examples of nonlinear Markov processes associated with Boltzmann type equations. Processes of this type describe the motion of a tagged molecule in a bath of infinitely many identical molecules. It arises as the limit as $N \to \infty$ of finite systems of N molecules in which there is an interaction which is symmetric in all pairs of particles. In this paper we consider another family of nonlinear Markov processes which arises in the study of reaction diffusion systems in chemical kinetics and population biology.

Example 2.1. The Brussel's Model of a Trimolecular Reaction

The Brussel's model is a model of a chemical reaction involving two reactants and a trimolecular reaction. Let $X(t)$, $Y(t)$, $t \geq 0$, denote the quantities of the two reactants in the system at time t. In the usual chemical notation the reaction is given by

$$A \rightleftarrows X$$

$$X + B \rightleftarrows Y + C$$

$$2X + Y \rightleftarrows 3X$$

$$X \rightleftarrows E$$

where A,B,C,E are auxiliary reactants whose quantities are held at fixed values throughout the experiment. Several versions of this model have been studied including (deterministic) nonlinear diffusion models and birth and death models. The diffusion approximation description of this model leads to the following pair of nonlinear stochastic differential equations:

$$(2.11) \quad dX(t) = [a-(b+1)X(t)+X^2(t)Y(t)]dt+g_1(X(t))dw_1(t)+D_1[E(X(t))-X(t)]dt,$$

$$dY(t) = [bX(t)-X^2(t)Y(t)]dt+g_2(Y(t))dw_2(t)+D_2[E(Y(t))-Y(t)]dt ,$$

where $a,b \geq 0$, $w_1(t)$, $w_2(t)$ are independent Wiener processes and g_1,g_2 are functions describing the density dependence of the fluctuations. The nonnegative real numbers D_1, D_2 denote the diffusivities of the chemicals X and Y respectively. The "nonlinear" terms $D_i[E(X(t))-X(t)]$ model the effects of spatial diffusion. The idea is that $X(t)$, $Y(t)$ denote the quantities of X and Y in a "small volume" which is exchanging mass by diffusion with the surrounding environment. The term $-D_1X(t)$ denotes the diffusion of molecules out of the small volume and the term $D_1E(X(t))$ denotes the diffusion of molecules into the small volume. The significance of the expectation "$E(X(t))$" is that it is assumed that the fluctuations in the number of incoming molecules are independent of the fluctuations in the small volume. This approximation is known as the mean field approximation and serves as a simple model exhibiting some of the same qualitative behavior as the original spatially distributed system. The bifurcation structure of the mean field model has been studied extensively (e.g. refer to Nicolis and Prigogine (1977)). The mean field model has been shown to exhibit limit cycle behavior, the formation of spatial structures and chaotic response to regular excitation.

Example 2.2. Model Problem: The Stochastic Logistic With Diffusion

For the purpose of this paper we require a nonlinear Markov model having a nontrivial qualitative behavior but one in which an exact solution is available. This will make it possible to evaluate the effectiveness of methods of approximate numerical solution. The model chosen is one described in Dawson (1980) and referred to as the stochastic logistic model. This model is given by the nonlinear stochastic differential equation

$$(2.12)\qquad dX(t) = (X(t) - X^2(t))dt + (\gamma X(t))^{\frac{1}{2}}dw(t) + \rho[E(X(t)) - X(t)]dt$$

where $\gamma > 0$, $w(t)$ is a standard Wiener process and $0 \leq \rho \leq 1$ is a measure of the rate of diffusion into a distinguished small volume as in Example 2.1.

The nonlinear Fokker-Planck equation corresponding to Equation (2.12) is

$$(2.13)\quad \partial p(t,x)/\partial t = -\partial/\partial x[(\rho\int yp(t,y)dy + ax - bx^2)p(t,x)] + \tfrac{1}{2}\gamma\partial^2/\partial x^2(xp(t,x))$$

$$= L^*p(t,x)\ .$$

We are interested in identifying the stationary probability distributions, p^*, for this process. One stationary distribution is

$$(2.14)\qquad p^*(dx) = \delta(x)\ .$$

If $\rho = 0$ (no diffusion), then 0 is an accessible boundary and (2.14) is the only stationary probability distribution. For the case $\rho = 1$, it is proved in Dawson (1980) that the steady state mean, $m^* = E_*(X)$, must be a root of the equation

$$(2.15)\qquad m(e) = e\ ,$$

where
$$m(e) \equiv [\Gamma(e/\gamma + \tfrac{1}{2})/\Gamma(e/\gamma)]^{\frac{1}{2}}.$$

It is possible to verify that

(i) $m(e)/e$ is a monotone decreasing function of e,

(ii) $\lim_{e\to\infty} m(e)/e = 0$, and

(iii) $\lim_{e\downarrow 0} m(e)/e = (\pi/\gamma)^{\frac{1}{2}}$.

Hence for $\gamma \geq \gamma_c = \pi$, the only root of Equation (2.15) is $e = 0$. However for $\gamma > \gamma_c$,

$$\lim_{e\downarrow 0} m(e)/e > 1\ ,$$

and therefore there is a second root, m^*, of Equation (2.15) which corresponds to a second stationary probability distribution.

An alternative approach to the study of the solution of Equation (2.12) is through the system of differential equations for the moments. Let

$$m_n(t) \equiv E(X^n(t)).$$

By application of Itô's Lemma it follows that the functions $m_n(t)$ satisfy the system of differential equations

$$(2.16)\qquad dm_1(t)/dt = m_1(t) - m_2(t)$$

$$dm_n(t)/dt = (n\rho m_1(t) + \tfrac{1}{2}n(n-1)\gamma)m_{n-1}(t) + n(1-\rho)m_n(t) - nm_{n+1}(t).$$

Given a nonlinear Markov model such as the stochastic logistic there are two basic questions to be answered. The first, briefly indicated above, is to describe the family of stationary probability measures for the process and in particular to describe the associated bifurcation structure as a function of the diffusivity, ρ , and the demographic temperature, γ . The second problem is that of computing numerical approximations to the nonlinear Markov process. There are two main approaches to the problem of numerical approximation. The first involves the approximate solution of the moment equations (2.16). However the system of equations (2.16) forms an "open hierarchy" in which the equation for the nth moment has a term involving the (n+1)st moment, $m_{n+1}(t)$. Various truncation schemes have been proposed; the simplest and most frequently used is the cumulant neglect method in which the (n+1)st and higher cumulants are set equal to zero. In this way the system of moment equations reduces to a closed nonlinear system of n equations. This method has been widely used in the study of the Brussel's model (e.g. Nicolis and Prigogine (1977)). This "closure" problem has also been extensively investigated in the study of turbulence (Leslie (1973)). Unfortunately as we indicate below the use of the cumulant neglect method (or variants of it) can lead to misleading results. The second approach to the approximate solution of stochastic differential equations is via a finite difference approximation to the nonlinear Fokker-Planck equation. The probabilistic aspects of this method with applications to nonlinear filtering and control has been developed by Kushner (1977). In recent years Galerkin methods such as finite element and collocation methods have been a serious alternative to finite difference methods for the solution of elliptic and parabolic partial differential equations (e.g. refer to Douglas and Dupont (1970)). In the next section we outline a Galerkin approach to the approximation of solutions to linear and nonlinear Fokker-Planck equations. In Section 4 we introduce a hybrid Galerkin-finite difference method for the approximate evaluation of function space integrals which combines the advantages of both the Galerkin method's ability to follow the nonlinearity and the finite difference method's ability to evaluate general function space integrals.

3. THE GAUSS-GALERKIN NUMERICAL APPROXIMATION

3.1. Objectives.

To begin the discussion of the Gauss-Galerkin numerical approximation to a Markov process let us list the objectives of such an approximation.

(i) The method should provide an approximation to the distribution $p(t,dx) = \mathcal{L}(X(t))$ in the sense of the approximation of probability measures.

(ii) The method should preserve as much of the structure of the problem as possible; in particular, a Markov process should be approximated by a Markov process.

(iii) The method should provide an approximate value of the expectation of a function space integral with respect to the function space measure induced by the Markov process.

(iv) The approximation should be carried out in a computationally efficient manner.

In this section we introduce a method, the Gauss-Galerkin Method which appears to fulfill these objectives.

3.2 The Gauss-Christoffel Approximation of Measures.

As stated above the first objective is to provide a numerical approximation to the probability distribution on R^1 , $p(t,dx) = \mathcal{L}(X(t))$. In most applications the probability measure is used to compute expectations. For this reason it is appropriate to view the approximation of the measure in the context of numerical quadrature formulas. Specifically the objective of our method is to provide an approximate Gaussian quadrature formula for the probability law $p(t,dx)$.

An n point Gauss-Christoffel approximation to the measure, μ, is given by

$$\mu \simeq \sum_{k=1}^{n} a_k \delta_{x_k} , \tag{3.1}$$

where the $\{\delta_{x_k}\}$ denotes atoms at the Gauss-Christoffel points $\{x_k\}$ and the $\{a_k\}$ denote the Gauss-Christoffel weights. The main classical results on Gaussian quadrature are summarized in the following theorem.

Theorem 3.1. a) The n points $\{x_k\}$ and n weights $\{a_k\}$ can be uniquely chosen so that

$$\int f(x)\mu(dx) = \sum_{k=1}^{n} a_k f(x_k) \tag{3.2}$$

holds for all polynomials of degree less than or equal to $(2n-1)$.

b) Let $\{P_m(x)\}$ be the family of orthogonal polynomials associated with the measure μ, that is, $P_m(x)$ is for each m a polynomial of degree m and

$$\int P_m(x)P_n(x)\mu(dx) = 0 \quad \text{if} \quad m \neq n . \tag{3.3}$$

Then the Gauss-Christoffel points $\{X_k; k=1,\ldots,n\}$ are the zeros of the polynomial $P_n(\cdot)$.

c) The Gauss weights are uniquely obtained as the unique solution of the equations:

$$\int f(x)\mu(dx) = \sum_{k=1}^{n} a_k f(x_k) \tag{3.4}$$

for all polynomials of degree less than or equal to $(n-1)$.

d) If $f \in C^{2n}(R)$, $f^{(2n)} \geq 0$, $f^{(2n)} \not\equiv 0$, then

$$\int f(x)\mu(dx) > \sum_{k=1}^{n} a_k f(x_k) . \tag{3.5}$$

Proof. Refer to Stroud (1974).

The theory for quadrature formulae in dimensions higher than one is less developed. However a comparable development of quadrature formulae with nonnegative weights has been developed on the basis of the fundamental theorem of Tchakaloff (cf. David (1967), Stroud (1971)).

In order to extend the application of the Gauss-Christoffel approximation to the study of Markov processes the following lemma on the approximation of the conditional probability law will be required.

Lemma 3.1. Let ν be a probability measure on R^1 and let $P(x,dy)$ be a Markov transition kernel, that is,

(3.6.a) for every $x \in R^1$, $P(x,\cdot)$ is a probability measure on R^1, and

(3.6.b) for every Borel set B, $P(\cdot,B)$ is Borel measurable function.

Let μ be a probability measure on R^1 defined by:

$$\mu(B) \equiv \int P(x,B)\nu(dx) \ . \tag{3.7}$$

Then there exists a set of n points and for every x a set of conditional (signed) weights $\{a_k(x)\}$ such that:

(i) $\int f(y)P(x,dy) = \sum_{k=1}^{n} f(x_k)a_k(x)$

for every polynomial, f, of degree less than or equal to $(n-1)$,

(ii) if $a_k \equiv \int a_k(x)\nu dx)$, then the a_k are the Gauss-Christoffel weights, that is

$\int f(y)\nu(dy) = \sum_{k=1}^{n} f(x_k)a_k$,

for all polynomials, f, of degree less than or equal to $(2n-1)$.

Proof. Taking a basis $\{f_m, m=1,2,\ldots,n\}$ for the polynomials of degree less than or equal to $(n-1)$ we obtain n equations for the conditional weights $\{a_k(x)\}$:

$$f_m(y)P(x,dy) = \sum_{k=1}^{n} f_m(x_k)a_k(x) \ , \quad m=1,\ldots,n \ . \tag{3.8}$$

The existence and uniqueness of a solution to this linear system of equations follows since the Vandermonde matrix

$$V = \begin{pmatrix} 1 & 1 & 1 & 1 & \ldots & 1 \\ x_1 & x_2 & x_3 & x_4 & & x_n \\ \cdot & \cdot & \cdot & \cdot & \cdot & \cdot \\ x_1^{n-1} & x_2^{n-1} & \ldots & & & x_n^{n-1} \end{pmatrix}$$

is non-singular provided that the Gauss-Christoffel points are distinct. But then

$$\int f_m(x)\mu(dx) = \int\int f_m(x)P(y,dx)\nu(dy) \tag{3.9}$$

$$= \int \sum_{k=1}^{n} f_m(x_k)a_k(y)\nu(dy) = \sum_{k=1}^{n} f_m(x_k)a_k \ ,$$

for $m=1,\ldots,n$. But according to Theorem 3.1, Part c, Equation (3.9) uniquely determines the Gauss-Christoffel weights. To complete the proof we note that the Gauss-Christoffel points $\{x_k\}$ are obtained in the usual manner.

Remark 3.1. In the terminology of numerical analysis the conditional weights obtained above correspond to a conditional Newton-Cotes quadrature formula. It is well known that signed weights of large magnitude when the number of points becomes large (cf. Krylov (1962)).

3.3 The Galerkin Equations for the Gauss-Christoffel Approximation

The basic idea behind the Gauss-Galerkin approximation is to solve dynamically for a set of n Gauss-Christoffel points and weights for $L(X(t))$ for each t. A set of differential equations for the Gauss-Galerkin points $\{x_k(t)\}$ and weights $\{a_k(t)\}$ are obtained by the method of Galerkin. For expository reasons we describe the approximation for a linear Markov process, the modification for a nonlinear Markov process is straightforward.

The Fokker-Planck equation

$$(3.10) \qquad \partial p(t,\cdot)/\partial t = L^* p(t,\cdot) \ ,$$

can be written in <u>weak form</u> as

$$(3.11) \qquad (\partial p(t)/\partial t, \phi) = (L^* p(t), \phi) \quad \text{for each } \phi \in C^2(R^1)$$

$$= (p(t), L\phi) \quad \text{since } L^* \text{ is the formal adjoint of } L \ .$$

The solution of Equation (3.10) is measure-valued, that is, for each t, $p(t) \in \underline{\underline{M}}(R^1)$, the space of non-negative measures on R^1.

Let $\underline{\underline{M}}_n(R^1)$ denote the subspace of $\underline{\underline{M}}(R^1)$ consisting of atomic measures having at most n atoms. Then any n-point Gauss-Christoffel approximation belongs to the subspace $\underline{\underline{M}}_n(R^1)$. Let $\underline{\underline{P}}_k(R^1)$ denote the space of all polynomials on R^1 of degree less than or equal to k.

The <u>n-point Gauss-Galerkin approximation</u> to $p(t,\cdot)$, denoted by $p_n(t)$ is given by a mapping from $[0,\infty)$ into $\underline{\underline{M}}_n(R^1)$ or alternately by a family of functions $\{x_k(t)\}$ and $\{a_k(t)\}$ specifying the Gauss-Christoffel points and weights, respectively, as a function of time. Let $\{f_m : m=1,\ldots,2n\}$ denote a basis of functions for $\underline{\underline{P}}_{2n-1}(R^1)$. Then the <u>Galerkin Equations</u> for the functions $x_k(t)$, $a_k(t)$, $k=1,\ldots,n$, are:

$$(3.12) \qquad d/dt\Big(\sum_{k=1}^{n} a_k(t) f_m(x_k(t))\Big) = \sum_{k=1}^{n} a_k(t)(Lf_m)(x_k(t)), \quad m=1,\ldots,2n \ .$$

This is a system of 2n differential equations for the 2n unknown functions $\{a_k(t)\}$, $\{x_k(t)\}$.

An alternate approach to the derivation of the Galerkin Equations is to begin with the stochastic differential equation

$$(3.13) \qquad dX(t) = e(X(t))dt + \sigma(X(t))dw(t), \ X(0) = x \ .$$

For $\phi \in C^2(R^1)$, Itô's Lemmas yields:

$$(3.14) \qquad d\phi(X(t)) = \phi'(X(t))dX(t) + \tfrac{1}{2}\phi''(X(t))\sigma^2(X(t))dt,$$ where ' denotes the derivative.

The solution, $X(t)$, of (3.13) induces a probability measure, P_x, on $C([0,\infty))$ which is characterized by the fact that for each $\phi \in C^2(R^1)$ with polynomial growth at infinity,

$$(3.15) \qquad \phi(X(t)) - \phi(X(s)) - \int_s^t \phi'(X(s))e(X(s))ds - \tfrac{1}{2}\int_s^t \phi''(X(s))\sigma^2(X(s))ds$$

is a P_x-martingale if $s \le t$. Taking expectations this implies that

(3.16) $$d/dt(E(\phi(X(t))) = E(\phi'(X(t))e(X(t))) + \tfrac{1}{2}E(\phi''(X(t))\sigma^2(X(t))).$$

Applying Equation (3.16) to the functions f_m , $m=1,\ldots,2n$, we again obtain the Galerkin Equations (3.12).

Let

(3.17) $$y_{1,m} \equiv a_m' , \quad y_{2,m} \equiv a_m x_m' , \quad 1 \le m \le n .$$

Then the Galerkin Equations (3.12) become

(3.18) $$\begin{pmatrix} A_{11} & A_{12} \\ A_{21} & A_{22} \end{pmatrix} \begin{pmatrix} y_1^T \\ y_2^T \end{pmatrix} = \begin{pmatrix} B_1 a^T \\ B_2 a^T \end{pmatrix}$$

where $a = (a_1,\ldots,a_n)$, a^T denotes the transpose of a and

$$B_1 = \begin{pmatrix} Lf_1(x_1) & Lf_1(x_2) & \cdots\cdots & Lf_1(x_n) \\ Lf_2(x_1) & Lf_2(x_2) & \cdots\cdots & Lf_2(x_n) \\ \cdots & \cdots & \cdots & \cdots \\ Lf_n(x_1) & Lf_n(x_2) & \cdots\cdots & Lf_n(x_n) \end{pmatrix}$$

$$B_2 = \begin{pmatrix} Lf_{n+1}(x_1) & Lf_{n+1}(x_2) & \cdots & Lf_{n+1}(x_n) \\ Lf_{n+2}(x_1) & Lf_{n+2}(x_2) & \cdots & Lf_{n+2}(x_n) \\ \cdots & \cdots & \cdots & \cdots \\ Lf_{2n}(x_1) & Lf_{2n}(x_2) & \cdots & Lf_{2n}(x_n) \end{pmatrix}$$

$$A_{11} = \begin{pmatrix} f_1(x_1) & f_1(x_2) & \cdots\cdots & f_1(x_n) \\ f_2(x_1) & f_2(x_2) & \cdots\cdots & f_2(x_n) \\ \cdots & \cdots & \cdots & \cdots \\ f_n(x_1) & f_n(x_2) & \cdots\cdots & f_n(x_n) \end{pmatrix}$$

$$A_{12} = \begin{pmatrix} f_1'(x_1) & f_1'(x_2) & \cdots\cdots & f_1'(x_n) \\ f_2'(x_1) & f_2'(x_2) & \cdots\cdots & f_2'(x_n) \\ \cdots & \cdots & \cdots & \cdots \\ f_n'(x_1) & f_n'(x_2) & \cdots\cdots & f_n'(x_n) \end{pmatrix}$$

$$A_{21} = \begin{pmatrix} f_{n+1}(x_1) & f_{n+1}(x_2) & \cdots & f_{n+1}(x_n) \\ f_{n+2}(x_1) & f_{n+2}(x_2) & \cdots & f_{n+2}(x_n) \\ \cdots & \cdots & \cdots & \cdots \\ f_{2n}(x_1) & f_{2n}(x_2) & \cdots & f_{2n}(x_n) \end{pmatrix}$$

$$A_{22} = \begin{pmatrix} f'_{n+1}(x_1) & f'_{n+1}(x_2) & \cdots & f'_{n+1}(x_n) \\ f'_{n+2}(x_1) & f'_{n+2}(x_2) & \cdots & f'_{n+2}(x_n) \\ \cdots & \cdots & \cdots & \cdots \\ f'_{2n}(x_1) & f'_{2n}(x_2) & \cdots & f'_{2n}(x_n) \end{pmatrix}.$$

The linear system of equations (3.18) has the solution:

$$\text{(3.19)} \qquad y_2^T = (A_{22} - A_{21}A_{11}^{-1}A_{12})^{-1}(B_2 - A_{21}A_{11}^{-1}A_{12}B_1)a^T$$

$$y_1^T = A_{11}^{-1}[B_1 - A_{12}(A_{22} - A_{21}A_{11}^{-1}A_{12})^{-1}(B_2 - A_{21}A_{11}^{-1}A_{12}B_1)]a^T .$$

Note that the matrices B_1, B_2, A_{11}, A_{12}, A_{21}, A_{22} are functions of the $\{x_k\}$ but do not depend on the values of the $\{a_k\}$. Therefore (3.19) yields a system of nonlinear differential equations for the Gauss-Christoffel points $\{x_k(t)\}$ coupled with a linear system of differential equations for the Gauss-Christoffel weights $\{a_k(t)\}$. The Gauss-Galerkin approximation to $p(t,\cdot)$ is obtained by the nuemrical solution of this system of ordinary differential equations.

Remark 3.1. The Gauss-Galerkin approximate points and weights obtained as the solution of the system of differential equations (3.19) should not be confused with the n-point Gauss-Christoffel approximation of the probability law $p(t,\cdot)$. In other words,

$$\hat{p}_n(t) = \sum_{k=1}^{n} a_k(t)\delta_{x_k(t)} , \quad \text{where } a_k(t), x_k(t) \text{ is the solution of (3.19),}$$

is an approximation to $p(t,\cdot)$ but in general it is not identical to

$$p_n^*(t) = \sum_{k=1}^{n} a_k^*(t)\delta_{x_k^*(t)} ,$$

where the $a_k^*(t)$, $x_k^*(t)$ are the Gauss-Christoffel weights and points associated with the probability law $p(t,\cdot)$.

3.4 Relation with the Hankel Truncation

In this section we restrict our attention to the case of a stochastic differential equation with polynomial coefficients whose solution is a non-negative process. In this case the Gauss-Galerkin method is equivalent to a truncation of the moment hierarchy.

Given a solution, $X(t)$, of the stochastic differential equation, let

$$m_n(t) \equiv E(X^n(t)) .$$

The Hankel determinants are defined as follows:

$$\text{(3.20)} \qquad \Delta_n \equiv \begin{vmatrix} m_0 & m_1 & \cdots & m_n \\ m_1 & m_2 & \cdots & m_{n+1} \\ \cdots & \cdots & \cdots & \cdots \\ m_n & m_{n+1} & \cdots & m_{2n} \end{vmatrix} ; \quad \Delta_n^{(1)} \equiv \begin{vmatrix} m_1 & m_2 & \cdots & m_{n+1} \\ m_2 & m_3 & \cdots & m_{n+2} \\ \cdots & \cdots & \cdots & \cdots \\ m_{n+1} & m_{n+2} & \cdots & m_{2n+1} \end{vmatrix}$$

The Hankel truncation of order (k+1) of the first type consists in setting:

(3.21)
$$\Delta_{k+1} = \Delta_{k+2} = \ldots. = 0 \ ,$$
$$\Delta^{(1)}_{k+1} = \Delta^{(1)}_{k+2} = \ldots. = 0 \ .$$

The Hankel truncation of order (k+1) of the second type consists in setting:

(3.22)
$$\Delta_{k+1} = \Delta_{k+2} = \ldots. = 0 \ ,$$
$$\Delta^{(1)}_{k} = \Delta^{(1)}_{k+1} = \ldots. = 0 \ .$$

Note that as a result of the Hankel truncation of the first type,

(3.23)
$$m_r = G_r(m_0, m_1, \ldots, m_{2k}), \quad r \geq 2k+1 \ ,$$

where $G_r(\cdot,\cdot,\cdot,\cdot)$ is a polynomial.

If the stochastic differential equation has polynomial coefficients, then the Hankel truncation yields a closed system of nonlinear differential equations for $m = (m_1,\ldots,m_{2k})$:

(3.24)
$$m' = Am + G(m_1,\ldots,m_{2k})$$

where A is a $2k \times 2k$ matrix and the $2k$ components of $G(\cdot,\cdot,\cdots,\cdot)$ are polynomials in $m_1,\ldots,m_{2k}$.

The source of the relationship between the Gauss-Galerkin method and the Hankel truncation method lies in the classical moment problem. Let $\{m_n : n = 0,1,2,\ldots\}$ be a sequence of nonnegative real numbers. The Stieltjes moment problem is to find a nonnegative measure, μ, on $[0,\infty)$ such that

(3.25)
$$m_n = \int_0^\infty x^n \mu(dx); \quad n = 0,1,2,3,\ldots$$

Theorem 3.2. (a) A necessary and sufficient condition for the existence of a solution of the Stieltjes moment problem is that:

(3.26)
$$\Delta_n \geq 0, \quad \Delta^{(1)}_n \geq 0, \quad n = 0,1,2,3,\ldots$$

(b) In order that there exist a solution whose support is not reducible to a finite set of points, it is necessary and sufficient that

(3.27)
$$\Delta_n = 0, \quad \Delta^{(1)}_n > 0, \quad n = 0,1,2,3,\ldots$$

(c) In order that there exist a solution consists of exactly (k+1) points distinct from 0, it is necessary and sufficient that

(3.28)
$$\Delta_0 > 0,\ldots,\Delta_k > 0, \quad \Delta_{k+1} = \Delta_{k+2} = \ldots = 0 \ ,$$
$$\Delta_0 > 0,\ldots,\Delta^{(1)}_k > 0, \quad \Delta^{(1)}_{k+1} = \Delta^{(1)}_{k+2} = \ldots = 0 \ .$$

(d) In order that there exist a solution whose support consists of exactly (k+1) points, one of them equal to 0, it is necessary and sufficient that

(3.29) $$\Delta_0 > 0, \ldots, \Delta_k > 0, \quad \Delta_{k+1} = \Delta_{k+2} = \ldots = 0 ,$$

$$\Delta_0^{(1)} > 0, \ldots, \Delta_{k-1}^{(1)} > 0, \; \Delta_k^{(1)} = \Delta_{k+1}^{(1)} = \ldots = 0 .$$

In the last two cases there is a unique solution to the Stieltjes problem.

Proof. See Shohat and Tamarkin (1950).

Parts (c) and (d) of Theorem 3.2 imply that the Hankel truncation of the moment hierarchy in the case of a stochastic differential equation with polynomial coefficients is equivalent to looking for an n-point atomic approximation as in the Gauss-Galerkin approximation. Thus in this case we have the option of solving the Gauss-Galerkin equations (3.19) or the Hankel closure of the moment hierarchy given by Equation (3.24).

3.5 Numerical Approximation of the Stochastic Logistic Model System

In this section we describe the application of the Gauss-Galerkin approximation to obtain a numerical approximation to the stochastic logistic nonlinear equation

(3.30) $$dX(t) = (X(t) - X^2(t))dt + (\gamma X(t))^{\frac{1}{2}} dw(t) + \rho(E(X(t)) - X(t))dt .$$

For $\rho = 1$, the exact value of the steady state mean, $m^*(\gamma)$ is given by Equation (2.15). The associated steady state probability density function is given by

(3.31) $$p(x) = c\, x^{(2m^*/\gamma)-1} \exp(-x^2/\gamma), \quad x > 0, \quad \gamma < \pi,$$

where c is a normalizing constant.

The Gauss-Galerkin method developed above can be applied, with minor modifications, to the nonlinear stochastic logistic model. We choose as the functions, $f_n(\cdot)$,

(3.32) $$f_n(x) = x^n , \quad \text{for } n = 1,2,3,\ldots$$

Since the coefficients in (3.30) are polynomials, we can apply either the Gauss-Galerkin equations (3.19) or the Hankel truncation (3.24).

Numerical Study 3.1. In the first numerical study Hankel truncations of both the first and second types involving three and four Gauss-Christoffel points were carried out. In implementing this method the Hankel truncation was carried out by using the algorithms for the Cholesky decomposition of the (positive) Hankel matrix. To briefly describe this recall that the Cholesky decomposition of an $N \times N$ matrix A is of the form $A = R \cdot R^T$ where R is a lower triangular matrix. The rows of the matrix R are computed according to the algorithm:

(3.33) for $k = 1, \ldots, N$

for $j = 1, 2, \ldots, k-1$,

$$r_{kj} = (a_{kj} - \sum_{p=1}^{j-1} r_{kp} r_{jp}) / r_{jj} ,$$

$$r_{kk} = (a_{kk} - \sum_{j=1}^{k-1} r_{kj}^2)^{\frac{1}{2}} .$$

In order to obtain m_{2n} as a polynomial in $m_1,\ldots,m_{2n-1}$ which is a consequence of the equation

$$\Delta_n = 0 ,$$

we modify the last step of the algorithm (3.33) for the case $k = 2n = N$ and set

$$r_{NN} = 0 \quad \text{which gives} \quad a_{NN} = \sum_{j=1}^{N-1} r_{Nj}^2 . \tag{3.34}$$

The resulting set of nonlinear ordinary differential equations were solved using a modified trapezoidal method. The solution was allowed to evolve to steady state and the limiting steady state values of the appropriate moments were computed.

Numerical Study 3.2. In the second numerical study a four point Gauss-Galerkin approximation (including one point fixed at the origin) was developed. The system of ordinary differential equations (3.19) for the Gauss-Christoffel points $\{x_k(t)\}$ and weights $\{a_k(t)\}$ were solved numerically by a standard method. The limiting steady state values were computed and from these the appropriate moments were computed.

Numerical Study 3.3. For the purposes of comparison the system of ordinary differential equations obtained from the cumulant neglect ($K_3 = 0$, $K_4 = 0$, or $K_5 = 0$) were solved using the Adams-Bashforth method.

Summary of Results.

The four point Gauss-Galerkin and the four point Hankel truncation approximations were in close agreement and the resulting values were within 5% of the exact values for $\rho = 1$, $\gamma < 2$. The results are plotted in Figure 3.1. The approximate solutions predicted the correct qualitative behavior and were stable near the critical value. Perhaps surprisingly the value of the critical parameter, γ_c, obtained by the four point approximation was within 5% of the exact value, $\gamma_c = \pi$. By comparison the cumulant neglect method performed well only for very small values of γ and the resulting system of equations exhibited unstable behavior near the critical parameter. The numerical results of 3 and 4 point Gauss-Galerkin approximations for the case $\rho = 0.5$ are plotted in Figure 3.2. In this case these two approximations are in close agreement except in the neighbourhood of the critical point.

Example of Rigorous Bound on the Solution. Note that the Gauss-Christoffel inequality (3.5) or Hankel inequality (3.26) also provide direct rigorous bounds for $m^*(\gamma)$. In the case $\rho = 1$, the Hankel inequalities $\Delta_1^{(1)} \geq 0$, $\Delta_2 \geq 0$, together with the moment equations (2.16) yield the bounds:

$$1 - \tfrac{1}{2}\gamma \leq m^*(\gamma) \leq 1 - \tfrac{1}{4}\gamma . \tag{3.35}$$

Remark 3.2. When the number of points used in the Gauss-Galerkin method is large, large matrices must be inverted and the conditioning of the problem becomes critical. Gautschi (1970) showed that the classical moment problem becomes increasingly ill-conditioned as the number of points increases. On the other hand he showed that on a finite interval, if the moments are replaced with the "modified moments" associated with a family of orthogonal polynomials such as the Tchebyscheff polynomials, the problem remains remarkably well-conditioned. In view of this a Gauss-Galerkin approximation with a large number of points should be carried out as follows. First a preliminary transformation:

$$Y(t) \equiv \exp(-\alpha X(t)) , \quad \alpha > 0 , \tag{3.36}$$

should be made. Secondly the Gauss-Galerkin Equations (3.19) should be set up with the choice of $\{f_n(\cdot)\}$ as a family of orthogonal polynomials on $[0,1]$. Finally, the resulting family of ordinary differential equations should be solved using a method suitable for a family of stiff equations.

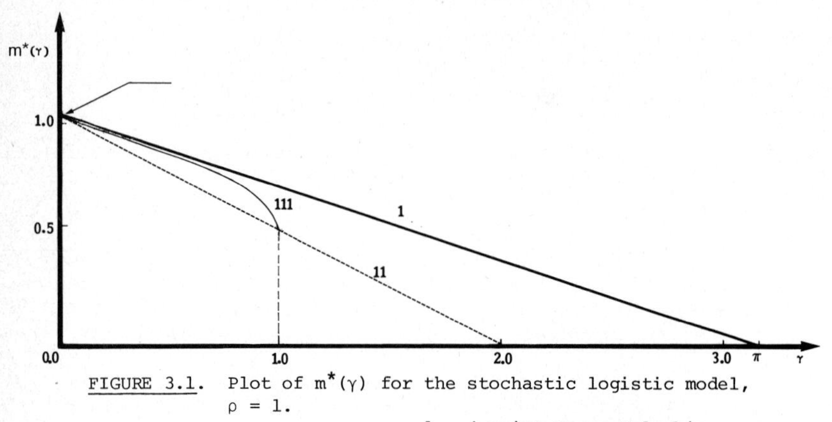

FIGURE 3.1. Plot of $m^*(\gamma)$ for the stochastic logistic model, $\rho = 1$.

Slope at $\gamma = 0$ is $-.25$

1. 4 point Gauss-Galerkin
11. 2 point Gauss-Galerkin
111. K_3 cumulant neglect

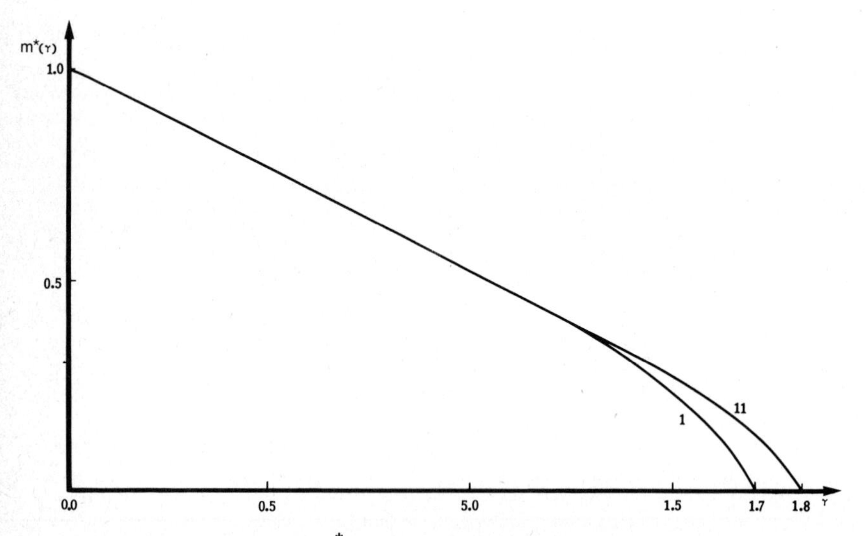

FIGURE 3.2. Plot of $m^*(\gamma)$ for the steady state solution of the Stochastic Differential Equation:

$$dX(t) = (0.5X(t) + 0.5E(X(t)) - X^2(t))dt + (\gamma X(t))^{\frac{1}{2}} dw(t)$$

CURVE 1: 3 point Gauss-Galerkin
CURVE 11: 4 point Gauss-Galerkin

4. GAUSS-GALERKIN APPROXIMATION OF FUNCTION SPACE INTEGRALS

In the previous section the Gauss-Galerkin approximation was developed to approximate the laws $L(X(t))$ for each t. However in probabilistic applications more is required in that it is the objective to approximate function space integrals. In this section we will develop a method of approximation of the function space integrals with respect to the function space probability measure induced by the Markov process, $X(t)$.

Consider the Gauss-Galerkin Equation (3.19). These equations can be rewritten in the form:

$$(4.1.a) \qquad x_k'(t) = F_k(a_1,\ldots,a_n;\ x_1,\ldots,x_n)\ , \quad \text{for } k = 1,2,\ldots,n\ ,$$

and

$$(4.1.b) \qquad a'(t) = a(t)Q$$

wehre $a(t) = (a_1(t),\ a_2(t),\ldots,\ a_n(t))$ and

$$(4.2) \qquad Q^T \equiv A_{11}^{-1}[B_1 - A_{12}(A_{22} - A_{21}A_{11}^{-1}A_{12})^{-1}(B_2 - A_{21}A_{11}^{-1}A_{12}B_1)]\ .$$

Note that the matrix Q depends on the $\{x_k\}$ but is independent of the $\{a_k\}$.

Equations (4.1) have the following interpretation. Equations (4.1.a) form a system of nonlinear ordinary differential equations describing the deterministic motion of the Gauss-Christoffel points $\{x_k(t)\}$. On the other hand Equations (4.1.b) are analogous to the forward Kolmogorov Equations for a nonlinear Markov jump process, $X_n(t)$, which for each t has state space $\{x_1(t),\ldots,x_n(t)\}$. However the latter is not a process in the usual sense; rather, it is a quasi-process in that the off-diagonal elements of the matrix, Q, can be negative. We make the assumption that 1 is in the span of $\{f_1,\ldots,f_n\}$ which implies that the matrix Q has zero row sums.

<u>Lemma 4.1</u>. Assume that the n-point Gauss-Galerkin Equations (3.19) have a bounded solution on the interval $[0,T]$. Then there is a quasi-process $X_n(t)$ such that

(i) for $t > 0$, $X_n(t)$ takes its values in the set $\{x_1(t),\ldots,x_n(t)\}$, and

(ii) there is a canonical version of $X_n(t)$ on $D[0,T]$ and an induced signed measure, $P^{\pm}_{n,\mu}$ on $D[0,T]$ of finite total variation such that

$$(4.3) \qquad \begin{aligned} &P^{\pm}_{n,\mu}(X_n(t) = x_i(t)) = a_i(t)\ , \quad \text{and} \\ &d/dt(P^{\pm}_{n,\mu}(X_n(t+s) = x_j \mid X_n(s) = x_i)\big|_{t=0} = Q(L(X(s)))_{ij}\ . \end{aligned}$$

Proof. Consider the finite dimensional distributions given by: for $0 \le t_1 < \ldots < t_k \le T$,

$$(4.4) \qquad \begin{aligned} P^{\pm}_{n,\mu}&(X(0) = x_{i_0}(\mu),\ X(t_1) = x_{i_1}(t_1),\ldots,X(t_k) = x_{i_k}(t_k)) \\ &= a_{i_0}(\mu)q(0,t;x_{i_0},x_{i_1})\ldots.q(t_{k-1},t_k;x_{i_{k-1}},x_{i_k}) \end{aligned}$$

where $\{a_i(\mu), x_i(\mu)\}$ are the Gauss-Christoffel weights and points for the measure μ. In (4.4), $q(s,t; x_i, x_j)$ is given by the solution of the matrix-valued differential equation

$$d/dt(q(s,t)) = q(s,t)Q(L(X(t))), \quad t \geq s, \tag{4.5}$$

$$q(s,s; x_i, x_j) = \delta_{ij} .$$

The consistency of the finite dimensional distributions (4.4) follows from Equation (4.5) which implies a Chapman Kolmogorov Equation as well as (4.3). $P^{\pm}_{n,\mu}$ is a signed probability measure in the sense that $\int P^{\pm}_{n,\mu}(d\nu) = 1$. The total variation $|P^{\pm}_{n,\mu}|$ is given by

$$|P^{\pm}_{n,\mu}| \equiv \sup \sum |P^{\pm}_{n,\mu}(X(0) = x_{i_0}, X(t_1) = x_1(t_1), \ldots, X(t_{k-1}) = x_{i_{k-1}}(t_{k-1}), \tag{4.6}$$

$$X(T) = x_{i_k}(T))|$$

$$\leq \exp\left(\int_0^T |Q(t)|dt\right),$$

where

$$|Q(t)| \equiv \sup_i \sum_{j \neq i} |Q(t)_{ij}| .$$

The proof that there is a canonical version on $D[0,T]$ proceeds in the same way as for an ordinary process (cf. Dynkin (1965)). Thus it suffices to prove that the quasi-Markov process $X_n(\cdot)$ is stochastically continuous. However the stochastic continuity follows from the fact that the solution to Equation (4.5) is continuous in t. Thus there exists a signed probability measure $P^{\pm}_{n,\mu}$ of finite total variation on $D[0,T]$ and the proof is complete.

Recall that the Markov diffusion process, $\{X(t) : t \geq 0\}$ on R^+ is identified with a family of probability measures, $\{P_\mu : \mu \in \underline{\underline{M}}\}$ on the space $C[0,T]$ for each $T > 0$. Let $F(\cdot)$ be a functional defined on the space $D[0,T]$. Then the Gauss-Galerkin approximation to a function space integral

$$I \equiv \int F(\omega)P_\mu(d\omega), \tag{4.7}$$

associated with the quasi-process $\{X_n(\cdot)\}$ is given by

$$I^{\pm}_n \equiv \int F(\omega)P^{\pm}_{n,\mu}(d\omega) . \tag{4.8}$$

In the next section it is shown that the approximation (4.8) is not suitable for numerical approximation when n becomes large. For this reason we introduce a second family of function space measures associated with the n-point Gauss-Galerkin approximation. This family is obtained by superimposing a nonlinear random walk on the Gauss-Galerkin points and combines elements of both the Galerkin and finite difference approaches. Consider a nonlinear Markov chain which lives on the Gauss-Galerkin points and which makes transitions from $x_i(t)$ to its nearest neighbours $x_{i-1}(t)$ and $x_{i+1}(t)$ (assuming that the $x_i(0)$ are chosen in increasing order). The infinitesimal transition matrix $Q^*(\mu)$ is defined by: for $\mu \in \underline{\underline{M}}_n(R^+)$, $\mu = \sum_{i=1}^{n} a_i \delta_{x_i}$,

(4.8.a) $Q^*(\mu)_{ij} = 0$ if $j \neq i, i-1, i+1$; $Q^*(\mu)_{1,0} = Q^*(\mu)_{n,n+1} = 0$;

(4.8.b) $Q^*(\mu)\underline{1} = 0$, where $\underline{1}$ denotes the unit column vector,

(4.8.c) $$Q^*(\mu)_{i,i+1}(x_{i+1}(\mu)-x_i(\mu))^2 + Q^*(\mu)_{i,i-1}(x_{i-1}(\mu) - x_i(\mu))^2$$

$$= \sum_{j\neq k} Q_{ij}(x_j(\mu) - x_i(\mu))^2, \quad i = 2,3,\ldots,n-1,$$

(4.8.d) $$Q^*(\mu)_{n,n-1}(x_{n-1}(\mu) - x_n(\mu))^2 + Q^*(\mu)_{12}(x_2(\mu) - x_1(\mu))^2$$

$$= \sum_{i=1,n} \sum_{j\neq i} (x_j(\mu) - x_i(\mu))^2 Q_{ij} ,$$

(4.8.3) $$\sum_{i=1}^{n} a_i Q^*(\mu)_{ij} = \sum_{i=1}^{n} a_i Q_{ij}, \quad j = 1,2,\ldots,n-1 .$$

It follows from (4.8.a) and (4.8.b) that the matrix $Q^*(\mu)$ is determined by (2n-2) parameters. Equations (4.8.c) and (4.8.d) together give (n-1) equations and (4.8.e) gives (n-1) equations. Thus (4.8) is a linear system of (2n-2) equations in (2n-2) unknowns. The equations for $Q^*(\mu)$ can be rewritten as

(4.9) $$AV = B$$

where V is the (3n-2)-vector given by

$$V^T = (Q^*_{11}, Q^*_{12}, Q^*_{21}, Q^*_{22}, Q^*_{23}, Q^*_{32}, Q^*_{33}, Q^*_{34}, \ldots, Q^*_{n,n-1}, Q^*_{n,n})$$

and

$$B^T = (\overbrace{0,0,\ldots,0}^{n}, b_2, \ldots, b_{n-1}, b_0, a^*_1, \ldots, a^*_{n-1}), \quad a^*_j = \sum_{i=1}^{n} a_i Q_{ij} ,$$

$$b_i = \sum_{j\neq i} Q_{ij}(x_j(\mu) - x_i(\mu))^2; \quad b_0 = \sum_{i=1,n} \sum_{j\neq i} (x_j(\mu) - x_i(\mu))^2 Q_{ij} ,$$

and $A = (A_1, A_2, A_3)^T$ where A_1 is $n \times m$ A_2 is $(n-1) \times m$, and A_3 is $(n-1) \times m$; $m=3n-2$

$$A_1 = \begin{pmatrix} 1\ 1\ 0\ 0 \ \cdots\cdots\cdots\cdots \\ 0\ 0\ 1\ 1\ 1\ 0\ 0 \ \cdots\cdots\cdots \\ 0\ 0\ 0\ 0\ 0\ 1\ 1\ 1\ 0 \ \cdots\cdots \\ \cdots\cdots\cdots\cdots\cdots\cdots\cdots \\ \cdots\cdots\cdots\ 0\ 0\ 1\ 1\ 1\ 0\ 0 \\ \cdots\cdots\cdots\ 0\ 0\ 0\ 0\ 0\ 1\ 1 \end{pmatrix}$$

$$A_2 = \begin{pmatrix} 0 & 0 & d_{21} & 0 & d_{23} & 0 & 0 & 0 \ \cdots\cdots\cdots\cdots\cdots \\ 0 & 0 & 0 & 0 & 0 & d_{32} & 0 & d_{34} \cdots\cdots\cdots\cdots\cdots \\ \cdots\cdots\cdots\cdots\cdots\cdots\cdots\cdots\cdots\cdots\cdots \\ \cdots\cdots\cdots\cdots\cdots\ 0 & d_{n-1,n-2} & 0 & d_{n-1,n} & 0 & 0 \\ 0 & d_{12} & 0 & 0 & 0 \ \cdots\cdots\cdots\ 0 & 0 & 0 & d_{n,n-1} & 0 \end{pmatrix}$$

where $d_{j,j+1} = (x_{j+1}(\mu) - x_j(\mu))^2$, $d_{j,j-1} = (x_{j-1}(\mu) - x_j(\mu))^2$.

$$A_3 = \begin{pmatrix} a_1 & 0 & a_2 & 0 & 0 & \cdots & & & & & & & \\ 0 & a_1 & 0 & a_2 & 0 & a_3 & 0 & \cdots & & & & & \\ 0 & 0 & 0 & 0 & a_2 & 0 & a_3 & 0 & a_4 & \cdots & & & \\ \cdots & & & & & & a_3 & 0 & a_4 & \cdots & & & \\ \cdots & & & & & & & & & & & & \\ \cdots & & & & & & & & a_{n-1} & 0 & 0 & 0 & 0 \\ \cdots & & & & & & a_{n-2} & 0 & a_{n-1} & 0 & a_n & 0 & \end{pmatrix}$$

If all the weights, a_j, are non-zero and all the x_j are distinct, then the matrix A is non-singular and Equation (4.9) can be solved for V. Note that at time 0 a modified initializing step is required and in this step Equation (4.8.a) is deleted. It can be verified that after the initial step the points are distinct and the weights are non-zero. Having solved (4.9) for the infinitesimal transition matrices, $Q^*(\cdot)$, a nonlinear Markov process is constructed as in Lemma 4.1. We denote by $X_n^*(\cdot)$ the resulting nonlinear Markov process and by $P_{n,\cdot}^*$ the resulting family of function space probability measures on $D[0,T]$. Note that the process $X_n^*(t)$ has the same marginal distributions as the quasi-process $X_n(t)$ because of assumption (4.8.e) and the same infinitesimal conditional variance structure as $X_n(t)$ because of the Assumption (4.8.d). Finally note that the process $X_n^*(\cdot)$ is nonlinear even in the case the original Markov process is linear. Given a functional $F(\cdot)$ on $D[0,T]$ the approximation to the function space integral (4.7) associated with the approximation $X_n^*(\cdot)$ is given by:

(4.10) $$I_n^* \equiv \int F(\omega) P_{n,\mu}^*(d\omega) \ .$$

It will be indicated in the next section that the approximation (4.10), in contrast to (4.8), is suitable for the numerical approximation of function space integrals.

5. THE QUESTION OF CONVERGENCE

In Section 4 we constructed two families of probability measures (signed) $P_{n,\cdot}^*$ and $P_{n,\cdot}^{\pm}$ on the space $D[0,T]$. In this section we briefly sketch an introduction to the analysis of these function space measures. In particular we discuss the question of the convergence:

(5.1) $$P_{n,\mu}^{\cdot} \to P_\mu \quad \text{as} \quad n \to \infty \ .$$

In the case of the signed measures $P_{n,\cdot}^{\pm}$, it can be shown that the total variation goes to infinity as $n \to \infty$ (cf. Krylov (1962),(1965)). For this reason the usual notion of weak convergence of probability measures is not suitable and we introduce a weaker notion of convergence. On the other hand the limit of the nonlinear random walks $X_n^*(\cdot)$ can be studied by the standard methods of probability theory and we will merely state the results. A detailed analysis of these function space measures will appear elsewhere.

Example 5.1. Consider the solution to the stochastic differential equation

(5.2) $$dX(t) = (aX(t) + e)dt + (\gamma X(t))^{\frac{1}{2}} dw(t) \ .$$

The hierarchy of moment equations for the solution of (5.2) is triangular and therefore the truncated systems can be solved easily. It follows that the Gauss-Galerkin points and weights $\{x_i(t)\}$, $\{a_i(t)\}$ associated with $X_n(t)$ coincide with the Gauss-Christoffel points and weights for the process $X(t)$. The same is true for the signed conditional weights and points so that the finite dimensional distributions of $X_n(t)$ converge to those of $X(t)$ in the sense that

$$(5.3) \qquad E^{\pm}_{n,\mu}(g_1(X_n(t_1)\ldots g_k(X_n(t_k))) \to E_\mu(g_1(X(t_1))\ldots g_k(X(t_k)))$$

as $n \to \infty$, for every k-tuple of polynomials $\{g_1,\ldots,g_k\}$. However the total variations of the signed finite dimensional distributions can go to infinity so that the finite dimensional distributions may not converge in the sense of weak convergence of probability distributions.

We next establish a convergence theorem for a more general situation. Consider the stochastic differential equation for a process on R^+:

$$(5.4) \qquad dX(t) = a(X(t))dt + \sigma(X(t))dw(t)$$

where $a(x)$ and $\sigma(x)$ are continuous functions which satisfy:

$$(5.5.a) \qquad a(0) > 0, \quad \sigma(0) = 0 ,$$

$$(5.5.b) \qquad a(x) \leq a_1 + a_2 x ,$$

$$(5.5.c) \qquad \sigma(x) \leq \sigma_1 x^{\frac{1}{2}} .$$

Then the moments $m_n(t)$ satisfy the equations:

$$(5.6) \qquad dm_n(t)/dt = nE(X^{n-1}(t)a(X(t))) + \tfrac{1}{2}n(n-1)E(X^{n-2}(t)\sigma^2(X(t)))$$

$$\leq na_1 m_{n-1}(t) + a_2 n m_n(t) + \tfrac{1}{2}n(n-1)\sigma_1^2 m_n(t) .$$

This implies that there exists constants $\{k_n : n \geq 1\}$ and $\theta_0 > 0$ such that:

$$(5.7) \qquad m_n(t) \leq k_n \quad \text{for} \quad 0 \leq t \leq T, \quad \text{and}$$

$$(5.8) \qquad \sum_{n=1}^{\infty} \theta^n k_n/n! < \infty \quad \text{for all} \quad \theta < \theta_0 .$$

Theorem 5.1. Let $X_N(t)$ denote the N-point Gauss-Galerkin approximation to the solution of the stochastic differential equation (5.4). Then for $0 \leq t \leq T$,

$$(5.9) \qquad L(X_N(t)) \to L(X(t)) \quad \text{as} \quad N \to \infty ,$$

in the sense of weak convergence of probability measures on R^+.

Proof. The N-point Gauss-Galerkin approximation is obtained by solving Equations (3.19) with the initial condition

$$L(X_N(0)) = \sum_{k=1}^{N} a_k(0)\, \delta_{x_k}(0)$$

where $\{a_k(0), x_k(0)\}$ denote the Gauss-Christoffel weights and points of $L(X(0))$, respectively. Let $m_{N,n}(t) \equiv E(X_N^n(t))$. Then by the Equation (3.16) it follows that for $n \leq 2N-1$,

(5.10) $$dm_{N,n}(t)/dt \le na_1 m_{N,n-1}(t) + a_2 n m_{N,n}(t) + \tfrac{1}{2}n(n-1)\sigma_1^2 m_{N,n-1}(t).$$

Therefore, for $n \le 2N-1$,

(5.11) $$m_{N,n}(t) \le k_n \quad \text{for} \quad 0 \le t \le T .$$

From this it follows that the collection $\{m_{N,n}(t) : 0 \le t \le T,\ N > \tfrac{1}{2}(n+1)\}$ is bounded and equicontinuous and therefore relatively compact in $C[0,T]$. Thus there exists a limiting family $\{m_{*,n}(t) : 0 \le t \le T\}$ of continuous functions. Since for each t, it is the limit of a sequence $M_{n,\cdot}(t)$ which satisfies (3.26) it follows that $m_{*,\cdot}(t)$ also satisfies (3.26). Therefore $m_{*,\cdot}(t)$ induces a probability measure $p_*(t)$ on R^+ for each $0 \le t \le T$ and the probability measures $L(X_N(t))$ converge weakly to the probability measures $p_*(t)$. Recall that for $n \le 2N-1$,

(5.12) $$m_{N,n}(t)-m_{N,n}(0) = \int_0^t (nE(X_N^{n-1}(s)a(X_N(s))) + \tfrac{1}{2}n(n-1)E(X_N^{n-2}(s)\sigma^2(X_N(s))))ds.$$

Using the fact that $X_N(t)$ converges in distribution to $X(t)$, the inequalities (5.11) and the bounded convergence theorem we can verify that for every n,

(5.13) $$m_{*,n}(t)-m_{*,n}(0) = \int_0^t (nE(X_*^{n-1}(s)a(X_*(s))) + \tfrac{1}{2}n(n-1)E(X_*^{n-2}(s)\sigma^2(X_*(s))))ds.$$

But from the uniqueness of the solution to the stochastic differential equation (5.4) it follows that

$$m_{*,n}(t) = E(X(t)) ,$$

and

$$p_*(t) = L(X(t)) ,$$

and the proof is complete.

Theorem 5.2. Assume that in addition to (5.5), the functions $a(x)$ and $\sigma^2(x)$ are polynomials in x. If $g_1,\ldots,g_k$ are polynomials, then

(5.14) $$\lim_{N\to\infty} E^{\pm}_{N,\mu}(g_1(X_N(t_1))\cdots\cdot g_k(X_N(t_k))) \equiv E_\mu(g_1(X(t_1))\cdots\cdot g_k(X(t_k))) .$$

Proof. The proof is by induction in k. Assume that the result is true for k. Then consider $h_{N,n}(t) \equiv E^{\pm}_{N,\mu}(g_1(X_N(t_1))\ldots g_k(X_N(t_k))\cdot X_N^n(t))$ for $t \ge t_k$. Then for $n \le N-1$, it follows from (4.3) that

(5.15) $$\begin{aligned} dh_{N,n}(t)/dt &= nE^{\pm}_N(E^{\pm}_N(X_N^{n-1}(t)a(x_N(t))\,|\,X_N(t_k))g_1(X_N(t_1))\ldots g_k(X_N(t_k))) \\ &\quad + \tfrac{1}{2}n(n-1)E^{\pm}_N(E^{\pm}_N(X_N^{n-2}(t)\sigma^2(X_N(t))\,|\,X_N(t_k))g_1(X_N(t_1))\ldots g_k(X_N(t_k))) \\ h_{N,n}(t_k) &= E^{\pm}_N(g_1(X_N(t_1))\ldots g_k(X_N(t_k))X_N^n(t_k)). \end{aligned}$$

We can also obtain bounds for the $h_n(t)$ of the type (5.11) which allows one to select a convergent subsequence. The limiting functions $h_n^*(t)$, $t \ge t_k$, then satisfy the equations (5.15) for all n. Then by the uniqueness of the solution to these equations (subject to the established bounds) it follows that

$$h_n^*(t) = E_\mu(g_1(X(t_1)).....g_k(X(t_k))X^n(t)) \quad \text{for} \quad t \geq t_k \ .$$

Hence the induction step is established. On the other hand the proof for $k = 1$ follows from Theorem 5.1. Hence the proof is complete.

Remark 5.1. It is important to note that the finite dimensional distributions need not converge in the sense of weak convergence of probability measures. The reason for this is the fact that the total variation of the finite dimensional signed measures can diverge as $N \to \infty$. This phenomena occurs even for measures on bounded intervals; in the latter case the discrete approximations are known as Newton-Cotes quadrature formulae. The behavior of the total variation and the nature of the convergence of Newton-Cotes formulae has been studied by V.I. Krylov (1962).

Nevertheless it is of interest to determine in what sense the function space measures, $P^{\pm}_{N,\mu}$ converge to P_μ. From the above comment it is clear that we cannot expect convergence in the sense of weak convergence of probability measures. However it is possible to establish the convergence

$$E^{\pm}_{N,\mu}(F) \to E_\mu(F) \quad \text{as} \quad N \to \infty, \tag{5.16}$$

for a class of functionals, F, on $D[0,T]$ called analytic functionals by Krylov (1965). These include the functionals of the form:

$$F(x(\cdot)) \equiv \sum_{k=1}^{K} \int_0^T \dots \int_0^T h_k(t_1,\dots,t_k)x(t_1)\dots x(t_k)dt_1\dots dt_k \ , \tag{5.17}$$

where $h_k(t_1,\dots,t_k)$ is a continuous function of $t_1,\dots,t_k$.

Finally, we come to the question of the convergence of the function space measures, $P^*_{n,\mu}$, associated with the nonlinear random walks defined by (4.8). This is described by the following result.

Theorem 5.3. Let $X_n^*(\cdot)$ denote the nonlinear Markov process defined by (4.8) and let $P^*_{n,\cdot}$ denote the family of probability measures on $D[0,T]$ associated with this process. Then

$$P^*_{n,\mu} \to P_\mu \quad \text{as} \quad n \to \infty \ , \tag{5.18}$$

in the sense of weak convergence of probability measures on $D[0,T]$.

Proof. The proof of this result is based on a modification of the standard proof of convergence of Markov chains to diffusions (cf. Stroock and Varadhan (1979; 11.2)) and is omitted.

Remark 5.2. According to (5.16) we can approximate the function space integral of an analytic functional by $E^{\pm}_{N,\cdot}(F)$. However this is not suitable for numerical approximation in view of the fact that the large total variation of the signed measure results in serious round-off errors. On the other hand Theorem 5.3 implies that the function space integral of a continuous functional can be well approximated by $E^*_{N,\cdot}(F)$. The proof of convergence for the non-linear stochastic logistic will also appear elsewhere.

6. REFERENCES

[1] Ames, W.F. (1977), Numerical Methods for Partial Differential Equations, Second Edition, Academic Press.

[2] Dahlquist, G. and Björck, A. (1974), Numerical Methods, Prentice Hall.

[3] David, P.J. (1967), Approximate integration rules with non-negative weights. In Lecture Series in Differential Equations, Session 7, Catholic University, 233-256.

[4] David, P.J. and Rabinowitz, P. (1975), Methods of Numerical Integration, Academic Press.

[5] Dawson, D.A. (1980). Qualitative behavior of geostochastic systems, J. Stochastic Processes and their Applications (10), 1-31.

[6] Douglas, J. and Dupont, T. (1970), Galerkin methods for parabolic problems, SIAM J. Numer. Anal. 4, 575-626.

[7] Dynkin, E.B. (1965). Markov Processes, Volumes I,II, Springer-Verlag.

[8] Fife, P.C. (1979), Mathematical Aspects of Reacting and Diffusing Systems, Springer Lecture Notes in Biomathematics 28.

[9] Gautschi, W. (1970), On the construction of Gaussian quadrature rules from modified moments, Math. Comp. 24, 245-260.

[10] Hebert, D.J. (1974), Nonlinear parabolic equations and probability, Bull. Amer. Math. Soc. 80, 965-969.

[11] Itô, K. and Watanabe, S. (1978), Introduction to stochastic differential equations. In Proceedings of the International Symposium on Stochastic Differential Equations, ed. K. Itô.

[12] Karlin, S. and Studden, W.J. (1966), Tchebysheff Systems with Applications in Analysis and Statistics, Interscience Publ.

[13] Krylov, V.I. (1962), Approximate Calculation of Integrals, translated by A.H. Stroud. Macmillan, New York, London.

[14] Krylov, V.I. (1965), Integration of analytic functionals with respect to distributions with alternating signs, Doklady Akad. Nauk. SSSR. 163(2), 289-292.

[15] Kushner, H.J. (1977), Probability Methods for Approximations in Stochastic Control and Elliptic Equations, Academic Press.

[16] Laurie, D.P. and Rolfes, L. (1979), Computation of Gaussian quadrature rules from modified moments, J. Comp. and Appl. Math. 5, 235-243.

[17] Leslie, D.C. (1973), Developments in the Theory of Turbulence, Clarendon Press.

[18] Lions, J.L. (1969), Quelques méthods de résolution des problèmes aux limites non-linéaires, Dunod. Gauthier Villars.

[19] Ladyzenskaja, O.A., Solonnikov, V.A. and Ural'ceva, N.N. (1968), Linear and Quasilinear Equations of Parabolic Type, Transl. Math. Monographs 23, Amer. Math. Soc.

[20] McKean, H.P. (1966), A class of Markov processes associated with nonlinear parabolic equations, Proc. N.A.S., U.S.A. 56, 1907-1911.

[21] Nicolis, G. and Prigogine, I. (1977), Self-organization in non-equilibrium systems, Wiley-Interscience.

[22] Ogawa, S. (1974). Processus de Markov en interaction et système semi-linéaire d'équations d'évolution, Ann. Inst. H. Poincaré B (10), 279-299.

[23] Shohat, J.A. and Tamarkin, J.D. (1950), The problem of Moments, Amer. Math. Soc.

[24] Stroud, A.H. (1981), Approximate Calculation of Multiple Integrals, Prentice-Hall.

[25] Stroud, A.H. (1974), Numerical Quadrature and Solution of Ordinary Differential Equations, Springer-Verlag.

[26] Stroock, D.W. and Varadhan, S.R.S. (1979), Multidimensional Diffusion Processes, Grund. der math. Wissenschaften 233, Springer-Verlag.

[27] Pridor, A. (1977), Estimation of moments for the numerical solution of transport problems, SIAM J. Numer. Anal. 14, 426-440.

KEY WORDS AND PHRASES: Nonlinear Markov process, nonlinear diffusion, Galerkin approximation, Gaussian quadrature, function space integral.

Research supported by the Natural Sciences and Engineering Research Council of Canada.

STATISTICS AND RELATED TOPICS
M. Csörgö, D.A. Dawson, J.N.K. Rao, A.K.Md.E. Saleh (eds.)

DONSKER CLASSES OF FUNCTIONS[1]

R. M. Dudley

Massachusetts Institute of Technology
Cambridge, Massachusetts, U.S.A.

Some central limit theorems for empirical measures (Ann. Prob. 6, 1978, 899-929; 7, 1979, 909-911) are extended from classes of sets to classes of functions. For uniformly bounded classes of functions the extension is straightforward. For possibly unbounded classes, the paper gives some simple results for sequences, and then the following main theorem: let $(X,\mathcal{A},P)$ be a probability space. Let $\mathcal{F}$ be a class of functions on X for which suitable measurability conditions hold. Let $F(x): = \sup\{|f(x)|:f \in \mathcal{F}\}$ and assume $F \in \mathcal{L}^p(X,\mathcal{A},P)$ for some $p > 2$. Assume that for some $M < \infty$ and some γ, $0 < \gamma < 1 - 2/p$, for $0 < \delta < 1$ there are functions $f_1,\ldots,f_m$ in $\mathcal{L}^1(X,\mathcal{A},P)$, where $m := m(\delta) \leq \exp(M\delta^{-\gamma})$, such that for each $f \in \mathcal{F}$ there are j and k with $f_j \leq f \leq f_k$ and $\int f_k - f_j \, dP \leq \delta$. Let P_n be the nth (random) empirical measure for P. Then $n^{\frac{1}{2}}(P_n - P)$ converges in law, with respect to uniform convergence on $\mathcal{F}$, to a limiting Gaussian process.

1. INTRODUCTION AND PRELIMINARIES

This section parallels sec. 1 of [7]; the emphasis here will be on definitions and results which are different for classes of functions than for classes of sets. Sec. 2 below treats sequences of functions, and Sec. 3 metric entropy with bracketing, corresponding to Secs. 2 and 5 of [7] respectively. Sec. 7 of [7] on Vapnik-Červonenkis classes is extended in [9], with a new formulation of measurability conditions.

Let $(X,\mathcal{A})$ be a measurable space and $\mathcal{F}$ a collection of real-valued measurable functions on X. For each $x \in X$ let $F(x) := F_{\mathcal{F}}(x) := \sup\{|f(x)| : f \in \mathcal{F}\}$. Then F will be called the <u>envelope function</u> for $\mathcal{F}$.

Let P be a probability measure on $\mathcal{A}$. Let $(X^\infty, \mathcal{A}^\infty, P^\infty)$ be a countable product of copies of $(X,\mathcal{A},P)$, with coordinates $X_j := X(j)$, $j = 1,2,\ldots,$ so that the X_j are independent, identically distributed random variables with values in X and distribution P. Let $P_n := (\delta_{X(1)} + \ldots + \delta_{X(n)})/n$, $n = 1,2,\ldots,$ where δ_x is the unit mass at x. Let $\nu_n := n^{\frac{1}{2}}(P_n - P)$, and for any signed measure ν let $\nu(f) := \int f \, d\nu$.

We will be concerned with the suprema of ν_n or its absolute value over collections of sets or functions. Pollard [12] has pointed out that to require X_j independent and identically distributed with law P does not specify the laws of such suprema. For example, let P be Lebesgue measure on $X = [0,1]$. Let A be a subset with inner measure 0 and outer measure 1. Let Q be the law induced on A by P. Let $\mathcal{F} := \{1_{\{x\}}: x \in A\}$. Then $\sup_{f \in \mathcal{F}} Q_1(f) = 1$ and $\sup_{f \in \mathcal{F}} P_1(f)$ is non-measurable, even though the identity map from A into $[0,1]$ is a random variable with distribution P. It was to avoid such non-uniqueness that above, X_j were taken to be coordinate functions on X^∞.

Let $\mathcal{L}^2(X,\mathcal{A},P)$ denote the set of all $\mathcal{A}$-measurable real-valued functions f on X with $\int f^2 dP < \infty$. (Functions equal P-almost everywhere are not identified here; the notation $L^2(X,\mathcal{A},P)$ is used when they are.) The usual pseudo-metric on $\mathcal{L}^2(X,\mathcal{A},P)$ will be written

$$e_P(f,g) := (\int (f-g)^2 dP)^{\frac{1}{2}} .$$

(Note that for $A,B \in \mathcal{A}$, $d_P(A,B) := P(A \Delta B) = e_P(1_A,1_B)^2$; thus d_P and e_P define different, although uniformly equivalent pseudo-metrics on $\mathcal{A}$.)

By the finite-dimensional central limit theorem, for any finite collection of functions $\{f_1,\ldots,f_m\} \subset \mathcal{L}^2(X,\mathcal{A},P)$, $\{\int f_j d\nu_n\}_{j=1}^m$ converges in law in $\mathbb{R}^m$ as $n \to \infty$ to $\{G_P(f_j)\}_{j=1}^m$ where G_P is a Gaussian process indexed by $\mathcal{L}^2(X,\mathcal{A},P)$, with mean 0 and covariance

$$EG_P(f)G_P(g) = \int fg\, dP - \int f\, dP \int g\, dP, \quad f,g \in \mathcal{L}^2(X,\mathcal{A},P) \quad .$$

We call $\mathcal{F}$ a $\underline{G_P\text{BUC class}}$ if and only if G_P on $\mathcal{F}$ can be realized with sample functions almost surely bounded and uniformly continuous for e_P. The following is proved just as [7, Prop.1.0].

1.1 <u>Proposition.</u> For any $\mathcal{F} \subset \mathcal{L}^2(X,\mathcal{A},P)$, $\mathcal{F}$ is a G_PBUC class if and only if its closure in L^2 is both a GB and a GC set.

<u>Definition.</u> A function G on a subset $\mathcal{F}$ of a real vector space V will be called <u>pre-linear</u> iff whenever $\Sigma_{1\le i\le m}\, a_i f_i = 0$, for real a_i, $f_i \in \mathcal{F}$, and $m = 1,2,\ldots,$ then $\Sigma_{1\le i\le m}\, a_i G(f_i) = 0$.

Clearly a pre-linear G on a set $\mathcal{F} \subset V$ extends uniquely to a linear function on the linear span of $\mathcal{F}$.

For any $\mathcal{F} \subset \mathcal{L}^2(X,\mathcal{A},P)$, let $C_{b,lin}(\mathcal{F},e_P)$ be the set of all bounded, pre-linear functions on $\mathcal{F}$ for which the extended linear function on the linear span of $\mathcal{F}$ is e_P-uniformly continuous on the symmetric convex hull of $\mathcal{F}$. Let $D_2(\mathcal{F},P)$ be the linear space of all functions $\phi + \psi$, where $\phi \in C_{b,lin}(\mathcal{F},e_P)$ and $\psi(f) = \Sigma_{1\le i\le m}\, a_i f(x_i)$ for some $m = 1,2,\ldots,$ real a_i, and $x_i \in X$.

We always have $P \in C_{b,lin}(\mathcal{F},e_P)$ and $P_n \in D_2(\mathcal{F},P)$, so $\nu_n \in D_2(\mathcal{F},P)$. If $\mathcal{F}$ is G_PBUC, then G_P can be taken to have values in $C_{b,lin}(\mathcal{F},e_P) \subset D_2(\mathcal{F},P)$ by a series representation ([6, Thm. 0.3(e)] or earlier Itô and Nisio [11, Thm.4.11]).

Note that all elements of $D_2(\mathcal{F},P)$ are pre-linear. On $D_2(\mathcal{F},P)$, or any space of bounded real functions on a set $\mathcal{F}$, we have the supremum norm

$$\|G\|_{\mathcal{F}} := \|G\|_\infty := \sup\{|G(f)| : f \in \mathcal{F}\} \quad .$$

(D_2 is in general not complete for $\|\cdot\|_\infty$.) Let $\mathcal{B}_b := \mathcal{B}_b(D_2)$ be the σ-algebra of subsets of $D_2(\mathcal{F},P)$ generated by the set of all open balls for $\|\cdot\|_\infty$ (with centers in the space).

We say $\mathcal{F}$ is <u>P-EM</u> (empirically measurable for P) iff for all n, $\omega \to (f \to \int f\, dP_n)$ is measurable from the measure-theoretic completion of $(X^\infty,\mathcal{A}^\infty,P^\infty)$ to $(D_2(\mathcal{F},P),\mathcal{B}_b)$.

Note that $\mathcal{F}$ is P-EM if it has a countable subset $\{f_m\}$ such that for each $f \in \mathcal{F}$, some subsequence of $\{f_m\}$ converges pointwise and in $\mathcal{L}^2(P)$ to f. Note also that $\mathcal{B}_b$ measurability of $\omega \to (f \to \int f\,dP_n)$ and $\omega \to (f \to \int f\,d\nu_n)$ are equivalent since $P \in C_{b,lin}(\mathcal{F},e_P)$.

Let $\mathcal{B}_c$ be the smallest σ-algebra of subsets of $D_2(\mathcal{F},P)$ for which all the evaluations $G \to G(f)$ are measurable, $f \in \mathcal{F}$. Let $\mathcal{B}_{bc}$ be the σ-algebra generated by $\mathcal{B}_b$ and $\mathcal{B}_c$. Although we may have $\mathcal{B}_b \subset \mathcal{B}_{bc}$ strictly, P-EM implies that the laws of P_n and ν_n are defined on $\mathcal{B}_{bc}$ since they are always defined on $\mathcal{B}_c$. In this paper, we will not need stronger measurability conditions such as "Pε-Suslin" [7] or "image Pε-Suslin" [9] although an example [10] shows that in some cases P-EM is not enough.

Definition. A P-EM class $\mathcal{F}$ will be called a P-Donsker class iff it is a G_PBUC class and $EH(\nu_n) \to EH(G_P)$ whenever H is a bounded real function on $D_2(\mathcal{F},P)$, continuous for $\|\cdot\|_\infty$ and $\mathcal{B}_{bc}$-measurable, where we use a realization of G_P with sample functions in $C_{b,lin}(\mathcal{F},e_P)$.

Note that a P-Donsker class $\mathcal{F}$, being G_P-BUC, is totally bounded for e_P, so it has a countable e_P-dense subset. Sample functions of G_P are a.s. uniformly continuous on such a subset by assumption, and thus extend uniquely to functions which, again by series ([6, Thm. 0.3(e)], [11, Thm. 4.1]) are in $C_{b,lin}(\mathcal{F},e_P)$. Since this space is separable for $\|\cdot\|_\infty$, $\mathcal{L}(G_P)$ is then defined on $\mathcal{B}_{bc}$.

Wichura [13, Thm. 2] and [5] imply that if $\mathcal{F}$ is Donsker and $\mathcal{A}_n$ are σ-algebras of subsets of $D_2(\mathcal{F},P)$ such that $\mathcal{B}_b \subset \mathcal{A}_n \subset \mathcal{B}$, where $\mathcal{B}$ is the σ-algebra of Borel sets generated by all the open sets for $\|\cdot\|_\infty$, and such that $\mathcal{L}(\nu_n)$ is defined on $\mathcal{A}_n$, then there is a probability space Ω_W with random variables T_n, $n \geq 0$, having values in $D_2(\mathcal{F},P)$, with $\mathcal{L}(T_n) = \mathcal{L}(\nu_n)$ on $\mathcal{A}_n$ for all $n \geq 1$ and $\sup_{f \in \mathcal{F}}|(T_n - T_0)(f)| \to 0$ as $n \to \infty$, where $\mathcal{L}(T_0) = \mathcal{L}(G_P)$ on $C_{b,lin}(\mathcal{F},e_P)$. Thus, as noted in [7, p.900], $\sup_{f \in \mathcal{F}}|(T_n - T_0)(f)|$ is measurable.

Since $\mathcal{L}(\nu_n)$ and $\mathcal{L}(G_P)$ are always defined on $\mathcal{B}_{bc}$ for a P-EM G_P-BUC class, the conclusion of Wichura's theorem for $\mathcal{A}_n \equiv \mathcal{B}_b$ here implies it for $\mathcal{B}_{bc}$. Thus in the definition of P-Donsker class, $\mathcal{B}_{bc}$ can be replaced equivalently by $\mathcal{B}_b$. We also have:

1.2 Theorem. A class $\mathcal{F} \subset \mathcal{L}^2(X,\mathcal{A},P)$ is a Donsker class if and only if it is P-EM and there is a probability space Ω_W with random variables T_n, $n \geq 0$, having values in $D_2(\mathcal{F},P)$, such that $\mathcal{L}(T_n) = \mathcal{L}(\nu_n)$ on $\mathcal{B}_{bc}$ for all $n \geq 1$ and

$$\lim_{n\to\infty} \sup_{f\in\mathcal{F}} |(T_n - T_0)(f)| = 0 \quad \text{a.s.}$$

For any σ-algebras $\mathcal{A}_n$ of Borel sets in $D_2(\mathcal{F},P)$ on which $\mathcal{L}(\nu_n)$ is defined for each n, we can take $\mathcal{L}(T_n) = \mathcal{L}(\nu_n)$ on $\mathcal{A}_n$, $n \geq 1$.

Let $\Omega := X^\infty$, $\Pr := P^\infty$, and let $\mathcal{E}$ be the completion of $\mathcal{A}^\infty$ under Pr. Then for any $A \subset \Omega$ the outer probability is defined as usual by

$$\Pr^*(A) := \inf\{\Pr(B) : B \supset A,\ B \in \mathcal{E}\}.$$

Given $\varepsilon > 0$ and $\delta > 0$, let

$$B_{\delta,\varepsilon} := \{h \in D_0(\mathcal{F},P) : \text{for some } f,g \in \mathcal{F},\ e_p(f,g) < \delta \text{ and } |h(f) - h(g)| > \varepsilon\}.$$

Here is a characterization of Donsker classes of functions, extending [7, Theorem 1.2]. We write $\nu_n(f) := \int f\,d\nu_n$.

1.3 Theorem. Given a probability space $(X,\mathcal{A},P)$ and a P-EM class $\mathcal{F} \subset \mathcal{L}^2(X,\mathcal{A},P)$, $\mathcal{F}$ is a Donsker class if and only if both

a) $\mathcal{F}$ is totally bounded for e_P, and

b) for any $\varepsilon > 0$ there is a $\delta > 0$ and an n_0 such that for $n \geq n_0$, $Pr^*\{\nu_n \in B_{\delta,\varepsilon}\} < \varepsilon$.

Proof. We follow the proof of [7, Thm. 1.2], replacing $\mathcal{C}$ by $\mathcal{F}$, sets A in $\mathcal{C}$ by functions h in $\mathcal{F}$, $\nu_n(A)$ by $\int h\, d\nu_n$, d_P by e_P, etc. Since $\mathcal{F}$ is totally bounded for e_P, hence bounded, we may assume, taking a constant multiple, that $\int f^2 dP < 1/4$ for all $f \in \mathcal{F}$ when replacing sets by functions in Lemma 1.3 of [7]. Noting the published correction toward the end of the proof (a further "Clarification" is available from the author), the proof in [7] then goes through with the indicated changes.

We next develop some stability properties of Donsker classes.

1.4 Corollary. Any P-EM subset $\mathcal{G}$ of a Donsker class $\mathcal{F}$ is Donsker.

Proof. This follows directly from 1.3. (It also follows from 1.2 if we take $\mathcal{A}_n \supset \mathcal{B}_{bc}$ on $D_2(\mathcal{F},P)$ to make the restriction map to $D_2(\mathcal{G},P)$ measurable.) Q.E.D.

1.5 Proposition. For any Donsker class $\mathcal{F}$ and finite set $\mathcal{G} \subset \mathcal{L}^2(X,\mathcal{A},P)$, $\mathcal{F} \cup \mathcal{G}$ is Donsker.

Proof. By induction we can take $\mathcal{G} = \{g\}$, $g \in \mathcal{L}^2$. Clearly $\mathcal{H} := \mathcal{F} \cup \{g\}$ is P-EM and totally bounded. Using Theorem 1.3, given $0 < \varepsilon < 1$ take $\delta > 0$ such that for $\mathcal{F}$, $Pr^*(B_{\delta,\varepsilon/2}) < \varepsilon/2$. Take $\gamma := \min(\delta, \varepsilon^2/2)$. Define $B^{\mathcal{H}}_{\alpha,\beta}$ as $B_{\alpha,\beta}$ for $\mathcal{H}$. If there is no $f \in \mathcal{F}$ with $e_P(f,g) < \gamma/2$, then $Pr^*(B^{\mathcal{H}}_{\gamma/2,\varepsilon}) < \varepsilon/2$. If there is such an f, choose one. Then for any $h \in \mathcal{F}$ with $e_P(g,h) < \gamma/2$, we have $e_P(f,h) < \gamma \leq \delta$. Using $|\nu_n(h-g)| \leq |\nu_n(h-f)| + |\nu_n(f-g)|$ and Chebyshev's inequality, we get

$$Pr^*(B^{\mathcal{H}}_{\gamma/2,\varepsilon}) \leq Pr^*(B^{\mathcal{F}}_{\delta,\varepsilon/2}) + Pr(|\nu_n(f-g)| \geq \varepsilon/2)$$

$$\leq \varepsilon/2 + \varepsilon/2 = \varepsilon, \quad \text{Q.E.D.}$$

1.6 Proposition. Let $\mathcal{F}$ be P-EM, $\mathcal{F} \subset \mathcal{L}^2(X,\mathcal{A},P)$, and let $\mathcal{G}$ be the symmetric convex hull of $\mathcal{F}$. Then $\mathcal{G}$ is a Donsker class if and only if $\mathcal{F}$ is.

Proof. By definition of D_2, there is a natural 1-1 map J of $D_2(\mathcal{F},P)$ onto $D_2(\mathcal{G},P)$ such that J^{-1} is restriction to $\mathcal{F}$. We have $\|Jh\|_{\mathcal{G}} = \|h\|_{\mathcal{F}}$ for all $h \in D_2(\mathcal{F},P)$. Thus J is an isomorphism for $\mathcal{B}_b$ and for $\mathcal{B}_{bc}$ measurability. Since J preserves ν_n, $\mathcal{G}$ is P-EM. Also J can be taken to preserve G_P.

If $\mathcal{F}$ is Donsker, take T_n from Theorem 1.2. Then JT_n serve for $\mathcal{G}$ there, so $\mathcal{G}$ is Donsker. The converse follows from 1.4. □

If $\mathcal{F}$ is Donsker and c is a constant, then it is straightforward that $\{cf: f \in \mathcal{F}\}$ is Donsker. If $\mathcal{F}$ is any P-EM class with the same symmetric convex hull as a class $\mathcal{H}$, then by 1.6 and its proof, $\mathcal{F}$ is Donsker if and only if $\mathcal{H}$ is.

Thus if $\mathcal{F}$ is Donsker, so are the following classes:

(1.7.1) $\{cf: f \in \mathcal{F},\ 0 \leq c \leq M\}$, $M < +\infty$;

(1.7.2) $\{cf: a \le c \le b,\ f \in F\}$, $a \le b$;

(1.7.3) $\{2f: f \in F\} \cup \{2g\}$, for any $g \in L^2$, by 1.5; hence

(1.7.4) $\{af + (2-a)g: f \in F,\ 0 \le a \le 2\}$; hence

(1.7.5) $\{f + g: f \in F\}$, for any fixed $g \in L^2$.

In [7], the Donsker property of a class C of sets was defined with respect to a function space

$$D_0(C,P) := \{\phi + j: j = \Sigma_{1 \le i \le n} a_i \delta_{x(i)}\ ,\ \phi \in C_b(C,d_P)\}$$

where C_b is the class of functions bounded and d_P-uniformly continuous on C . For the corresponding class $F = \{1_A;\ A \in C\}$, we can regard $C_{b,lin}(F,e_P)$, with its additional pre-linearity and extended continuity properties, as a subspace of $C_b(C,d_P)$.

Clearly C is totally bounded for d_P if and only if F is for e_P. Since Theorem 1.3(b) above and Theorem 1.2(b) of [7] both use $Pr := P^\infty$ on X^∞ , they are equivalent here. If $\sup_{A \in C} |(\nu_n - G)(1_A)|$ is Pr-completion measurable for each $G \in D_0(C,P)$, it is a fortiori for $G \in D_2(F,P)$. Thus if C is P-EM as defined in [7], then F is P-EM as defined here above. Hence if C is Donsker as in [7], so is F as defined here. The following is now clear:

1.8 Proposition. If C is a P-EM class of sets [7] and $F = \{1_A: A \in C\}$, then C is a Donsker class of sets [7] if and only if F is a Donsker class of functions.

Whether F P-EM or even Donsker implies C P-EM I must leave open here.

Next, here is a general way to extend the Donsker property from a class of sets to a related uniformly bounded class of functions.

1.9 Theorem. Let $G := \{1_A: A \in C\}$ be a Donsker class of functions, $M < \infty$, and

$$F := \{f: X \to [-M,M],\ f^{-1}([a,b[) \in C \text{ whenever } a < b\}.$$

Then F is a Donsker class.

Proof. Let H be the set of all finite linear combinations $\Sigma b_i g_i$ where $g_i \in G$ and $\Sigma |b_i| \le M$. Then by 1.6 and 1.7.1, H is also a Donsker class.

By 1.7.5 with $g \equiv M$ or $g \equiv -M$ we can replace the range $[-M,M]$ equivalently by $[0,2M]$. Then by 1.7.1, we can replace $[0,2M]$ by $[0,1]$.

Then for any $f \in F$, $n = 1,2,\ldots$, and $j = 1,\ldots,n$, let $A(n,j) := f^{-1}([j/n,2[)$ and $f_n(x) := [nf(x)]/n$, where $[y]$ is the greatest integer $\le y$, so $\sup_x |(f-f_n)(x)| \le 1/n$. Then

$$f_n = \Sigma_{1 \le j \le n} 1_{A(n,j)}/n\ .$$

All such f_n belong to the H above with $M = 1$. Each functional $f \to (\int f d\nu_n)(\omega)$, or $f \to T_n(f)(\omega)$ as in Theorem 1.2 for H , is (for fixed ω) in $D_2(H,P)$, and thus uniformly continuous in $h \in H$ with respect to $\|\cdot\|_\infty$. These T_n thus extend uniquely to $\|\cdot\|_\infty$-continuous functions on the $\|\cdot\|_\infty$-closure of H , which includes F. These extensions, restricted to F , are in $D_2(F,P)$, and converge uniformly on F as $n \to \infty$ since they do on H . For any $j \in D_2(F,P)$,

$$\sup_{f \in F} |(\nu_n - j)(f)| = \sup_{h \in H} |(\nu_n - j)(h)| .$$

Thus F is P-EM, completing the proof, using 1.2.

2. SEQUENCES OF FUNCTIONS

In [7, sec.2], it is shown that for any sequence $\{A_m\}$ of measurable sets with $\Sigma(P(A_m)(1 - P(A_m)))^r < \infty$ for some $r < \infty$, $\{A_m\}$ is a Donsker class for P, and that in case $\{A_m\}$ are independent for P this condition is also necessary. Let $A(m) := A_m$. We consider extensions of the results just quoted to sequences of functions $\{f_m\}$, first to $f_m = c_m 1_{A(m)}$ for numbers c_m. Four easy propositions will be proved.

First, suppose $c_m \to \infty$ and $p_m := P(A_m) \to 0$ as $m \to \infty$. If $\{A_m\}$ are independent for P, then for each n and j, let E_{nj} be the event $\{c_j \nu_n(A_j) \geq 1\}$. For j large enough (n fixed), $c_j \geq 2n^{\frac{1}{2}}$ and $p_j \leq 1/(2n)$, so $Pr(E_{nj}) \geq np_j/2$. If $\Sigma\, p_j = +\infty$, then by the Borel-Cantelli lemma $Pr(\cup_{j \geq m} E_{nj}) = 1$ for all m and n. By Theorem 1.3 this means $\{c_j 1_{A(j)}\}$ cannot be a Donsker class, no matter how slowly $c_j \to \infty$. If $p_m \not\to 0$ the same is clearly true. Thus we have:

2.1 Proposition. For any sequence $0 < c_m \to +\infty$ and independent sets $A_m := A(m)$ for P such that $\Sigma\, P(A_m) = +\infty$, $\{c_m 1_{A(m)}\}_{m=1}^{\infty}$ is not a Donsker class.

Letting $\|f\|_2 := (\int f^2 dP)^{\frac{1}{2}}$, this gives:

2.2 Corollary. For $r > 2$, $\Sigma_{m=1}^{\infty} \|f_m\|_2^r < \infty$ does not imply $\{f_m\}$ is a Donsker class.

Let $\sigma_P^2(f) := \int f^2 dP - (\int f\, dP)^2 \leq \|f\|_2^2$. Then for each n, $\mathrm{var}(\int f\, d\nu_n) = \sigma_P^2(f)$. By (2.2), the following gives the best exponent $r = 2$:

2.3 Theorem. If $\Sigma_m \sigma_P^2(f_m) < +\infty$, e.g. if $\Sigma \|f_m\|_2^2 < +\infty$, then $\{f_m\}$ is a P-Donsker class.

Proof. Since $\{f_m\}$ is countable, the measurability property (P-EM) holds. We have $E \int f_m d\nu_n = 0$ for all m and by Chebyshev's inequality, for any $\varepsilon > 0$ and m,

$$\Sigma_{j \geq m} Pr(|\int f_j d\nu_n| \geq \varepsilon) \leq \Sigma_{j \geq m} \sigma_P^2(f_j)/\varepsilon^2 \to 0$$

as $m \to \infty$. Thus by Theorem 1.3, the result follows, Q.E.D.

Let $A_j := A(j)$ now be disjoint measurable sets and c_j constants. Let $F(\{A_j\},\{c_j\})$ be the set of all finite sums $\Sigma_{j \in F} c_j 1_{A(j)}$ over all finite sets F.

2.4 Theorem. The class $F(\{A_j\},\{c_j\})$ is Donsker if and only if $\Sigma c_j (P(A_j))^{\frac{1}{2}} < \infty$.

Proof. Durst and Dudley [10, Theorem 3.1] prove this for $c_j \equiv 1$. The general case can be proved in the same way.

3. METRIC ENTROPY WITH BRACKETING

Definition. For $\varepsilon > 0$, let $N_I(\varepsilon, F, P)$ be the least m such that for some measurable functions $f_1, \ldots, f_m$ (not necessarily in F), for each $f \in F$ there are j and $k \leq m$ with $f_j(x) \leq f(x) \leq f_k(x)$ for all x and $\int f_k - f_j\, dP < \varepsilon$.

Note that then the envelope $F(x) \leq \max(|f_1|,\ldots,|f_m|)(x)$, taking e.g. $\varepsilon = 1$. On the other hand we can always assume that $|f_j| \leq F$ for all j. Note also that N_I is an extension of N_I as defined for sets [7, p. 914], since if $f \leq 1_A \leq g$ and $\int g - f\, dP < \varepsilon$, $B := \{f > 0\}$, and $C := \{g \geq 1\}$, then $B \subset A \subset C$ and $P(C \setminus B) < \varepsilon$.

Here is the main result.

3.1 Theorem. Suppose F is a P-EM class with envelope $F = F_F \in L^p(X,A,P)$ for some $p > 2$, and that for some γ, with $0 < \gamma < 1 - 2/p$, and some $M < \infty$, $N_I(\varepsilon,F,P) \leq \exp(M\varepsilon^{-\gamma})$ for ε small enough. Then F is a P-Donsker class.

Remark. As $p \to \infty$, the condition on γ approaches $\gamma < 1$; if F is a collection of indicator functions of sets, the theorem for $\gamma < 1$ follows from [7, Theorem 5.1], noting the $N_I(\delta^2)$ there. For $\gamma = 1$, it appears that the theorem fails (F need not be G_PBUC), specifically where F is the collection of indicator functions of convex sets in $\mathbb{R}^3$ and P is Lebesgue measure on the unit cube, combining results of Bronštein [3, Theorem 3] with [8] (for details, see (3.14) below).

Proof of Theorem 3.1. For each $i = 1,2,\ldots,$ let $A_i := A(i) := \{x : F(x) \leq 2^i\}$. Then since $F \in L^p$, we have $P(X\setminus A_i) = o(2^{-pi})$ as $i \to \infty$.

If $n \leq 2^{ip}$, then we have

$$\Pr\{X_j \notin A_i \text{ for some } j = 1,\ldots,n\} \leq nP(X\setminus A_i) \leq 2^{ip}P(X\setminus A_i) \to 0$$

as $i \to \infty$. For each n let $r(n)$ be the least integer r such that $2^{rp} \geq n$. Then

$$\Pr(P_n(A_{r(n)}) = 1) \to 1 \quad \text{as} \quad n \to \infty. \tag{3.2}$$

On the other hand, for $r := r(n)$,

$$n^{\frac{1}{2}} \int_{X\setminus A(r)} F\, dP \leq \int_{X\setminus A(r)} F^{1+(p/2)} dP \tag{3.3}$$

$$\leq \int_{X\setminus A(e)} F^p\, dP \to 0 \quad \text{as} \quad n \to \infty.$$

Let $\varepsilon > 0$ be fixed, $\varepsilon \leq 1$. Then for each $i = 1,2,\ldots,$ let $\tau_i := \varepsilon/2^{ip/2}$. Choose $f_{i1},\ldots,f_{iN(i)}$, where $N(i) := N_I(\tau_i,F,P)$ and $\{f_{ik}\}_{k=1}^{N(i)}$ serve as f_k in the definition of N_I. By assumption,

$$N(i) \leq \exp(2^{ip\gamma/2}M) \text{ for } i \text{ large enough, some } \gamma,\ 0 < \gamma < 1-2/p, \tag{3.4}$$
$$\text{and some } M = M(\varepsilon) < \infty.$$

For each $g \in F$ take functions f_{ij} and f_{ik} as above, $1 \leq j := j(i,g) \leq N(i)$, $1 \leq k := k(i,g) \leq N(i)$, such that $f_{ij} \leq g \leq f_{ik}$ and $\int f_{ik} - f_{ij}\, dP \leq \tau_i$. We can always assume that $|f_{is}| \leq F$ for all s. Set $h_{is} := f_{is}$ on A_i and $h_{is} := 0$ on $X\setminus A_i$. Then $h_{ij} \leq g \leq h_{ik}$ on A_i and $\int h_{ik} - h_{ij}\, dP \leq \tau_i$. For each $g \in F$, we get a sequence of funtions $g_i := h_{ik(i,g)}$, $i = 1,2,\ldots,$ such that $\|g_i\|_\infty \leq 2^i$, $\int_{A(i)}(g_i - g)^2 dP \leq 2^{i+1}\tau_i = \varepsilon 2^{1+i-ip/2}$, and for $i_o(\varepsilon)$ large enough, we have for all $i \geq i_o(\varepsilon)$ that (3.4) holds, $\int_{X\setminus A(i)} g_{i+1}^2 dP \leq 4^{i+1}P(X\setminus A_i)$

$\leq 2^{(2-p)i}$, and $2^{i-ip/2} < \varepsilon$; then

$$2^{(2-p)i} \leq 2^{i-ip/2}\varepsilon, \quad i \geq i_o(\varepsilon) ,$$

and $\int_{X\backslash A(i)} (g_{i+1} - g_i)^2 \, dP \leq 2^{i-ip/2}\varepsilon$. Next,

$$\int_{A(i)} (g_{i+1} - g_i)^2 dP \leq 2 \int_{A(i)} (g_i - g)^2 dP + 2 \int_{A(i)} (g_{i+1} - g)^2 \, dP$$

$$\leq 2^{i+3} \tau_i = 2^{3+i-ip/2} \varepsilon ,$$

so $\int (g_{i+1} - g_i)^2 dP \leq 2^{4+i-ip/2} \varepsilon , \quad i \geq i_o(\varepsilon)$.

Let $H_i := \{h_{ij} : j=1,\ldots,N(i)\}$. For $i > 1$ let $\mathcal{D}_i$ be the set of all differences $f = \phi - \psi$ such that ϕ and ψ belong to $H_i \cup H_{i+1}$ and $\int |f| dP \leq 3\tau_i$. Then $\|f\|_\infty \leq 2^{i+2}$ and $\int |f|^2 dP \leq 2^{i+4}\tau_i = 2^{4+i-ip/2}\varepsilon$. Note that always $g_{i+1} - g_i \in \mathcal{D}_i$. By (3.4), for $i_1(\varepsilon)$ large enough, with $i_1(\varepsilon) \geq i_o(\varepsilon)$, we have for some $M_1 := M_1(\varepsilon)$,

$$\text{card}(\mathcal{D}_i) \leq 4N(i+1)^2 \leq 4 \exp(2^{1+(i+1)p\gamma/2} M) \tag{3.5}$$

$$= 4 \exp(2^{ip\gamma/2} M_1), \quad i \geq i_1(\varepsilon) .$$

Next, Bernstein's inequality (Bennett [1]) gives for each $f \in \mathcal{D}_i$, if $n \geq 2^{pi}$,

$$P_{f,i,n} := \Pr\{|\int f \, d\nu_n| > i^{-2}\}$$

$$\leq 2 \exp(-i^{-4}/(2^{5+i-ip/2}\varepsilon + i^{-2}2^{i+3}/n^{1/2}))$$

$$\leq 2 \exp(-1/(i^4 2^{5+i-ip/2}\varepsilon + i^2 2^{3+i-ip/2})) .$$

For any α,β such that $p\gamma/2 < \alpha < \beta < (p/2) - 1$, and for $i_2(\varepsilon) \geq i_1(\varepsilon)$ large enough, we have $P_{f,i,n} \leq \exp(-2^{i\beta})$ if $i \geq i_2(\varepsilon)$, $f \in \mathcal{D}_i$, and $n \geq 2^{pi}$.

Thus by (3.5), for $i_3(\varepsilon) \geq i_2(\varepsilon)$ large enough,

$$\Pr\{|\int f \, d\nu_n| > i^{-2} \text{ for some } f \in \mathcal{D}_i\} \tag{3.6}$$

$$\leq \exp(-2^{i\alpha}), \; i \geq i_3(\varepsilon), \; n \geq 2^{pi} .$$

We have by (3.2) and (3.3) that for some $n_o(\varepsilon)$ large enough, the outer probability

$$\Pr^*\{|\int_{X\backslash A(r(n))} f \, d\nu_n| > \varepsilon \text{ for some } f \in F\} < \varepsilon , \tag{3.7}$$

$$n \geq n_o(\varepsilon) .$$

Take $i(\varepsilon) \geq i_3(\varepsilon)$ large enough so that

(3.8)
$$\Sigma_{i \geq i(\varepsilon)} i^{-2} < \varepsilon, \quad \text{and}$$
$$(\exp(-2^{i(\varepsilon)\alpha}))/\{1 - \exp(-2^{i(\varepsilon)\alpha}(2^{\alpha} - 1))\} < \varepsilon .$$

For any $g \in F$, and n large enough so that $r(n) > i(\varepsilon)$, we have

$$|\int g - g_{i(\varepsilon)} d\nu_n| \leq |\int g - g_{r(n)} d\nu_n| + |\Sigma_{j=i(\varepsilon)}^{r(n)-1} \int g_{j+1} - g_j \, d\nu_n| .$$

By (3.6) we have, since $j < r(n)$ implies $2^{jp} < n$,

(3.9)
$$\Pr\{|\int g_{j+1} - g_j d\nu_n| > j^{-2} \text{ for some } j,\ i(\varepsilon) \leq j < r(n), \text{ and some } g \in F\}$$
$$\leq \Sigma_{j \geq i(\varepsilon)} \exp(-2^{j\alpha}) < \varepsilon$$

by a geometric series and (3.8). Using (3.6) and (3.8), we have

(3.10) $$\Pr\{|\int f \, d\nu_n| > \varepsilon \text{ for some } f \in \mathcal{D}_i\} < \varepsilon, \ i \geq i(\varepsilon), \ n \geq 2^{pi},$$

and hence in particular for $i = r(n) - 1$ and $n \geq n_1(\varepsilon)$ for some large enough $n_1(\varepsilon) \geq n_o(\varepsilon)$.

Now let $r := r(n)$; on A_r, for some j and k,

$$f_r := h_{rj} \leq g \leq h_{rk} := g_r, \quad \text{with } \int g_r - f_r dP \leq \tau_r = 2^{-rp/2},$$

so that

(3.11) $$0 \leq n^{\frac{1}{2}} \int_{A(r)} g_r - g \, dP \leq n^{\frac{1}{2}} \tau_r \leq \varepsilon, \quad \text{and}$$

(3.12)
$$0 \leq n^{\frac{1}{2}} \int_{A(r)} g_r - g \, dP_n \leq n^{\frac{1}{2}} \int g_r - f_r \, dP_n$$
$$\leq \int g_r - f_r \, d\nu_n + \varepsilon .$$

Then for each $n \geq n_1(\varepsilon)$, there is an event B_n from (3.7), (3.9) and (3.10), such that $\Pr(B_n) < 3\varepsilon$, and for any $g \in F$, and $\omega \notin B_n$, we have:

$$|\int_{X \backslash A(r)} g \, d\nu_n| \leq \varepsilon \quad \text{by (3.7)} ;$$

by (3.10) for $i = r(n) - 1$, (3.11) and (3.12),

$$|\int_{A(r)} g_r - g \, d\nu_n| \leq 3\varepsilon ,$$

and by (3.8) and (3.9), $|\int g_r - g_{i(\varepsilon)} d\nu_n| < \varepsilon$, so that

(3.13) $$|\int g - g_{i(\varepsilon)} d\nu_n| < 5\varepsilon, \quad \omega \notin B_n, \ n \geq n_1(\varepsilon), \quad g \in F .$$

Now let $i := i(\varepsilon)$ and $\delta := \delta(\varepsilon) := \tau_i$. Take any $g, j \in F$ with $\int |g-j|^2 \, dP < \delta^2$, so that $\int |g-j| dP < \tau_{i(\varepsilon)}$. Then take g_i as defined previously for g, and likewise j_i for j. We apply (3.13) for $j - j_i$ as well as $g - g_i$. Now g_i and j_i both belong to H_i, and

$$\int |g_i - j_i| dP \le \int_{A(i)} |g_i - g| + |g - j| + |j_i - j| dP \le 3\tau_i ,$$

by choice of g_i and j_i. Thus $g_i - j_i \in D_i$. So for $i := i(\varepsilon)$, $n \ge 2^{pi}$, and $\omega \notin B_n$, we have using (3.10), $|\int g_i - j_i d\nu_n| < \varepsilon$. Combining this with (3.13) for g and j gives $|\int g - j \, d\nu_n| < 11\varepsilon$, $n \ge n_1(\varepsilon)$, $\omega \notin B_n$, $g, j \in F$, and if $\int |g-j| dP < \delta(\varepsilon)$. Thus Theorem 1.3 applies and, letting $\varepsilon \downarrow 0$, Theorem 3.1 is proved, Q.E.D.

Remark. De Hardt [4, Lemma 1] noted that for a uniformly bounded class F of measurable functions on $\mathbb{R}^k$ with $N_I(\delta, F, P) < \infty$ for all $\delta > 0$, almost surely

$$\lim_{n\to\infty} \sup_{f \in F} |\int f \, d(P_n - P)| = 0$$

(Glivenko-Cantelli theorem uniformly on F).

Let $Co(I^k)$ denote the collection of convex subsets of the unit cube I^k in $\mathbb{R}^k$. Let λ be Lebesgue measure on I^k. For collections of sets rather than functions, N_I is defined with $\subset$ rather than $\le$ [7, sec.5].

3.14 Proposition. For any $k \ge 2$ and some $M := M_k < \infty$, $N_I(\varepsilon, Co(I^k), \lambda) \le \exp(M/\varepsilon^{(k-1)/2})$ for $0 < \varepsilon \le 1$.

Proof. Let $N_h(\varepsilon, C)$ denote the smallest number of sets in a class C needed to approximate all sets in C within ε for the Hausdorff metric h. E. Bronštein [3, Thm. 3] proved that for some $N := N_k$, $k \ge 2$, $N_h(\varepsilon, Co(I^k)) \le \exp(N/\varepsilon^{(k-1)/2})$.

For any set $F \subset \mathbb{R}^k$ and $\varepsilon > 0$ let

$$F^\varepsilon := \{y: |y-x| < \varepsilon \text{ for some } x \in F\},$$

$${}_\varepsilon F := \{y: |y-x| < \varepsilon \text{ implies } x \in F\}.$$

Then ${}_\varepsilon F \subset F \subset F^\varepsilon$ and $h(F^\varepsilon, F) \le \varepsilon$. Thus for some constant A_k and any convex $F \subset I^k$, $(F^\varepsilon \backslash F) \le A_k \varepsilon$ [2, p.41,5].

Given $x \in F$ convex, let

$$d_j(x, F^c) := \inf\{|y_j - x_j|: y \notin F, \ y_i = x_i \text{ for all } i \ne j\}.$$

If $d_j(x, F^c) \ge k^{\frac{1}{2}}\varepsilon$ for all j, then $x \in {}_\varepsilon F$. Let

$$F_j := \{x \in F \backslash {}_\varepsilon F: d_j(x, F^c) < k^{\frac{1}{2}}\varepsilon\}.$$

Then $\cup_{j=1}^k F_j = F \backslash {}_\varepsilon F$. Since F and ${}_\varepsilon F$ are convex, $\{y_j: y \in F_j, \ y_i = x_i, \ i \ne j\}$ consists of at most four intervals, for any x, with total length at most $2k^{\frac{1}{2}}\varepsilon$. Thus by Fubini's theorem, $\lambda(F \backslash {}_\varepsilon F) \le \Sigma_{j=1}^k \lambda(F_j) \le k \cdot 2k^{\frac{1}{2}}\varepsilon$.

Now suppose $C, D \in C_o(I^k)$ and $h(C,D) < \varepsilon$. Then clearly $D \subset C^\varepsilon$. To show that ${}_{\varepsilon}C \subset D$, suppose not, and take $x \in {}_{\varepsilon}C \backslash D$. Take a hyperplane H through x with D on one side. There is a point on the other side, in C, at distance ε from H and hence at distance $\geq \varepsilon$ from D, a contradiction. So ${}_{\varepsilon}C \subset D \subset C^{\varepsilon}$. For some constant $J_k < \infty$, we have $\lambda(C^{\varepsilon} \backslash {}_{\varepsilon}C) \leq J_k \varepsilon$ for all $C \in Co(I^k)$. Thus Bronštein's result on approximation of sets in the Hausdorff metric implies (3.14), Q.E.D.

Then, taking $k = 3$, we get the result mentioned in the Remark after the statement of 3.1.

Acknowledgement. Many thanks to Kenneth Alexander for some corrections.

4. REFERENCES

[1] Bennett, G., Probability inequalities for the sum of independent random variables, J. Amer. Statist. Assoc. 57 (1962) 33-45.

[2] Bonnesen, T., and Fenchel, W., Theoreie der Konvexen Körper (Springer, Berlin, 1934).

[3] Bronštein, E.M., ε-entropy of convex sets and functions, Sibirskii Mat. Zh. 17 (1976) 508-514 = Siberian Math. J. (English translation) 17 393-398.

[4] DeHardt, J., Generalizations of the Glivenko-Cantelli theorem, Ann. Math. Statist. 42 (1971) 2050-2055.

[5] Dudley, R. M., Measures on non-separable metric spaces, Illinois J. Math. 11 (1967) 449-453.

[6] Dudley, R.M., Sample functions of the Gaussian process, Ann. Probability 1 (1973) 66-103.

[7] Dudley, R.M., Central limit theorems for empirical measures, Ann. Probability 6 (1978) 899-929; Correction, ibid. 7 (1979) 909-911.

[8] Dudley, R.M., Lower layers in $\mathbb{R}^2$ and convex sets in $\mathbb{R}^3$ are not GB classes, Lecture Notes in Math. 709 (1979) 97-102.

[9] Dudley, R.M., Vapnik-Červonenkis Donsker classes of functions (preprint), to appear in Les aspects statistiques et les aspects physiques des processus gaussiens (Colloque C.N.R.S., St. Flour, 1980).

[10] Durst, Mark, and Dudley, R.M., Empirical processes, Vapnik-Červonenkis classes and Poisson processes, to appear in Probability and Mathematical Statistics 1 (1980, Wrocław, Poland).

[11] Itô, Kiyosi, and Nisio, Makiko, On the convergence of sums of independent Banach space valued random variables, Osaka J. Math. 5 (1968) 35-48.

[12] Pollard, David, Limit theorems for empirical processes, preprint (1980).

[13] Wichura, M.J., On the construction of almost uniformly convergent random variables with given weakly convergent image laws, Ann. Math. Statist. 41 (1980) 284-291.

Room 2-245, M.I.T.
Cambridge, Mass. 02139 U.S.A.

FOOTNOTE

[1] This research was partially supported by National Science Foundation Grant MCS-79-04474 (U.S.A.).

AMS 1970 subject classifications: Primary 60F05; Secondary 60B10, 60G17, 28A05, 28A40.

Key words and phrases: central limit theorems, empirical measures, Donsker classes, metric entropy with bracketing.

STATISTICS AND RELATED TOPICS
M. Csörgö, D.A. Dawson, J.N.K. Rao, A.K.Md.E. Saleh (eds.)

CAUSAL CALCULUS OF BROWNIAN FUNCTIONALS, AND ITS APPLICATIONS

Takeyuki Hida

Department of Mathematics
Nagoya University
Chikusa-ku, Nagoya
464 JAPAN

We discuss a causal calculus of Brownian functionals in terms of white noise $\dot{B}(t)$. As an application of the analysis we propose stochastic partial differential equations which involve the partial derivatives such as $\partial/\partial\dot{B}(t)$ and multiplication by $\dot{B}(t)$. Some examples show that those equations characterize Brownian functionals as well as stochastic processes of particular type.

1. BACKGROUND

The purpose of the present note is to propose Stochastic Partial Differential Equations for Brownian functionals. A Brownian functional, that is a function of a Brownian motion $\{B(t, \omega), t \in R, \omega \in \Omega(P), P$ a probability measure$\}$, may be expressed in the form

$$\phi(\dot{B}(t, \omega), t \in T), \tag{1}$$

where $\dot{B}$ is the time derivative of the Brownian motion : $\dot{B}(t, \omega) = dB(t, \omega)/dt$ the white noise.

In order to carry out the analysis of such functionals we first concretize them as members of the Hilbert space $(L^2) \equiv L^2(\mathcal{S}^*, \mu)$, where $\mathcal{S}^*$ is the dual space of the Schwartz space $\mathcal{S}$ and where μ is the measure of white noise introduced on $\mathcal{S}^*$ by the characteristic functional

$$C(\xi) = \exp[-\tfrac{1}{2}\|\xi\|^2], \qquad \xi \in \mathcal{S}. \tag{2}$$

Then, almost every element x of $\mathcal{S}^*$ (with respect to μ) is viewed as a sample function of $\dot{B}(t)$, so that $\phi(x) \in (L^2)$ may be thought of as a realization of a Brownian functional of the form (1) with finite variance.

The Hilbert space (L^2) admits the Wiener-Itô decomposition :

$$(L^2) = \sum_{n=0}^{\infty} \oplus \mathcal{H}_n, \tag{3}$$

where $\mathcal{H}_n$ is formed by the multiple Wiener integrals of degree n. (For details see [1, Chapt. 4].)

A powerful tool from analysis is the integral representation of Brownian functionals. The representation is given by the transformation $\mathcal{T}$ on (L^2) :

$$(\mathcal{T}\phi)(\xi) = \int_{\mathcal{S}^*} e^{i\langle x,\xi\rangle} \phi(x)d\mu(x), \qquad \phi \in (L^2). \tag{4}$$

The vector space $\mathbb{F} = \{(\mathcal{T}\phi)(\xi);\ \phi \in (L^2)\}$ is topologized so as to be isomorphic to (L^2) under the transformation $\mathcal{T}$. Indeed, the space $\mathbb{F}$ is made to be a reproducing kernel Hilbert space with reproducing kernel $C(\xi-\eta)$, $(\xi, \eta) \in \mathcal{S}\times\mathcal{S}$, and we have

$$\mathbb{F} \cong (L^2) \qquad \text{under } \mathcal{T}. \tag{5}$$

Once $\mathcal{T}$ is restricted to the subspace $\mathcal{H}_n$, we are given the following expression.

$$\begin{cases} (\mathcal{T}\phi)(\xi) = i^n C(\xi)U(\xi), & \phi \in \mathcal{H}_n, \\ U(\xi) = \int_{R^n}\cdots\int F(u_1,\cdot\cdot, u_n)\xi(u_1)\cdot\cdot\xi(u_n)du_1\cdots du_n, & \end{cases} \tag{6}$$

where F is a member of $\widehat{L^2(R^n)} = \{F \in L^2(R^n);\ F \text{ is symmetric}\}$. In addition, the mapping

$$\phi \longrightarrow F \in \widehat{L^2(R^n)}, \qquad \phi \in \mathcal{H}_n,$$

is surjective, and we have

$$\|\phi\|_{(L^2)} = \sqrt{n!}\,\|F\|_{L^2(R^n)}. \tag{7}$$

The representation (6) of $\phi \in \mathcal{H}_n$ in terms of $\widehat{L^2(R^n)}$-function is called the <u>integral representation</u>. The functional $U(\xi)$ in (6) is said to be the U-functional associated with ϕ. For the representation of a general $\phi \in (L^2)$ we make use of the expansion $\phi = \sum_n \phi_n$, $\phi_n \in \mathcal{H}_n$, to have a sequence $\{F_n;\ n \geq 0\}$ of $\widehat{L^2(R^n)}$-functions. The U-functional associated with ϕ in this case is simply to be the sum of the U-functionals associated with the ϕ_n's.

2. GENERALIZED BROWNIAN FUNCTIONALS

We are interested in developing a causal calculus of Brownian functionals, where the evolution in time is explicitly taken into account. The basic idea is to take $\{\dot{B}(t)\}$ to be the system of variables of Brownian functionals in question. The advantage to do so is that $\dot{B}(t)$ is associated with the time $\underline{t}$ and that the $\dot{B}(t)$ itself changes as $\underline{t}$ goes by. However there is a disadvantage that almost all sample functions of $\dot{B}(t)$ are generalized functions, so that we meet some difficulty when we form nonlinear functions of the $\dot{B}(t)$. Still we should like to deal with elementary nonlinear functionals such as polynomials in $\dot{B}(t)$'s and as exponential functions of the $\dot{B}(t)$.

The above observation leads us to introduce generalized Brownian functionals. Under the set-up established in the last section, $\dot{B}(t)$ is realized by a member, say $\underline{x}$, of $\mathcal{S}^*$ and the space $\mathcal{H}_n^{(-n)}$ of generalized functionals of $\underline{x}$ (generalized Brownian functionals) of degree $\underline{n}$ is defined by the following diagram :

$$
\begin{array}{ccccc}
③\ \widehat{H^{-\frac{n+1}{2}}(R^n)} & \cong & ④\ \mathcal{H}_n^{(-n)} & \cong & \mathcal{F}_n^{(-n)} \\
\cup\!\uparrow & & \cup\!\uparrow & & \cup\!\uparrow \\
\widehat{L^2(R^n)} & \cong & \mathcal{H}_n & \cong & \mathcal{F}_n \\
\cup\!\uparrow & & \cup\!\uparrow & & \cup\!\uparrow \\
①\ \widehat{H^{\frac{n+1}{2}}(R^n)} & \cong & ②\ \mathcal{H}_n^{(n)} & \cong & \mathcal{F}_n^{(n)}
\end{array}
$$

where $\widehat{H^m(R^n)} = H^m(R^n) \cap \widehat{L^2(R^n)}$, $m > 0$, $H^m(R^n)$ being the Sobolev space on R^n of order $\underline{m}$, and $\widehat{H^{-m}(R^n)}$ is the dual space of $\widehat{H^m(R^n)}$ with $\widehat{H^0(R^n)}^* = \widehat{H^0(R^n)} = \widehat{L^2(R^n)}$. The logical steps in the diagram are ①→② ③→④ (with arrows ①→③ and ②→④). The spaces $\mathcal{F}_n$ and $\mathcal{F}_n^{(\pm n)}$ of U-functionals correspond to $\mathcal{H}_n$ and $\mathcal{H}_n^{(\pm n)}$, respectively.

The sum $\sum_n \oplus \mathcal{H}_n^{(n)} = (L^2)^+$ is the space of test functionals, and $(L^2)^-$ the dual space of $(L^2)^+$ is the space of <u>generalized Brownian functionals</u>.

<u>Example</u>. Hermite polynomials in $\dot{B}(t)$ (or in $x(t)$, $x \in \mathcal{S}^*$). Since $\underline{x}$ is a generalized function, monomial of the form $x(t)^n$ has no meaning at all, but $H_n(x(t); \frac{1}{dt})$, $H_n(t, \sigma^2)$ being the Hermite polynomial, does have meaning in our sense and the associated U-functional is $U(\xi) = \frac{1}{n!}\xi(t)^n$.

Set

$$M_m(dt) = m! H_m(x(t); 1/dt)dt,$$

and set

$$
\prod_{j=1}^{k} \bullet M_{n_j}(dt_j) = \begin{cases} \prod_j M_{n_j}(dt_j), & t_j\text{'s are different,} \\ 0, & \text{otherwise.} \end{cases} \tag{8}
$$

Then $\prod_{j=1}^{k} \bullet M_{n_j}(dt_j)$ plays the role of a random measure. (See [1, Chapt. 8] and [2]). In fact, such a product is called a generalized random measure. An $\mathcal{H}_n^{(-n)}$-functional which is expressed as an integral with respect to a generalized random measure given by (8) with $\Sigma n_j = n$ is called a <u>normal functional</u> of degree $\underline{n}$. A sum of normal functionals of various degrees is simply called a normal functional.

<u>Example</u> of a normal functional of degree $\underline{n}$.

$$
\phi(x) = \int_{R^k} \cdots \int G(u_1, \cdots, u_k) M_{n_1}(du_1) \bullet \cdots \bullet M_{n_k}(du_k), \quad \Sigma n_j = n. \tag{9}
$$

<u>Remark</u> 1. In the expression (9) above, the kernel should be regarded as an

$\widehat{H^{-(n+1)/2}(R^n)}$-function, although (9) looks like a functional as if there were an $L^2(R^k)$-kernel.

Remark 2. An $\mathcal{H}_n$-functional may be expressed in the form (9) by taking $k = n$, $n_1 = \cdots = n_k = 1$ and G in $\widehat{L^2(R^n)}$.

3. DIFFERENTIAL CALCULUS

Since each $\dot{B}(t)$ can be thought of as a variable of generalized Brownian functionals, we are naturally led to introduce the differential operator $\partial/\partial\dot{B}(t)$. Differentiation with respect to $\dot{B}(t)$ should describe the variation of a generalized Brownian functional when $\dot{B}(t)$ is changed into $\dot{B}(t) + \delta\dot{B}_1(t)$, where $\{\dot{B}_1(t)\}$ is a white noise independent of $\{\dot{B}(t)\}$. The variation

$$\delta\phi(\dot{B}) = \text{the linear part of } \phi(\dot{B} + \delta B_1(t)) - \phi(\dot{B})$$

would be represented by the variation of the U-functional corresponding to ϕ. It is therefore quite reasonable to define the operator $\partial/\partial\dot{B}(t)$ as follows :

Definition. Let $\phi(x)$ be an $(L^2)^-$-functional and let $U(\xi)$ be the corresponding U-functional. If $U(\xi)$ has the functional derivative $U'_\xi(t)$ in the Fréchet sense, and if there exists an $(L^2)^-$-functional ϕ' whose U-functional is $U'_\xi(t)$, then ϕ is a said to be differentiable at t with respect to $\dot{B}(t)$ (or with respect to $x(t)$) and the ϕ' is denoted by $\frac{\partial}{\partial\dot{B}(t)}\phi$ (or by $\frac{\partial}{\partial x(t)}\phi$). We often denote the operator simply by ∂_t.

The following assertion is straightforward.

Proposition 1. i) The domain of the operator ∂_t includes $\sum_n \mathcal{H}_n^{(n)}$ (the algebraic sum).

ii) Let ϕ be in $\mathcal{H}_n^{(n)}$ and let F be its kernel. Then we have

$$\partial_t\phi = n\int_{R^{n-1}}\cdots\int F(u_1,\cdots, u_{n-1}, t)M_1(du_1)\bullet\cdots\bullet M_1(du_{n-1}).$$

(For notations see Remark 2)

iii) If ϕ is given by the formula (9) with continuous G, then we have

$$\partial_t\phi = \sum_j n_j H_{n_j-1}(x(t); \tfrac{1}{dt})\int_{R^{k-1}}\cdots\int G(u_1,\cdots, \overset{j}{\downarrow}{t}, -u_k) \tag{10}$$

$$M_{n_1}(du_1)\bullet\cdots\overset{j}{\downarrow}{*}\cdots\bullet M_{n_k}(du_k).$$

($\overset{j}{\downarrow}{*}$ means $M_{n_j}(du_j)$ is missing.)

As a triviality we have

$$\phi \in \mathcal{H}_n^{(-n)} \quad \text{implies} \quad \partial_t \phi \in \mathcal{H}_{n-1}^{(-n+1)}.$$

The adjoint operator ∂_t^* is defined by

$$\langle \partial_t^* \phi, \psi \rangle = \langle \phi, \partial_t \psi \rangle \qquad \text{for any} \quad \psi \in (L^2)^+,$$

where $\langle \, , \, \rangle$ is the canonical bilinear form connecting $(L^2)^+$ and $(L^2)^-$. We now introduce two notations. For $F \in \widehat{L^2(R^n)}$ and the delta function δ_t,

$$(\delta_t * F)(u_1, \cdots, u_{n-1}) = F(u_1, \cdots, u_{n-1}, t),$$

and

$$(\delta_t \hat{\otimes} F)(u_1, \cdots, u_{n+1}) = \text{symmetrization of the tensor product}$$

$$\delta_t(u_{n+1}) \otimes F(u_1, \cdots, u_n).$$

We then introduce <u>multiplication</u> by $\dot{B}(t)$ (or by $x(t)$, $x \in \mathcal{S}^*$). If the limit (in the $(L^2)^-$-topology)

$$\lim \frac{\langle x, \chi_\Delta \rangle}{\Delta} \phi(x)$$

exists as $\Delta \to \{t\}$, Δ containing <u>t</u>, then ϕ is said to be in the domain of multiplication and the above limit is denoted by $\dot{B}(t) \cdot \phi$ or by $x(t) \cdot \phi(x)$.

The following proposition is due to Kubo and Takenaka [3].

<u>Proposition</u> 2. Let ϕ be in $(L^2)^+$, and let $\{F_n;\, n \geq 0\}$ be the sequence of the kernels of the integral representation. Then,

i) the sequences of the kernels of $\partial_t \phi$ and $\partial_t^* \phi$ are

$$\{n\delta_t * F_n;\, n \geq 1\} \quad \text{and} \quad \{\delta_t \hat{\otimes} F_n;\, n \geq 0\}, \tag{11}$$

respectively.

ii) In particular

$$\partial_t : \mathcal{H}_n^{(n)} \longrightarrow \mathcal{H}_{n-1}^{(n-1)},$$

$$\partial_t^* : \mathcal{H}_n^{(n)} \longrightarrow \mathcal{H}_{n+1}^{-(n+1)}.$$

iii) We have

$$[\partial_t, \partial_s^*] \equiv \partial_t \partial_s^* - \partial_s^* \partial_t = \delta_{t-s} I. \tag{12}$$

iv) The product $x(t)\phi(x)$ is defined for $\phi \in \mathcal{H}_n^{(n)}$ and we have

$$x(t)\phi(x) = \partial_t^* \phi(x) + \partial_t \phi(x). \tag{13}$$

The proof is easy but needs a comment about the domain. (For details, see [3].)

Remark 3. In physics terminology ∂_t is the annihilation operator, while ∂_t^* is the creation operator.

Second order differential operators

$$\partial_t \partial_s \quad \text{and} \quad \partial_t^{(2)}$$

are defined by using the second order functional derivatives $U''_{\xi\eta}(t, s)$ and and $U''_{\xi^2}(t)$, respectively. The Laplacian operator can also be defined by

$$\Delta = \int \partial_t^{(2)} dt$$

and is in agreement with the Lévy's Laplacian. (See P. Lévy [4, Part III].)

4. STOCHASTIC PARTIAL DIFFERENTIAL EQUATIONS

We discuss several examples of an equation involving the operators ∂_t, ∂_t^* and multiplication by $\dot{B}(t)$. Before doing so, three remarks are in order.

i) Multiple Wiener integrals should not be viewed as definite integrals, but as polynomials in the $\dot{B}(t)$'s.

ii) The operator ∂_t, when applied to a generalized functional, acts not quite like an ordinary differential operator on $L^2(R^n)$. Its action, however, is modified so as to correspond to the functional derivative.

iii) When we deal with some kind of a generalized functional, it is often convenient to regard it as a renormalized one of a relevant functional having intuitive meaning. Examples can be seen in the paper [5] by L. Streit and the author.

We are now ready to give examples.

1) Frequency modulation. Given a Gaussian process $X(t) = \int F(t, u)\dot{B}(u)du$, F being a Volterra kernel, then we can realize it as a process living in $\mathcal{H}_1$, call it $\phi(t)$. Set

$$\psi(t) = \exp[i\phi(t)], \tag{14}$$

and divide it by the mean to have

$$\tilde{\psi}(t) = \psi(t)/E(\psi(t)).$$

Then we have

$$\partial_s \tilde{\psi}(t) = iF(t, s)\tilde{\psi}(t). \tag{15}$$

Conversely, the equation (15) together with $E\tilde{\psi}(t) = 1$ characterizes the frequency modulation of the form (14).

Remark 4. If, in particular, we set $iF(t, u) = f(u)\chi_{(-\infty,t]}(u)$, then $\tilde{\psi}(t)$ is a martingale, and (15) is a characterization of the martingale.

2) Start with the renormalization of a formal functional of the form $\exp[ic\int \dot{B}(u)^2 du]$. The renormalized functional is realized by an $(L^2)^-$-functional ϕ whose U-functional is

$$U(\xi) = \exp[ic'\int \xi(u)^2 du], \qquad c' = \frac{c}{1-2ic}. \tag{16}$$

We obviously have

$$\partial_t \phi = 2ic' \partial_t^* \phi. \tag{17}$$

Conversely, if we put additional assumptions that ϕ is a normal functional and that $\langle \phi, 1 \rangle = 1$, then the solution to the equation (17) has to be an $(L^2)^-$-functional with U-functional (16).

Since $\partial_t^* + \partial_t = \dot{B}(t)\cdot$, we are given an equivalent equation

$$\partial_t \phi = 2ic\dot{B}(t)\phi. \tag{17'}$$

If, in particular, $\underline{c}$ is taken to be $i/2$, then

$$\partial_t \phi = - \dot{B}(t)\phi \qquad (\text{or } = x(t)\phi). \tag{17''}$$

Remark 5. The last formula suggests to us that we may think of the renormalization of $\exp[-\frac{1}{2t}\int \dot{B}(u)^2 du]$ as being the continuous analogue of the Gauss kernel up to constant.

3) The generating function of the Hermite polynomials $H_{2n}(B(t); t)$, $n \geq 0$, is

$$\phi(t) = \sum_{n=0}^{\infty} \beta^{2n} H_{2n}(B(t); t), \qquad \beta \text{ real}. \tag{18}$$

It is realized in (L^2) by taking $B(t) = \langle x, \chi_{[0,t]} \rangle$. The stochastic partial differential equation that characterizes the stochastic process $\phi(t)$, $t \geq 0$, is

$$\partial_{s_1} \partial_{s_2} \phi(t) = \begin{cases} \beta^2 \phi(t), & s_1, s_2 \leq t, \\ 0, & \text{otherwise}, \end{cases}$$

with additional conditions

$$E\phi(t) = 1, \qquad E(\partial_s \phi(t)) \equiv 0.$$

A systematic approach of the theory of stochastic partial differential equations will be reported in a separate paper.

5. REFERENCES

[1] T. Hida, Brownian motion. (Iwanami Pub. Co. 1975). English translation, (Springer-Verlag, Applications of Math. 11, 1980).
[2] ———, Analysis of Brownian functionals. (Carleton Mathematical Lecture Notes no. 13, 2nd edition 1978, Carleton Univ., Ottawa).
[3] I. Kubo and S. Takenaka, Calculus on Gaussian white noise, I, II. (Proc. Japan Academy. (to appear)
[4] P. Lévy, Problème concrets d'analyse fonctionnelle. (Gauthier-Villars, Paris, 1951).
[5] L. Streit and T. Hida, Generalized Brownian functionals and Feynman integrals. (to appear)

STATISTICS AND RELATED TOPICS
M. Csörgö, D.A. Dawson, J.N.K. Rao, A.K.Md.E. Saleh (eds.)

WHITE NOISE ANALYSIS AND AN APPLICATION TO STOCHASTIC DIFFERENTIAL EQUATIONS IN HILBERT SPACE

Yoshio Miyahara

Faculty of Economics
Nagoya City University
Mizuhocho, Mizuhoku, Nagoya
JAPAN

We discuss the so-called bilinear stochastic differential equations in a Hilbert space by means of white noise analysis.

§0. Introduction.

The theory of stochastic processes in a Hilbert space has been developed by many authors (A. V. Balakrishnan [1,2], Yu. L. Daletskii [4], D. A. Dawson [5], Z. Haba [9], R. Marcus [7], Y. Miyahara [17], B. L. Rozovskii [18], A. Shimizu [19], M. Yor [20]). In particular the so-called bilinear stochastic differential equations have been extensively studied.

In this paper we shall discuss two stochastic differential equations

$$dX_t = -\hat{\omega}X_t dt + X_t \cdot RdB_t \tag{1}$$

and

$$\begin{cases} dX_t = Y_t dt \\ dY_t = -\hat{\omega}Y_t dt + X_t \cdot RdB_t \end{cases} \tag{2}$$

on $H = L^2[0, \pi]$ (for the precise definitions of these equations, see §2 (2.2) and (2.14)). Equation (1) can be regarded as a realization (or a modification) of the formal equation

$$\frac{\partial X(t, \sigma)}{\partial t} = -\hat{\omega}X(t, \sigma) + X(t, \sigma)\cdot w(t, \sigma)$$

where $w(t, \sigma)$ is a multi-parameter (Gaussian) white noise. On the other hand the equation (2) is a realization of the formal equation

$$\frac{\partial^2 X(t, \sigma)}{\partial t^2} = -\hat{\omega}\,\frac{\partial X(t, \sigma)}{\partial t} + X(t, \sigma)\cdot w(t, \sigma).$$

We shall investigate these equations in §2 by means of the theory of white noise analysis which has been developed by T. Hida and others (T. Hida [11], [12], I. Kubo [13], Y. Miyahara [17]).

Our main results are Theorems 2.1, 2.4, and 2.5. Theorem 2.1 gives a system of equations in terms of the kernels of the integral representation of X_t (that is, the solution of the equation (1)). Theorems 2.4 and 2.5 give the equations which determine the covariance functions of X_t^n (= the n-th component of the Wiener-Itô decomposition of X_t) and X_t respectively.

§1. Stochastic integrals and multiple Wiener integrals.

In this section we shall give definitions of some basic concepts and summarize some known results (for details, see [20], [11], [17]).

1. Assume that a probability space (Ω, F, P) and an increasing family of σ-fields F_t, $t \geqq 0$, $F_t \subset F$, are given. Let H be a real separable Hilbert space.

Definition 1.1. (M. Yor [20]). A mapping $B_t(h, \omega)$: $[0, \infty)\times H\times\Omega \to R^1$ is called a cylindrical Brownian motion (abb. c.B.m.) on H if it satisfies the following conditions:

(i) $B_0(h, \cdot) = 0$ and $B_t(h, \cdot)$ is F_t-adapted.

(ii) For any $h \in H$, $h \neq 0$, $\frac{1}{\|h\|} B_t(h, \cdot)$ is a one-dimensional Brownian motion.

(iii) For any $t \in [0, \infty)$ and $\alpha, \beta \in R^1$, and $h, k \in H$, the following formula holds

$$B_t(\alpha h + \beta k) = \alpha B_t(h) + \beta B_t(k), \qquad (P\text{-a.s.}).$$

We will define stochastic integrals with respect to the c.B.m. B_t. Let $\phi(t, \omega)$ be a F_t-adapted measurable function from $[0, \infty)\times\Omega$ into H such that

$$E[\int_0^t \|\phi(s)\|^2 ds] < \infty \qquad \text{for any } t > 0,$$

where E[ ] denotes the expectation with respect to P.

Definition 1.2. The stochastic integral $\int_0^t \langle \phi(s), dB_s\rangle$ of ϕ is the martingale given by

$$\int_0^t \langle \phi(s), dB_s\rangle = \sum_{n=1}^{\infty} \int_0^t (\phi(s), e_n) dB_s(e_n),$$

where $\{e_n; n = 1, 2, \ldots\}$ is an orthonormal base in H.

Given two Hilbert spaces H and K, we denote by $\sigma_2(H, K)$ the Hilbert space consisting of all Hilbert-Schmidt operators from H into K. Let $\Phi(t, \omega)$ be a $\sigma_2(H, K)$-valued F_t-adapted function defined on $[0, \infty)\times\Omega$ into $\sigma_2(H, K)$ such that

$$E[\int_0^t \|\Phi(s)\|^2_{\sigma_2(H,K)} ds] < \infty.$$

Definition 1.3. The stochastic integral of Φ is the K-valued martingale M_t which is uniquely determined by

$$(y, M_t)_K = \int_0^t \langle \Phi^*(s)y, dB_s\rangle, \quad y \in K,$$

and it is denoted by $\int_0^t \Phi(s) dB_s$, where $\Phi^*(s)$ is the dual operator of $\Phi(s)$.

Remark 1.1. It is easy to verify

$$E[|\int_0^t \langle \phi(s), dB_s\rangle|^2] = E[\int_0^t \|\phi(s)\|^2 ds]$$

and

$$E[\|\int_0^t \Phi(s)dB_s\|^2] = E[\int_0^t \|\Phi(s)\|^2_{\sigma_2(H,K)} ds].$$

A stochastic integral equation on K is an equation of the form

$$X_t = X_0 + \int_0^t a(X_s)ds + \int_0^t G(X_s)dB_s, \tag{1.1}$$

where a and G are mappings such that a: $K \to K$, G: $K \to \sigma_2(H, K)$. For simplicity, we write the equation (1.1) in the form of stochastic differential equation

$$dX_t = a(X_t)dt + G(X_t)dB_t. \tag{1.2}$$

If we consider the case $a(X_t) = AX_t$, where A is an unbounded linear operator with the domain D(A) of dense in K, then we have an evolutionary stochastic differential equation

$$dX_t = -AX_t dt + G(X_t)dB_t. \tag{1.3}$$

2. We start with a Gelfand triple

$$\mathcal{E} \subset \mathcal{H} = L^2(D\times T) \subset \mathcal{E}^*,$$

where D is a domain in R^d, $T = (-\infty, \infty)$ (the time space), $\mathcal{E}$ is a nuclear space and $\mathcal{E}^*$ is the dual space of $\mathcal{E}$.

Definition 1.4. A (Gaussian) white noise on a real Hilbert space $H = L^2(D)$ or a (Gaussian) white noise with parameter space D×T is the probability space $(\mathcal{E}^*, \mathcal{B}, \mu)$ whose characteristic function is given by

$$C_\mu(\eta) = \int_{\mathcal{E}^*} e^{i\langle\eta,\omega\rangle} d\mu(\omega) = \exp\left\{-\frac{1}{2}\|\eta\|^2\right\},$$

$$\|\eta\|^2 = \int_{D\times T} |\eta|^2 dx,$$

where $\langle\ ,\ \rangle$ is the canonical bilinear form connecting $\mathcal{E}$ and $\mathcal{E}^*$.

In this paper we adopt the white noise space $(\mathcal{E}^*, \mathcal{B}, \mu)$ as the basic probability space, and we denote by $\mathcal{B}_t$ the σ-field generated by $\{\langle\eta, \omega\rangle ; \eta \in \mathcal{E}, \omega \in \mathcal{E}^*, \mathrm{supp}\{\eta\} \subset D\times(-\infty, t]\}$.

A c.B.m. can be formed from a white noise. In fact a mapping given by $B_t(\xi) = \langle \xi\otimes\chi_{[0,t]}, \omega\rangle$, $\xi \in H$, is a c.B.m. on H, where $\chi_{[0,t]}$ is the indicator function of [0, t].

3. We introduce the Hilbert space $(L^2)_D = L^2(\mathcal{E}^*, \mu)$ of all real valued functionals of white noise having finite variances. Then we obtain Wiener-Itô decomposition of $(L^2)_D$:

$$(L^2)_D = \sum \oplus \mathcal{H}_n \cong \sum \oplus \sqrt{n!}\ \hat{L}^2((D\times T)^n), \quad \text{under } \tau, \tag{1.4}$$

$$\|\phi\|_{(L^2)_D} = \sqrt{n!}\ \|\tau\phi\|_{L^2((D\times T)^n)} .$$

For $F \in L^2((D\times T)^n)$ we can define the multiple Wiener integral $I(F)$ of F in such the way as $I(F)$ has the following properties:

(i) If $F = \eta_1 \otimes \cdots \otimes \eta_n$, $\eta_j = \xi_j \otimes \zeta_j$ ($\{\xi_j\}$ and $\{\zeta_j\}$ are orthogonal systems respectively), then $I(F) = \int \cdots \int \zeta_1(t_1) \cdots \zeta_n(t_n) dB_{t_1}(\xi_1) \cdots dB_{t_n}(\xi_n)$, where the right hand is the usual (finite dimensional) multiple Wiener integral of degree n.

(ii) $I \cdot \tau$ = identity.

Let $F \in L^2((D\times T)^n)$. Then for fixed $(x_2, \ldots, x_n)$ and t_1, F can be identified with an element of $H = L^2(D)$. Hence the stochastic integral $\int \langle F, dB_{t_1} \rangle$ is well-defined by Definition 1.3, where $B_t(\xi) = \langle \xi \otimes \chi_{[0\wedge t, 0\vee t]}, \omega \rangle$. Repeating this procedure n-times, we are finally given the iterated stochastic integral $\hat{I}(F)$ of F. It can be proved that $I(F) = n!\, \hat{I}(F)$ for $F \in \hat{L}^2((D\times T)^n)$.

We next consider the Hilbert space $L^2(\mathcal{E}^* \to K)$ of all K-valued functionals of white noise with finite variance. The Wiener-Itô decomposition of $L^2(\mathcal{E}^* \to K)$ is as follows:

$$(1.5)\qquad L^2(\mathcal{E}^* \to K) = \sum_{n=1}^{\infty} \oplus \mathcal{H}_n(K) \cong \sum \oplus \sqrt{n!}\; \sigma_2(\hat{L}^2((D\times T)^n), K)$$

$$\cong \sum \oplus \sqrt{n!}\; \hat{L}^2((D\times T)^n \to K), \qquad \text{under } \tau^*,$$

$$\|\Phi\|_{L^2(\mathcal{E}^* \to K)} = \sqrt{n!}\; \|\tau^*\Phi\|_{L^2((D\times T)^n \to K)}$$

For $F \in L^2((D\times T)^n \to K)$ the multiple Wiener integral $I(F)$ of F and the iterated stochastic integral $\hat{I}(F) = \int \cdots \int_{t_1 \leqq \cdots \leqq t_n} F dB_{t_1} \cdots dB_{t_n}$ can be well defined, and they have the following properties:

(i) For $\psi \in K$, $(\psi, I(F))_K = I(F^*\psi)$ and $(\psi, \hat{I}(F))_K = \hat{I}(F^*\psi)$ under the identification of $L^2((D\times T)^n \to K)$ with $\sigma_2(L^2((D\times T)^n), K)$.

(ii) $I \cdot \tau^*$ = identity.

(iii) $I(F) = n!\, \hat{I}(F)$ for $F \in \hat{L}^2((D\times T)^n \to K)$.

Definition 1.5. Let $X \in L^2(\mathcal{E}^* \to K)$ and $X = \sum \oplus X^n$, $X^n \in \mathcal{H}_n(K)$. Then $\tau^* X^n \in \hat{L}^2((D\times T)^n \to K)$ is called the kernel of degree n of the integral representation of X.

§2. Application to stochastic differential equations.

We are interested in the so-called bilinear equations formally described as

$$(*)\qquad \frac{\partial X(t, \sigma)}{\partial t} = -\hat{\omega} X(t, \sigma) + X(t, \sigma) w(t, \sigma), \quad t \geqq 0, \quad \sigma \in [0, \pi]$$

and

$$(**)\qquad \frac{\partial^2 X(t, \sigma)}{\partial t^2} = -\hat{\omega}\, \frac{\partial X(t, \sigma)}{\partial t} + X(t, \sigma) w(t, \sigma),$$

where $\hat{\omega} = \sqrt{-\Delta} = \sqrt{-\frac{\partial^2}{\partial\sigma^2}}$ and $w(t,\sigma)$ is a 2-parameter white noise. The precise definitions of which will be given later.

Put $H = L^2([0, \pi])$. Let $\Delta = \frac{d^2}{d\sigma^2}$ be the Laplacian on H with the Neumann boundary condition. Then Δ is a non-positive self-adjoint operator for which $\{-j^2: j = 0, 1, \ldots\}$ and $\{\xi_0 = \frac{1}{\sqrt{\pi}}, \xi_j = \sqrt{\frac{2}{\pi}} \cos j\sigma: j = 1, 2, \ldots\}$ form the eigensystem. Hence we can define the operator $\sqrt{-\Delta}$ and we put $\hat{\omega} = \sqrt{-\Delta}$. $\hat{\omega}$ is a non-negative self-adjoint operator on H, and $\{j: j = 0, 1, \ldots\}$ and $\{\xi_j: j = 0, 1, \ldots\}$ form the eigensystem of it.

We can now regard the stochastic differential equation

$$(2.1) \qquad dX_t = -\hat{\omega}X_t dt + X_t \cdot dB_t, \qquad t \geqq 0$$

as a realization of the formal equation (*), where $X_t \cdot$ denotes the multiplicative operator and B_t is the c.B.m. on H given by $B_t(\xi) = \langle \xi \otimes \chi_{[0\wedge t, 0\vee t]}, \omega \rangle$.

It is known that Equation (2.1) has no solution on H, and not even on a Hilbert scale (see [17] §5). Accordingly, we consider a modified equation

$$(2.2) \qquad dX_t = -\hat{\omega}X_t dt + X_t \cdot R dB_t,$$

where R is an operator given by

$$(Rf)(\sigma) = \int R(\sigma, \sigma')f(\sigma')d\sigma'.$$

It is easy to see that for $X \in H$, $X \cdot R$ is a Hilbert-Schmidt operator from H to H and that the mapping $X \to X \cdot R \in \sigma_2(H)$, $X \in H$, is Lipschitz continuous ($\|X \cdot R\|_{H-S} \leqq c\|X\|$). Therefore the equation (2.2) has a unique solution in H.

Let X_t be the solution of (2.2) in H and let $\{\Phi_n(t): n = 0, 1, \ldots\}$ be the kernels of the integral representation of X_t. Our first problem is the determination of the kernels $\{\Phi_n\}$. Before we do this, we must prepare two lemmas.

<u>Lemma 2.1</u>. The kernel of degree n of $\int_0^t -\hat{\omega}X_s ds$ is

$$\int_{t_n}^t -\hat{\omega}\Phi_n(s; x_1, \ldots, x_n; \cdot)ds \quad \text{for } 0 \leqq t_1 \leqq \cdots \leqq t_n \leqq t.$$

<u>Proof</u>. By definition, X_t is represented in the form

$$X_t = \sum_{n=0}^{\infty} n! \int \cdots \int_{0 \leqq t_1 \leqq \cdots \leqq t_n \leqq t} \Phi_n(t) dB_{t_1} \cdots dB_{t_n}$$

From this it follows that

$$\int_0^t -\hat{\omega}X_s ds = \sum n! \int_0^t \left\{ \int \cdots \int_{0 \leqq t_1 \leqq \cdots \leqq t_n \leqq s} -\hat{\omega}\Phi_n(s) dB_{t_1} \cdots dB_{t_n} \right\} ds$$

$$= \sum n! \int \cdots \int_{0 \leqq t_1 \leqq \cdots \leqq t_n \leqq t} \left\{ \int_{t_n}^t -\hat{\omega}\Phi_n(s) ds \right\} dB_{t_1} \cdots dB_{t_n}.$$

This proves the lemma. (Q.E.D.)

Lemma 2.2. Let $S \in L^2(T\times\mathcal{E}^* \to \sigma_2(H))$ and assume that $S(t, \omega)$ satisfies:

(i) $S(t, \omega)$ is $\mathcal{B}_t$-adapted.

(ii) $S(t, \cdot) \in L^2(\mathcal{E}^* \to \sigma_2(H))$ and Wiener-Itô decomposition of $L^2(\mathcal{E}^* \to \sigma_2(H))$ gives

$$S(t) = \sum_{n=0}^{\infty} S_n(t), \qquad S_n(t) \in \mathcal{H}_n(\sigma_2(H)),$$

$$S_n(t) \cong S_n(t; x_1, \ldots, x_n; *),$$

where $S_n(t) \in L^2((\widehat{D\times T})^n\times D \to H)$ and $D = [0, \pi]$.
Then the $\mathcal{H}_{n+1}(H)$-component of $\int_0^t S(s)dB_s$ is equal to $\int_0^t S_n(s)dB_s$ and its kernel is given by

$$\frac{1}{n+1}\chi_{[0,t]}(t_{n+1})S_n(t_{n+1}; x_1, \ldots, x_n; *_{n+1}), \quad \text{for } t_1 \leqq \cdots \leqq t_{n+1}. \tag{2.3}$$

Proof. See [17] Lemma 5.1. (Q.E.D.)

Corollary 2.1. The kernel of degree n+1 of $\int_0^t X_s\cdot RdB_s$ is given by

$$\frac{1}{n+1}\chi_{[0,t]}(t_{n+1})\sum_{j=0}^{\infty}\xi_j\left(\int_0^{\pi}\Phi_n(t_{n+1}; x_1, \ldots, x_n; \sigma)R(\sigma, *_{n+1})\xi_j(\sigma)d\sigma\right)$$
$$\text{for } t_1 \leqq \cdots \leqq t_{n+1}.$$

Proof. It is easy to see that the n-th component $(X_s\cdot R)_n$ of $X_s\cdot R$ is

$$(X_s\cdot R)_n(x_1, \ldots, x_n; \cdot) = \sum_{j=0}^{\infty}\xi_j(\cdot)\left(\int_0^{\pi}\Phi_n(s; x_1, \ldots, x_n; \sigma)R(\sigma, \cdot)\xi_j(\sigma)d\sigma\right).$$

Applying Lemma 2.2 to $X_s\cdot R$, we obtain the result. (Q.E.D.)

We are now ready to state our first theorem.

Theorem 2.1. The system of the kernels $\{\Phi_n : n = 0, 1, \ldots\}$ of X_t satisfies the following functional equations

$$\Phi_0(t) = X_0 + \int_0^t -\hat{\omega}\Phi_0(s)ds, \tag{2.4}$$

$$\Phi_{n+1}(t) = \chi_{[0,t]}\int_{t_{n+1}}^{t} -\hat{\omega}\Phi_{n+1}(s; x, \ldots, x_{n+1}; \cdot)ds +$$
$$+ \frac{1}{n+1}\chi_{[0,t]}(t_{n+1})\sum_{j=0}^{\infty}\xi_j\left(\int_0^{\pi}\Phi_n(t_{n+1}; x_1, \ldots, x_n; \sigma)R(\sigma, *_{n+1})\xi_j(\sigma)d\sigma\right),$$
$$\text{for } t_1 \leqq t_2 \leqq \cdots \leqq t_{n+1}, \quad n = 0, 1, \ldots .$$

Proof. This theorem follows immediately from Lemma 2.1 and Corollary 2.1. (Q.E.D.)

Remark 2.1. Equation (2.4) is equivalent to

$$\Phi_0(t) = T_t X_0 \tag{2.5}$$

$$\Phi_{n+1}(t) = \frac{1}{n+1} \chi_{[0,t]}(t_{n+1}) \sum_{j=0}^{\infty} \xi_j e^{-j(t-t_{n+1})} \times$$

$$\times \left(\int_0^{\pi} \Phi_n(t_{n+1}; x_1, \ldots, x_n; \sigma) R(\sigma, x_{n+1}) \xi_j(\sigma) d\sigma\right),$$

$$\text{for } t_1 \leqq \cdots \leqq t_{n+1},\ n = 0, 1, \ldots .$$

Together with Equation (2.4), we get

Theorem 2.2. Equation (2.4) has a unique solution, $\{\Phi_n: n = 0, 1, \ldots\}$ which satisfies

$$(2.6) \qquad \int_{t_1 \leqq \cdots \leqq t_n} \cdots \int \|\Phi_n(t)\|^2 dx_1 \cdots dx_n \leqq c_0 \frac{(\pi\gamma^2 t)^n}{(n!)^3}, \quad n = 0, 1, \ldots,$$

where $c_0 = \|X_0\|^2$ and $\gamma = \max|R(\sigma, \sigma')|$.

Proof. Using the formula (2.5), we can prove (2.6) inductively. (Q.E.D.)

Remark 2.2. From (2.6) it follows that

$$E[\|I(\Phi_n)\|_H^2] = n! \int \cdots \int \|\Phi_n\|^2 dx_1 \cdots dx_n$$

$$= (n!)^2 \int_{t_1 \leqq \cdots \leqq t_n} \cdots \int \|\Phi_n\|^2 dx_1 \cdots dx_n \leqq c_0 \frac{(\pi\gamma^2 t)^n}{n!}$$

Therefore we know that $X(t) = \sum_{n=0}^{\infty} I(\Phi_n(t))$ converges in $L(\mathcal{E}^* \to H)$ and that

$$E[\|X(t)\|^2] = \sum_{n=0}^{\infty} E[\|I(\Phi_n(t))\|^2] \leqq c_0 e^{\pi\gamma^2 t}.$$

We shall next see that Equation (2.2) is equivalent to a system of (non-random) ordinary differential equations on H.

For $\eta \in \mathcal{E}$ put

$$U^{(n)}(t; \eta) = \int \cdots \int \Phi_n(t; x_1, \ldots, x_n; \cdot) \eta^{n\otimes}(x_1, \ldots, x_n) dx_1 \cdots dx_n,$$

and put

$$U(t; \eta) = \sum_{n=0}^{\infty} U^{(n)}(t; \eta).$$

Then it is easily seen that the system $\{U(t; \eta): \eta \in \mathcal{E}\}$ determines $\{\Phi_n(t; x_1, \ldots, x_n), n = 0, 1, 2, \ldots\}$ completely. Without loss of generality, we may assume that $\{\eta: \eta = \xi \otimes \zeta \in \mathcal{E}\}$ is dense in $\mathcal{E}$. We therefore conclude that $\{U(t; \eta): \eta = \xi \otimes \zeta \in \mathcal{E}\}$ determines $\{\Phi_n(t); n = 0, 1, 2, \ldots\}$ completely.

Theorem 2.3. For $\eta = \xi \otimes \zeta \in \mathcal{E}$, $U(t; \eta)$ satisfies the following equation on H

$$(2.7) \qquad \frac{dU(t; \eta)}{dt} = -\hat{\omega} U(t; \eta) + \zeta(t) G(\xi) U(t; \eta), \quad t > 0,$$

$$U(0; \eta) = X_0,$$

where $G(\xi)$ is a bounded linear operator on H given by

$$G(\xi)h = (R\xi)\cdot h, \qquad h \in H.$$

Proof. Using (2.4), we can prove

$$U^{(n+1)}(t;\eta) = \int_0^t -\hat{\omega}U^{(n+1)}(s;\eta)ds + \int_0^t \zeta(s)G(\xi)U^{(n)}(s;\eta)ds,$$

$$n = 0, 1, 2, \ldots .$$

Therefore we get

$$U(t;\eta) = \sum_{n=0}^{\infty} U^{(n)}(t;\eta) = U^{(0)}(t;\eta) + \sum_{n=0}^{\infty} U^{(n+1)}(t;\eta)$$

$$= X_0 + \int_0^t -\hat{\omega}\Phi_0(s)ds + \sum_{n=0}^{\infty} \int_0^t -\hat{\omega}U^{(n+1)}(s;\eta)ds$$

$$+ \sum_{n=0}^{\infty} \int_0^t \zeta(s)G(\xi)U^{(n)}(s;\eta)ds$$

$$= X_0 + \int_0^t -\hat{\omega}U(s;\eta)ds + \int_0^t \zeta(s)G(\xi)U(s;\eta)ds.$$

Thus the proof is completed. (Q.E.D.)

Let X_t^n be the $\mathcal{H}_n(H)$-component of X_t (i.e. $X_t^n = I(\Phi_n(t)) = n!\ \hat{I}(\Phi_n(t))$).

Definition 2.1. A function $V_n(t;\sigma,\sigma')$ which satisfies the condition:

$$E[(X_t^n, f)(X_t^n, g)] = \int_0^\pi \int_0^\pi V_n(t;\sigma,\sigma')f(\sigma)g(\sigma')d\sigma d\sigma',$$

for any $f, g \in H$, is called the covariance function of X_t of degree n.

For the simplicity, from now on we assume that the initial value X_0 is a constant function ($X_0(\sigma)$ = const.).

Theorem 2.4. There exist covariance functions $V_n(t;\sigma,\sigma')$ of X_t of degree n, n = 0, 1, ..., and they satisfy the following equations

(2.8) $$V_0(t;\sigma,\sigma') = c^2,$$

$$V_{n+1}(t;\sigma,\sigma') = \int_0^t \int_0^\pi \int_0^\pi Q(t-s;\sigma,\sigma';\tau,\tau')\tilde{R}(\tau,\tau')V_n(s;\tau,\tau')d\tau d\tau' ds,$$

$$n = 0, 1, \ldots,$$

where $Q(s;\sigma,\sigma';\tau,\tau') = \sum_{j,k=0}^{\infty} e^{-(j+k)s}\xi_j(\sigma)\xi_k(\sigma')\xi_j(\tau)\xi_k(\tau')$ and $\tilde{R}(\tau,\tau') = \int_0^\pi R(\tau,*)R(\tau',*)d*$.

Remark 2.3. The function $Q(s;\sigma,\sigma';\tau,\tau')$ has the following properties:

(i) $$Q(0;\sigma,\sigma';\tau,\tau') = \delta(\sigma-\tau,\sigma'-\tau').$$

(ii) Q is continuous on $(0, \infty)\times[0, \pi]\times[0, \pi]$.

(iii) $\int_0^\pi \int_0^\pi Q(s; \sigma, \sigma'; \tau, \tau')d\tau d\tau' = 1$.

(iv) $Q(s; \sigma, \sigma'; \tau, \tau') \geqq 0$.

(v) $\int_0^\pi \int_0^\pi Q(s; \sigma, \sigma'; \rho, \rho')Q(u; \rho, \rho'; \tau, \tau')d\rho d\rho' = Q(s+u; \sigma, \sigma'; \tau, \tau')$ (Chapman-Kolmogorov).

Remark 2.4. Let $\{\tilde{T}_t: t \geqq 0\}$ be a semi-group of operators on $H\times H$ given by

$$(\tilde{T}_t F)(\sigma, \sigma') = \int_0^\pi \int_0^\pi Q(t; \sigma, \sigma'; \tau, \tau')F(\tau, \tau')d\tau d\tau'.$$

Then the equation (2.8) is expressed in the form

$$V_{n+1} = \int_0^t \tilde{T}_{t-s}(R\cdot V_n)ds. \tag{2.9}$$

It can be proved that $\tilde{T}_t = \begin{bmatrix} T_t & 0 \\ 0 & T_t \end{bmatrix}$ and that the infinitesimal generator $-\tilde{\omega}$ of $\{\tilde{T}_t\}$ is given by $-\tilde{\omega} = \begin{bmatrix} -\hat{\omega} & 0 \\ 0 & -\hat{\omega} \end{bmatrix}$.

Proof of Theorem 2.4. By definition,

$$\begin{aligned} &E[(X_t^n, f)(X_t^n, g)] \\ &= (n!)^2 \int_0^\pi \int_0^\pi \{\int \cdots \int_{t_1 \leqq \cdots \leqq t_n} \Phi_n(t; x_1, \ldots, x_n; \sigma)\Phi_n(t; x_1, \ldots, x_n; \sigma') \\ &\qquad dx_1 \cdots dx_n\} f(\sigma)g(\sigma')d\sigma d\sigma'. \end{aligned}$$

Therefore we obtain

$$V_n(t; \sigma, \sigma') = (n!)^2 \int \cdots \int_{t_1 \leqq \cdots \leqq t_n} \Phi_n(t; x_1, \ldots, x_n; \sigma)\Phi_n(t; x_1, \ldots, x_n; \sigma')dx_1 \cdots dx_n.$$

Using (2.5), we obtain

$$\begin{aligned} &V_{n+1}(t; \sigma, \sigma') \\ &= \int_0^t \int_0^\pi \int_0^\pi Q(t-s; \sigma, \sigma'; \tau, \tau')\tilde{R}(\tau, \tau')V_n(s; \tau, \tau')d\tau d\tau' ds \\ &= \int_0^t \tilde{T}_{t-s}(\tilde{R}\cdot V_n(s))ds. \end{aligned}$$

(Q.E.D.)

Corollary 2.2. Let $v_n(t) = \sup\{V_n(u; \sigma, \sigma'): 0 \leqq u \leqq t, \sigma, \sigma' \in [0, \pi]\}$. Then $v_n(t)$ satisfies

$$v_n(t) \leqq c^2 \frac{(\pi\gamma^2)^n t^n}{n!}, \qquad n = 0, 1, \ldots . \tag{2.10}$$

Proof. Assume that (2.10) is true for n. Then by the formula (2.8)

$$|V_n(u;\sigma,\sigma')| \leqq \int_0^t \int_0^\pi \int_0^\pi Q(u-s;\sigma,\sigma';\tau,\tau')|\tilde{R}(\tau,\tau')V_n(s;\tau,\tau')|d\tau d\tau' ds$$

$$\leqq \pi\gamma^2 c^2 \frac{(\pi\gamma^2)^n}{n!}\int_0^n\{\int_0^\pi\int_0^\pi Q(u-s;\sigma,\sigma';\tau,\tau')d\tau d\tau'\}s^n ds$$

$$= c^2 \frac{(\pi\gamma^2)^{n+1}}{n!}\int_0^u s^n ds$$

$$= c^2 \frac{(\pi\gamma^2)^{n+1}u^{n+1}}{(n+1)!} .$$

Hence

$$v_{n+1}(t) \leqq c^2 \frac{(\pi\gamma)^{n+1}t^{n+1}}{(n+1)!} .$$

Thus (2.10) has been proved by induction. (Q.E.D.)

Theorem 2.5. Put $V(t;\sigma,\sigma') = \sum_{n=0}^{\infty} V_n(t;\sigma,\sigma')$. Then $V(t;\sigma,\sigma')$ converges on $[0,\infty)\times[0,\pi]\times[0,\pi]$ and satisfies

$$|V(t;\sigma,\sigma')| \leqq c^2 e^{\pi\gamma^2 t}, \tag{2.11}$$

$$V(t;\sigma,\sigma') = c^2 + \int_0^t \tilde{T}_{t-s}(\tilde{R}\cdot V(s))ds. \tag{2.12}$$

Proof. The inequality (2.11) follows from (2.10). Since $\sum_{n=0}^{\infty} V_n(t;\sigma,\sigma')$ converges, (2.12) follows from (2.9). (Q.E.D.)

Remark 2.5. The function $V(t;\sigma,\sigma')$ denotes the covariance function of X_t. By (2.12), $V(t;\sigma,\sigma')$ satisfies

$$\frac{\partial V(t;\sigma,\sigma')}{\partial t} = -\tilde{\omega}V + \tilde{R}\cdot V, \qquad t > 0.$$

Remark 2.6. When $t \to \infty$, $V(t)$ does not converges (with exceptional cases, e.g. $c = 0$ or $R = 0$). From this we know that the equation (2.2) has no invariant measure.

We shall next discuss Equation (**). This equation is equivalent to

$$\frac{\partial X(t,\sigma)}{\partial t} = Y(t,\sigma), \tag{**'}$$

$$\frac{\partial Y(t,\sigma)}{\partial t} = -\hat{\omega}Y(t,\sigma) + X(t,\sigma)\cdot w(t,\sigma).$$

Therefore we can regard the stochastic differential equation

$$dX_t = Y_t dt, \tag{2.13}$$

$$dY_t = -\hat{\omega}Y_t dt + X_t\cdot dB_t$$

as a realization of the formal equation (**). Unfortunately Equation (2.13) has no solution on H for the same reason as that discussed in the beginning of this section for Equation (2.1). Thus we are led to a modified equation

$$dX_t = Y_t dt \tag{2.14}$$

$$dY_t = -\hat{\omega}Y_t dt + X_t\cdot RdB_t,$$

where R is the same operator given in (2.2).

The kernels of the integral representations of X_t and Y_t, are given by the following theorems.

Theorem 2.6. Let $\{\Phi_n(t): n = 0, 1, \dots\}$ and $\{\Psi_n(t): n = 0, 1, \dots\}$ be the kernels of X_t and Y_t respectively. Then the following system of equations is satisfied:

$$
(2.15) \quad
\begin{aligned}
\Psi_0(t) &= T_t Y_0, \\
\Phi_0(t) &= X_0 + \int_0^t \Psi(s)ds,
\end{aligned}
$$

$$\Psi_{n+1}(t) = \chi_{[0,t]}(t_{n+1}) \int_{t_{n+1}}^t -\hat{\omega}\Psi_{n+1}(s)ds$$

$$+ \frac{1}{n+1} \chi_{[0,t]}(t_{n+1}) \sum_{j=0}^{\infty} \xi_j \int_0^{\pi} \Phi_n(t_{n+1}; x_1, \dots, x_n; \sigma) R(\sigma, x_{n+1}) \xi_j(\sigma) d\sigma,$$

$$\Phi_{n+1}(t) = \int_{t_{n+1}}^t \Psi_{n+1}(s)ds, \qquad n = 0, 1, \dots .$$

Proof. This theorem can be proved in the same way as the proof of Theorem 2.1. (Q.E.D.)

Remark 2.7. Equation (2.15) is equivalent to

$$
(2.16) \quad
\begin{aligned}
\Psi_0(t) &= T_t Y_0, \\
\Phi_0(t) &= X_0 + \int_0^t \Psi_0(s)ds, \\
\Psi_{n+1}(t) &= \frac{1}{n+1} \chi_{[0,t]}(t_{n+1}) \sum_j \xi_j e^{-j(t-t_{n+1})} \times \\
&\quad \times \int_0^{\pi} \Phi_n(t_{n+1}; x_1, \dots, x_n; \sigma) R(\sigma, x_{n+1}) \xi_j(\sigma) d\sigma, \\
\Phi_{n+1}(t) &= \int_{t_{n+1}}^t \Psi_{n+1}(s)ds, \qquad n = 0, 1, \dots .
\end{aligned}
$$

Using (2.15) or (2.16), we can obtain $\Psi_0, \Phi_0, \Psi_1, \Phi_1, \dots, \Psi_n, \Phi_n, \dots,$ inductively.

Theorem 2.7. The kernels $\{\Phi_n(t): n = 0,1,\dots\}$ and $\{\Psi_n(t): n = 0,1,\dots\}$ satisfy

$$
(2.17) \quad
\begin{aligned}
&\|\Psi_0(t)\|^2 \leqq c_2^2, \\
&\|\Phi_0(t)\|^2 \leqq 2(c_1^2 + c_2^2 t^2), \\
&\int \cdots \int_{t_1 \leqq \cdots \leqq t_n} \|\Psi_n(t)\|^2 dx_1 \cdots dx_n \leqq 2^{-2n+3} (\pi\gamma^2)^n (c_1^2 + c_2^2 t^2) \frac{t^{4n-3}}{(n!)^4}, \\
&\int \cdots \int_{t_1 \leqq \cdots \leqq t_n} \|\Phi_n(t)\|^2 dx_1 \cdots dx_n \leqq 2^{-2n+2} (\pi\gamma^2)^n (c_1^2 + c_2^2 c^2) \frac{t^{4n}}{(n!)^4 (n+1)},
\end{aligned}
$$

where $c_1 = \|X_0\|$ and $c_2 = \|Y_0\|$.

Proof. Two inequalities

$$\|\Psi_0(t)\|^2 \leqq c_2^2$$

and $$\|\Phi_0(t)\|^2 \leqq 2(c_1^2 + c_2^2 t^2)$$

are immediate. From (2.16) it follows that

(2.18) $$\int\cdots\int_{t_1\leqq\cdots\leqq t_{n+1}} \|\Psi_{n+1}(t)\|^2 dx_1 \cdots dx_{n+1}$$

$$\leqq \frac{\pi\gamma^2}{(n+1)^2}\int_0^t\{\int\cdots\int_{t_1\leqq\cdots\leqq t_n}\|\Phi_n(t_{n+1})\|^2 dx_1\cdots dx_n\}dt_{n+1},$$

and

(2.19) $$\int\cdots\int_{t_1\leqq\cdots\leqq t_{n+1}} \|\Phi_{n+1}(t)\|^2 dx_1 \cdots dx_{n+1}$$

$$\leqq t^2\int_0^t\{\int\cdots\int_{t_1\leqq\cdots\leqq t_{n+1}\leqq s}\|\Psi_{n+1}(s)\|^2 dx_1\cdots dx_{n+1}\}ds.$$

Using (2.18) and (2.19), we can prove (2.17) by induction. (Q.E.D.)

Corollary 2.3. The solution X_t of the equation (2.14) satisfies

(2.20) $$E[\|X_t\|^2] \leqq c(c_1^2 + c_2^2 t^2)e^{(\pi\gamma^4/4)t^4}.$$

Now that we have determined the kernels of X_t and Y_t, we are able to carry on the same analysis on the (2.14) as we have done on (2.2).

Let

$$U_X^{(n)}(t;\eta) = \int\cdots\int\Phi_n(t)\eta^{n\otimes}dx_1\cdots dx_n, \qquad \eta\in\mathcal{E}$$

and

$$U_X(t;\eta) = \sum_{n=0}^{\infty} U_X^{(n)}(t;\eta).$$

Theorem 2.8. For $\eta = \xi\otimes\zeta\in\mathcal{E}$, $U_X(t;\eta)$ satisfies

(2.21) $$\frac{d^2U_X(t)}{dt^2} = -\hat{\omega}\frac{dU_X(t)}{dt} + \zeta(t)G(\xi)U_X(t), \qquad t>0,$$

$$U_X(0) = X_0, \qquad \left.\frac{dU_X}{dt}\right|_{t=0} = Y_0,$$

where $G(\xi)h = (R\xi)\cdot h$ (which is the same operator given in Theorem 2.3).

Theorem 2.9. The covariance functions $\{V_n^X(t;\sigma,\sigma'): n = 0, 1, \ldots\}$ of the solution X_t of (2.14) satisfy

(2.22) $$V_{n+1}^X(t;\sigma,\sigma') = \int_0^t \tilde{\hat{T}}_{t-s}(\tilde{R}V_n^X(s))ds, \qquad n = 0, 1, \ldots,$$

where $\tilde{\hat{T}}_u$ is the operator on $H\times H$ given by $\tilde{\hat{T}}_u = \begin{bmatrix}\hat{T}_u & 0\\ 0 & \hat{T}_u\end{bmatrix}$, $\hat{T}_u = \int_0^u T_s ds$, and where

$\tilde{R}(\tau, \tau') = \int_0^\pi R(\tau, x)R(\tau', x)dx$.

Corollary 2.4. The covariance function $V^X(t; \sigma, \sigma') = \sum_{n=0}^{\infty} V_n^X(t; \sigma, \sigma')$ of X_t satisfies

$$V^X(t; \cdot, \cdot) = V_0^X(t; \cdot, \cdot) + \int_0^t \tilde{\tilde{T}}_{t-s}(\tilde{R}\cdot V^X(s))ds. \tag{2.23}$$

Since the proofs of the above theorems can be carried out in the same manner as those of Theorem 2.3, 2.4 and Corollary 2.2, we omit the details.

§3. Concluding remarks.

In §2 we have investigated only the special case $D = [0, \pi]$. But our methods can be applied to the more general cases, for example, the cases where $D = R^d$ or $D \subset R^d$.

Recently D. A. Dawson and H. Salehi [7] have investigated an equation

$$\frac{\partial X(t, x)}{\partial t} = \frac{1}{2}\Delta X(t, x) + \sigma X(t, x)W'(t, x), \quad t \geq 0, \quad x \in R^d. \tag{3.1}$$

This equation is equivalent to the stochastic differential equation

$$dX_t = \frac{1}{2}\Delta X_t dt + X_t \cdot RdB_t \tag{3.2}$$

on $H = L^2(R^d)$ under our formulation of the problems, and we can apply the theory developed in the previous sections to Equation (3.2).

§4. References

[1] A. V. Balakrishnan, Stochastic Optimization Theory in Hilbert Spaces-1, Applied Mathematics and Optimization, Vol.1, No.2 (1974), 97-120.

[2] ———, Stochastic Bilinear Partial Differential Equations, Proc. U.S.-Italy Conference on Variable Structure Systems, Oregon, 1974.

[3] R. F. Curtain and P. L. Falb, Stochastic Differential Equations in Hilbert Space, J. Differential Equations, Vol.10 (1971), 412-430.

[4] Y. Daletskii, Infinite Dimensional Elliptic Operators and Parabolic Equations Connected with them, Uspekhi. Math. Nauk. t.22 (1967).

[5] D. A. Dawson, Stochastic Evolution Equation, Mathematical Biosciences, Vol.15 (1972), 287-316.

[6] ———, Stochastic Evolution Equations and Related Measure Processes, J. Multivariate Analysis, Vol.5 (1975), 1-55.

[7] D. A. Dawson and H. Salehi, Spatially Homogeneous Random Evolutions, to appear.

[8] I. M. Gelfand and N. Ya. Vilenkin, Generalized Functions, Vol.4, Academic Press, 1964.

[9] Z. Haba, Functional Equations for Extended Hadrons, J. Math. Phys., Vol.18, No.11 (1977), 2133-2137.

[10] ——— and J. Lukierski, Stochastic Description of Extended Hadrons, to appear.

[11] T. Hida, Analysis of Brownian Functionals, Carleton Math. Lecture Notes No.13, 2nd ed. Ottawa, Canada, 1978.

[12] ———, Brownian Motion, Springer, 1980.

[13] I. Kubo, Hida Calculus on Gaussian White Noise, Nagoya Univ. Lecture Notes, 1979.

[14] R. Marcus, Parabolic Itô Equations, Trans. Amer. Math. Soc., Vol.198 (1974), 177-190.

[15] Y. Miyahara, Stochastic Differential Equations in Hilbert Space,"OIKONOMIKA" (Nagoya City University, Japan), Vol.14, No.1 (1977), 37-47.

[16] ———, Stability of Linear Stochastic Differential Equations in Hilbert Space, in Information, Decision and Control in Dynamic Socio-Economics (ed. by H. Myoken), Bunshindo/Kinokuniya, Tokyo, 1978, pp.237-252.

[17] ———, Infinite Dimensional Langevin Equation and Fokker-Planck Equation, to appear in Nagoya Math. J., Vol.81 (1981).

[18] B.L. Rozovskii, On Stochastic Differential Equations with Partial Derivatives, Math. Sb. 96 (1975), 314-341.

[19] A. Shimizu, Construction of a Solution of Linear Stochastic Evolution Equations on a Hilbert Space, in Proceedings of the International Symposium on Stochastic Differential Equations, Kyoto, 1976, pp.385-395.

[20] M. Yor, Existence et Unicité de Diffusions à Valeurs dans un Espace de Hilbert, Annales de l'Institut Henri Poincare, Section B., Vol.10, No.1 (1974), 55-88.

STATISTICS AND RELATED TOPICS
M. Csörgö, D.A. Dawson, J.N.K. Rao, A.K.Md.E. Saleh (eds.)

ON THE INCREMENTS OF STOCHASTIC PROCESSES AND THE RECONSTRUCTION OF THEIR DISTRIBUTIONS

Josef Steinebach

University of Düsseldorf[1)]

Erdős and Rényi (1970) pointed out a connection between maximum increments of partial sum sequences and their underlying distributions. This review paper reports on some generalized versions having applications to the solution of so-called "stochastic geyser problems". Some convergence rate results are also given.

1. THE STOCHASTIC GEYSER PROBLEM

Rényi (1962) introduced the following problem:

Consider a sequence $S_1' = S_1 + R_1$, $S_2' = S_2 + R_2$, ... of observations, where $S_1, S_2, \ldots$ are the partial sums of an i.i.d. sequence $\{X_i\}_{i=1,2,\ldots}$ and $R_1, R_2, \ldots$ are random error terms. For sake of simplicity suppose that the errors are bounded, i.e. $|R_n| \leq r_n$ $(n = 1,2,\ldots)$. The question is, under what conditions on the magnitude r_n of the error terms R_n is it possible to determine the underlying distribution function $F(x) = P(X_1 \leq x)$ with probability 1 (w.p.1), even if only an erroneous realization $S_1', S_2', \ldots$ is known?

The question was posed by Rényi for the case $S_n' = [S_n]$, where $[\cdot]$ denotes the integer part, and it is known as the "stochastic geyser problem". The title comes from the following little story: Somewhere on an island there lives a lonely man. His only companion is a geyser erupting at random times. The man is interested in the distribution of the inter-occurence-times X_i between two eruptions of his geyser. But having no watch or other facility to measure the time exactly, he only records the number of eruptions day by day, thus getting a sequence $N_1, N_2, \ldots$. Supposing that the man lives infinitely long and hence obtains an infinite sequence, is this sequence sufficient to determine the d.f. $F(x) = P(X_i \leq x)$? Realizing that there is a one-to-one correspondence between the sequences $N_1, N_2, \ldots$ and $[S_1], [S_2], \ldots$, this leads to Rényi's problem, assuming that the X_i's are i.i.d. random variables.

There are several solutions to this problem, depending on the magnitude r_n of the error terms:

1) In case of $r_n \equiv 0$ (or $r_n = o(1)$) an immediate positive answer is given by the Glivenko-Cantelli-theorem which yields

$$\lim_{n \to \infty} \sup_x |F_n(x) - F(x)| = 0 \quad \text{w.p.1} ,$$

where $F_n(x)$ denotes the empirical d.f. based on the first n observations. But if $r_n \to \infty$ $(n \to \infty)$ the Glivenko-Cantelli-theorem cannot be used any longer to approximate F w.p.1 and the problem becomes more difficult. Bártfai (1966) was the first to give an answer for the case $r_n = o(\log n)$:

2) If the moment-generating function $\phi(\tau) = E\exp(\tau X_1)$ is finite in some interval $|\tau| < \tau_1$, then it is possible to estimate the cumulants of X_1 recursively w.p.1, thus getting the m.g.f., which in turn determines the underlying distribution.

Bártfai conjectured that for $r_n = \varepsilon \log n$ the answer becomes negative, i.e. the d.f. cannot be determined any longer w.p.1. His conjecture was confirmed by a strong approximation theorem of Komlós, Major and Tusnády (1975/76) who proved:

3) If $EX_1 = 0$, $EX_1^2 = 1$ and $\phi(\tau) < \infty$ $(|\tau| < \tau_1)$, then there exists a sequence $Y_1, Y_2, \ldots$ of standard normal variables in such a way that

$$|S_n - T_n| = O(\log n) \quad \text{w.p.1},$$

where $S_n = X_1 + \ldots + X_n$, $T_n = Y_1 + \ldots + Y_n$ $(n = 1,2,\ldots)$. This means, a sequence $S_1', S_2', \ldots$ with $S_n' = S_n + \varepsilon \log n$ could behave like a sequence of partial sums of normal variables which makes it impossible to recognize the actually underlying distribution.

Hence $\log n$ appears to be a critical order for the magnitude of error terms in the stochastic geyser problem. This fact was used by Erdős and Rényi (1970) to give a remarkable and completely different new solution to the problem at hand, derived from - what they called - a new law of large numbers.

2. THE ERDŐS-RÉNYI LAW OF LARGE NUMBERS

Let $S_1, S_2, \ldots$ be partial sums corresponding to a non-degenerate i.i.d. sequence $X_1, X_2, \ldots$, satisfying $\phi(\tau) = E\exp(\tau X_1) < \infty$, $\tau \in (0, \tau_1)$. Then the Erdős-Rényi "new law of large numbers" is as follows:

Theorem 1 (see [8]). Under the above assumptions, we have for $a \in A = \{\phi'(\tau)/\phi(\tau) : \tau \in (0,\tau_1)\}$ and $C = C(a)$ such that $\exp(-1/C) = \inf_\tau \phi(\tau)\exp(-\tau a) = \rho(a)$, that

$$\lim_{N\to\infty} \max_{0\le n\le N-[C\log N]} \frac{S_{n+[C\log N]} - S_n}{[C\log N]} = a \quad \text{w.p.1.} \tag{1}$$

Now, if $r_n = o(\log n)$, assertion (1) still holds, even if the values S_n are replaced by the erroneous values S_n'. Hence, also a realization $S_1', S_2', \ldots$ determines the functional dependence $C = C(a)$ w.p.1 on some interval of a-values. A solution of the stochastic geyser problem then follows from the following

Lemma (see [8]). The functional dependence between a and $C = C(a)$ in Theorem 1 determines the d.f. of the X_i's uniquely.

For the proof just note that

$$1/C = -\log\rho(a) = \sup_\tau (\tau a - \log\phi(\tau))$$

is the conjugate of the strictly convex cumulant-generating function $\log\phi(\tau)$. By convex analysis we then also know $\log\phi(\tau)$, given its conjugate. But $\log\phi(\tau)$ resp. $\phi(\tau)$ uniquely determines the underlying distribution.

The proof of Theorem 1 is mainly based on three facts:

i) For each K, $\{S_{n+K} - S_n\}_{n=0,1,2,\ldots}$ is a stationary sequence;

ii) For each K, $\{S_{(i+1)K} - S_{iK}\}_{i=0,1,2,\ldots}$ is an i.i.d. sequence;

iii) $\lim\limits_{K\to\infty} P(S_K \geq Ka)^{1/K} = \rho(a)$ for all $a \in A$.

The latter relation follows from the existence of m.g.f. ϕ in some interval by Chernoff's (1952) large deviation theorem. Since there is a strong connection between an exponential tail behaviour of the partial sums and existence of the underlying m.g.f., it turns out that $\phi(t)$ being finite in some interval is also necessary for the Erdős-Rényi law in the following sense:

Theorem 2 (see [16]). If $\phi(\tau) = \infty$ for all $\tau > 0$, then

$$\text{(2)} \qquad \limsup_{N\to\infty} \max_{0\leq n\leq N-[C\log N]} \frac{S_{n+[C\log N]} - S_n}{[C\log N]} = \infty \quad \text{w.p.1}$$

for each constant $C > 0$.

Hence the Erdős-Rényi law really depends upon an exponential behaviour of the sequence $\{S_n\}_{n=0,1,2,\ldots}$ in the sense of iii).

On the other hand, an exponential tail behaviour can be found in rather general situations. For examples see Bahadur (1971, Theorem 2.2) or the following result, both using moment-generating function techniques:

Theorem 3 (see [12]). Let $\{T_n\}_{n=1,2,\ldots}$ be a sequence of real-valued random variables on (Ω, A, P) satisfying:

i) $\phi_n(\tau) = Ee^{\tau T_n} < \infty$ for all $\tau \in (0,\tau_1)$,

ii) $\lim\limits_{n\to\infty} \frac{1}{n}\log\phi_n(\tau) = g(\tau)$ for all $\tau \in (\tau_0,\tau_1)$, $0 \leq \tau_0 < \tau_1$,

iii) $g(\tau)$ is strictly convex on (τ_0,τ_1).

Then, for any sequence $\{a_n\}_{n=1,2,\ldots}$ with $\lim\limits_{n\to\infty} a_n = a \in \{g_+'(\tau) : \tau \in (\tau_0,\tau_1)\}$, we have

$$\text{(3)} \qquad \lim_{n\to\infty} P(T_n \geq na_n)^{1/n} = \inf_\tau \exp\{g(\tau) - \tau a\} = \exp\{g(\bar\tau) - \bar\tau a\},$$

where $\bar\tau$ is the unique solution of $a = g_+'(\bar\tau)$ and $g_+'(\tau)$ denotes the right derivative of the convex function g at τ.

Other rather general theorems on exponential large deviation behaviour have recently been proved by Groeneboom, Oosterhoff and Ruymgaart (1979) and Bahadur and Zabell (1979), using different techniques.

Since the Erdős-Rényi law of large numbers turned out to provide a principle for determining an unknown exponential large deviation rate $\rho(a)$ w.p.1 from a single realization $S_1,S_2,\ldots$ of a partial sum sequence, even if error terms are present, it might be asked whether such a principle could hold true in more general situations involving exponential rates. And a next question would be, what properties of the underlying distributions could be recognized via such a general principle? We report on some answers in our next section.

3. SOME OTHER ERDŐS-RÉNYI TYPE LAWS

In a recent paper S. Csörgő (1979a) realized that points i), ii) and iii), mentioned in Section 2, are indeed the essential facts for the proof of Theorem 1. And forgetting about the special structure of partial sums he proved two rather general Erdős-Rényi type laws, one of which may be stated as follows:

Theorem 4 (see [6], [7]). Let $X_1, X_2, \ldots$ be an i.i.d. sequence taking values in some measurable space (X, L) and let T_n be real-valued test-statistics depending upon $X_1, X_2, \ldots$ only through the first n observations, i.e. $T_n = T_n(X_1, \ldots, X_n)$. Define "generalized increments" $T_{n,K}$, setting $T_{n,K} = T_K(X_{n+1}, \ldots, X_{n+K})$, $n = 0,1,2,\ldots$; $K = 1,2,\ldots$. Suppose that

$$\text{(iii)} \qquad \lim_{K\to\infty} P(T_K \geq Ka)^{1/K} = \rho(a)$$

exists for a's in some interval (a_o, a_1), where ρ is a strictly decreasing function with $0 < \rho(a) < 1$. Then, for $a \in (a_o, a_1)$ and $C = C(a)$ such that $\exp(-1/C) = \rho(a)$, we have

$$(4) \qquad \lim_{N\to\infty} \max_{0\leq n\leq N-[C\log N]} \frac{T_{n,[C\log N]}}{[C\log N]} = a \quad \text{w.p.1} .$$

Of course, Theorem 4 generalizes Theorem 1. The proof is again essentially based on the facts that

i) for each K, $\{T_{n,K}\}_{n=0,1,2,\ldots}$ is a stationary sequence,

ii) for each K, $\{T_{iK,K}\}_{i=0,1,2,\ldots}$ is an i.i.d. sequence,

iii) $\lim_{K\to\infty} P(T_{0,K} \geq Ka)^{1/K} = \lim_{K\to\infty} P(T_K \geq Ka)^{1/K} = \rho(a)$.

As in Erdős and Rényi's (1970) original proof, the main idea is to approximate

$$\max_{0\leq n\leq N-[C\log N]} T_{n,[C\log N]}$$

by

$$\max_{i=0,1,\ldots,[N/C\log N]-1} T_{i[C\log N],[C\log N]} ,$$

which is a maximum of independent terms.

Bahadur (1971, Theorem 7.2) pointed out that an exponential behaviour like iii) of a sequence of test statistics under the null hypothesis can be used to determine "exact slopes" even under alternatives, if only the statistics are strongly consistent, which is often quite easy to verify. But usually it is non-trivial to determine the rate $\rho(a)$.

On the other hand, Theorem 4 again provides a principle for determining the unknown rate ρ from a single realization $X_1, X_2, \ldots$, even if some error terms are involved. In view of this fact, S. Csörgő (1979b) proposes a comparison of the efficiency of different sequences of test statistics making use of the connection between their exact slopes and their Erdős-Rényi maxima in the sense of Theorem 4, avoiding the exact computation of the underlying slopes.

Although Theorem 4 provides a rather general and very useful principle to determine exponential rates from sequences of observations, it still depends upon an i.i.d. situation, which turns out to be unnecessary. There are other typical situations, not too far away from the i.i.d. case of course, but substantially different, also providing certain Erdős-Rényi type laws. Consider, for instance, waiting-times in a queuing system of type G/G/1. Let X_n denote the difference between service-time B_n for the nth customer and interarrival-time A_n between n^{th} and $(n+1)^{th}$ customer. Then the waiting-time W_n of the $(n+1)^{th}$ customer is given by

$$W_o \equiv 0, \quad W_n = [W_{n-1} + X_n]^+ \quad (n = 1,2,\ldots) ,$$

where $[\cdot]^+$ denotes the positive part. Setting again $S_o = 0$, $S_n = X_1 + \ldots + X_n$ we have

$$W_n = \max_{k=0,\ldots,n} (S_n - S_k) \quad (n = 0,1,2,\ldots).$$

Suppose that $X_1, X_2, \ldots$ is a non-degenerate i.i.d. sequence with

$$EX_1 < 0, \quad \phi(\tau) = E \exp(\tau X_1) < \infty \quad (\tau \in [0,\tau_1)) .$$

Moreover, let $\tau_o \in (0,\tau_1)$ be such that $\phi(\tau_o) = 1$. Then, using Theorem 3, it can be proved that

$$\lim_{n\to\infty} P(W_n \geq na)^{1/n} = \rho(a) ,$$

where

$$\rho(a) = \begin{cases} e^{-\tau_o a} , & a \in (0, a_0] \\ \inf_\tau \phi(\tau) \exp(-\tau a), & a \in (a_0, a_1) , \end{cases}$$

$$a_o = \phi'(\tau_o) , \quad a_1 = \sup_{\tau < \tau_1} \phi'(\tau)/\phi(\tau) .$$

Hence we have again exponential large deviation rates for the sequence $\{W_n\}_{n=0,1,2,\ldots}$, but the increments

$$W_{n,K} = W_{n+K} - W_n , \quad n = 0,1,2,\ldots; \quad K = 1,2,\ldots ,$$

are neither stationary nor independent. Nevertheless an Erdős-Rényi type law is still available.

Theorem 5 (see [15]). Let $\{W_n\}_{n=0,1,2,\ldots}$, $\rho(a)$, a_o, a_1 be as defined above. Then, for $a \in (0,a_1)$ and $C = C(a)$ such that $\exp(-1/C) = \rho(a)$, we have

$$\lim_{N\to\infty} \max_{0\leq n\leq N-[C \log N]} \frac{W_{n+[C \log N]} - W_n}{[C \log N]} = a \quad \text{w.p.1} . \tag{5}$$

Main idea of proof is to select, for each N, a suitable sequence $0 = \nu_{0,N} < \nu_{1,N} < \ldots$ of random indices such that

$$\{W_{\nu_{i,N}+[C\log N]} - W_{\nu_{i,N}}\}_{i=0,1,2,\ldots}$$

is an i.i.d. sequence and, for some $0 < \delta < 1$,

$$\sum_{N=1}^{\infty} P(L_N \le (1-\delta)[N/[C \log N]]) < \infty ,$$

where $L_N = \#\{\nu_{i,N} : \nu_{i,N} \le N - [C \log N]\}$.

Then

$$\max_{0 \le n \le N-[C\log N]} (W_{n+[C\log N]} - W_n)$$

is approximated by

$$\max_{\nu_{i,N} \le N-[C\log N]} (W_{\nu_{i,N}+[C\log N]} - W_{\nu_{i,N}}) ,$$

which is now a maximum of a random number of i.i.d. terms.

Having proved Theorem 5, a "stochastic geyser problem for waiting-times" could be added, because, if $-\log \rho(a)$ in Theorem 5 possesses a strictly convex part, the functional dependence $C = C(a)$ again determines the m.g.f. $\phi(\tau)$ from a single realization $W_0, W_1, W_2, \ldots$ of the waiting-time process. And as before, the realization may even be observed erroneously up to an order $o(\log n)$.

Similar results are valid under a renewal-type situation. Consider, for instance, a sequence $X_1, X_2, \ldots$ of so-called "failure times" in a renewal process, i.e. $X_1, X_2, \ldots$ is a sequence of i.i.d. nonnegative random variables. Let N_t denote the number of renewals occuring up to time t, which means

$$N_t = \max\{n \ge 0 : S_n \le t\} \quad (t \ge 0).$$

Assuming that the X_n's are non-degenerate and using Theorem 3 it can be shown that the renewal counting process $\{N_t\}_{t \ge 0}$ has an exponential tail behaviour in the following sense:

For a's in a certain interval (a_0, a_1), we have

$$\lim_{t\to\infty} P(N_t \ge ta)^{1/t} = \inf_{\xi} \exp g(e^{\xi}) - \xi a\} = \rho(a) ,$$

where $g(x) = \lim_{t\to\infty} \frac{1}{t} \log Ex^{N_t}$ (for details see [17] or [18]).

Moreover, if $\phi(\tau)$ denotes the m.g.f. of the X_n's, which is finite at least for all $\tau < 0$, it holds that

$$g\left(\frac{1}{\phi(\tau)}\right) = -\tau \quad (\tau < 0) .$$

But again, the increments $N_{t+K} - N_t$ of the process are stationary and independent only in the special case of exponentially distributed failure times leading to a Poisson process, otherwise not. Similarly to the waiting-time process however, an Erdős-Rényi type law is still available:

Theorem 6 (see [17], [18]). Let $\{X_n\}_{n=1,2,\ldots}$, $\{N_t\}_{t\geq 0}$, and $\rho(a)$ be as defined above. If the X_n's are non-degenerate and $EX_1^3 < \infty$, then we have, for $a \in (a_o, a_1)$ and $C = C(a)$ such that $\exp(-1/C) = \rho(a)$,

$$\text{(6)} \qquad \lim_{T\to\infty} \sup_{0\leq t\leq T-C\log T} \frac{N_{t+C\log T} - N_t}{C \log T} = a \quad \text{w.p.1} .$$

Remark 1. More general versions are available, e.g. for first-passage-times of random walks or secondary processes associated with the counting process $\{N_t\}_{t\geq 0}$. The methods of proof are similar (see [17], [18]).

Remark 2. Using some simple estimations, it can be verified that Theorem 6 holds in a discrete version too; this means

$$\text{(6')} \qquad \lim_{N\to\infty} \max_{0\leq n\leq N-[C\log N]} \frac{N_{n+[C\log N]} - N_n}{[C \log N]} = a \quad \text{w.p.1} ,$$

(see [18]).

The latter version enables us to solve a "stochastic geyser problem" also for the renewal counting process, because once again a realization $N_1, N_2, \ldots$ of the process, observed in discrete time and probably erroneously observed up to an order $o(\log n)$, still determines the functional dependence $C = C(a)$ in Theorem 6. Since $\frac{1}{C} = \sup_\xi \{\xi a - g(e^\xi)\}$, convex analysis yields the convex function $g = g(e^\xi)$ by knowing its conjugate. Finally, the special form of g, i.e. $g(\frac{1}{\phi(\tau)}) = -\tau$ $(\tau < 0)$, allows us to determine $\phi = \phi(\tau)$, which again characterizes the underlying distribution. Hence the problem is solved.

The proof of Theorem 6 takes advantage of the fact that, for each T, a sequence $0 = \tau_{0,T} < \tau_{1,T} < \ldots$ of random times can be selected such that

$$\{N_{\tau_{i,T}+[C\log T]} - N_{\tau_{i,T}}\}_{i=0,1,2,\ldots}$$

is an i.i.d. sequence, and, for some $0 < \delta < 1$, and T being integer-valued,

$$\sum_{T=1}^{\infty} P(M_T \leq (1-\delta)T/C \log T) < \infty ,$$

where $M_T = \#\{\tau_{i,T} : \tau_{i,T} \leq T - C \log T\}$. Simply take the $\tau_{i,T}$'s as suitable renewal points. Thus an approximation by a random number of i.i.d. terms is possible again.

Theorems 5 and 6 both follow the same ideas which can be stated as a rather general principle:

Theorem 7. Let $\{T_{n,K}\}_{\substack{n=0,1,2,\ldots \\ K=1,2,\ldots}}$ be a double sequence of real-valued random random variables on (Ω, A, P) satisfying:

(i) $\limsup_{K\to\infty} P(T_{n,K} \geq Ka)^{1/K} \leq \rho(a)$ uniformly in n, for a's in some interval (a_0, a_1), where ρ is a strictly decreasing function with $0 < \rho(a) < 1$;

(ii) there exists a sequence $0 = \nu_{0,N} < \nu_{1,N} < \ldots$ of random indices such that, for each $N = 1,2,\ldots$ and $C > 0$,

$$\{T_{\nu_{i,N};[C \log N]}\}_{i=0,1,2,\ldots}$$

is an i.i.d. sequence, and, for some $0 < \delta < 1$,

$$\sum_{N=1}^{\infty} P(L_N \leq (1-\delta)[N/C \log N]) < \infty ,$$

where $L_N = \#\{\nu_{i,N} : \nu_{i,N} \leq N - [C \log N]\}$;

(iii) $\lim_{K\to\infty} P(T_{0,K} \geq Ka)^{1/K} = \rho(a)$.

Then, for $a \in (a_0, a_1)$ and $C = C(a)$ such that $\exp(-1/C) = \rho(a)$, we have

$$\lim_{N\to\infty} \max_{0\leq n\leq N-[C \log N]} \frac{T_{n,[C \log N]}}{[C \log N]} = a \quad \text{w.p.1} . \tag{7}$$

Clearly, Theorem 7 generalizes Theorem 4; simply use a deterministic sequence $\nu_{i,N} = i[C \log N]$ $(i = 0,1,2,\ldots)$ of indices and stationarity and independence of the "generalized increments" defined in Theorem 4. However, our Theorem 7 is mainly designed for cases where, although stationarity and independence fails to hold, at least an approximation by a suitable random subsequence of i.i.d. "increments" is available.

4. CONVERGENCE RATES IN ERDŐS-RÉNYI LAWS

Recent results of Révész (1980) for Wiener processes showed that convergence rate statements were available in the original Erdős-Rényi law under normal distributions. Stimulated by Révész's (1980) results we proved in [5] that such statements hold true for other distributions too:

Theorem 8 (see [5]). Under the assumptions of Theorem 1 we have

$$\lim_{N\to\infty} \left\{ \max_{0<n<N-[C \log N]} \frac{S_{n+[C \log N]} - S_n}{[C \log N]^{\frac{1}{2}}} - [C \log N]^{\frac{1}{2}} a \right\} = 0 \quad \text{w.p.1} . \tag{8}$$

In other words, assertion (8) means that

$$\left| \max_{0 \leq n \leq N-[C \log N]} \frac{S_{n+[C \log N]} - S_n}{[C \log N]} - a \right| = o((\log N)^{-\frac{1}{2}}) \quad \text{w.p.1} ,$$

and, from Révész's (1980) paper it can also be seen that this convergence rate cannot be improved to get better rates like $o((\log N)^{-(\frac{1}{2}+\delta)})$, say, for any $\delta > 0$.

For the proof, a more exact large deviation expansion than that of iii) in Section 2 is needed. In [5] it is shown that

iii')
$$P(S_K \geq Ka + K^{\frac{1}{2}}\varepsilon) \leq \rho^K(a) \exp(-\delta K^{\frac{1}{2}}) ,$$

$$P(S_K > Ka - K^{\frac{1}{2}}\varepsilon) \geq A\, \rho^K(a) \exp(\delta K^{\frac{1}{2}}) ,$$

for any $\varepsilon > 0$ and K sufficiently large will do, where $\delta = \delta(\varepsilon) > 0$, $A = A(\varepsilon) > 0$.

Now, in more general situations, like those of Theorem 4 or even Theorem 7, similar improvements of the Erdős-Rényi laws are available whenever the rough exponential rate assumption

(iii)
$$\lim_{K \to \infty} P(T_{0,K} \geq Ka)^{1/K} = \rho(a)$$

can be replaced by an improved condition

iii')
$$P(T_{0,K} \geq Ka) \leq \rho^K(a) \exp(-\delta K^{\frac{1}{2}}) ,$$

$$P(T_{0,K} > Ka) \geq A\, \rho^K(a) \exp(\delta K^{\frac{1}{2}}) ,$$

or something similar, as mentioned after Theorem 8.

Such an improved condition is fulfilled in the renewal-type situation of Theorem 6, noting that we have the following identity

$$P(N_t \geq ta) = P(S_{[ta]} \leq t) ,$$

which means that the estimations for partial sums, as given in [5], can be used here, too. Therefore we obtain the following improved version of Theorem 6:

Theorem 9. Under the assumptions of Theorem 6 we have

$$\text{(9)} \qquad \lim_{T \to \infty} \left\{ \sup_{0 \leq t \leq T - C \log T} \frac{N_{t+C \log T} - N_t}{(C \log T)^{\frac{1}{2}}} - (C \log T)^{\frac{1}{2}} a \right\} = 0 \quad \text{w.p.1} .$$

It would be interesting to know whether convergence rate statements are also available in other situations, for instance, in some of the series of examples given by S. Csörgő (1969a) in his comprehensive paper [6].

5. REFERENCES

[1] Bahadur, R.R. (1971). Some Limit Theorems in Statistics. Regional Conference Series in Appl. Math. No. 4, SIAM, Philadelphia.

[2] Bahadur, R.R., Zabell, S. (1979). Large deviations of the sample mean in general vector spaces. Ann. Probability 7, 587-621.

[3] Bártfai, P. (1966). Die Bestimmung der zu einem wiederkehrenden Prozess gehörenden Verteilungsfunktion aus den mit Fehlern behafteten Daten einer einzigen Realisation. Studia Sci. Math. Hungar. 1, 161-168.

[4] Chernoff, H. (1952). A measure of asymptotic efficiency for tests of a hypothesis based on the sum of observations. Ann. Math. Statist. 23, 493-507.

[5] Csörgő, M., Steinebach, J. (1980). Improved Erdős-Rényi and strong approximation laws for increments of partial sums. Carleton Mathematical Series 166, 19 pp. (Preprint).

[6] Csörgő, S. (1979a). Erdős-Rényi laws. Ann. Statist. 7, 772-787.

[7] Csörgő, S. (1979b). Bahadur efficiency and Erdős-Rényi maxima. Submitted to Sankhyā Ser. A.

[8] Erdős, P., Rényi, A. (1970). On a new law of large numbers. J. Analyse Math. 23, 103-111.

[9] Groeneboom, P., Oosterhoff, J., Ruymgaart, F.H. (1979). Large deviation theorems for empirical probability measures. Ann. Probability 7, 553-586.

[10] Komlós, J., Major, P., Tusnády, G. (1975). An approximation of partial sums of independent rv's and the sample df. I. Z. Wahrscheinlichkeitstheorie verw. Geb. 32, 111-131.

[11] Komlós, J., Major, P., Tusnády, G. (1976). An approximation of partial sums of independent rv's and the sample df. II. Z. Wahrscheinlichkeitstheorie verw. Geb. 34, 33-58.

[12] Plachky, D., Steinebach, J. (1975). A theorem about probabilities of large deviations with an application to queuing theory. Period. Math. Hungar. 5, 343-345 (1976).

[13] Rényi, A. (1962). A sztochasztikus gejzir. Publ. Math. Inst. Hungar. Acad. Sci. 7, 643.

[14] Révész, P. (1980). How to charactierize the asymptotic properties of a stochastic process by four classes of deterministic curves? Carleton Mathematical Series 164, 34 pp. (Preprint).

[15] Steinebach, J. (1976). Das Gesetz der grossen Zahlen von Erdős-Rényi im Wartemodell G/G/1. Studia Sci. Math. Hungar. 11, 459-466 (1980).

[16] Steinebach, J. (1978a). On a necessary condition for the Erdős-Rényi law of large numbers. Proc. Amer. Math. Soc. 68, 97-100.

[17] Steinebach, J. (1978b). A strong law of Erdős-Rényi type for cumulative processes in renewal theory. J. Appl. Prob. 15, 96-111.

[18] Steinebach, J. (1981). The stochastic geyser problem for first-passage times. J. Appl. Prob. (to appear).

---

1) Work done while the author was on leave from University of Düsseldorf and a Visiting Scientist at Carleton University, Ottawa, supported by NSERC grants of M. Csörgő, D.A. Dawson and J.N.K. Rao and by DFG-grant Ste 306/1.

AMS 1979 subject classifications. Primary 60F15; Secondary 60F10, 60G50, 60K05, 60K25.

Key words and phrases. Increments of processes, Erdős-Rényi laws, stochastic geyser problem, strong approximations, large deviations.

TITLES OF CONTRIBUTED PAPERS

M. Ahsanullah (University of Brazil) and *A.B.M.L. Kabir* (Carleton University): A Likelihood Ratio Test of Equality of Means in a Bivariate Normal Distribution.

D. Bellhouse (University of Western Ontario): Optimal Randomization for Experiments in which Autocorrelation is Present.

M. Brown (Florida State University): On Choice of Variance for the Log Rank Test.

Y. Chikuse (Univeristy of Pittsburgh and Kagawa University, Japan): Robustness of a Test Statistic in Gauss-Markov Model.

T.D. Dwivedi and Y.P. Chaubey (Concordia University): Moments of Ratio of Two Positive Definite Quadratic Forms in Normal Variables.

R. Fischler (Carleton University): How to Find the "Golden Number" without Really Trying.

A. de Fontany (Communications, Canada): Seasonal Analysis: from Concepts to Models.

R.D. Gupta (University of Saskatchewan): Calculation of Zonal Polynomials of $n \times n$ Positive Definite Symmetric Matrices.

P.H. Peskun (York University): Theoretical Tests for Choosing the Parameters of the General Mixed Linear Congruential Pseudo Random Number Generator.

Prajnesu (Birkbeck College, London): Stochastic Models for Biological Population Growth.

S. Raman (University of Ottawa) *and C.L. Chiang* (University of California, Berkley): Explicit Solutions of Kolmogorov Equations for Multiple Transition Probability.

D. St. Richards (University of West Indies): Differential Operators Associated with Zonal Polynomials.

V. Shryskov and D.E. Stout (Rider College, New Jersey): The Homogeneity Problem in Statistics.